Ultrastructural Investigations on the Pituitary-Gonadal Axis

D. A. Bhiwgade • Sasikumar Menon

Ultrastructural Investigations on the Pituitary-Gonadal Axis

Springer

D. A. Bhiwgade
School of Biotechnology
and Bioinformatics
DY Patil Deemed-To-Be University
Navi Mumbai, Maharashtra, India

Sasikumar Menon
Institute for Advanced Training &
Research in Interdisciplinary Sciences
Ramnarain Ruia Autonomous College
Mumbai, Maharashtra, India

ISBN 978-981-99-3275-7 ISBN 978-981-99-3276-4 (eBook)
https://doi.org/10.1007/978-981-99-3276-4

This Springer imprint is published by the registered company Springer Nature Singapore Pte Ltd.
The registered company address is: 152 Beach Road, #21-01/04 Gateway East, Singapore 189721, Singapore

Paper in this product is recyclable.

"Subtle interrelationships between cells and tissues that comprise the Hypothalamic—Pituitary—Gonadal axis reflect the complex interconnection between our body and nature."—
Dr. D. A. Bhiwgade

Foreword

This interesting compilation of an explorative journey about the *Ultrastructural Investigations on the Pituitary-Gonadal Axis*, authored by Dr. D.-A. Bhiwgade and Dr. Sasikumar Menon, is an essential reference guide to everyone who is involved in advanced studies or research in the field of endocrinology and reproduction. The authors have painstakingly compiled and collated results from their various research publications and investigations made by them over several decades. This volume contains an elaborate study of important aspects of many features of cellular architecture with reference to their functional status. It also brings out the significance of structural variations in comparison to the normal with those related to different phases of growth, development and some pathological and clinical manifestations.

Every cell, in a biological system, is meticulously designed as an advanced manufacturing unit of a massive live workshop to suit the varied functional and life requirements. These units are not only independent but also interdependent on physiological, metabolic and genetic inputs of the biological system. These subtle interactions result in perfect functional coordination, between them, without any derangements to aid in the best outcome for growth and development of the biological entity.

The endocrine system forms an interesting part of this subtle physiological and functional web of life. Endocrine glands are unique as they secrete specific hormones as needed and the synthesis of hormones is done by specific cells with distinct ontogeny. The functional aspects of such glands are initially studied with light microscopy using thin sections stained specifically for various kind of cells in the gland. With the advancement of research techniques, electron microscopic investigations are initiated for ultrastructural evaluation of the cytoarchitecture and associated cellular functions. This enables a clear and better understanding of their involvement in the production of hormones.

It is this scientific acumen which stimulated the authors to undertake intensive investigations, spanning over three decades. Their perseverant efforts enabled them in reaping the incentives of original results which they have complied in this volume as a ready reference for anyone who has interest in the field of research.

The authors present here results of their in-depth studies using transmission electron microscopic investigations undertaken on animals with different reproductive strategies related to sperms, fertilization, zygote formation and implantation in the uterus. Their observations related to changes in various layers and cellular patterns in the endometrium, the development pattern and the type of implantation of embryo are quite comprehensive. The formation and functions of the placental layers are different in different animals studied by them. The observations are compared with those already available and with those seen in the normal and some pathological aspects of the cellular function.

The growth and development of every endocrine gland is governed by the stimulating hormones with a distinctive feedback mechanism that is perfectly synchronized with growth, synthesis and hormone action. The involvement of various controlling factors, their stimulation on the functional and synthetic aspects of the endocrine glands and their cells have been very well brought out.

Apart from these aspects, the animal studies with treatment of several synthetic, anti-gonadal and contraceptive hormonal preparations are used to evaluate the responses of testis, epididymis, placenta, corpus luteum and endocrine glands of the anterior pituitary at various phases of their activity. Comparison of experimental observations with the normal state has been used to understand the cellular modifications after such experimental treatments.

The authors have specifically dealt with the embryo, its implantation, its development, the foeto-placental relationship in different placental barriers and the modifications at various cellular compartments in the uterus at every stage of development. The significance of such changes has been discussed clearly by appropriately quoting relevant publications. Finally, the authors and their team of workers have also put forth their own views and interpretations.

The modifications of specifically designed cells, their components or subcellular and microcellular organelles that result due to stimulatory functional variations or due to selected treatment schedules have been discussed with reference to biochemical, histochemical, immunochemical and physiological changes. The significances of these changes have been discussed in comparison with the normal developmental patterns.

I found this detailed and comparative treatise to be a clear representation of in-depth investigations and interpretations of changes in cellar functions and responses seen in normal and experimental animals. This book provides a horde of information useful as good reference repository for future research. Ideally, this volume should find a place in every teaching and research establishment, especially those involved in Biological Sciences.

A devoted endocrinologist, Prof. Govindarajulu, is better known as founder of the Department of Endocrinology at the University of Madras, Tamil Nadu, India. With a vast research experience spanning more than 55 years, he has been the research supervisor for 50 doctoral research scholars. With more than 150 research publications, Prof. Govindarajulu has presented his research work in many national and international seminars and conferences, both in India and abroad. A keen endocrinologist, Prof. Govindarajulu, has received grants from several Governmental agencies like the Indian Council of Medical Research, the Department of Science and Technology, Government of India, and other prestigious institutions like the Ford Foundation and the World Health Organization. He is a fellow of the Society for Reproduction and Endocrinology and has been a visiting-scientist at the St. Mary's Hospital, London, UK. During his tenure at the University of Madras, he had held several important administrative posts as the Registrar of the University, as Member of the senate and as the Secretary of the University. He continues to be actively involved in educational activities as the trustee of Sriram Educational Trust at Chennai, Tamil Nadu, India.

Peranaidu Govindarajulu M.Sc., Ph.D., F.S.R.E

Preface

Experimental methods are backbone of biological investigations to discover body responses to xenobiotics and to understand how a biological system manages stress. Biochemical evaluations on body fluids and tissue samples provide a broad understanding of the status of the physiological milieu associated with a pathology. It is the light microscopic evaluation that enables us to pinpoint nature of lesion at the tissue level. Electron microscopic studies very minutely unravels the cytoarchitectural basis of a pathological lesion. This provides a different dimension in understanding not only the aetiology of pathological condition, but one could also predict the progression of pathology. Ultrastructural studies have added newer perspective to existing knowledge about many diseases and pathological conditions. In several clinical conditions, electron microscopy enables clinicians to diagnose specific nature of the disease looking at the cytoarchitectural changes and prescribe a better treatment strategy.

Microscopic studies not only supplement biochemical observations but also correlate the biochemical changes in relation to changes in cytological structures. Ultrastructural studies on normal tissues provide invaluable data related to the arrangement and interaction between different cell types that make up the tissue. Establishing the interesting correlation between anatomy and physiology of a tissue system enables us in understanding not only the biological role of a tissue system but also its evolutionary lineage in related animal orders.

Functional relationships of complex system of neuroendocrine cells in the pituitary have been an area of great interest to cytologists, endocrinologists and physiologists. Ultrastructural characterization of different pituitary cell types, that is reported in this book, will be of interest to physiologists and endocrinologists alike. Light microscopic studies when correlated with electron microscopic observations can provide interesting interpretations and this book provides our experiences in this field. The differential staining techniques that have been fine-tuned at our laboratory provided us with a great tool in understanding distribution and abundance of pituitary cell types across various animal orders. Further, the application of electron microscopy enabled us in understanding changes in internal cytoplasmic structures, especially the secretory granules, endoplasmic reticulum and mitochondria. The identification features of different pituitary cell types like FSH, LH and prolactin cells, as explained in this book, will enable many cytologists in understanding and evaluating cytoarchitectural changes correlated with physiological activity of the gland.

The complex relation between the pituitary—gonadal axis and the placental system have always intrigued endocrinologists and developmental biologists. Over several vertebrate orders, pituitary—gonadal axis remains the significant endocrinological regulator that influences reproductive physiology. With the evolution of placental system of nourishment of foetus within the maternal body, a new dimension gets added to this complex physiological regulatory relationship. Whereas the pituitary —gonadal axis regulates the development and sustenance of placenta, the placenta itself functions as an endocrine organ while maintaining the integrity of maternal and foetal environments. Thus, cytoarchitecture of specialized tissue systems and arrangement of different cell types within the placenta can provide valuable insights into the physiology of trans-placental transport. The comparison of various placental system with differential involvement of foetal and maternal tissues has been possible by studies on different families of Indian Chiropterans. Chiropterans exhibit different placental types within the same mammalian order. Our electron microscopic studies have revealed interesting structural differentiations in foetal and maternal tissues during developmental stages of the placental barrier in different Chiropteran families. This will be a valuable insight for understanding placental failures in women under different clinical conditions.

Chiropterans with their unique adaptations in reproductive behaviour provide an interesting array of placental types within various bat families. Reproductive biologists have been working on various aspects of Chiropteran reproductive anatomy, physiology and behaviours. Our laboratory has been equally intrigued by their findings and have been fascinated by the structural variations in placenta seen within various families of the single order. For a period of over three decades, we applied electron microscopy to investigate different pituitary and placental samples from several species of mammals to understand the phylogenetic trend and possible physiological roles of different cell types that constituted the placental system. This book is a record for all interesting findings that we came across while peering hours and hours over the visualization screen of electron microscope. When we encountered some unusual cytoarchitectural arrangements, the obvious questions that we pondered over ranged from; why such variations? and what significance does it have in relation to the physiology? With scanty literature available in this area, it became very important to share our findings with scores of biologists who are engaged in finding explanations to many complexities of the pituitary—gonadal axis and the mammalian placenta. This book will indeed help those in their quest to understand physiological roles played by pituitary and placenta in reproduction. It is our hope that this book will not only assist in answering many questions but will also serve as an atlas of endocrine cells at the ultrastructural level. Across its pages, in this book, we have collated, classified, described and illustrated various observations that we made on cellular organelles and inclusions that reflect physiological state of the tissue.

The book includes several hand drawn illustrations to highlight important cytoarchitectural features in a cell type and arrangement of cell types in the placental system. These illustrations have been carefully compiled from several electron micrographs so that cellular features are made visible in a single plane to make it easier to understand the cytoarchitecture. Many hours of painstaking sketching and corrections have gone into the making of illustrations that are included in the book. These illustrations have been made to reflect the cytoarchitecture, as close to the actual observations as possible. While the illustrations of different pituitary types have been developed based on several electron microscopic observations of the same cell type, the illustrations of placenta are based on many electron microscopic observations recorded at different angles and at different regions of the same placenta. Many of these observations have already been published, as cited in the text, but several have been reproduced in the book which have not reached the public domain. This book is an attempt to compile all observations that were made at our laboratory, to provide a more holistic overview of ultrastructural characteristics which otherwise is not feasible in a typical research paper due to constraints on figures and words. The book, we are sure, will serve as unique guidance to more in-depth studies in future.

Comparative ultrastructural evaluations on pituitary cell types and placental barrier across different animal species is the main focus of this book. This approach enables the reader to appreciate the wide array of structural and functional diversity in ultrastructural features within the tissue. A comprehensive list of relevant references is provided for all chapters that will enable readers to get better perspective of the cytoarchitecture, especially for investigating such areas in other species or a pathologic condition. The book will also help the larger group of researchers who have been recently initiated into the field of clinical pathology and need a broader understanding of electron microscopy which will be useful in pursuing their specific areas of interest.

The realm of electron microscopy largely dwells on ultrastructural details which are beyond the reach of light microscopy. The changes in ultrastructural features like the cilia, polyribosomes, endoplasmic reticulum, cell membrane, etc. recorded under electron microscopy have changed our perception of cellular physiology and pathology. The writing of this book is a self-imposed mission for us, mainly because the efforts that we attempted at our laboratory in recording ultrastructural details of the pituitary and the placental barrier have led us to understand some hitherto unknown facts about this complex tissue systems. We felt that writing a book would be the best and most comprehensive way to bring this information to the scientific fraternity. The placental barrier in a complex manner, many facets of which are yet to be understood, establishes cooperation between two genetically different individuals the foetus and the mother. It is the good and efficient foeto-maternal cooperation that is established through the placental barrier that enables viviparity to be sustained successfully till term. This book, therefore, covers ultrastructural observations that is correlated with the development, morphology and evolution of foetal membranes and accessory maternal elements in different types of placental barrier. Well-illustrated figures have been provided to make ultrastructural details of placental barrier easily understood by beginners too.

Maharashtra, India

D. A. Bhiwgade
Sasikumar Menon

Acknowledgements

The genesis of this book is mainly because of the guidance and sincere dedication that I saw in my teacher and research guide Late. Dr. Gopalakrishnan, Professor in Zoology at the Department of Zoology, Institute of Science, Nagpur, Maharashtra, India. His knowledge and perseverance in reproductive biology and mammalian embryology was infectious and my interest in the subject was solely due to his influence. Coming from a family with humble background, for me, Dr. Gopalakrishnan was a father figure in research and he convinced me that being sincere, working hard under proper guidance will always bear fruit. This book is a testimony to Dr. Gopalakrishnan's prophecy. Having completed my education under trying circumstances, to migrate to a megapolis like Mumbai and complete most of the work cited in this book was a dream come true.

My tryst with electron microscopy started at Jaslok Hospital and Research Centre in Mumbai, India. Late Dr. Arun Chitale, who headed the department of Electron Microscopy at the hospital, was a well-known Clinical Microscopist who is known for his diagnostic acumen both in India and abroad. Under his guidance, I started my first investigations in electron microscopy and my pursuance of ultrastructural studies continued for more than three decades. I am equally indebted to all my research students for their involvement in the work without which such a voluminous work could not have been possible.

This book is a unique compilation of observations and interpretations made by my team of researchers and records more than four decades of explorations into the fascinating cellular relationships that exists within the pituitary-gonadal axis. The book brings out some of the most intriguing structural adaptative radiations within a single order of mammals that could throw light into the evolutionary history of placenta among the modern eutherians. The subtility of the cellular interdependence and relationships seen within the ultrastructural characteristics of placental components would be vital in understanding several clinical conditions of failed gestation in human.

This book is a lifetime mission for me, and I would remain ever gracious to all those associated with its conception, development and production. It is difficult to cite individually all those who helped me in these initial years of research and during the actual production of the book. Many friends and acquaintances have provided specimens from different parts of India and their contributions have been significant to the advancement of scientific knowledge. The technical skills of the staff at the Department of Electron

Microscopy of Jaslok Hospital needs special mention since, it is their professional support that reflects in the quality of Electron micrographs. Mr. Babu Sarode (Nagpur), Mr. Vijay Kate, Mr. Dilip Khanaskar, Mr. Shivaji Bhosle and Mr. Ramesh Mahadik have put forth sincere efforts in processing and preparing ultrathin sections of resin embedded tissue blocks to get the correct perspective of the tissue. They have spent long hours in the dark room to print the correctly toned black and white prints from the negatives, some of which needed special attention due to the poor quality of image. Digital Technology made their work easier in the later years of the research.

Mr. Sawant and Mr. Siddiqi from the Institute of Science, Mumbai, India have contributed with their skills in photography and draftsmanship to provide drawings to represent various layers, membranes and different intracellular features. Their technique of using stripling and hatching helped in making original drawings from rough sketches to illustrate different ultrastructural features. I extend my sincere thanks to my friends and colleagues for their help and also to the skill and dedication of my co-workers, technicians and my laboratory support staff. In this context, the help rendered by Mr. Gorivale in collection and maintenance of animals needs special mention.

Many colleagues and research students have read through the manuscript and provided useful criticisms and suggestions. Most of them have been incorporated and if any errors still remain, those are solely my responsibility. My sincere acknowledgment goes to Mr. Sunil Berade (D. Y. Patil College) and Mr. Vinod Patil for patiently typing and formatting the handwritten manuscript into a printer-friendly layout due to which the publishing work could be expedited. The book could see the light of day, in its present form, mainly due to the skill, cooperation and professionalism of the editorial staff at Springer Nature.

One of the reasons that made the journey of this book so memorable and enjoyable was my wife, Jyoti Bhiwgade who has been a constant source of encouragement and support. As my true lifetime companion, she not only tolerates my assiduous involvement at the laboratory but even manages the family so well that my absence is rarely felt.

D. A. Bhiwgade
Mumbai,
2022

Publication of this book is an endeavour that evolved from my deep sense of indebtedness to my guide, Dr. D. A. Bhiwgade, without whom, the gates to the Eden of research would not have been opened to me. Becoming a scientist had always been my dream which I cherished from my school days and Dr. Bhiwgade should be credited for enabling me to realize that dream. His passion and his focused approach were mainly the inspiration for me to go off the beaten track while selecting my doctoral research topic on male contraceptives. His mastery in histological techniques was mainly responsible for me and my colleagues at our lab to delve into electron microscopy, which even today, remains a bastion of some endowed research centres in the country. I remember the cold-shoulder and rebuke that our team received from some of the established research centres during our early stint into this area. Pushed to the wall, we approached the Jaslok Hospital and Research

Centre, Mumbai, India. The encouragement and guidance of late Dr. Arun Chitale at the hospital was singularly responsible for making this challenging plunge to be not only enlightening but resulted in a great sense of accomplishment for our entire team. Dr. Arun Chitale was an expert clinical pathologist who specialized in electron microscopic diagnosis. His opinions were sought by clinicians from various parts of the country and the special interest that he took in our explorative work on pituitary-gonadal axis was infective. His successor Dr. (Mrs.) Shaila Khubchandani also encouraged our work in a similar manner.

The array of placental adaptations seen among chiropterans is what made our work both challenging and interesting. This book which is a compilation of our observations and findings would provide histologists with a ready reckoner for identifying ultrastructural characteristics of many placental pathologies.

I, finally, acknowledge the emotional support, inspiration and constant encouragement that was extended to me by my family, friends and colleagues throughout my explorative journey which culminated in the publication of this book.

Dr. Sasikumar Menon
Mumbai
2022

Research Scholars

(who worked under the guidance of Dr. D. A. Bhiwgade)

Dr. L.T. Gulhane	Histopathological study of the pituitary gland of the Indian five striped palm squirrel, *Funambulus pennanti* (Wroughton).	(1979)
Dr. A.G. Mankar	Cytology of adenohypophysis in some Indian bats.	(1980)
Dr. A.P. Manekar	Studies on the effects of certain steroid hormones on pituitary, testis and associated accessory structure of bat, *Rousettus leschenaulti* (Desmarest). 	(1986)
Dr. D.G. Senad	Studies on the effects of certain steroid hormones on pituitary, gonads, thyroid and adrenal of bat, *Rousettus leschenaulti* (Desmarest).	(1986)
Dr. (Mrs.) R.S. Jawade	Involvement of steroid hormones in reproduction of male and effect on mucopolysaccharides of female genital tract in bat, *Rousettus leschenaulti* (Desmarest).	(1986)
Dr. (Mrs.) C.S. Katoley	Histoenzymological studies in the female reproductive tract of the Indian fruit bat, *Rousettus leschenaulti* (Desmarest) during different phases of the sexual cycle.	(1987)
Dr. (Mrs.) H.N. Menon	A histoenzymological study : Effect of some contraceptive agents on the male reproductive organs of the bat *Rousettus leschenaulti* (Desmarest). 	(1987)

(continued)

Dr. S.N. Menon	Enzyme biochemistry and histological study : Effect of some contraceptive agents on the male reproductive organs of the bat, *Rousettus leschenaulti* (Desmarest). 	(1987)
Dr. V.N. Magare	Comparative anatomy and cytology of the pituitary gland in some marine teleost and electron microscopy of the adenohypophysis during gonad maturation in *Johnius belangerii*. 	(1988)
Dr. Indu Nair	Contraceptive agent (Gossypol): Ultrastructural and biochemical changes in the reproductive organs of rat. 	(1989)
Dr. (Ms.) A.B. Singh	Ultrastructural development of chorioallantoic placental barrier in the Miniopterus bat, *Miniopterus schreibersii fuliginosus* (Hodson).	(1989)

(continued)

Dr. K.M. Avari	Steroidal contraception: Effect of Depot medroxy- progesterone acetate and testosterone enanthate on the male reproductive system of albino rat. 	(1990)
Dr. (Mrs.) Vrinda Joshi	The cells of adenohypophysis and their electron microscopic study. 	(1990)
Dr. A. Mandal	Ultrastructural studies on the chorioallantoic placenta and pituitary in the Indian mouse tailed bat, *Rhinopoma microphyllum*.	(1991)
Dr. (Mrs.) M.N. Kulkarni	Clomiphene citrate: Light and electron microscopic observations on male reproductive organs of bat, *Rousettus leschenaulti*. 	(1991)
Dr. (Mrs.) V.L. Pereira	Studies on the histochemistry of the placenta and the ultrastructure of the yolk sac in some Indian Chiroptera.	(1991)

(continued)

Dr. V.M. Thakur	Endocrinology of reproduction in Bonnet monkey. *Macaca radiata*. 	(1991)
Dr. P.B. Kamble	Experimental analysis of pituitary-gonad-thyroid-adrenal in Bonnet monkey.	(1992)
Dr. A. Kothari	Ultrastructural studies on interhemal membrane in some mammals. 	(1993)
Dr. (Mrs.) C.S. Panse	Ultrastructural studies of interhemal membrane in two species of mammals. 	(1993)
Dr. P.D. Raut	Endocrinological approach to male fertility in Bonnet monkey, *Macaca radiata*: Testosterone enanthate. 	(1993)

(continued)

Dr. (Mrs.) V.A. Thakur	Comparative electron microscopy of the hemochorial Placenta.	(1994)
Dr. (Mrs.) A. J. Patel	GnRH: Endocrinological approach to control of male fertility in Bonnet monkey Macaca radiate 	(1994)
Dr. A.Y. Bellare	Effects of medroxyprogesterone acetate and testosterone enanthate on the hormones, biochemistry and histopathology of the white male rat. 	(1995)
Dr. (Ms.) Utkarsha Mantri	Functional classification of cell types of the anterior pituitary gland accomplished by electron microscopy. 	(1995)

(continued)

Dr. (Mrs.) K. D. Bansal	Endocrinological approach to male fertility in monkey *Macaca radiata*: DMPA + TE. 	(1995)
Dr. (Mrs.) Vinda Sakpal Manjramkar	Studies on ultrastructure and biochemical aspects of mammalian epididymis. 	(1995)
Dr. H.P. Mishra	Experimental analysis of pituitary gland of *Clarias batrachus*: An electron microscopic study.	(1995)
Dr. (Mrs.) S.N. Date	Endocrinology of reproduction: Endocrine interaction during the different phases of female reproductive cycle in the Indian Fruit bat, *Rousettus leschenaulti*. 	(1996)
Dr. (Mrs.) S.S. Banerjee	Ultrastructural studies of the placenta in some Indian bats and the lipid profile during ageing in the Indian fruit bat, *Rousettus leschenaulti*.	(1996)

(continued)

Dr. (Ms.) Vaishali M. Kadam	Endocrinology of male reproduction. Can CPA + TE combination regimen prove to be an efficacious male contraceptive in the near future? 	(1998)
Dr. Manoj R. Borkar	Ultrastructural Studies: Identification of anterior pituitary cell in some mammals. 	(1999)
Dr.(Ms.) Deepthi Uthaman	Studies on histology of the oviduct and biochemical analysis of blood serum metabolites during different phases of the female reproductive cycle in the Indian fruit bat, *Rousettus leschenaulti*. 	(1999)

(continued)

Dr.(Mrs.) Jyotsna Mahaley	Implantation and ultrastructure of placenta in Indian Tylonycteris bat *Tylonycteris pachypus*. 	(1999)
Dr. (Mrs.) Saswati Bhattacharya	Endocrinology of reproduction in female Wistar rats : Facts about Depo-Provera.	(1999)
Dr. (Ms.) Suvarna Rawal	Effects of progestogenic compounds on reproductive organs in female albino rat: Histological and clinical approach. 	(2000)
Dr. (Ms.) Pratibha G. Posam	Effect of female contraceptives on clinical biochemistry and ultrastructure of female reproductive organs in rats.	(2000)
Dr. (Ms.) Neera T. Shetty	Genetic Study in High-Risk Pregnancy. 	(2000)
Dr. (Ms.) Smita A. Savant	Ultrastructural changes in the corpus luteum of bats during the reproductive cycle.	(2002)

(continued)

Dr. Vinod S. Narayane	Development towards male contraception : Comprehensive investigation on CPA +TE combination regimen and ultrastructure of pituitary gland in some mammals. 	(2002)
Dr. S.M. Deshbhratar	Structure and development of the Inter-Haemal membrane of the vespertilionid bat, *Scotophilus heathii* & Pteropodidae bat, *cynopterus sphinx gangeticus* : An Ultra-Structural Study 	(2002)
Dr. (Mrs.) Manda Thakur	Reproduction in male fruit bat, *Rousettus leschenaulti* and effect of anticancer drug on the endocrine glands and reproductive organs of male white rat. 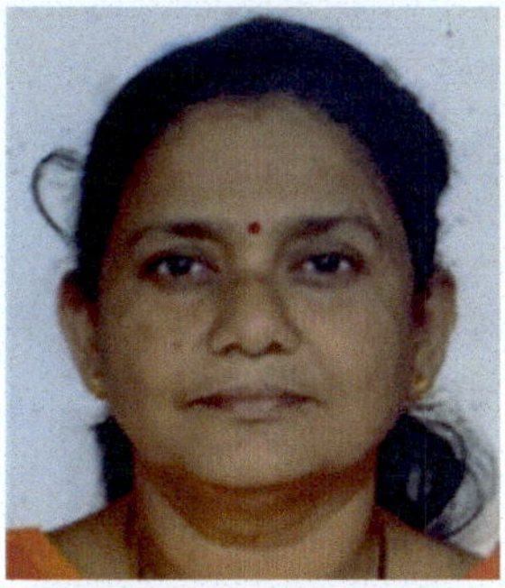	(2002)
Dr. (Mrs.) Tejashree.V. Shanbhag	Endocrinology of Reproduction in Vertebrates 	(2003)

(continued)

Dr. Mata Deen Mishra	Effect of anticancer drug doxorubicin on cellular and molecular changes associated with the testicular and epididymal tissue of rat.	(2004)
Dr. (Ms.) Pratibha R. Kamble	Ultrastructural and antioxidants status of hepatic and renal tissue following the treatment of anticancer drugs Cisplatin and Etoposide in male rat. 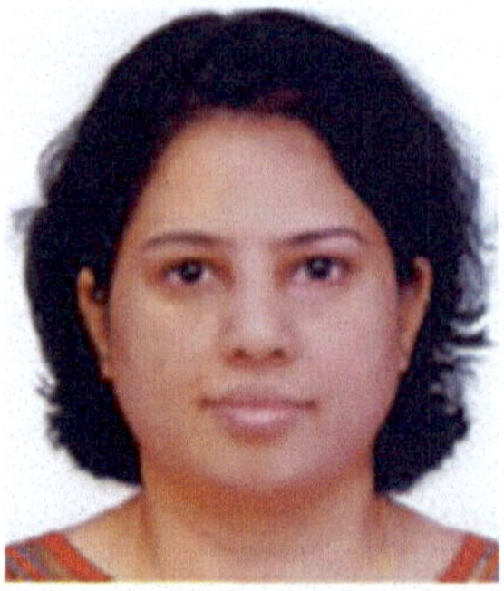	(2004)
Dr. (Ms.) Thayyil Liji Appu	Analysis of the alternations in testis and epididymis of Wistar rats following Etoposide treatment using transmission electron microscopy, biochemical and molecular techniques.	(2006)
Dr. (Ms.) Manisha S. Kayande	Observations on the fine structure of the Yolk Sac in three species of bats. 	(2009)
Dr. Rajivkumar Shah	Study of adenosine deaminase in HIV & tuberculosis patients.	(2010)
Dr. (Mrs.) Rekha Bhagwat	Assessment of newer cardiac markers as independent risk factors of CVD in Indian population.	(2010)
Dr. (Ms.) Tarika Sonawane	Effect of vincristine and embelin on steroidogenic pathway of male Wistar rats. 	(2013)
Dr. (Ms.) Shine Devarajan	Molecular modelling studies on the confirmation and specificity of Amyloid Beta (AB42) protein towards understanding the aggregation and apoptosis in Alzheimer's disease.	(2015)

List of students awarded M.Sc. (By Research)
under the supervision of Prof. D. A. Bhiwgade

Ms. V.V Akolkar	Endocrinology of Reproduction : Electron microscopy of anterior pituitary gland in bat, *Rousettus leschenaulti*. 	(1986)
Ms. L.S. Rohatgi	Fine structure of placental barrier in bat, *Taphozous melanopogon*. 	(1993)
Ms. V.M. Kadam	Studies on the effects of certain chemical compounds on the testicular structure of mammals.	(1995)
Ms. Ranjana A. Gaonkar	Ultrastructural cytology of anterior pituitary in *Bandicota bengalensis*.	(1998)

Contents

About the Authors

D. A. Bhiwgade M.Sc., Ph.D., D.Sc. Coming from a very humble background of a village near Nagpur, Maharashtra, India, he climbed the heights of academic research by perseverance and sustained dedication. He completed his doctoral research on mammalian embryology in 1973 for which he received a special award from the University for completing doctoral degree in record time. He received the post-doctoral degree of D.Sc. (Doctor of Science or *Vigyan Pandit*) in 1989 for his research in Mammalian Embryology, Endocrinology and Male Contraception. With more than four decades of involvement in academics and research, he attained superannuation as the Professor and Head, Department of Biotechnology and Bioinformatics at the Padmashree D. Y. Patil University, Belapur, Navi Mumbai, India. Before that, he was the Professor and Head, at the Department of Life science, University of Mumbai, India. Before taking up the responsibilities at the University of Mumbai, he was Head of the Department of Zoology, at The Institute of Science, Mumbai, India. In his academic career, he successfully supervised 50 research scholars for their doctoral work. He is a recipient of Paras Das Jain Gold Medal for his valuable contributions to the field of research in Life Sciences.

Sasikumar Menon M.Sc., Ph.D. An adherent Zoologist at heart, his passion for interdisciplinary approach is evident in all his research work. He received his doctoral degree in 1987, on male contraception, under the supervision of Dr. D. A. Bhiwgade. After his doctoral research, he shifted his interests to drug development and drug toxicology with special emphasis on bioanalytical investigations. He is the Director of Institute for Advanced Research in Interdisciplinary Sciences, at Sion, Mumbai, India. He is associated with Ramnarain Ruia College, as the Associate Professor in the Department of Pharma Analytical Sciences. Formerly, he was the Reader in Zoology, at the Ramnarain Ruia College, Mumbai, India. In his academic career of more than three and half decades, he has successfully guided 24 research scholars for their doctoral degrees. He continues to guide research students and is active in drug development programmes. He has been associated with several research projects funded by Government of India and various pharmaceutical companies.

Ultrastructure of Anterior Pituitary, Testis, and Epididymis in Mammals

1

General Introduction

'Here in this well concealed spot, almost to be covered by a thumb nail, lies the very mainspring of primitive existence; vegetative, emotional and reproductive'.

In these striking words, Harvey Cushing described the pituitary body, which is one of the smallest of the endocrine glands but with a complex task of integration assigned to it. Rapid advances have been made in the last two decades in our knowledge of pituitary structure and function following the application of immunochemistry and electron microscopy. Until some decades ago, what used to be the uncharted sea of endocrines strewn with hypotheses, is now known to be unified by the influence of a single gland, the pituitary gland. The integrative role of the pituitary was prompted by the knowledge that the anterior lobe of this gland was found to be secreting certain trophic hormones, which when released into the circulation would eventually stimulate the activity of the endocrine glands. The number of such trophic hormones was then a matter of speculation but it was soon realized that these chemicals were endocrine gland specific in their stimulatory action.

It was later confirmed that the release of anterior pituitary trophic hormones depended on the secretion of certain hypothalamic 'neurohormones' or 'releasing factors', which were carried down the infundibular stalk in the hypophysial portal system. These chemicals were finally carried through the portal venous system flowing from the median eminence of the hypothalamus to the anterior lobe of the pituitary to control the output of trophic hormones. Thus, it was eventually accepted that the pituitary gland is only an intermediary in these endocrine chains of events.

In the days when haematoxylin-eosin was a satisfying stain, the structure of the adenohypophysis was found to be simple and consisted of three cell types according to their tinctorial affinities. The adenohypophysial cells were identified as either acidophils, basophils or chromophobes (Thorn 1901; Gemelli 1907). However this phase was short lived and newer staining methods emerged not only giving us new pictures and terminologies.

Noteworthy among these novel methods were the nomenclature using Greek alphabets proposed by many authors (Bailey and Davidoff 1925). Subsequently, electron microscopists also used the nomenclature of Greek alphabets (Barnes 1962; Herlant, 1964). Some light microscopists used nomenclature suggestive of the endocrine function of the cells, e.g. gonadotroph and thyrotroph (Purves and Griesbach 1951a, b). Histochemical technique of Pearse (1948, 1949), Herlant (1949, 1950), Halmi (1950), Herlant and Racadot (1957), Ezrin et al. (1958), and Paget and Eccleston (1960) permitted further subdivision of chromophil cell types.

The introduction of McManus periodic acid Schiff (PAS) method of staining with Orange G

D. A. Bhiwgade, S. Menon, *Ultrastructural Investigations on the Pituitary-Gonadal Axis*,
https://doi.org/10.1007/978-981-99-3276-4_1

as a counter stain provided a new histological picture of the anterior pituitary more closely related to the biological and hormonal activity of the cells. Ezrin and his associates (1958) modified this technique further into Iron PAS technique and distinguished 5 types of cells. In essence, this phase of pituitary research was associated with developing methods of staining reactions that conferred to each cell type a characteristic tinge (El Etreby et al. 1972; Ruitenberg et al. 1974). With the advent of electron microscopy, the identification of anterior pituitary cells was based on the ultrastructural features of the cell. The incredibly high resolving power of the electron microscope made possible a morphometric study of cells and it was realized that the size and shape of the cell's secretory granules were the important makers of cell identity (Fernandez-Moran and Luft 1949). With the improvement of sectioning techniques, the early fifties saw extensive use of the electron microscope in pituitary cell identification. Rinehart and Farquhar (1953) are given the credit for having recognized the great potential of the electron microscope for studying pituitary cytology. Reports on fine structural cytology of pituitaries of various animals as well as human pituitary specimen from surgeries started pouring in (Barnes 1962; Kobayashi 1966; Gale 1972; Kovacs et al. 1977).

Electron microscopy was thus found to be suitable for identification of pituitary cell types as well as for assessing their functional state. One serious flaw, however, was apparent in course of these studies. It was realized that due to overlapping in the sizes of secretory granules and/or lack of distinguishing features; a varying number of cells could not be categorized by electron microscopy. This was one problem that electron microscopic study was plagued with, until it was categorically decided that an arbitrary classification of cells was impractical and that it would be more appropriate to rely on a set of ultrastructural features before ascertaining the cell identity (Horvath and Kovacs 1988). Another error to avoid was the extrapolation of findings from one species to another.

A relatively new finding that repeatedly called the attention of pituitary cytologists was the existence of plurimorphous cell communities in the normal and adenomatous pituitary of rodents & men (McComb et al. 1981; Horvath et al. 1974; Scheithauer et al. 1986). By virtue of the fact such plurihormonal cells produce more than one hypophysial hormone, the utility of conventional electron microscopy in cell identification was seriously undermined thereby creating new problems in classification and semantics. The co-localization of two or more hormones in one such plurihormonal cell rendered the 'One cell one hormone' theory obsolete and raised new questions concerning correlation between fine structure and hormone differentiation. It is at this time that the immunohistochemistry revolutionized specific cell identification (Hsu et al. 1981; Kovacs et al. 1981; Nakane 1970; Sternberger et al. 1970). Its fine structural application, immunoelectron microscopy (IEM) launched the study of pituitary into a yet higher orbit taking some of uncertainties out of cell identification (Moriarty 1973; Pelletier et al. 1978; Asa et al. 1988). Besides IEM made it possible to study the ultrastructural mechanisms of hormone production and of its transfer and release from the cell (Kurosumi 1991).

Some workers have opined that the anterior pituitary cells undergo transdifferentiation of anterior pituitary cells, which can be confirmed by applying techniques of IEM. However, such an approach is prohibitively expensive and routine electron microscopy still remains the most appropriate and effective tool in the classification of anterior pituitary, if used with certain interpretatory precautions. Perusal of literature shows that electron microscopy has contributed most significantly to the wealth of information on pituitary cytology. As a tool of investigation, electron microscopy has an edge over other routine methods of pituitary cell identification, as it can decide, with precision, the extent of activity of the cell by giving the observer an access to the fine structural characteristics of the cell. The cell identity can then be ascertained by correlating the structure and function.

The approach outlined in this book is aimed at integrating the form and function of adenohypophysial cells of some mammals. This work goes beyond a simple statement of observation relating to the size, shape, position or other morphological features. Here, an attempt has been made to correlate the structural and functional aspects of the cell simultaneously, thereby examining the cell as a whole. Such a subjective approach not only facilitates confirmation of cell identity but also yields valuable information regarding its functional state. Furthermore, it needs to be emphasized that such a correlation of morphological evidence with functional criteria is particularly useful in light of prevailing confusions regarding bi/pleurihormonal cells.

The implication of the foregoing statements is that the electron microscopic picture of a particular cell type would vary with the functional state of the same. Stimulating effects would result in proliferative changes. On the contrary, suppressive effects would show a cytostatic action. Thus, the cell has fine structural correlates for its functional state that show considerable variations among species. Such species specificity of these fine structural correlates therefore necessitates a comparative study of adenohypophysial cells in different animals or among different species with a common major taxonomic affiliation; such as different mammals.

Another useful approach for interpretation of the functional cytology of the pituitary is to correlate the ultrastructure of its cells with different phases of the reproductive cycle. This rationale has been adopted by many workers to investigate the adenohypophysis of various mammals viz. mouse (Barnes 1962; Gomez-Dumm and Echave-Llanos 1972), Syrian hamster (Dekker 1967), mink (Murphy and James 1976), rabbit (Young et al. 1967) and bat (Bhiwgade et al. 1989).

Hence in the first part of this work, mammals showing cyclic reproductive behaviour have been selected for the ultrastructural study of their anterior pituitary cells. The rationale behind the design of the experimental study is that the cellular organization can be sufficiently altered by exogenous influences. Such alterations lead to a proportionate imbalance in the functional state of the cell. This rationale carried further implies that experimentally upsetting an important endocrine function could well reflect in the ultrastructure of the corresponding pituitary cell by way of a feedback mechanism. This contention coupled with the fact that the adenohypophysis is an assemblage of fine structural equivalent of endocrine differentiation justifies, the experimental design of these investigations. In fact many electron microscopists have used this knowledge for a functional classification of the anterior pituitary cells based on the results of experimental morphology (Farquhar and Rinehart 1954a, b; Stefaneanu and Kovacs 1991; Lloyd 1991; Saeger 1992; Mantri 1994).

In the second part of our work, it has been experimentally verified that the cells of the anterior pituitary show fine structural alterations following drug treatment. The drugs administered to the experimental animals in this study, fall under four categories. They are synthetic antioestrogenic (tamoxifen citrate), oestrogenic (oestradiol valerate), progestogenic (progesterone and norethisterone heptanoate) and a combination regimen of progestogen and oestrogen (levonorgestrel + ethinyl oestradiol).

In our studies, a special emphasis has been placed on the gonadotroph (GTH) and prolactin (PRL) cells for obvious reasons. Also, we have not categorized the GTH cells into FSH and LH subtypes, since the dualist concept has been completely turned around by immunohistochemical data, indicating that FSH and LH are co-localized in the same cell (Nakane 1970; Phifer et al. 1973; Inoue and Kurosumi 1984).

The reason for selecting a rat as an experimental animal for our study is because it is the most commonly used laboratory animal and the basic morphological features of its pituitary cells as well as their response to endocrinological stimuli have been extensively studied and better defined than in any other species, including man (Kurosumi 1986; Nakane 1970; Costoff 1973; Moriarty 1973; Kurosumi and Fujita 1974; Girod 1976, 1977). The new developments in pituitary cytology suggest that modifications are required in our perception of pituitary cells, the implications of these new insights are entirely unknown. However, beyond the theoretical considerations, knowledge of fine structural cytology of the mammalian pituitary may provide valuable information to the clinician, concerning pathogenesis, prognosis and therapeutic responsiveness of the lesions.

Basis of Classification of Different Cell Types: An Update

The methodologies adopted for the identification and classification of anterior pituitary cells have gradually evolved from simple tinctorial methods, through immunocytochemical methods, to the use of electron microscope. The amalgamation of the last two aforementioned methods has given us yet another technique known as Immunoelectron microscopy, which has now set aside a few anomalies in cell identification, that could not be resolved by conventional transmission electron microscopy.

Immunoelectron microscopy authenticated the doubt expressed by some of the earlier workers, based on their cytochemical studies, that some cells of the anterior pituitary secrete more than one hormone. Thus, bihormonal and plurihormonal cells were demonstrated in the anterior pituitary. This technique thus offered specificity of cell identification and confirmed the identifications formerly proposed by conventional methods.

Nevertheless, conventional transmission electron microscopy still remains one of the most intensively used technique for cell identification as it permits the identification of cells on the basis of their cytoarchitectural characteristics. Over the last several decades numerous ultrastructural studies on the anterior pituitary cells of different animals have been done. Consolidation of the data on the morphology of the cell as influenced by the reproductive status and experimental manipulation has resulted in a general consensus regarding the fine structural criteria for cell identification.

1.1 Growth Hormone Cell (GH Cell)/Somatotroph (STH)

1.1.1 Introduction

This is the most prevalent cell type of the hypophysial pars anterior, with a well-characterized fine structure (Costoff 1973; Girod 1976; Kurosumi 1986; Moriarty 1973). About 50% of the pituitary cell population is composed of GH cells. Ultrastructurally, the growth hormone cell gives the impression of a large round cell with a variable shape and position of its nucleus. Though a remarkable stability of its ultrastructure has been claimed (Horvath and Kovacs 1988; Saeger 1992), cyclic and pathological influences do alter their morphology. Such influences have been described, in pregnant and lactating rats (Salazar 1963; Cinti et al. 1985). Analysis of morphometric differences between sparsely granulated adenomatous GH cell and their normal counterparts with immunoelectron microscopy was instrumental in detecting specific labelling for growth hormone over the secretory granules of this cell in rat (Kurosumi 1986). A heavy granulation of this cell has been a consistent observation made by many workers.

Granulation pattern of this cell shows alteration in accordance with the reproductive state as can be realized from the increase in secretory granule content reported in various pregnant animals (Barnes 1962, 1963; Smith and Farquhar 1966; Young et al. 1967; Dekker 1967; Costoff 1973; Weman 1974; Young and Chaplin 1975; Bhiwgade et al. 1989). A transient fall in the number and the size of the secretory granules following surgical and chemical thyroidectomy (Salazar 1963; Stratmann et al. 1972; Horvath and Kovacs 1988; Bhiwgade et al. 1989; Saeger 1992). An interesting observation made by Horvath & Kovacs is the transdifferentiation of GH cell into TSH cell in thyroidectomized animals (Yoshimura et al. 1973a, b; Horvath and Kovacs 1988).

1.1.2 Light Microscopic Observations

In the various histochemical techniques used the secretory granules of these non-mucoid and acidophilic cells are stained with only acid dyes. They are stained orange with Orange G and yellow with Martius yellow and blue-green with Luxol Fast Blue (Fig. 1.1) in Periodic Acid Schiff/Orange G, Martius Scarlet Blue, Luxol Fast Blue/Periodic Acid Schiff/Orange G and Carmoisine L/Orange G/Aniline blue/Acid alizarin blue and Crossmon. They are negative to basic dyes Periodic Acid Schiff, Alcian Blue, Aldehyde Thionin, Aldehyde Fuschin and Aniline blue. Most of these cells are round or oval in shape with a diameter of 7–10 μ. These cells are distributed throughout the gland in the form of cluster of 3–4 cells or sometimes they occur singly. Their plasma membrane in clearly defined and the cytoplasm appears homogenous owing to the uniformly distributed fine secretory granules. Nuclei of these cells are small and round or oval with a diameter of 3–5 μ. They are either centrally placed or eccentric in position. The chromatin material is in the form of numerous clumps and a Golgi zone is observed in juxtanuclear position. A single nucleolus is often observed in the nucleus.

1.1.3 Electron Microscopic Observations

During active breeding season in the anterior pituitary of female garden lizard (*Calotes versicolor*), the growth hormone and prolactin hormones secreting cells are the most frequently cells observed. The GH cell is characteristically rounded or ovoid and typically arranged in groups. They are easily identified by the presence of a large number of very dense, spherical secretory granules that are scattered throughout the cytoplasm. The perigranular membrane is closely attached to the granules which present a sharp outline. The endoplasmic reticulum appears in small rows throughout the cytoplasm. The mitochondria are round and scattered throughout the cytoplasm. Apart from a slightly increased extent of granulation during the active oestrous phase in the female, these cells appear to undergo most conspicuous changes during anoestrous phase (Figs. 1.2 and 1.3).

The growth hormone (GH) cells are the most frequently encountered cells in the pars distalis. The GH.

The cellular organization is observed in oestrous females of the bat, *Rhinopoma microphyllum* (Fig. 1.4), late pregnant stage of bat, *Rhinolophus rouxi* (Fig. 1.5) and late pregnant stages of bats *Cynopterous sphinx* (Fig. 1.6) and *Hipposideros fulvus* (Fig. 1.7). The cell is round to polygonal in shape with an eccentrically placed nucleus. The highly electron dense secretory granules are round to oval in shape scattered throughout the cytoplasm. The Rough Endoplasmic Reticulum (RER) is lamellar, found at one pole of nucleus near the cell membrane. The Golgi zone is well developed with flattened sacs and round vesicles. They also contain a few mitochondria exhibiting prominent cristae.

In anterior pituitary of active male bat, *Rousettus leschenaulti*, the cell possesses granulation on one side, whereas the remaining cytoplasm is relatively clear of granules. Also, some ruptured mitochondria are present (Fig. 1.8).

These cells are polygonal in male rat (Fig. 1.9) and the secretory granules are round and electron dense. Lamellar RER profiles are observed with the RER cisternae arranged in parallel arrays.

In anterior pituitary of male monkey (Fig. 1.10), the secretory granules are of varying electron density. Four to five short arrays of lamellar RER are found near the cell membrane. Few mitochondria are discernible though mostly ruptured.

Apparently, drug treatment does not bring about demonstrable changes in the ultrastructure of this cell. In DMPA (depot medroxyprogesterone acetate) and Ovral (Ethinyl oestradiol + Levonorgestrel) treated female rats, this cell type is oval to elongated with centrally placed nucleus. Electron-dense secretory granules of a well-developed lamellar RER are found at the cell periphery near the nucleus (Figs. 1.11 and 1.12).

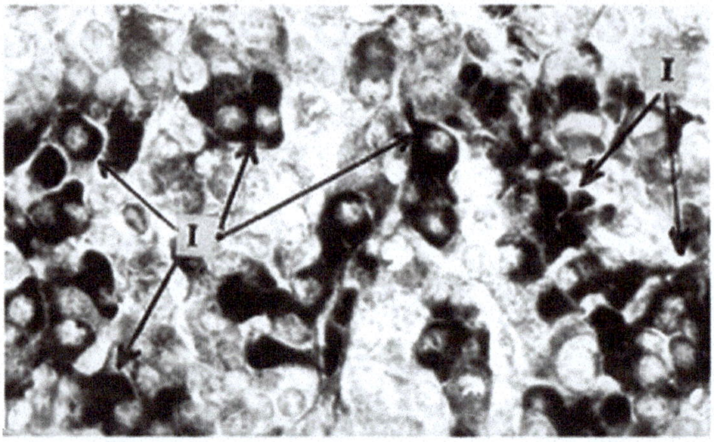

Fig. 1.1 Pars anterior of normal male bat, *Rousettus leschenaulti* showing STH cells (I arrows) stained with Luxol Fast Blue/Periodic Acid Schiff /Orange G [X 500] cell is characteristically rounded or ovoid and typically arranged in groups. They are easily identified by the presence of a large number of very dense, spherical secretory granules that are scattered throughout the cytoplasm. The perigranular membrane which presents a sharp outline is usually closely attached to the granule. However, some diffused granules can also be found the cell membrane can be clearly seen by separating the cytoplasm of one cell from that of another. The endoplasmic reticulum usually appears as a series of cisternae, arranged in parallel rows and stacked at one side of the cell. The exterior of this membranous system is richly studded with RNA particles. The Golgi complex may occupy a small zone near the nucleus or be extended over a large cytoplasmic area. The mitochondria look like long rods and are scattered throughout the cytoplasm, lipid droplets are also encountered

1.1.4 Discussion

The main basis for ultrastructural characterization of this cell type is the presence of characteristic, very dense, uniform secretory granules, absence of perigranular membrane, centrally located nucleus and inconspicuous Golgi apparatus (Salazar 1963; Herlant, 1964; Weman 1974; Murphy and James 1976).

The growth hormone cells (GH cells) are numerous, well distributed and clearly identifiable. Ultrastructurally, the GH cells are generally ovoid or polygonal and possess a slightly eccentric, spherical and mostly euchromatic nucleus, often with a fairly prominent dense nucleolus. The cytoplasmic storage granules are spherical, with a high electron density. The Golgi complex is inconspicuous in normal pituitary but is well marked in the pars distalis during pregnancy. The RER is not extensive but a few short cisternae are present in the cytoplasm. A subpopulation of GH cells also contains a varying number of small secretory granules intermingled with the larger ones. IEM consistently detects specific labelling for GH over the secretory granules (Kurosumi 1986).

The secretory activity of the GH cell rises with advancing pregnancy indicated by an increased number of mature, electron-dense granules, a well-developed Golgi complex and a moderate number of slightly hypertrophied mitochondria. Similar observations have been reported for the mouse (Barnes 1962, 1963), rat (Smith and Farquhar 1966), rabbit (Young et al. 1967), hamster (Dekker 1967), rat (Costoff 1973), mink (Weman 1974), deer (Young and Chaplin 1975) and bat (Bhiwgade et al. 1989). These observations are in accordance with the results of the cell population study of the dynamics in the rat adenohypophysis during full-term pregnancy which shows that the GH cells composed 47% of all the anterior pituitary cells in non-lactating rats (Poole and Kornegay 1982).

The ultrastructural morphology of the GH cell has been reported by Cinti et al. (1985) to be similar in the lactating and non-lactating homologs rats. A process of hormone release has been suggested for the pituitary basophils in

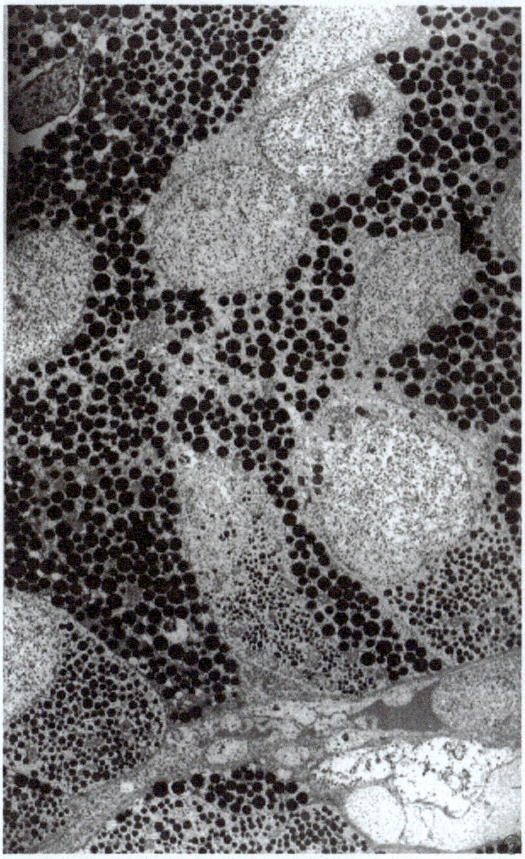

Fig. 1.2 EM of anterior pituitary of female garden lizard (*Calotes versicolor*) during breeding season. Note several somatotroph cells with dense large, round secretory granules. It is not uncommon to see coalesced clumps of secretion in most cells so extensive occupancy of secretory granules within the cytoplasm that other organelles are obscure or pushed to a relatively small part of the cytoplasm. Discreet nucleus with fairly uniform chromatin spread and occasional nucleolus is seen [X 4000]

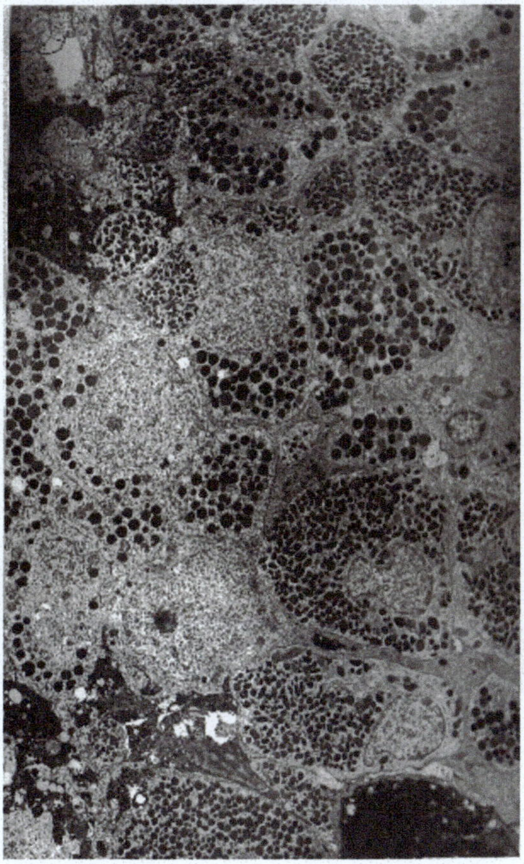

Fig. 1.3 EM of anterior pituitary of female garden lizard (*Calotes versicolor*) during active breeding season showing the mixed population of STH and LTH cells [X 4000]

rats (Herlant 1963), for the GH cells in mice after hepatectomy (Echave-Llanos et al. 1971) and in male mice under normal conditions (Gomez-Dumm and Echave-Llanos 1972).

1.2 Prolactin Cell (PRL Cell)

1.2.1 Introduction

This cell may be rightly called the 'Pituitary Cytologist Cindrella' since, several features of the prolactin cell have made it a favourite cell type for detailed morphological analysis of the secretory process. The percentage of PRL cells in the pituitary is between 10% and 30%. In the normal functional state, this cell cannot be distinguished from the GH cells by conventional stains. Despite the prevailing heterogeneity, the typical PRL cell, however, can be readily distinguished from the other cell types by ultrastructural morphology alone.

Such unmistakable fine structural features offer a distinct advantage in any study requiring an easy identification of individual cells. IEM studies of PRL cells of the rat (Nakane 1975; Parson and Erlandsen 1974), the guinea pig (Beauvillain et al. 1977), the cow (Dacheux and Dubois 1976) and sheep pituitaries (Parry et al. 1978, 1979) corroborate its fine structural characteristics. The cell as seen under the electron microscope is large, polygonal or irregular in shape with its nucleus showing a moderate amount of dense chromatin

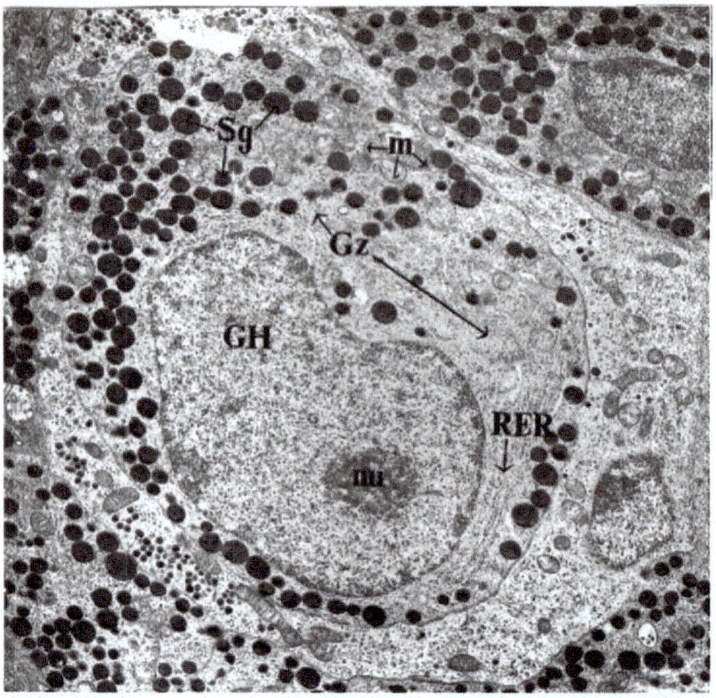

Fig. 1.4 High-power electron micrograph of the anterior pituitary of female bat, *Rhinopoma microphyllum*, during the oestrous phase showing GH cell. Note prominent nucleolus (nu), highly electron dense round secretory granules (Sg) and peripherally arranged lamellar cisternae of RER (RER). Also, note the Golgi zone (Gz) comprising of sac and vesicles [X 8500]

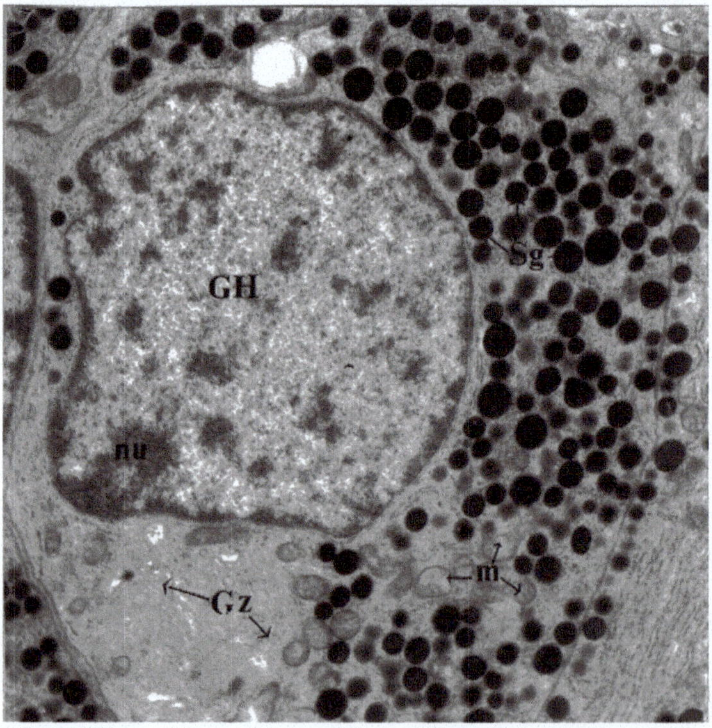

Fig. 1.5 Electron micrograph of anterior pituitary of pregnant female bat, *Rhinolopus rouxi*, showing GH cell with numerous electron-dense secretory granules (Sg), mitochondria (m) and Golgi zone (Gz) [X 10,000]

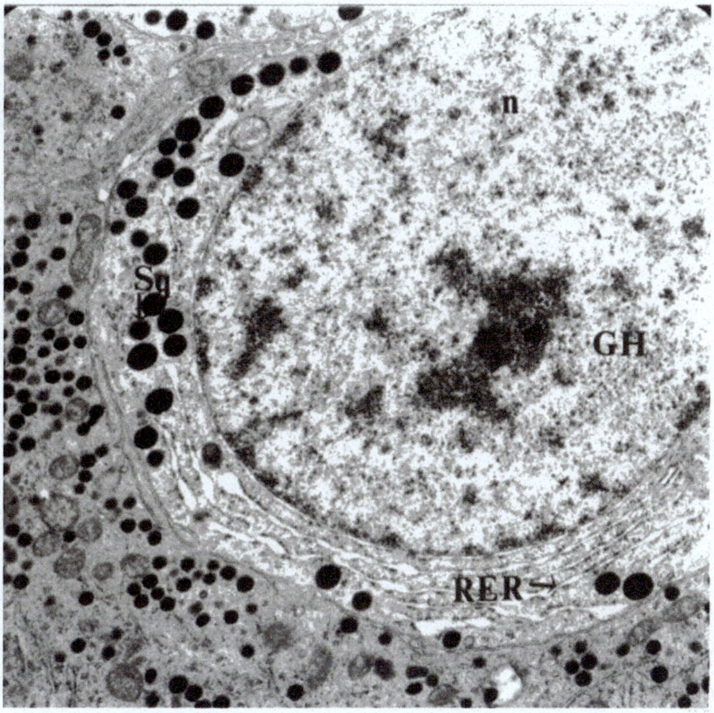

Fig. 1.6 Growth Hormone cell of a pregnant female bat, *Cynopterous sphinx*, showing large nucleus (n), peripheral arrangement of RER and round secretory granules (Sg) [X 16, 000]

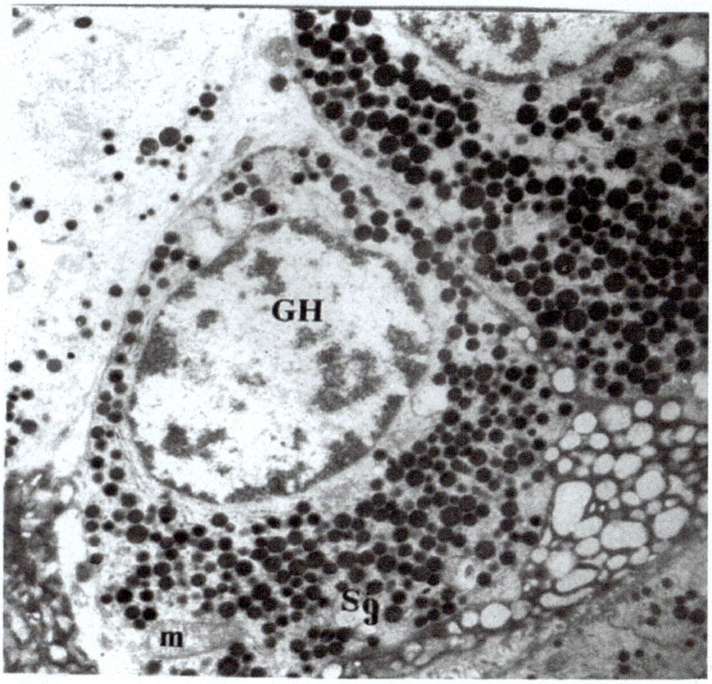

Fig. 1.7 Electron micrograph of pituitary of female bat, *Hipposideros fulvus*, showing densely granulated Growth Hormone cells. The secretory granules (Sg) cover the entire cytoplasmic area and a few mitochondria (m) [X 6500]

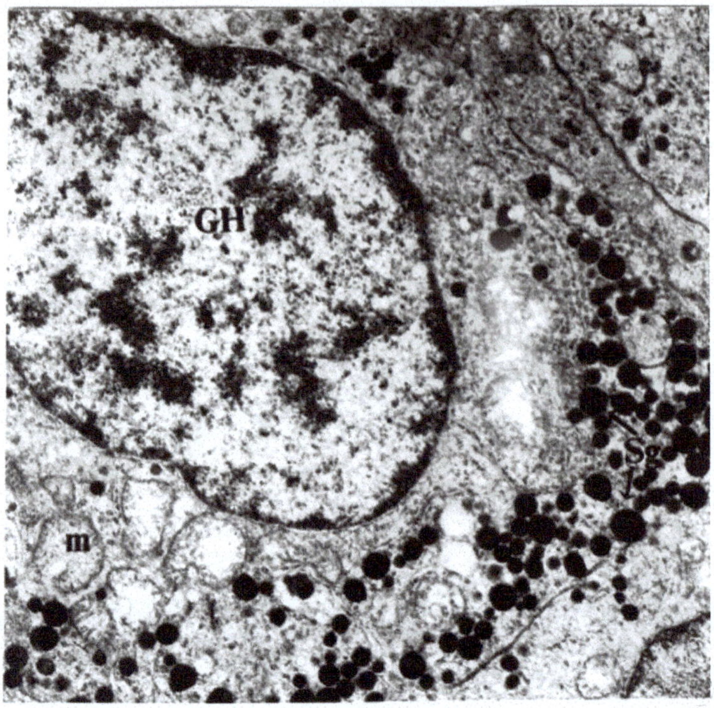

Fig. 1.8 Growth Hormone cell of anterior pituitary of male bat, *Rousettus leschenaulti*, showing round secretory granules (Sg) and numerous mitochondria (m) [X 8000]

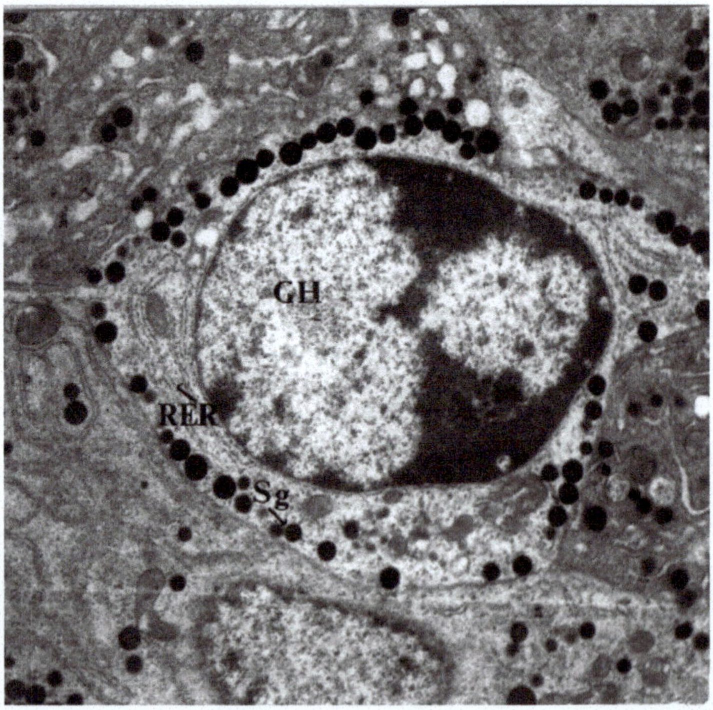

Fig. 1.9 Growth Hormone cell of anterior pituitary of male rat, note highly electron-dense secretory granules (Sg) and few parallel arrays of lamellar RER near the cell periphery [X 8500]

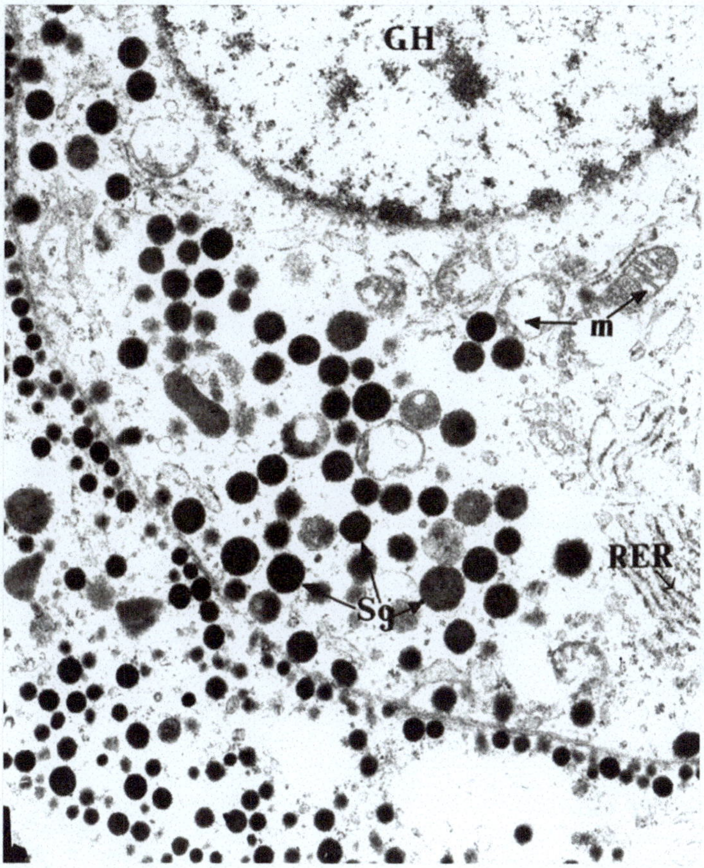

Fig. 1.10 High-power electron micrograph of the Growth Hormone cell of male monkey Macaca radiata. Note the round secretory granules of variable density (Sg), few short arrays of lamellar Rough Endoplasmic Reticulum (RER) and few ruptured mitochondria (m) [X 13,000]

and a nucleolus. Young et al. (1967) have recorded an increasing indentation of the nucleus in lactating females. Similar observations are also made by Murphy and James (1976).

This cell is extremely sensitive and it is possible to alter the intracellular processes of prolactin synthesis by physiological changes and experimental manipulations (McComb and Kovacs 1978). Such alterations are reflected in the changed morphology of the cell. A cyclicity of morphological alterations of the cell synchronized with the reproductive cycle has been demonstrated in different animals. Such altered fine structure is most pronounced in pregnant and lactating females (Shiino et al. 1972; Horvath and Kovacs 1988; Saeger 1977; Asa et al. 1984; Bhiwgade et al. 1989; Mantri et al. 1995; Borkar 1998).

Several authors have recorded a correspondence between the ultrastructure and the stage of prolactin secretion (Hymer et al. 1961; Young et al. 1967; Shiino et al. 1972; Merchant-Larois and Mena 1982). The whorls of concentric RER Nebenkern are a prominent feature of stimulated PRL cells during pregnancy and lactation (Farquhar and Rinehart 1954a, b; Hymer et al. 1961; Shiino et al. 1972; Cinti et al. 1985). Similarly, drug-induced fine structural alterations have also been studied (Hymer et al. 1961). In this context, it is now well known that oestrogens act as a strong stimulus for these cells and increase the synthesis and output of its hormone (Saeger 1977; Stefaneanu and Kovacs 1991; Lloyd 1991; Nogami 1984). Thus, the proliferative effect of oestrogens on the PRL cells could be of immense value in confirming the identity of the cell. The cytoarchitecture of such stimulated cells is similar to that of the cells during pregnancy (Saeger 1977, 1992; Bhiwgade et al. 2001).

The potentially important marker in the identification of these cells is the dense and spherical as well as irregular, secretory granules seen within the Golgi sacculi (Horvath and Kovacs 1988;

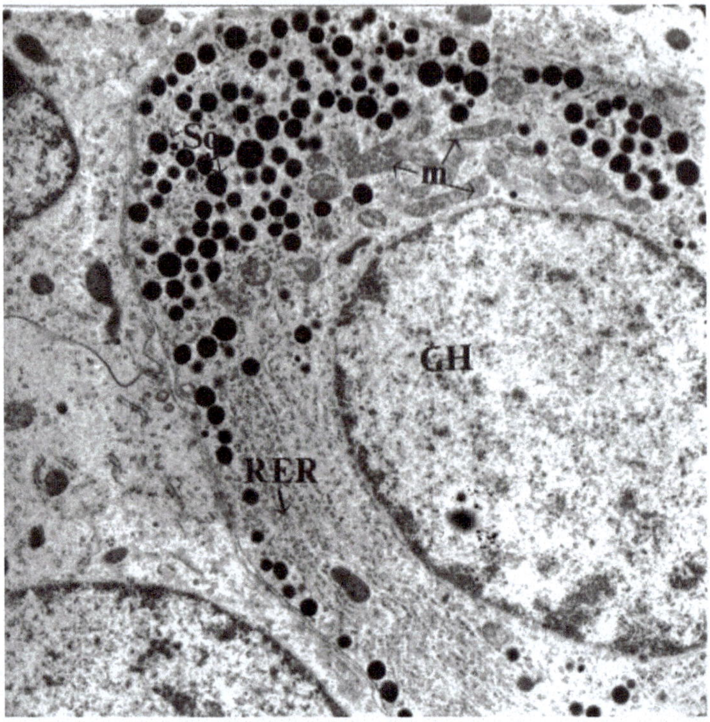

Fig. 1.11 Growth Hormone cell of anterior pituitary of DMPA treated female rat. Note the electron-dense round secretory granules (Sg), numerous mitochondria (m) and Rough Endoplasmic Reticulum situated near the periphery in the form of short parallel arrays (RER) [X 13,000]

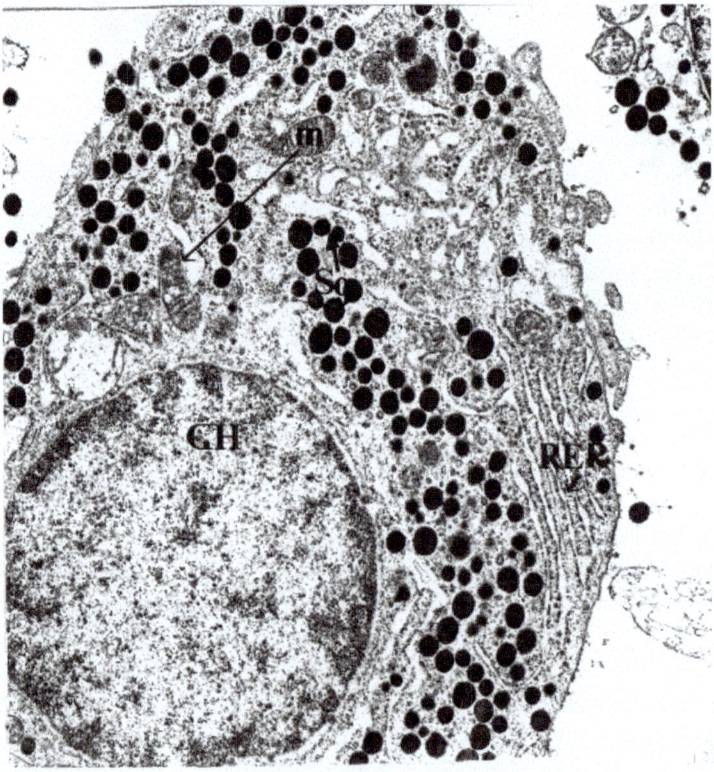

Fig. 1.12 Growth Hormone cells of Ethinyl oestradiol + Levonorgestrel treated rat showing numerous secretory granules (Sg), few mitochondria (m) and slightly dilated Rough Endoplasmic Reticulum cisternae which are found near the cell periphery (RER) [X 13,000]

Young et al. 1967; Dekker 1967). A reliable way of identifying this cell in the lactating female is reported by Shiino et al. (1972) in rats. Murphy and James (1976) in mink and Bhiwgade et al. (1989) in bat. All these workers have made an important concordant observation that in the lactating females isolated from their litter, accumulation of secretory granules is seen in the prolactin cells.

Pleomorphic nature of the mature secretory granules, as an identity of this cell, is confirmed in many rodent species (Dekker 1967; Gomez-Dumm and Echave-Llanos 1972; Murphy and James 1976) and rabbits (Young et al. 1967). In humans pleomorphism of prolactin secretory granules has been verified by Paiz and Hennigar (1970), Lawzewitsch et al. (1972), Kovacs et al. (1977), McComb and Kovacs (1978) and Horvath and Kovacs (1988). But polymorphic secretory granules as a criterion to identify lactotrophs are being reconsidered in the light of IEM findings. There is a strong argument that granular pleomorphism seen in rats (Pasteels 1963a; Smith and Farquhar 1966) and in humans (Herlant, 1964; Pasteels 1963b; Pasteels et al. 1973) is a matter of morphological heterogeneity of this cell type. Diversity in the shape and size of prolactin secretory granules is observed in the pituitary of guinea pig (Beauvillain et al. 1977), bovine and porcine pituitaries (Dacheux and Dubois 1976; Dacheux 1980).

Mammosomatotrophs, the cells secreting both GH and PRL, are a common pituitary cell type in a variety of animals as evidenced by studies with rats (Porter et al. 1990; Ishibashi and Shiino 1989a), musk shrew (Ishibashi and Shiino 1989a), bats (Ishibashi and Shiino 1989b), bovine pituitary (Kineman et al. 1991) and men (Bendayan 1982, 1984; Asa et al. 1988).

It is necessary to record that these bihormonal cells resemble the somatotrophs and have only subtle structural differences with the latter (Asa et al. 1988). Therefore, morphological detectability of mammosomatotrophs is a matter of antigen preservation in the mature pituitary.

Some carefully designed investigations on the cytogenesis of human foetal adenohypophysis have been carried out to record the structural differences between the mammosomatotrophs and the somatotrophs (Asa et al. 1988). Ultrastructurally, the mammosomatotroph is a large, round or polygonal cell with a round nucleus harbouring clumped heterochromatin and prominent nucleolus. The RER and Golgi are well developed. Perhaps the most significant feature of this bihormonal cell is its many large and pleomorphic granules. In such cells, immunogold technique has revealed co-localization of both PRL and GH.

Despite the firm denial of bihormonal cells in the rat pituitary, mammosomatotrophs are common (Boockfor and Frawley 1987; Frawley et al. 1985; Leong et al. 1985; Nikitowich-Winer et al. 1987). Such bihormonal cells also exist in mouse adenohyposis (Sasaki and Iwama 1989) and are smaller than PRL cells.

1.2.2 Light Microscopic Observations

The secretory granules of these cells are stained purple, red, pale yellow and gold tone. They are Alcian Blue—negative, Aldehyde Thionin—negative, weakly Periodic Acid Schiff—positive and exhibit slightly greater affinity towards Orange G in the combined staining procedures. The LTH cells are erythrosinophilic with the Cleveland—Wolfe's trichrome procedure, Carminophilic with Dowson and Friedgood's technique, Fuchsinophilic with Crossmon's technique (Fig. 1.13), pale yellow or gold tone with Luxol Fast Blue/Periodic Acid Schiff/Orange G procedure, red or purple with Martius—scarlet blue/green procedure, reddish orange with Brookes staining. Staining with Periodic Acid Schiff—Orange G alone or in combination with complex basic dyes (Alcian Blue, Aldehyde Fuschin and Aldehyde Thionin) sometimes results in reddish-orange colour of these cells (apparently due to their weak affinity for Periodic Acid Schiff). The LTH cells are distinguished easily from other acidophils (STH cells) with Crossman's (1937), Martius Scarlet Blue and Carmoisine L/Orange G/Light Green techniques. These cells most commonly occur in clusters but are also observed scattered. These cells are round or oval shaped

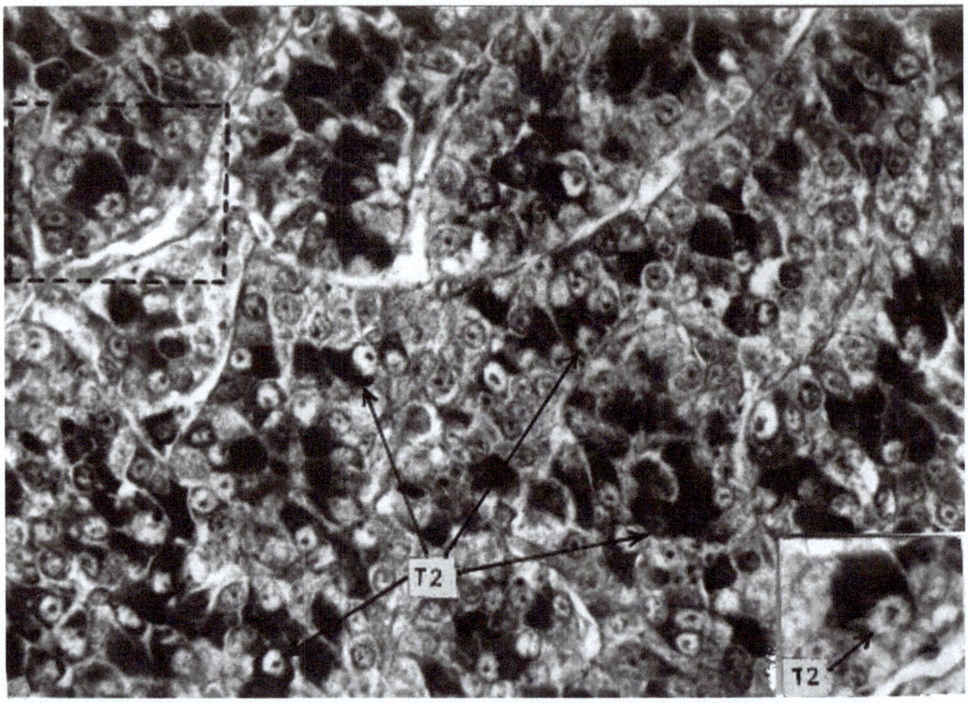

Fig. 1.13 Anterior pituitary of a female squirrel at 2 weeks of post-partum showing fuchsinophilic LTH cells (T2). Crossmon's stain. [X 250] Inset: Enlargement of the marked area

but are slightly irregular and larger in size. The cell boundaries are distinct. The nucleus is prominent and often eccentric. The cytoplasm of these cells appears homogeneous because of the evenly distributed fine granules. The cells undergo hypertrophy, hyperplasia and increased secretory activity at the onset of pregnancy. Even during early and mid-pregnancy they constitute the majority of cells in the pars anterior and in the lactating female they occupy most of the pars anterior. The signs of cellular hyperactivity such as hypertrophy and nucleus, presence of an enlarged Golgi zone and intense cytoplasmic granulation are conspicuous. In pregnant and lactating females the type-2 cells are densely granulated. These cells are large, round and are numerous in pregnant and lactating bats, *Rousettus leschenaulti*, which are Carmoisine L-positive LTH cells (Figs. 1.14 and 1.15) and dark green gonadotrophs cells.

Electron Microscopic Observations in Normal Cyclical Animals

In the pituitary of the female garden lizard (*Calotes versicolor*) during active breeding period, the PRL cells are numerous and occupy the entire anterior pituitary along with some STH cells. Polymorphic secreting granules are mostly observed. Mature secretory granules are mostly polymorphic in nature and distributed throughout the cytoplasm. The RER is the most prominent organelle and is arranged in parallel rows with frequent Nebenkern formations. The mitochondria are very well developed with cristae. The Golgi zone is also prominent (Figs. 1.16, 1.17, 1.18 and 1.19).

The PRL cell type is the most studied cell type of pars anterior in mammals. The electron microscope reveals elongated cell bodies, moderately developed rough endoplasmic reticulum (RER), and many spherical or irregularly shaped granules in the physiologically quiescent female. The cell

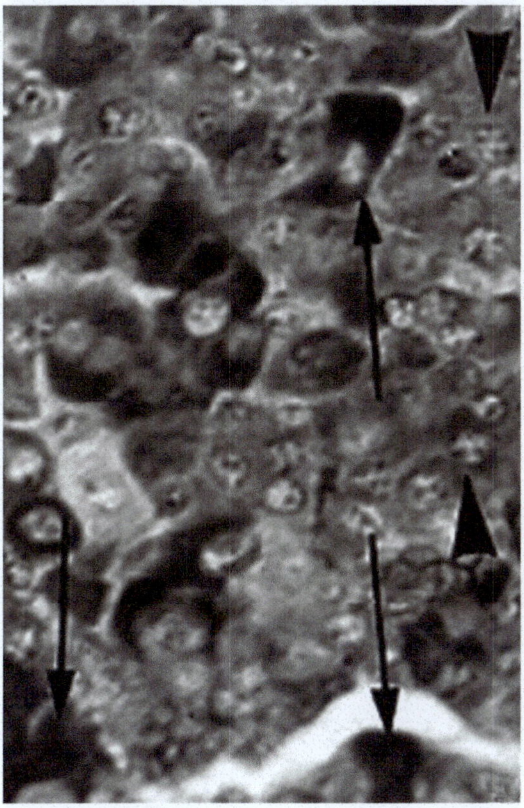

Fig. 1.14 Anterior pituitary of a lactating female bat, *Rousettus leschenaulti* Carmoisine L—positive LTH cells (arrow heads) dark-green gonadotrophs (arrows) Carmoisine L/Orange G/Light Green [X 600]

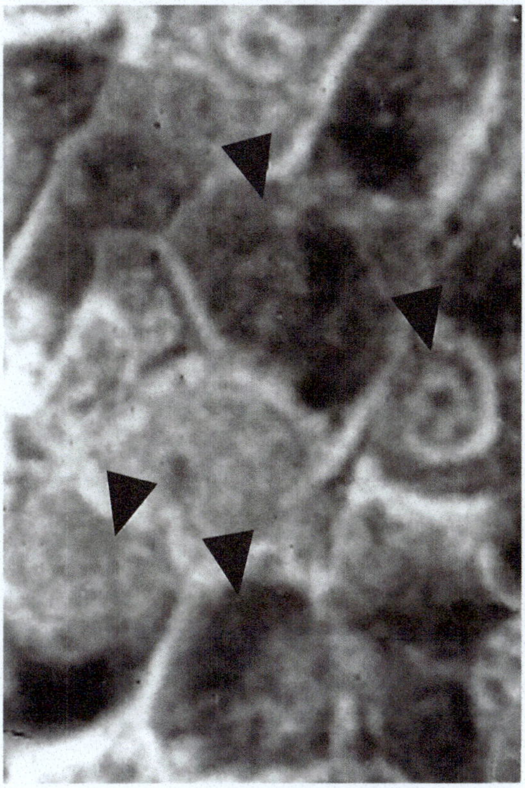

Fig. 1.15 Anterior pituitary of a lactating female bat, *Rousettus leschenaulti* Carmoisine L-positive LTH cells (arrow heads) [X 800]

with signs of higher secretory activity contains a well-developed RER with the formation of 'nebenkern', a prominent Golgi apparatus, and fewer and smaller granules. Exocytoses are frequently observed. In the period of advanced pregnancy and lactation, there is maximum physiologic stimulation of prolactin cells following which they transform into pregnancy cells which are characterized by their large elongated size with sparse granulation.

The PRL cell of oestrous female rat is polygonal in shape with an eccentrically placed nucleus. The cell contains a moderate number of pleomorphic granules, parallel arrays of lamellar RER, a well-developed Golgi zone (Gz), a few irregular mitochondria and lysosomes, which are in the form of degenerating secretory granules and multivesicular bodies (Fig. 1.20). In *Suncus murinus* (Fig. 1.21), the PRL cell includes a lobed nucleus with moderate chromatin material. The cytoplasm consists of sparse granulation with pleomorphic granules, the RER manifests in the 'nebenkern' form which is at one pole in most of the cells. The Gz comprises dilated sacculi and numerous vesicles which is near the nucleus. The mitochondria are few, irregular in shape with conspicuous cristae and lysosomal activity is seen in few cells.

In the Bonnet monkey (*Macaca radiata*) (Figs. 1.22 and 1.23) during early pregnancy the PRL cell contains a well-developed juxtanuclear Gz, with dilated sacculi and the luminal ends consist of granules in the stages of early formation. Enlarged mitochondria with inconspicuous cristae are observed near the Gz. The mature secretory granules are found aligned in a single

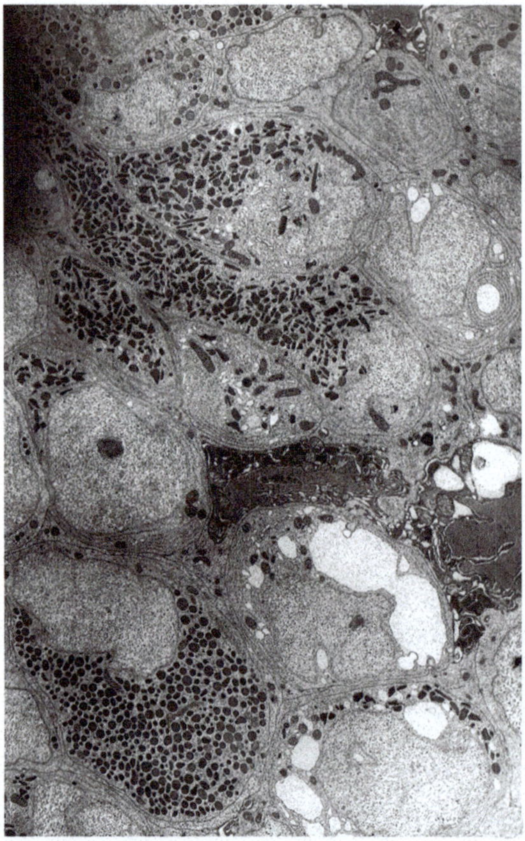

Fig. 1.16 Low-power electron micrograph showing numerous prolactin cells (PRL) in the pituitary of female *Calotes versicolor* during active breeding season. Note the dense polymorphic secretory granules. Also, note several mitochondria. For the first time, the polymorphism of secretory granules is shown in this reptilian model in the evolutionary hierarchy of amniotes. The morphology of the secretory granular deviate from its conventional round profile and assume various shapes filling much of the cellular matrix [X 5000]. Unpublished electron micrographs from Dr. Bhiwgade

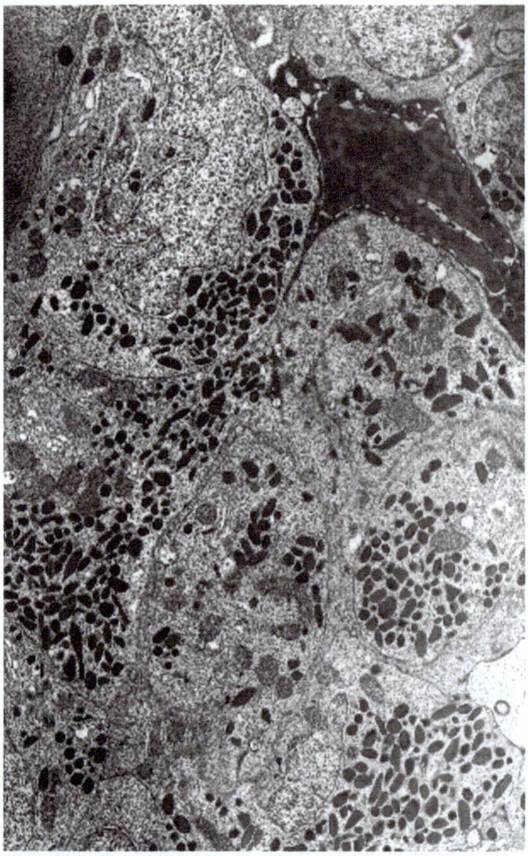

Fig. 1.17 Enlarged part of the PRL cell of female garden lizard (*Calotes versicolor*) during active breeding season. Note several PRL cells with polymorphic secretory granules and several elongated to round mitochondria [X 8000]. Unpublished electron micrographs from Dr. Bhiwgade

file arrangement along the cell membrane indicating that discharge of secretion is imminent.

In the pituitaries of the early pregnant *Suncus murinus* (Fig. 1.24), the cell picture is dominated by large, hypertrophied prolactin cells with cytological appearances of active secretion. The cytoplasm contains a well-developed rough-surfaced endoplasmic reticulum with the formation of nebenkern, a prominent Golgi zone, and fewer and smaller granules. Exocytoses are frequently observed.

During the period of advanced pregnancy, there is a maximal physiologic stimulation of prolactin cells, which transform into pregnancy cells and are characterized by their large, elongated size, well-developed RER and sparse secretory granules.

In various species of bats viz. *Hipposideros* and *Megaderma* (Figs. 1.25, 1.26 and 1.28) and in the monkey (*Macaca radiata*) (Fig. 1.29) and *Suncus (*Fig. 1.27*)* during late pregnancy, the PRL cells contain abundant RER indicative of stepped-up synthetic activity. This stimulation is further supplemented by the presence of concentric nebenkern in *Hipposideros* and *Suncus.* In Bonnet monkey, the RER is in the form of

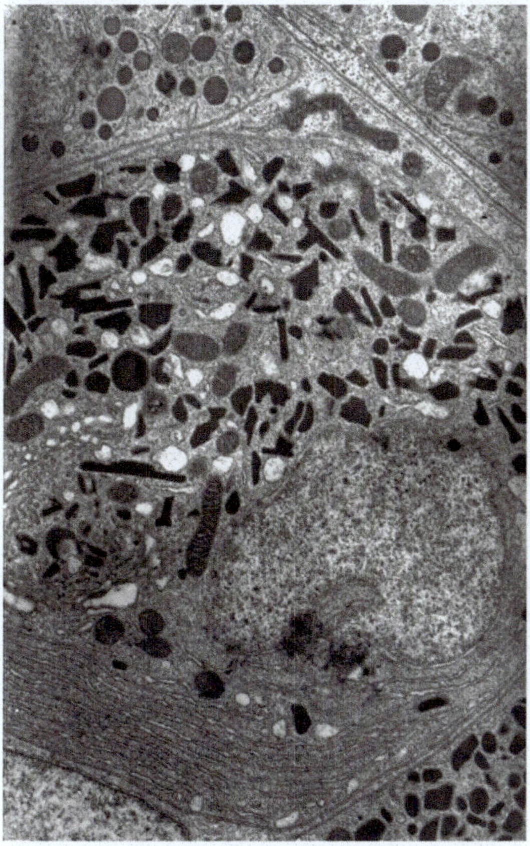

Fig. 1.18 Enlarged part of the PRL cells of anterior pituitary of female garden lizard (*Calotes versicolor*) during active breeding season. Note several dense polymorphic secretory granules, several elongated mitochondria and rough parallel endoplasmic reticulum. The sigmoid puzzle configuration of the secretory granules imparts amozaic tile design to the cell. The orderly arrangement sequence of nucleus, Golgi zones and mitochondria in the centre and the parallel array of rough endoplasmic reticulum convey a stratified impression [X 15,000]. Unpublished electron micrographs from Dr. Bhiwgade

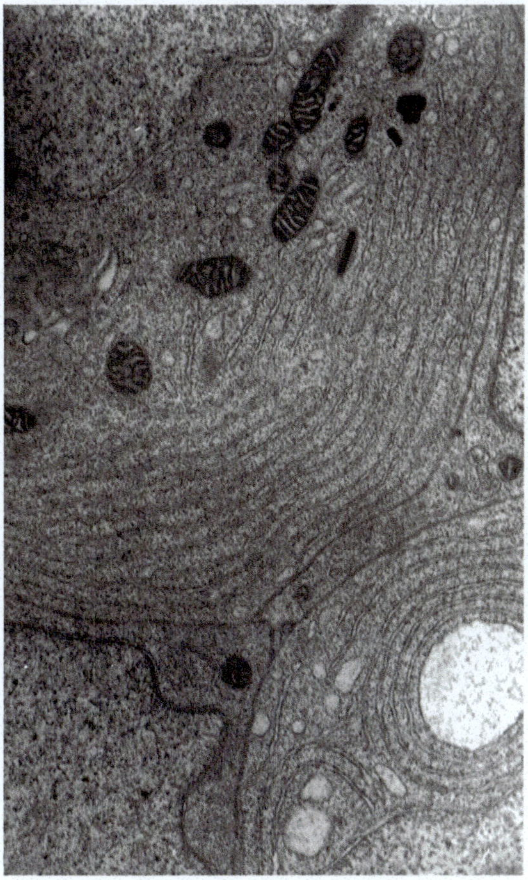

Fig. 1.19 Enlarged part of the prolactin cells of anterior pituitary of female garden lizard (*Calotes versicolor*) during active breeding season. Note the Nebenkern formation of RER and well developed mitochondria [X 15,000]. Unpublished electron micrographs From Dr. Bhiwgade

parallel lamellae whereas in *Megaderma* (Fig. 1.28) the RER is in extensive parallel arrays as well as stages transitional between parallel lamellae and concentric nebenkern. An increase in the number of RER cisternae indicates hypertrophy of the organelle. The granules are large, electron-dense and round in *Hipposideros, Megaderma* and *Suncus;* whereas the secretory granules of *Cynopterous* are large and pleomorphic. In *Hipposideros*, the peripheral alignment of secretory granules under the cell membrane prompts the speculation that the cell is also in the active secretory state. Very well-developed multiple Golgi zones are formed with many dilated sacculi and vesicles, in the vicinity of which are found many immature granules in case of Bonnet monkey (Fig. 1.29). The cytoplasmic volume is markedly increased. The cell becomes elongated. The RER is extremely developed showing highly organized parallel cisternae studded with many ribosomes. The Golgi apparatus is distinctly enlarged with slightly dilated sacculi and increased numbers of immature granules. In contrast, the mature secretory granules are very sparse and relatively small. Lysosomes are scanty, small and rich in content.

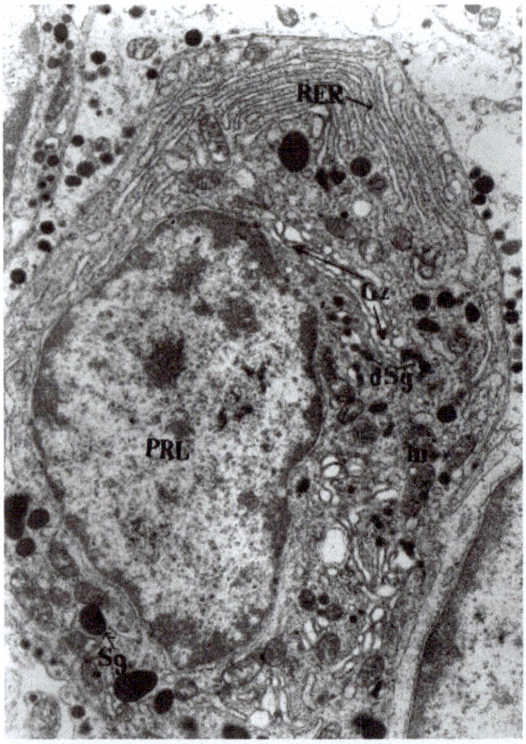

Fig. 1.20 A typical prolactin cell of anterior pituitary of female rat. This cell is characterized by existence of large polymorphic secretory granules (Sg). Note the distribution of compact parallel arrays of Rough Endoplasmic Reticulum (RER). Immature, small polymorphic granules juxtanuclear Golgi zones are often present in the Golgi region (Gz); also few mitochondria (m) are present [X 10,000]. Unpublished electron micrographs from Dr. Bhiwgade

Mitochondria are also sparse, oval or elongated. The cytoplasm is characterized by numerous mitochondria, some of which are ruptured as in case of *Cynopterous* (Fig. 1.30) whereas in case of *Suncus* (Fig. 1.27) a few mitochondria with a homogenous matrix are discerned. Presence of lysosomes probably denote the degradation of excessive hormone in *Hipposideros* (Fig. 1.26).

The granulation pattern in PRL cells in the late pregnant bat, *Taphozous melanonopogon* (Fig. 1.31) is an apparent deviation from its counterparts as in other species. In this species, the granulation is dense and the granules are pleomorphic. Such an ultrastructural peculiarity with respect to granulation may be assumed to be in anticipation of a full-term pregnancy during which the secretory activity is enhanced.

The lactating females with attached young ones of bat, *Pteropus giganteus* (Figs. 1.32 and 1.33) and *Rhinopoma microphyllum show* (Fig. 1.34) the PRL cell exhibits many fine structural similarities and differences. There is an eccentrically placed nucleus. In *Pteropus*, the cytoplasm contains parallel arrays of ER studded with ribosomes (Fig. 1.33). The extensively developed multiple Golgi complex is made up of flattened sacs, vacuoles and small vesicles. Here, the Golgi zone is also constituted by a few concentrically placed sacculi and few vesicles (Fig. 1.32). These observations coupled with a sparse granulation are suggestive of the cell having reached a peak of secretory activity. In case of *Rhinopoma microphyllum* (Fig. 1.35), the PRL cell demonstrates nebenkern formation of RER within whose core contains newly synthesized hormone material in the form of granules. The granules are packed in Golgi stacks and hence immature granules are found in the vicinity of Gz. Extrusion of granules as observed here indicates that the secretion is on, and will result in sparse granulation. Numerous mitochondria are also observed.

The PRL cells of untreated animals are large, round and polymorphic. The nuclei are round and moderately rich in chromatin. The RER is arranged regularly in parallel stacks of membranes. Round-shaped mitochondria with foliate cristae are scattered throughout the cytoplasm. The Golgi apparatus consists of 3–5 stacks of flat cisternae. The mature secretory granules are electron-dense and are surrounded by membranes.

Prolactin cells are stimulated by different drugs and hormones. The stimulation is mostly the effect of decreased dopamine inhibition, as the direct effect of increased secretion of prolactin releasing factor (PRF).

Remarkable changes are observed in the hypophysis after treatment with the following drugs / or hormones. There is a striking increase in the number of PRL cells showing characteristics of stimulated secretory cells.

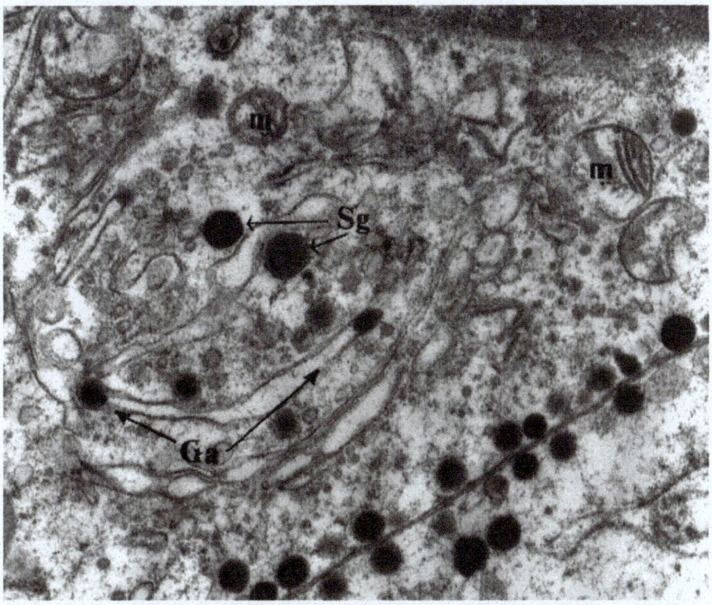

Fig. 1.21 Electron micrograph of PRL cells in an early pregnant female, musk shrew (*Suncus murinus*). Note the lamellar rough endoplasmic reticulum (RER) arranged in parallel arrays, few secretory granules (Sg) and homogenous mitochondria (m) present in the cell, and dilated sacs of Golgi apparatus (Ga) [X 6500]. Unpublished electron micrographs from Dr. Bhiwgade

The stimulated PRL cell (due to decreased dopamine suppression) has an ultrastructure similar to that of the cell during pregnancy. The cytoplasmic volume is markedly increased. The cell becomes elongated, the nuclei may be slightly lobated and contain more than one nucleoli. The rough endoplasmic reticulum is extremely developed showing highly organized parallel membranes with many ribosomes. The Golgi apparatus is distinctly enlarged with slightly dilated sacculi and increased numbers of immature granules. In contrast, the mature secretory granules are very sparse, slightly pleomorphic and relatively small.

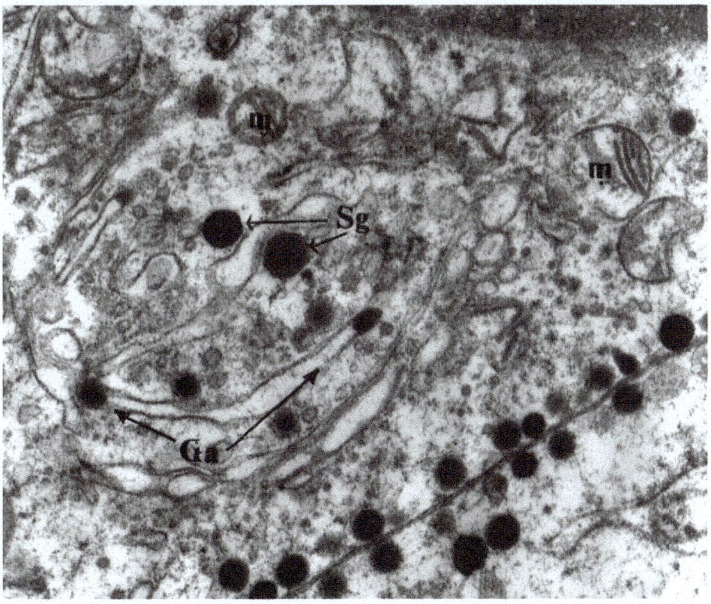

Fig. 1.22 High-power electron micrograph of early pregnant female monkey (*Macaca radiata*) showing PRL cell with a well-developed Golgi apparatus (Ga) with immature granules forming in the Golgi sacs. Note round secretory granules (Sg) and numerous hypertrophied mitochondria (m) [X 26,000]. Unpublished electron micrographs from Dr. Bhiwgade

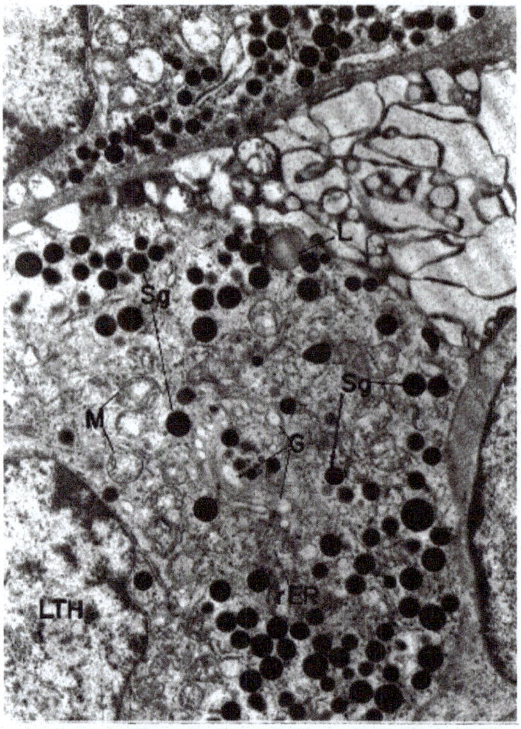

Fig. 1.23 Electron Micrograph of early pregnant female monkey (*Macaca radiata*) showing prolactin cell. Note well-developed Golgi zone (G), large secretory granules and lysosome (L) and few rough endoplasmic reticulum (RER) [X 26,000]. Unpublished electron micrographs From Dr. Bhiwgade

1.2.3 Discussion on Normal Prolactin Cells in Cyclical Animals

The prolactin cell presents as a large cell body, which is polygonal to elongated in shape, with an eccentrically placed nucleus, variable number of round to polymorphic highly electron-dense secretory granules, typical lamellar RER and a juxtaposed Golgi apparatus. Our observations on PRL cells of various species corroborate with the findings of Hymer et al. (1961), Aumuller et al. (1978), Horvath and Kovacs (1988) in rats; Dekker (1967) in hamsters and in various experimental animals (Saeger 1992). PRL cells showed polygonal or elongated shapes with small secretory granules, oval or polygonal cells with medium-sized round and polymorphic granules. PRL cells, which sometimes embrace a gonadotroph show the appearance of a cup-shaped cell.

Although an important criterion for identifying the PRL cells has been the occurrence of large and pleomorphic secretory granules, it does not necessarily serve as an absolute marker (Nogami 1984). Our study shows that the PRL cell granules varied from spherical to polymorphic, and from small to large in size after being packaged in the Golgi region.

Fig. 1.24 Low-power electron micrograph of AP of female musk shrew (*Suncus murinus*) during early pregnancy showing many PRL cells. Note the marked abundance of highly organized RER with 'nebenkern' (RER) with prominence of Golgi zone, large polymorphic secretory granules and numerous mitochondria [X 3500]. Unpublished electron micrographs from Dr. Bhiwgade

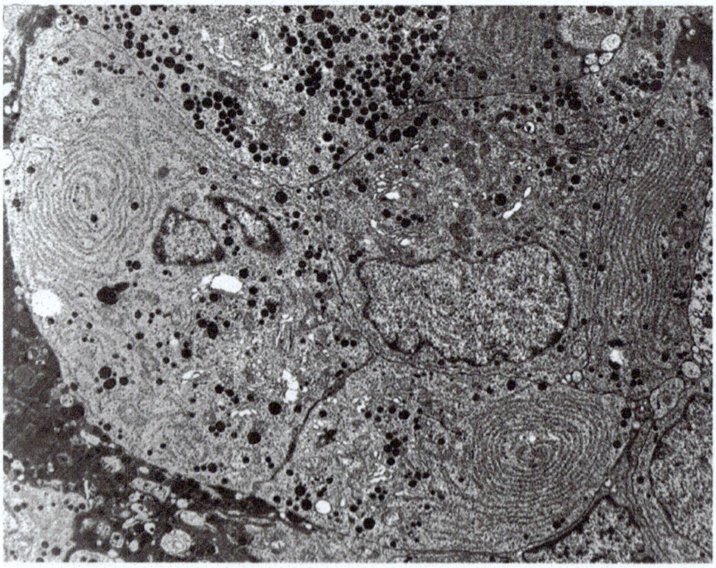

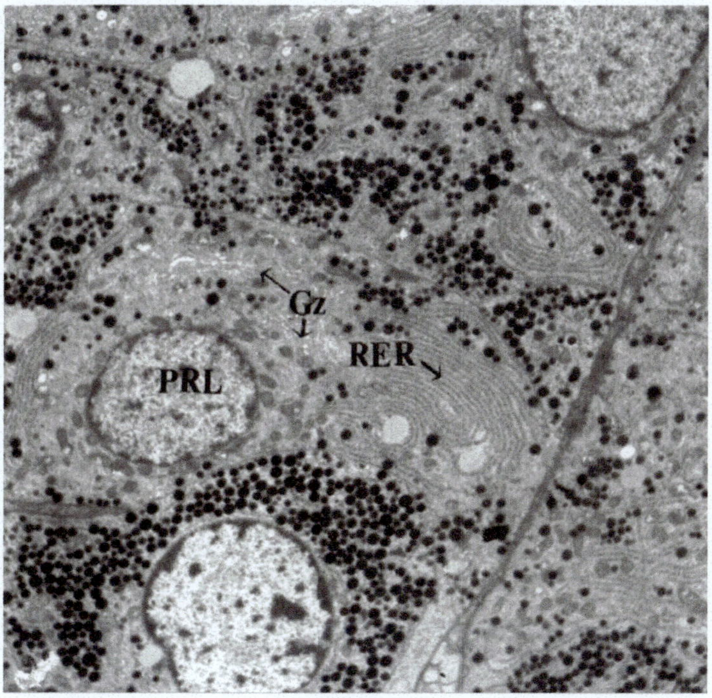

Fig. 1.25 Anterior pituitary of late pregnant female bat, *Hipposideros fulvus fulvus*. Note large PRL cell and well-developed parallel arrays of RER and Golgi zone (Gz) [X 6500] Unpublished electron micrographs from Dr. Bhiwgade

In our observations, the pituitaries of some species during oestrous period show a great increase in the prolactin cells with respect to RER, mitochondria and Golgi apparatus as seen in rats and Suncus. These morphological results are in good agreement with the findings of El

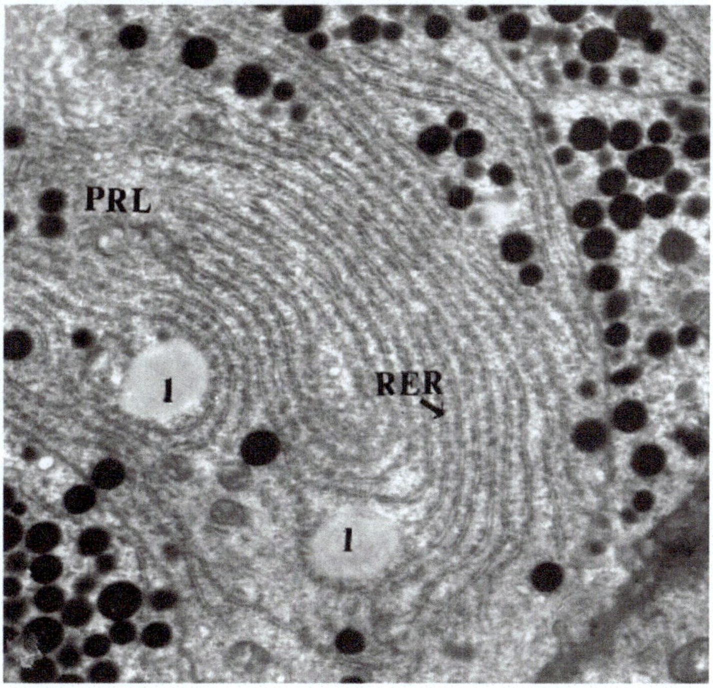

Fig. 1.26 High-power electron micrograph of the same PRL cell of *Hipposideros fulvus fulvus* in Fig. 1.13. Note extensively developed parallar rough endoplasmic reticulum (RER) and lipid droplets (l) initiation of nebenkern with core of lipid inclusions [X 6000]. Unpublished electron micrographs from Dr. Bhiwgade

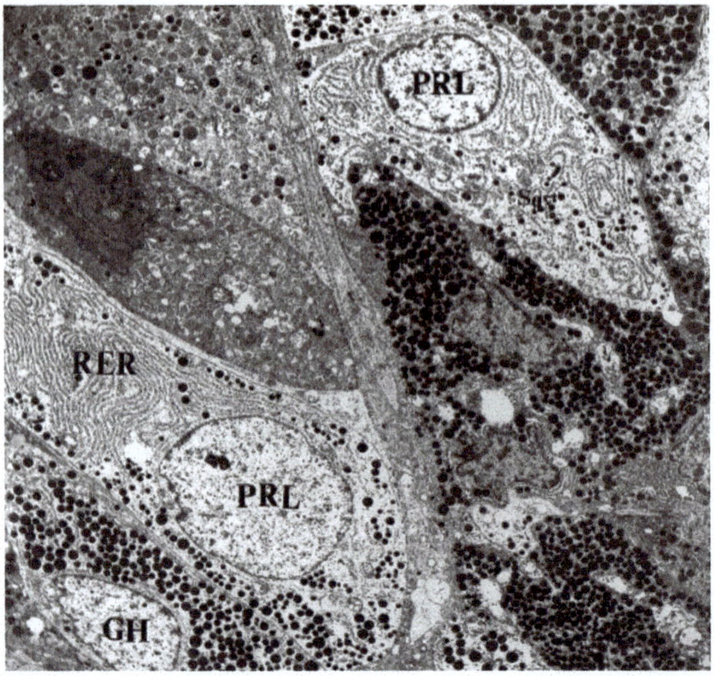

Fig. 1.27 Anterior pituitary of late pregnant female *Suncus murinus*. Note the PRL cells along with lamellar rough endoplasmic reticulum (RER) occupying most of the cytoplasmic area, secretory granules (Sg) present mostly near the cell periphery [X 5000]. Unpublished electron micrographs from Dr. Bhiwgade

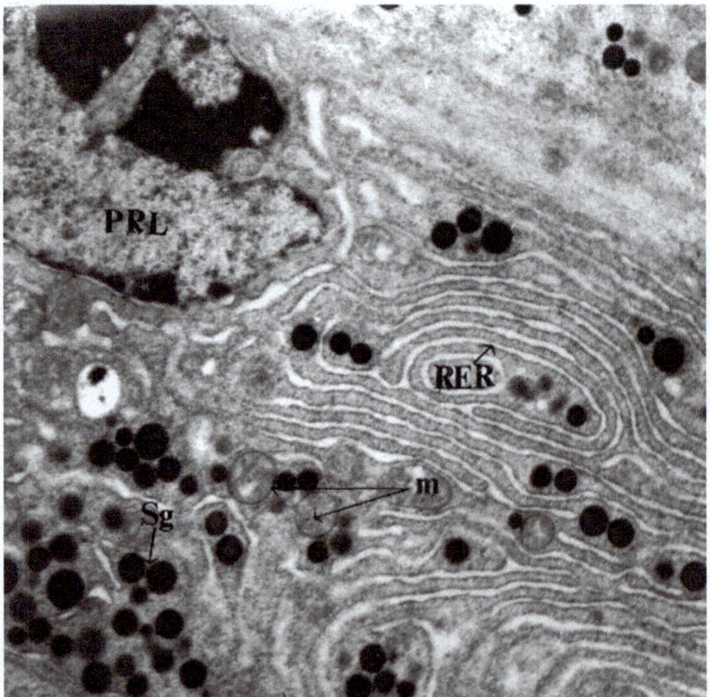

Fig. 1.28 PRL cell of anterior pituitary of bat, *Megaderma lyra lyra*, during late pregnancy. Note rough endoplasmic reticulum (RER) with stages transitional between parallel lamellar and concentric nebenkern (RER). Also note the mitochondria with inconspicuous matrix (m) and large, round, electron-dense secretory granules (Sg) [X 13,000]. Unpublished electron micrographs from Dr. Bhiwgade

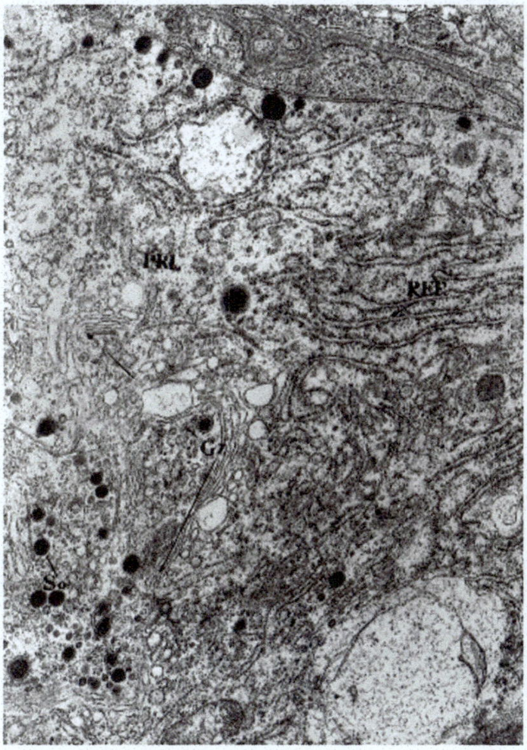

Fig. 1.29 PRL cell of anterior pituitary of the late pregnant female Bonnet monkey (*Macaca radiata*). Note dilated cisternae with numerous ribosomes of lamellar rough endoplasmic reticulum (RER), enlarged Golgi zone (Gz) and presence of secretory granules (Sg) and a pro-secretory granules near the Golgi apparatus [X 6000]. Unpublished electron micrographs from Dr. Bhiwgade

Etreby and Gunzel (1974) in rat, dog and monkey during oestrous phase. These observations are supported by the findings of Sar and Meites (1967) and Amenomori et al. (1979), who by radioimmunoassay were able to detect elevated prolactin levels in the oestrous rat.

During pregnancy, the PRL cells show marked aspects of synthetic and secretory activity such as hypertrophied Golgi complexes, abundant RER membranes, sometimes with Nebenkern formation; active formation of granules and their extrusions as seen in bats, *Rhinopoma microphyllum*, *Cynopterus*, *Hipposideros*, *Megaderma*, *Taphozous*, in Bonnet monkey and in *Suncus*. The cellular synthetic and secretory activity is seen more distantly heightened as the stage of pregnancy advances. This can be explained by the increasing PRL levels during the advancing stages of pregnancy. Our morphometric observations are in agreement with those made by Farquhar and Rinehart (1954a, b), Hymer et al. (1961), Shiino et al. (1972)) and Cinti et al. (1985) in rats and by Dekker (1967) in hamsters during pregnancy. During oestrous, pregnancy and lactation in beagle bitches, there is an increase in the number and activity of PRL cells (El Etreby et al. 1972). Similar results are obtained in men, rat, bat, mole, cat and badger (Herlant, 1964; pregnancy. During oestrous, pregnancy and lactation in beagle bitches, there is an increase in the number and activity of PRL cells (El Etreby et al. 1972). Similar results are obtained in men, rat, bat, mole, cat and badger (Herlant, 1964; Purves 1966; Herlant & Pasteels, 1967 and 1971). During pregnancy, the PRL levels in the serum are reported to be high and reach a peak as the stage of pregnancy advances (Murr et al. 1974). They further stated that the PRL levels become low on the last day of pregnancy and again rise after parturition.

In our study, a higher secretory activity is observed in lactating females with attached young ones of *Rhinopoma microphyllum* and *Pteropus giganteus* when the cell exhibits abundant lamellar RER, well-developed Golgi apparatus and scarce secretory granules. Similar observations have been shown by Farquhar and Rinehart (1954a, 1954b), Hymer et al. (1961), Shiino et al. (1972) and Cinti et al. (1985) in rat, Bhiwgade et al. (1989) in bat and El Etreby et al. (1972) in beagle bitches. Our morphological observations are in good agreement with the quantitative data reported by Murr et al. (1974), Gudelsky et al. (1981) and Crosby & Block, (1980), who were able to measure the increased PRL levels in the serum of lactating rats and women. It is known that lactating rats have a high concentration of serum prolactin and a high pituitary PRL level (Shiino et al. 1972). Prolactin cells of animals in this group showed many immature secretory granules in the Golgi area, well-developed RER and a low number of mature secretory granules in the peripheral cytoplasm. Generally, these morphological phenomena are believed to represent the several phases in active synthesis and packaging of protein hormones.

1.2.4 Electron Microscopic Observation in Experimental Animals

Ultrastructure of PRL cell of normal female rat—A large, elongated cell with large and eccentrically placed nucleus. The nucleus shows a distinct nucleolus and marginated chromatin. The most conspicuous feature of this cell is its abundant, with polar disposition but well-organized RER, displayed in loose lamellar arrays in the peripheral cytoplasm. Other noticeable fine features are the juxtanuclear Golgi zone. Condensing secretion (forming secretory granules) is often seen in the core of this Golgi zone. Mitochondrial inclusion in this cell is heavy, most of these being elongated with normal matricial density and clear cristae. Occasionally, a solitary multivesicular body is noticed distal to the Golgi zone (Fig. 1.20).

1. Effect of Tamoxifen Citrate

The prolactin cell is elongated large, pear-shaped and conspicuous due to its significantly altered fine structure. The cell morphology is varied. The euchromatic nucleus is round, indented and lobated (Fig. 1.26), with a dense chromatin margination and condensed configuration. The enlargement of the cell has resulted in an increased cell volume. The rich concentration of polyribosomes has imparted an increased density to the cytoplasm of the cell. Though the secretory granules are severely depleted, Type I aggregating secretory granules are seen (Figs. 1.36 and 1.37a) along with frequent exocytosis (Fig. 1.37a).

The RER has been noticeably augmented and graceful arrays of its lamellae arranged in stacks of 5–6 cisternae are commonly observed (Fig. 1.37a). Many of these have diffused non-discernible membranes and secretion-filled lumen. Formation of dense lamina has occurred

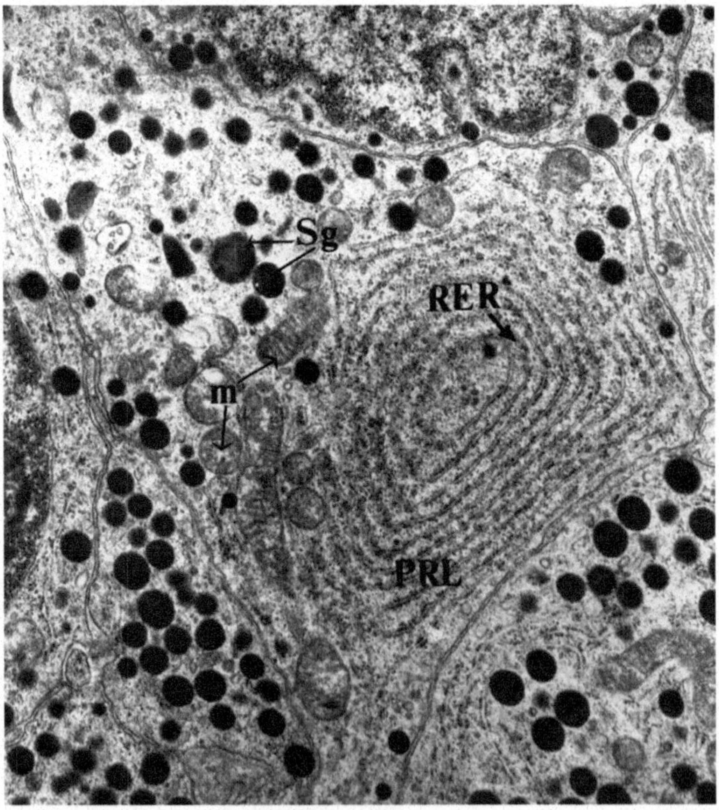

Fig. 1.30 PRL cell of female bat, *Cynopterus*, during late pregnancy. Note large secretory granules, hypertrophied mitochondria and well-developed RER [X 13,000]. Unpublished electron micrographs from Dr. Bhiwgade

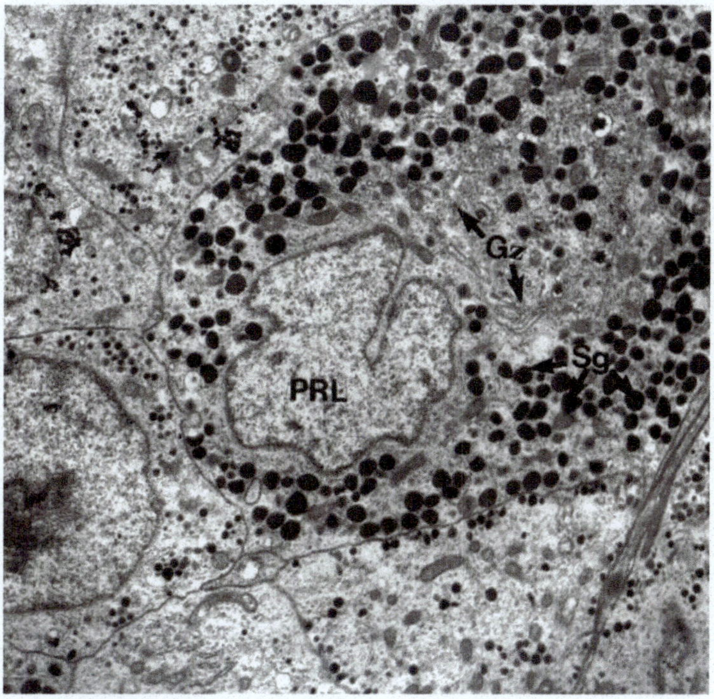

Fig. 1.31 Anterior pituitary of bat, *Taphozous melanopogon*, during late pregnancy. Note the PRL cell with numerous polymorphic secretory granules (Sg) and enlarged Golgi zone (Gz) and an irregular invaginated nucleus [X 13,000]. Unpublished electron micrographs from Dr. Bhiwgade and Dr. Mantri

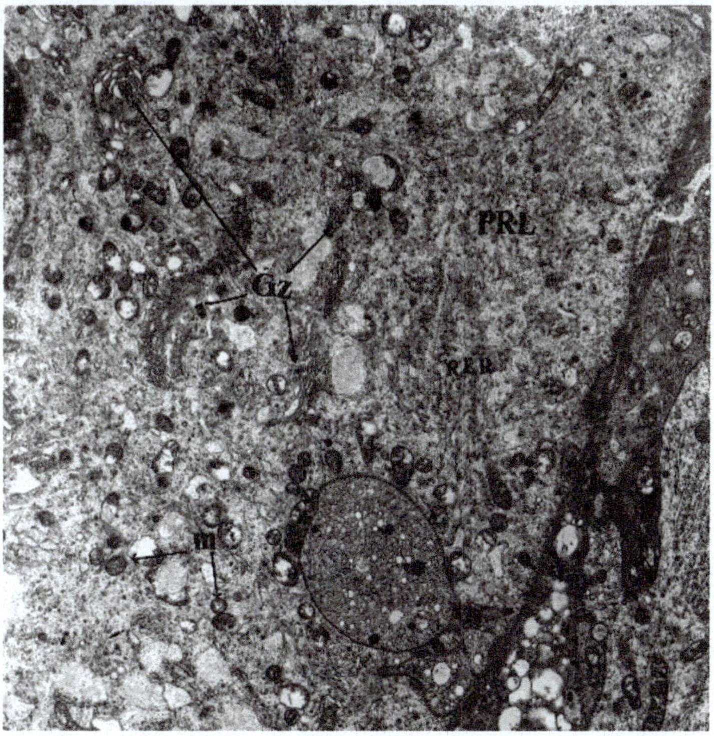

Fig. 1.32 Anterior pituitary of lactating female bat, *Pteropus giganteus*, with attached young one. The PRL cell is characterized by multiple Golgi zones (Gz), rough endoplasmic reticulum (RER) with numerous mitochondria (m) and a complete lack of secretory granules [X 5000]. Unpublished electron micrographs from Dr. Bhiwgade and Dr. Mantri

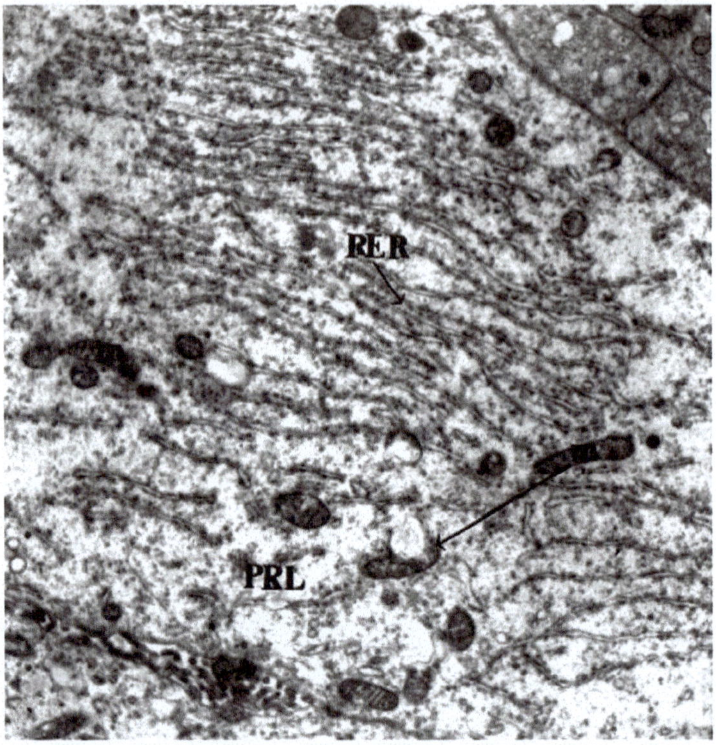

Fig. 1.33 EM of lactating female bat, *Pteropurs giganteus*, showing extensively developed parallel arrays of lamellar RER and few mitochondria (m) [X 10,000]. Unpublished electron micrographs from Dr. Bhiwgade and Dr. Mantri

due to apposing strands of confronting cisternae (Fig. 1.37b). Dilated RER cisternae have undergone irregular whirling (Fig. 1.37a and b) and their cores often include secretory granules as well as degenerated mitochondria with woolly focal densities (Fig. 1.38a).

A noteworthy feature of this cell is the formation of secretion granules within the RER cisternae (Fig. 1.37a). Such precocious formation of visible secretory product attests the intense secretory activity of this cell. The Golgi character is suggestive of 'work hypertrophy' as the Golgi fields have increased in size and number. The aggressive dilated Golgi sacs have resulted in a considerable distortion of the Golgi zone (Fig. 1.38a), with an unusually increased number of Golgi vesicles. In the Golgi area, many small and immature secretory granules are surrounded by smooth membranes (Figs. 1.36 and 1.38b) along with aggregating prosecretory granules of Types I and II (Fig. 1.38b). This represents the classical stage of 'packaging'.

The cluster of round and elongated mitochondria facing the 'trans' side of Golgi zone has undergone hypertrophy (Fig. 1.38b) and show enlargement of matrix chamber and decreased density due to loss of matrix caused by breaks in the walls. The degenerative changes shown by the mitochondria include a deficit of cristae, woolly focal densities and presence of myelin figures (Fig. 1.38b). Lysosomal activity has been remarkably heightened as confirmed by the presence of light multivesicular bodies and lysosomes (Fig. 1.36).

2. Effect of oestradiol Valerate

At the ultrastructure level, the prolactin cell of the oestradiol valerate-treated female rat has undergone hypertrophy and hyperplasia. The relatively small nucleus varies between spherical and irregular shape with an angular or a distinctly crenated margin and a distinct eccentrically placed nucleonema (Fig. 1.39). The densities of the nucleus and the cytoplasm have increased markedly due to the numerous free ribosomes and polyribosomes. The cytoplasm is expansive and has resulted in a decreased

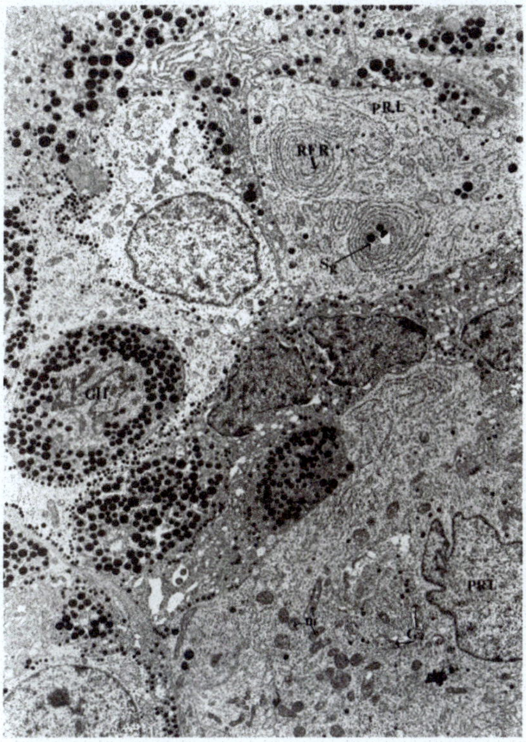

Fig. 1.34 Low-power EM of lactating female bat, *Rhinopoma microphyllum*, shows GH, PRL, FSH and ACTH cells in anterior pituitary. STH oval-shaped concentrically placed nucleus, PRL—completely degranulated, with RER nebenkern patterns, FSH completely degranulated hypertrophied Golgi, ACTH—polygonal eccentric nucleus [X 3000]. Unpublished electron micrographs from Dr. Bhiwgade and Dr. Mantri

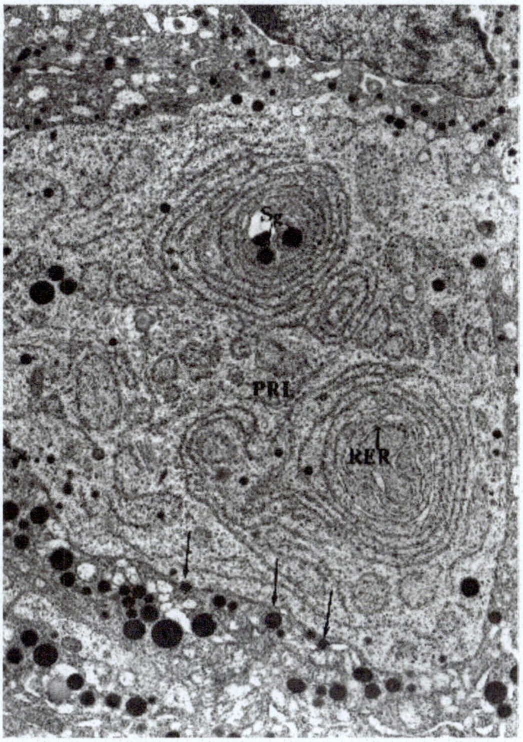

Fig. 1.35 High-power EM of the PRL cell of the lactating female bat, *Rhinophoma microphyllum*. Note the RER arranged in concentric nebenkern and granule extrusions (arrows) and secretory granules in the core of concentric 'nebenkerns'. Genesis of RER nebenkern is evident in this electron micrograph's extensive occupying of the cytoplasm [X 13,000]. Unpublished electron micrograph from Dr. Bhiwgade

nucleocytoplasmic ratio. The granulation is characteristically sparse and it is of interest to note that many forming secretory granules in the Golgi region and exocytosis are common occurrences in these cells. There is a phenomenal increase in the rough endoplasmic reticular membranes that have occupied the cytoplasm extensively along with the frequently encountered 'nebenkern' formations constituted by 12–14 concentrically arranged RER membranes (Figs. 1.39 and 1.40). The RER shows the most dynamic changes following oestrogen treatment and is well visualized due to its intensely stained, attached ribosomes (Fig. 1.41). The attached and free ribosomes have been obviously associated with the bizarre formations of the endoplasmic reticulum (Fig. 1.41).

A remarkable adenomatous feature of these cells is the concentric membranes that present themselves as rounded masses, their core showing focal vacuolar dilatations with sequestrated cytoplasm and mitochondria (Fig. 1.42). Some nebenkerns are myelinated and consequently distorted (Fig. 1.42). Some cells show a morphological variant of RER in the form of confronting cisternae, where two such cisternae are intimately applied to each other that a 'dense lamina' is formed between their confronting membranes (Fig. 1.41). This dense lamina is an electron-dense material formed due to the disintegration of the ribosomes trapped between the confronting cisternal membranes. There is simultaneous development of cystically dilated concentric cisternae, with their reticular

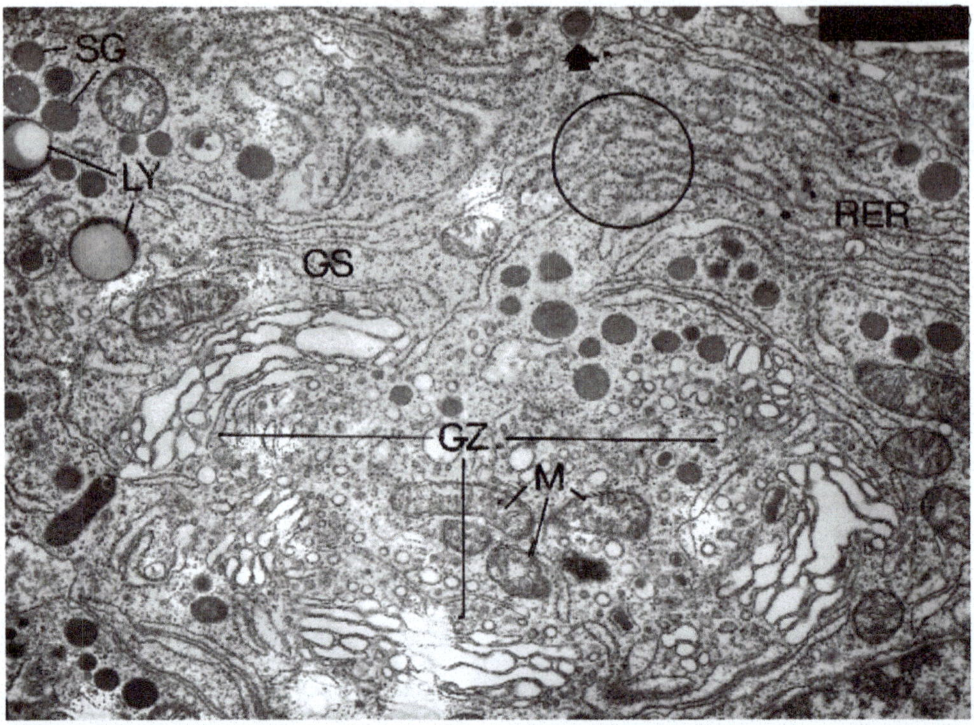

Fig. 1.36 The Golgi zone (GZ) and the surrounding cytoplasm of the prolactin cell of 24 weeks Tamoxifen citrate treated female albino rat is highly magnified in this electron micrograph. Note that some of the RER cisternae have their lumen engorged with an electron opaque secretion, resulting in reduced discernibility of their membranes (encircled). Also noteworthy is the formation of secretory granule (SG) within the RER cisternae (bold arrow). The core of the Golgi zone (GZ) shows forming of secretory granules and mitochondria. (M). Lysosomes (LY) have increased n number [X 13,000]

nature retained (Fig. 1.42). A concomitant increase in the Golgi elements has made the Golgi fields very prominent, the saccules of some of which have undergone strong dilatation and distortion, resulting in a vacuolated appearance of the cell. The mitochondria are grossly enlarged with breaks in the wall, disintegrated cristae and lucent matrix (Figs. 1.41 and 1.42). Some of these have also undergone myelination (Fig. 1.41). There are numerous large lysosomes that have fused with the secretory granules (Figs. 1.39 and 1.40) and light multivesicular bodies have been developed in addition to the annular lipid inclusions with lucent centres (Fig. 1.40). The ultrastructural features indicate that a sparsely granulated PRL cell adenoma has resulted.

3. Effect of Progesterone

After progesterone treatment, the cell has become elongated and irregular in shape with a small-to-moderately sized, round, predominantly euchromatic and a centrally placed nucleus, often with bordered clumps of dense chromatin. Distinct and paired nucleoli have appeared in the nuclei of some cells of treated animals. The cytoplasmic density has been accentuated and the cytoplasm is marked by a pronounced scarcity of secretory granules (Fig. 1.43a). The extremely few but typically large and pleomorphic secretory granules that are present in this degranulated cell, are loosely scattered, whereas numerous polymorphic secretory granules of Types II and III have started aggregating in the Golgi cores (Figs. 1.43b, c and 1.44). RER has increased in amount, solitary and parallel strands of which

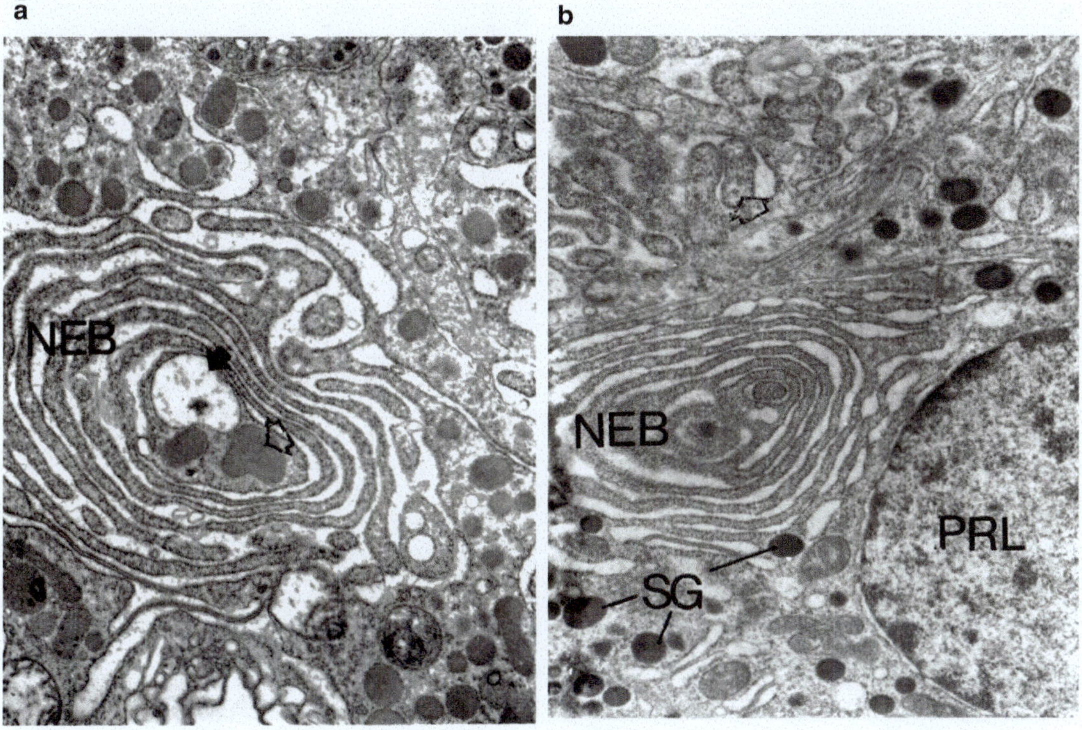

Fig. 1.37 (**a**, **b**) High-power electron micrographs of small portions of Prolactin cells of female albino rats treated with Tamoxifen citrate for 24 weeks. Note the irregular whorling of the rough endoplasmic reticular cisternae resulting in 'nebenkern' formation. Such nebenkern cores (**a**) include secretory granules (hollow arrow) as well as degenerated mitochondria (bold arrow). Vesicular profiles of rough endoplasmic reticulum are also seen (hollow arrow—**b**). Few mitochondria (M) are present [X 16,000]

show dilated terminal ends (Fig. 1.43b). Concentric arrangement of 10–12 RER cisternae has resulted in 'nebenkern', at the centre of which a small portion of the cytoplasm containing formed secretory granules and granular membranes with empty cores is sequestrated (Fig. 1.44). A slight degree of vesiculation has developed in some cisternae and the cytoplasm contains several pairs of closely apposed confronting cisternae with electron-dense lamina (Fig. 1.43c).

A prominent cup-shaped, crescentric or round hypertrophied Golgi zone, with several stacks of smooth surfaced sacs is placed close to the nucleus and its core contains numerous aggregating prosecretory granules. Mitochondrial associations with the Golgi zone have been markedly accentuated (Fig. 1.43b and c), and those outside the Golgi zone have suffered degenerative changes, like myelination and loss of matrix. Lysosomal degradation of the secretory aggregates has begun within and outside the Golgi zone (Fig. 1.43a). Multivesicular bodies are developed in the cytoplasm, opposite the forming face of the Golgi complex, in addition to the lysosomes containing electron-dense material and lipid droplets (Fig. 1.43a).

4. Effect of Norethisterone Heptanoate

After treatment with norethisterone heptanoate, the prolactin cell has undergone ultrastructural alterations suggestive of a stimulated state. These large, round or polymorphic cells have round, irregularly shaped nuclei with moderately rich chromatin content and a distinct nucleolus. The cytoplasmic volume has increased, thereby reducing the nucleo-

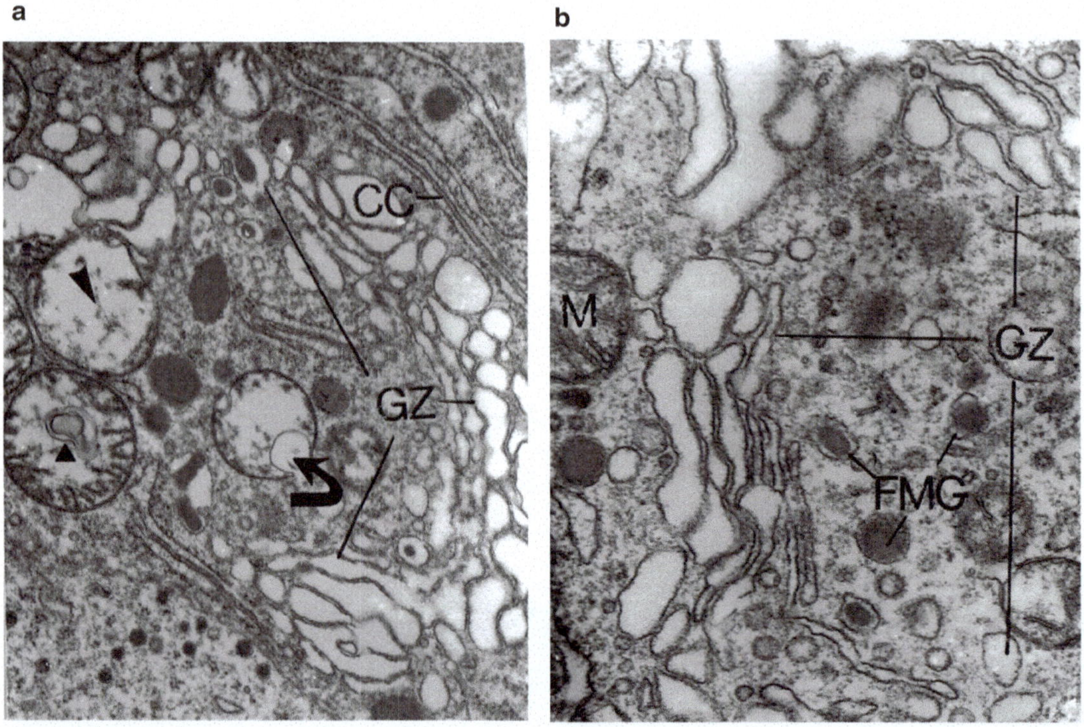

Fig. 1.38 (**a**, **b**) Highly distorted Golgi zones (GZ) of the prolactin cells of female albino rats treated for 24 weeks with Tamoxifen citrate. A marked hypertrophy, dilatation and distortion of the Golgi elements is evident. Observe the long strands of confronting cisternae (CC) immediately outside the Golgi zone (**a**). Note the mitochondria (M) clustered on the 'trans' side of the Golgi zone (**b**). Mitochondria association with the Golgi zone is a strategic cytoarchitectural feature of this cell. Mitochondria (**a**) are round, with matricial lucency (arrow head), wall breaks (curved arrow) and often show myelination (triangle). Also, see the forming secretory granules (FMG) in the core of the Golgi zone (**b**) [X 25,000]

cytoplasmic ratio. Aggregating prosecretory granules in various stages of development are encountered in the Golgi zones of these cells. Type-I secretory granules are found either within the Golgi saccules or in close association with them (Fig. 1.45a). The relatively large Type-II granules are typically polymorphic (Fig. 1.45b). Type-III variety are intermediate in shape and are not as variable as Type-II (Fig. 1.45c), whereas the Type-IV variety is more or less regularly rounded or ovoid (Fig. 1.45d). Densely packed groups of mature pleomorphic secretory granules are encountered within the cytoplasm. The elevated amount of prosecretory granules and the increased size of the juxtanuclear Golgi zone are the most conspicuous features of this cell.

The Golgi zone generally has a circular profile with substantially dilated components. The granular endoplasmic reticular profiles have undergone augmentation and exist in three distinct morphotypes. They form parallel lamellar arrays of extensively developed cisternae (Fig. 1.45b), or are irregularly scattered (Fig. 1.45a). The focal proliferation of the RER has resulted in a very characteristic formation of 'nebenkern', consisting of 7–8 continuous circular profiles of rough and partially granulated endoplasmic reticulum, with clusters of secretory granules in their core (Fig. 1.45d). A few confronting cisternae have also been formed (Fig. 1.45c). In some cells the RER has undergone modifications into a system of irregular sacs and vesicles, which have been distended

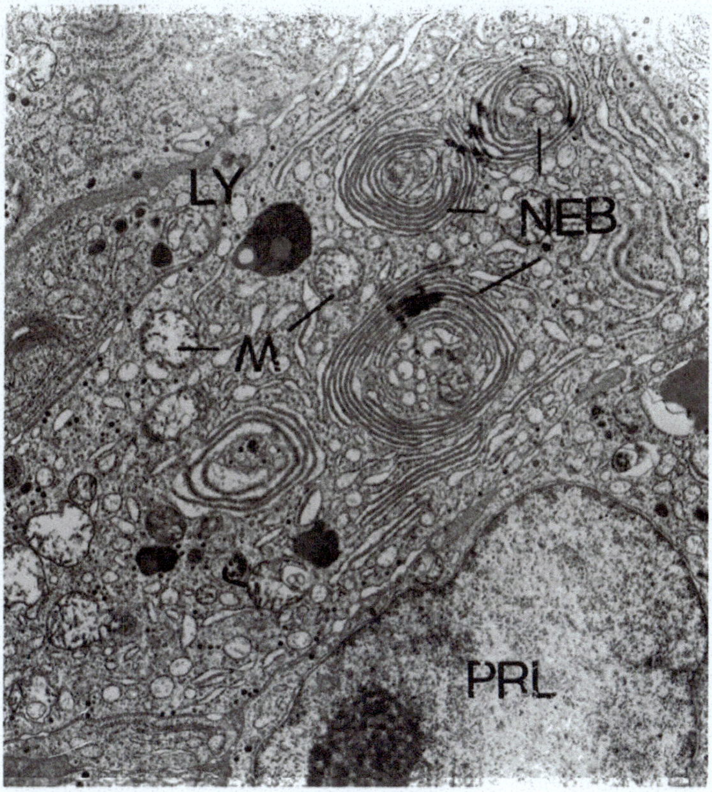

Fig. 1.39 High-power electron micrographs of the prolactin cells (PRL) of female albino rats treated with Oestradiol valerate for 24 weeks. This hyperplastic prolactin cell shows a large, irregular nucleus with a distinct, eccentric nucleolus. See the extensive 'Nebenkern' formations (NEB). Mitochondria (M) show light matrix and loss of cristae. Lysosomal aggregates (LY) are also seen (X, 8000)

with proteinaceous secretory material (Fig. 1.46). Intracisternal sequestration of the cytoplasm has occurred within the numerous papuliferous and a few vesicular profiles of RER (Figs. 1.45d, and 1.46a, b). Numerous intermediate vesicles have begun budding off from the RER stacks towards the Golgi complex (Fig. 1.46b). Detectable increase has occurred in the size and number of mitochondria in all these cells, most of which are specially concentrated near the Golgi zone (Figs. 1.45d, and 1.46a, b). The breaks in their walls are accountable for matricial lucency and many mitochondria have also suffered a loss of cristae and have undergone myelination. Lysosomal aggregates have also appeared in the cytoplasm (Figs. 1.45d and 1.46b). The cell displays a stimulated state engendered by chronic administration of the drug.

5. Effect of Levonorgestrel + Ethinyl Oestradiol

The cell shows a significantly altered cell picture at the ultrastructural level, following levonorgestrel + ethinyl oestradiol treatment. The large, evenly round and euchromatic nucleus of the cell shows a slightly dense chromatin near the nuclear margin and a distinct nucleolus. The extensive, electron-dense and sparsely granulated cytoplasm is enriched with free ribosomes, which usually present as rosettes, besides those attached to the membranes of the RER (Fig. 1.47a). The electron-dense secretory granules do not have specific enough morphology and vary between spherical, oval and elongated shapes. The Golgi cores of almost all the cells show Types-I and II varieties of granules (Fig. 1.47a and b). The formed spherical granule content has been released between the lateral borders of the cell resulting in 'misplaced exocytosis' (Fig. 1.48). The RER is arranged in flattened parallel arrays outside the Golgi Zone and forms confronting cisternae in some cells (Figs. 1.47b and 1.48). In

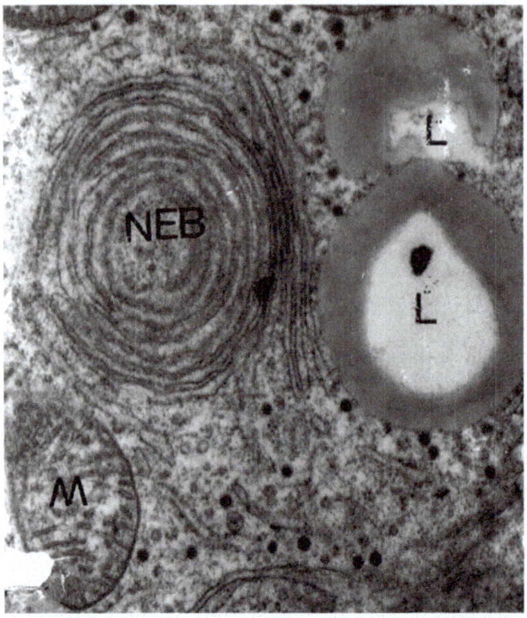

Fig. 1.40 See the annular lipid inclusions with lucent centres (L), in the vicinity of a 'Nebenkern' (NEB) that comprises 8 concentric cisternae of rough endoplasmic reticulum. An elliptical mitochondria (M) close to the nebenkern, shows matricial granules and a large break in its wall [X 25,000]

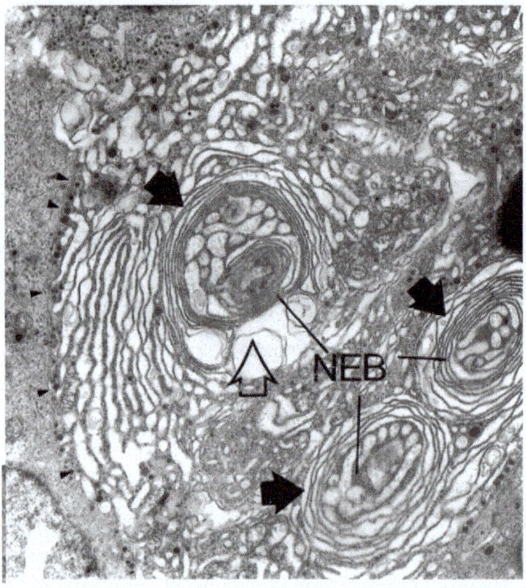

Fig. 1.41 This sparsely granulated adenomatous prolactin cell shows abundant rough endoplasmic reticulum and extensively formed 'Nebenkerns' (NEB) some of which are myelinated. The nebenkern cores show engorged cisternae filled with secretion (bold arrows) Note the lining up of the secretory granules beneath the cell membrane, as well as granule extrusions (small arrows) (X 8000)

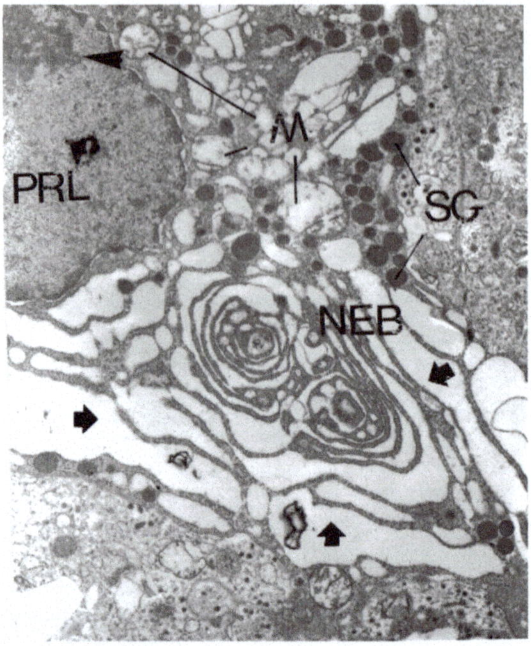

Fig. 1.42 Note the pronounced ultrastructural alterations that have occurred in this cell. The nucleus of this cell is distinctly crenated and shows an eccentric reticulum nucleolus (arrow head). A few secretory granules (SG) are peripherally displaced by the extensive and cystically dilated rough endoplasmic reticular cisternae (bold arrows). 'Nebenkerns' (NEB) have also been formed (X 10,000)

some cases, the cisternae are moderately to strongly dilated and have assumed cup-shaped configurations at their terminal ends (Fig. 1.47b).

Vesiculated RER is not rare. In one cell just outside the Golgi zone, a lamellar cup-shaped structure is in the process of wrapping around a mitochondrion to sequestrate it. The juxtanuclear Golgi zone is crescentric, circular or in the form of figure '8' (Fig. 1.47b) and has undergone hypertrophy and distortion characteristic of a tumorous cell (Fig. 1.48). The number of Golgi units have increased remarkably as a result of 'Work hypertrophy' (Figs. 1.47b and 1.48). The presence of forming secretory granules of Type-I and II varieties within the Golgi cores lend further support to this (Figs. 1.47b and 1.48). Though the hypertrophy of the Golgi complex is marked, it has not resulted in a distortion beyond a standard recognizable pattern. Many short, spherical or sinuous, enlarged mitochondria with normal lucency and cristae are intimately associated

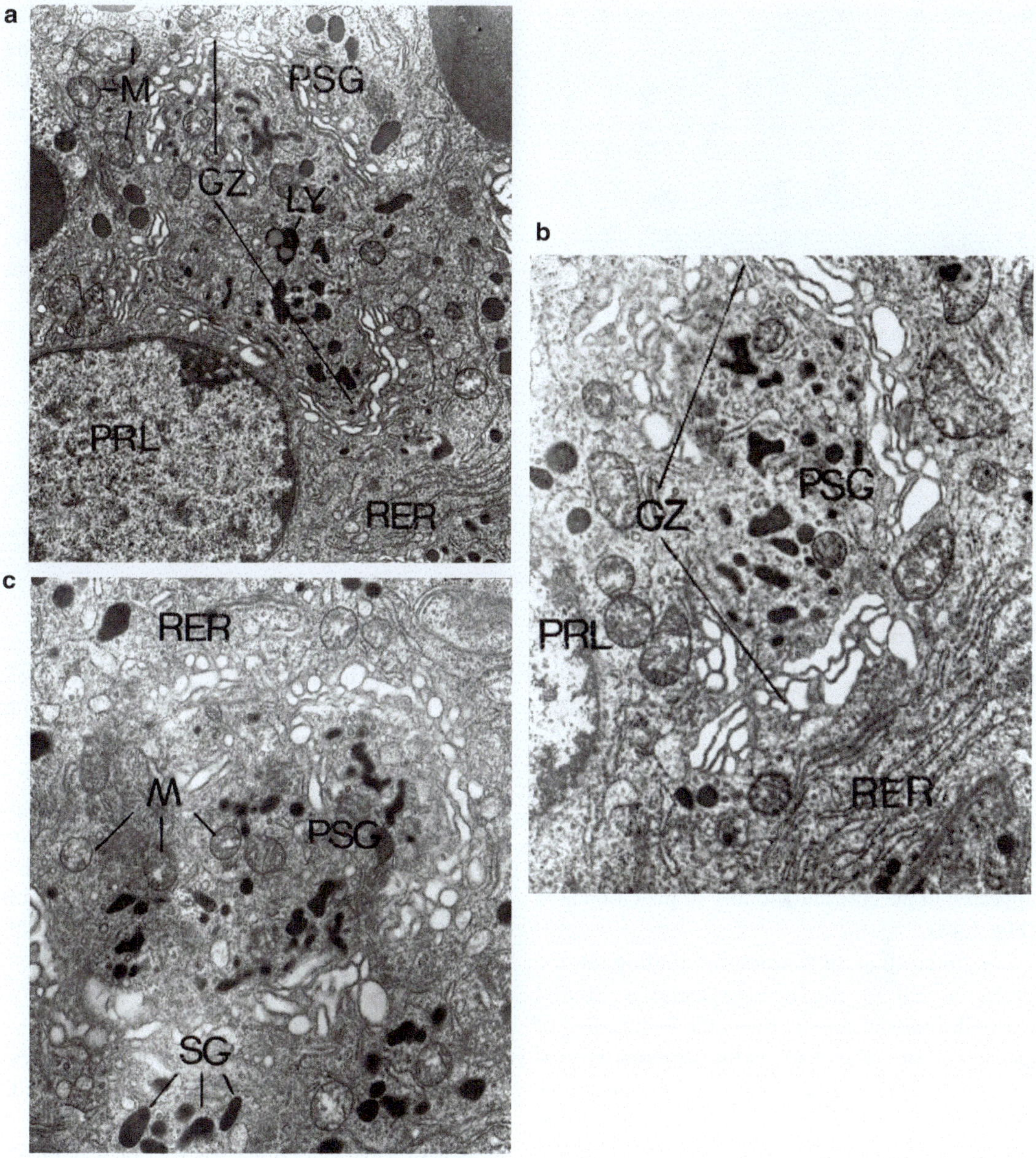

Fig. 1.43 (**a**) High-power electron micrographs of prolactin cell (PRL) or portions thereof, of female albino rats following 24 weeks of Progesterone treatment. The nucleus of this cell shows a distinct nucleolus (bold arrow) and marginated chromatin clumps. Note the loosely scattered cisternae of rough endoplasmic reticulum (RER; **a** and **b**). Also, note the juxta-nuclear Golgi zone (GZ) with substantially dilated sacs. The cores of the Golgi zones show many aggregating prosecretory granules (PSG). (**b**) The lysosomal aggregates (LY) in the Golgi zone are suggestive of crinophagy. Note the mature pleomorphic secretory granules (SG). (**c**) Mitochondria (M) grouped mainly around the Golgi zone. X 10,000 X 13,000 X 13,000, respectively

with the Golgi complex, particularly opposite its forming face. A solitary or a few mitochondria occupy the Golgi core routinely (Fig. 1.47b). Lysosomal activity has been augmented and many large-sized lysosomal aggregates have

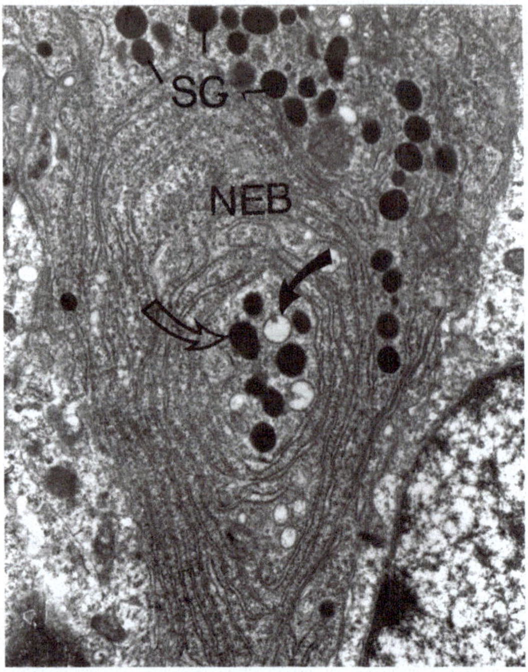

Fig. 1.44 This electron micrograph highlights the juxtanuclear disposition of the rough endoplasmic reticulum. This arrangement probably marks the beginning of 'Nebenkern' (NEB). The core of this early nebenkern includes a few secretory granules (hollow arrow). The empty spaces seen here (bold arrow) are the granular membranes with empty cores [X 13,000]

appeared in the core and vicinity of the Golgi area (Fig. 1.48).

An interesting ultrastructural finding of the prolactin cell of the levonorgestrel + ethinyl oestradiol-treated female rat is the occurrence of an oligocilium (Fig. 1.49). This structure is visible outside the Golgi zone near the secretory granule, at the end of an ergastoplasmic confronting cisternal pair. The cell picture confirms neuroendocrinoma.

6. Effect of depo medroxy progesterone acetate (DMPA) (Antifertility)

After DMPA treatment in female rats (Figs. 1.50a, b, and 1.51a, b), the cytoplasm of the PRL cell shows organelles in abundance which are associated with protein synthesis and packaging. The RER is arranged in the form of parallel lamellae, and RER's proximity to the Golgi aids in the packaging of secretory granules within the latter. The cytoplasm contains a comparatively larger Golgi zone situated near one pole of nucleus, and secretory granules are small to large. The smaller ones are immature or forming secretory granules called as Type I granules. Type.

II and Type III are evidently larger and are found scattered throughout the cytoplasm. The cell also shows lysosomal activity as an effort to rid the cell of excessive hormones.

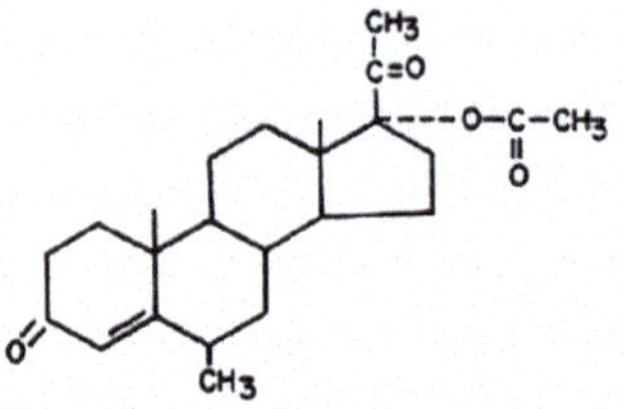

Depo Medrox y Progesterone Acetate

7. Effect of R-06. 5403 (Antiovulatory)

After R-06 treatment (Fig. 1.52), the PRL cell shows Nebenkern formation of RER with granules in its core and the secretory granules are pleomorphic. Coalesced secretory granules are found within the concave face of the Golgi. The sacculi of Golgi show dilation indicating the organelle being in its active state. The cell also shows hypertrophied round mitochondria with ruptured cristae. Lysosomes are scanty and small.

8. Effect of ovral (Antiovulatory) (oestrogen + progesterone)

After ovral (Ethinyl oestradiol + Levonorgestrel treatment) to female rat the PRL cells (Figs. 1.53 and 1.54) appear basically the same as that of DMPA counterpart except that in the former there is a greater abundance of lysosomal bodies. The RER displays two different patterns in the same cell (i) parallel flattened cisternae located in the peripheral cytoplasm and (ii) dilated cisternae in the vicinity of the Golgi complex. Polymorphous secretory granules are found dispersed between the Golgi cisternae. There is a well-developed Golgi apparatus with numerous flat stacks of Golgi cisternae, Golgi vesicles and in the vicinity of whose are observed immature secretory granules of Types II and III

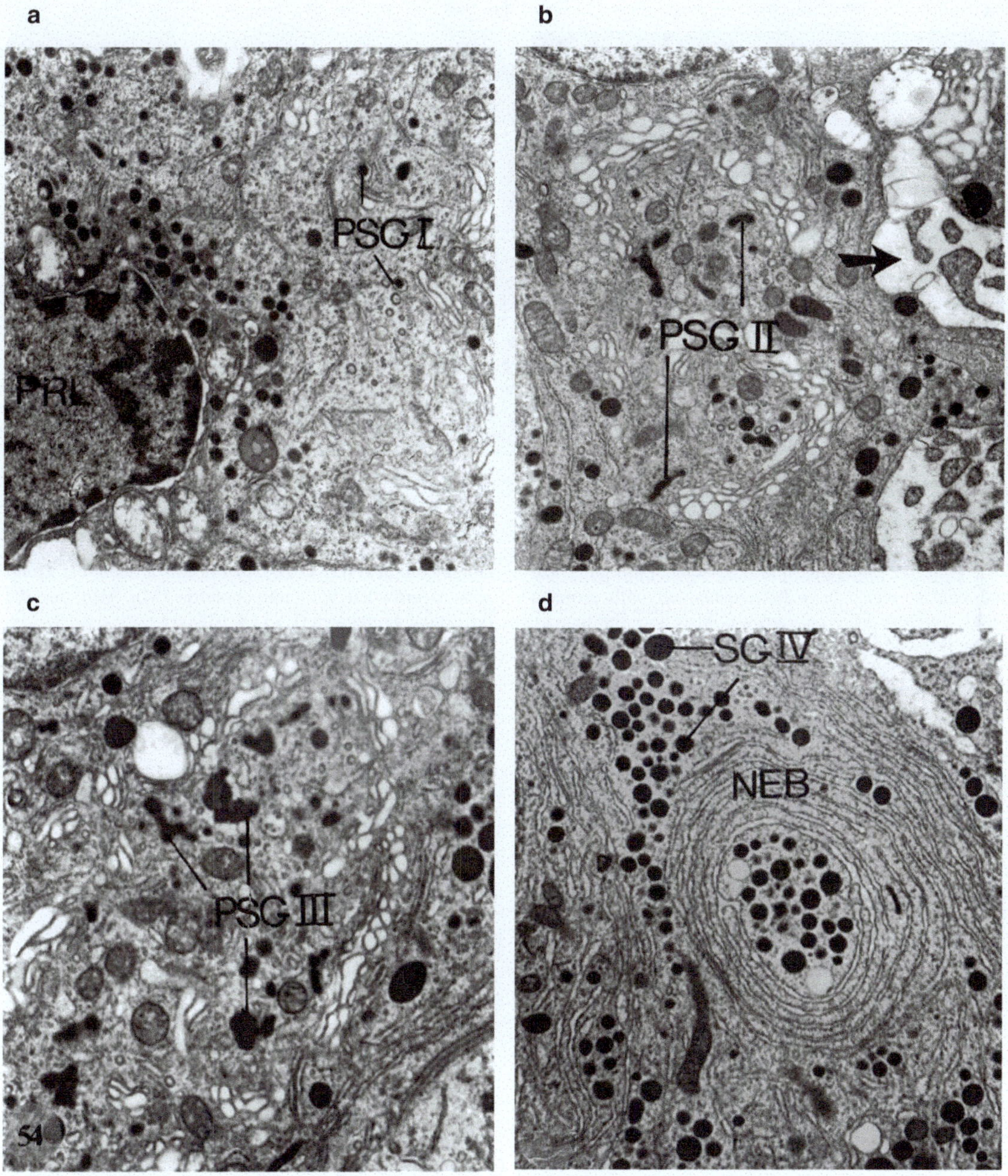

Fig. 1.45 High-power electron micrographs of the prolactin of female albino rats treated with Norethisterone heptanoate for 24 weeks. Note the irregular nucleus with peripheral dense chromatin (**a**). Also note the progressive formation of secretory granules passing through four stages. Type I prosecretory granules (PSG I—**a**) are small, spherical and have a distinct limiting membrane. Type II prosecretory granules (PSG II—**b**) are typically polymorphic. Type III variety (PSG III—**c**) is intermediate in shape. The pleomorphic electron-dense mature secretory granules of type IV (SG IV—**d**) occur freely in the cytoplasm or may be sequestered in the nebenkern cores (NEB). (**a**) [X 13,000] (**b**) [X 13,000] (**c**) [X 13,000] (**d**) [X 8000]

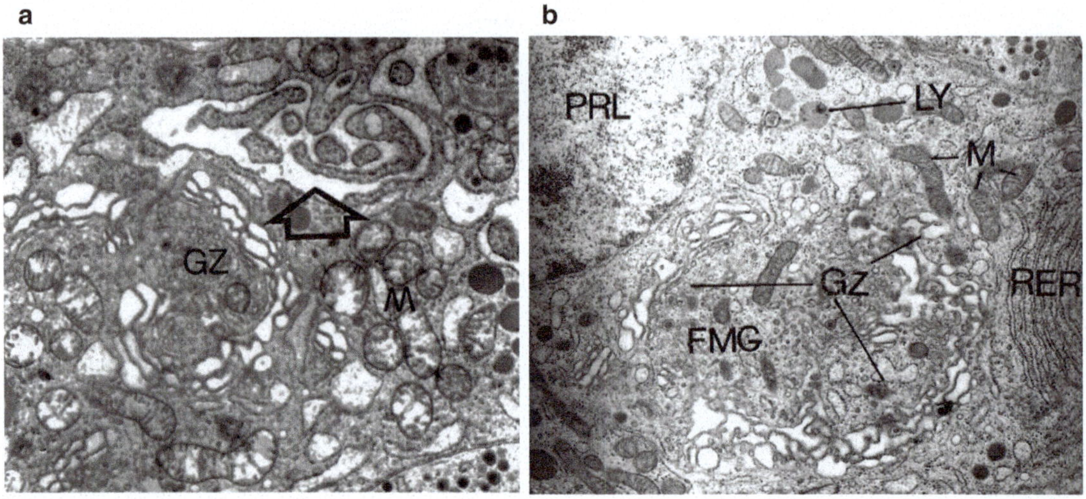

Fig. 1.46 High-power electron micrographs of prolactin cells of Norethisterone heptanoate treated female albino rats (**a**). In this cell, the rough endoplasmic reticulum has undergone modifications into a system of irregular sacs and vesicle, which have been distended with proteinaceous secretory maternal (arrow). Also note the increase in the size and number of mitochondria (M) that are concentrated around the Golgi zone (GZ) [X 13,000] (**b**) Note the numerous intermediate vesicles building off from the stacks of rough endoplasmic reticulum towards the Golgi complex (GZ). The cytoplasm around the Golgi zone is crowded with mitochondria (M). Also, see the lysosomal aggregates (LY) [X 13,000]

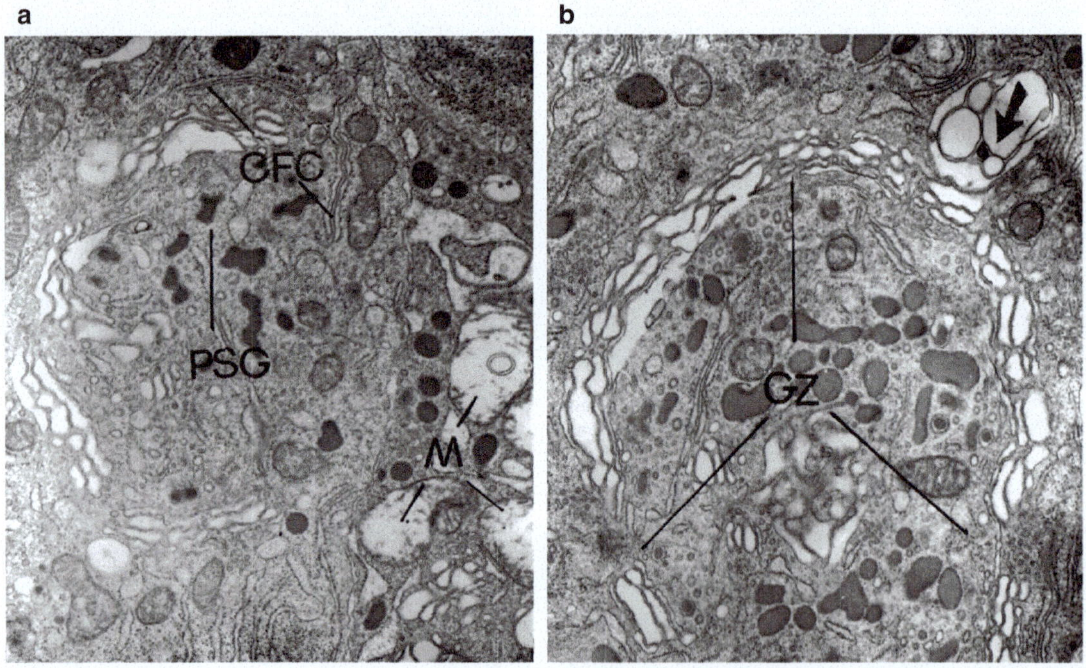

Fig. 1.47 High-power electron micrographs of the Golgi zones (GZ) of prolactin cells of the female albino rats treated with levonorgestrel + ethinyl oestradiol for 24 weeks. See the presence of prosecretory granules (PSG) in the Golgi core (Fig. 1.41). Some Golgi elements have been distorted (bold arrow—**b**). Also, see the confronting cisternae (CFC) outside the Golgi zone (**a**). (**a**) [X 16,000] (**b**) [X 16,000]

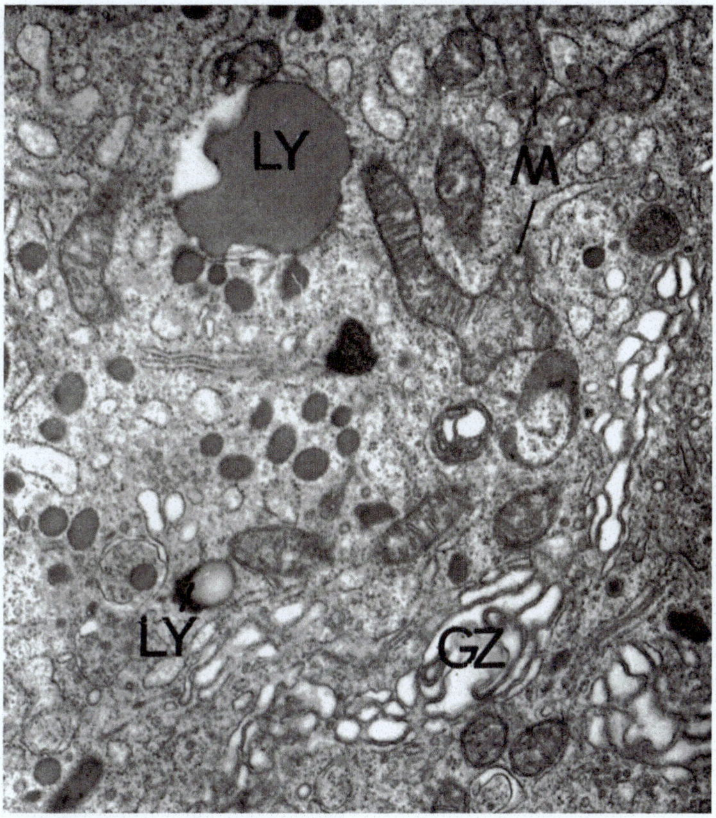

Fig. 1.48 High power of electron micrograph showing the cytoplasmic portion outside the Golgi zone (GZ) of a prolactin cell of a female albino rat treated with levonorgestrel + ethinyl oestradiol for 24 weeks. Note the intimate association between the mitochondria (M) and the Golgi zone. Large lysosomal aggregates (LY) have appeared in the vicinity of the Golgi zone [X 16,000]

which after amalgamation formed mature Type IV secretory granules. There are numerous round-shaped mitochondria scattered throughout the cytoplasm.

9. Effect of DMPA + TE in male rats (Antifertility)

In the DMPA + TE treatment for 17 weeks in male rats, remarkable changes are observed in the prolactin cell (Fig. 1.55a and b). A striking increase in the number of PRL cells showing characteristics of stimulated secretory cells. The chromatin is densely aggregated against the nuclear envelope which exhibits numerous pores (Fig. 1.55a). The well-developed RER may display parallel flattened cisternae in the vicinity of the Golgi complex. Polymorphous secretory granules are dispersed between the Golgi cisternae. The mitochondria become irregular and present a homogenous matrix.

1.2.5 Discussion on Observations in Experimental Animals

The PRL cell has been a cell of choice for electron microscopic investigations, on account of its unmistakable ultrastructural features as well as its susceptibility to a wide range of endogenous and exogenous influences. In fact, it is very obvious from our observations that practically all the nine drug regimens have had a profound influence on the structure and consequently on the function of this cell as well.

The prolactin cells of female rats treated with tamoxifen citrate, oestradiol valerate, progesterone and levonorgestrel + ethinyl oestradiol combination, have increased electron density as a common feature. The enlargement of cells is conspicuous in the tamoxifen citrate and oestradiol valerate regimens. The increased electron density is obviously due to the numerous free ribosomes accumulated in the cytoplasm and is suggestive of

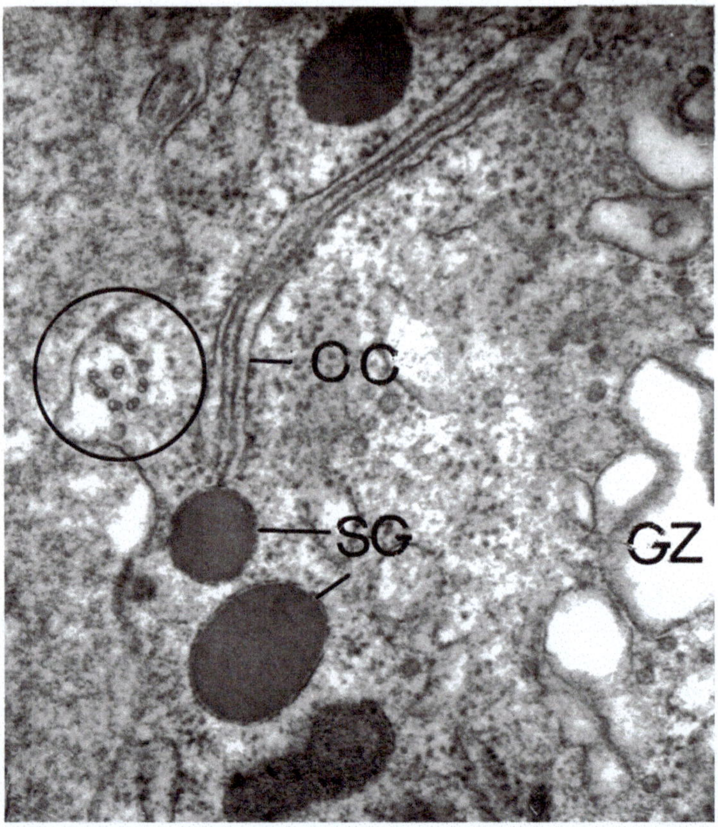

Fig. 1.49 Note the occurrence of an oligocilium (encircled) near the secretory granule (SG) at the end of an ergastoplasmic confronting cisternae (CC) [X 40,000]

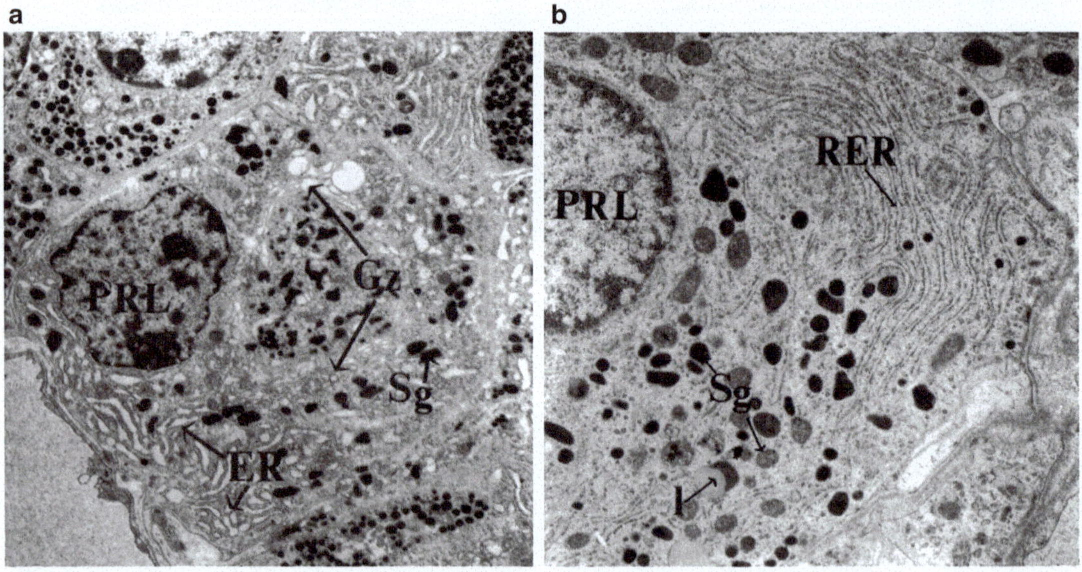

Fig. 1.50 (**a**, **b**) Low-power EM of PRL cell after DMPA treatment in female rats showing enlarged Golgi zone, polymorphic secretory granules and dilated lamellar cisternae of RER. (**a**) [X 13,000] (**b**) [X 13,000]

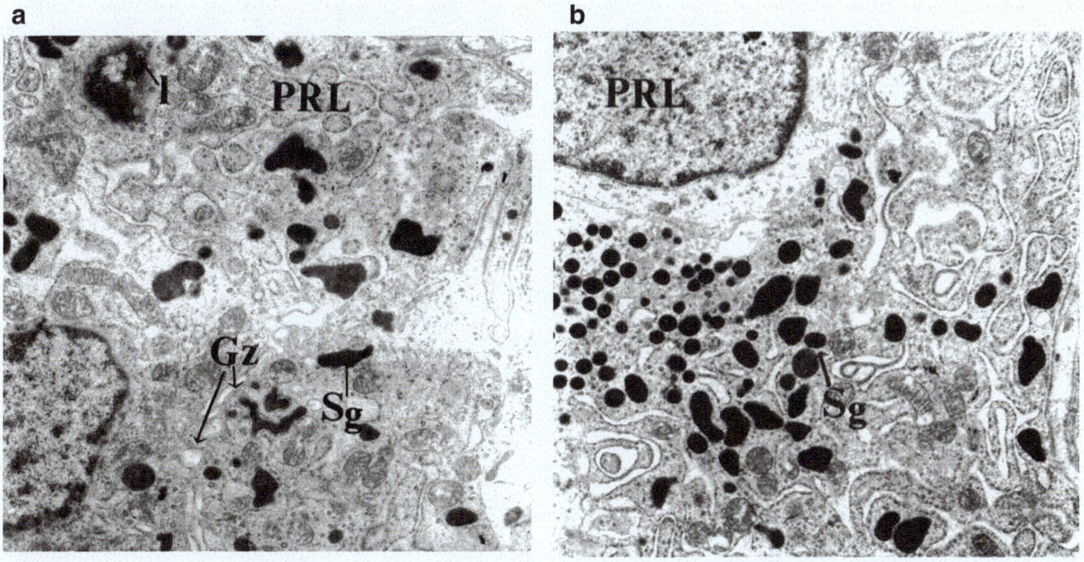

Fig. 1.51 (**a**, **b**) AP of female rat after 30 days DMPA treatment. Note the PRL cell showing the type of appearance on the basis of endoplasmic reticulum, polymorphic secretory granules (Sg) and Golgi zone (Gz) and lysosomes (L). (**a**) [X 13,000] (**b**) [X 13,000]

endocrine hyperactivity (Stefaneanu and Kovacs 1991). The stimulated prolactin cells showing signs of involution have been reported following Pimozide treatment which also shows dense cytoplasm due to the detachment of ribosomes from the RER (Santolaya et al. 1979).

The increased cell size of tamoxifen and oestradiol valerate-treated rat PRL cells, endorse their stimulated state and confirm hypertrophy. The markedly increased cytoplasmic volume of the stimulated prolactin cells under oestrogenic influence has been documented by Horvath and Kovacs (1988) and Saeger (1992). The large euchromatic nuclei seen in tamoxifen, progesterone and levonorgestrel + ethinyl oestradiol-treated rats, have been similarly reported by Stefaneanu and Kovacs (1991) in the estrone-treated female rats. It is interesting to note that

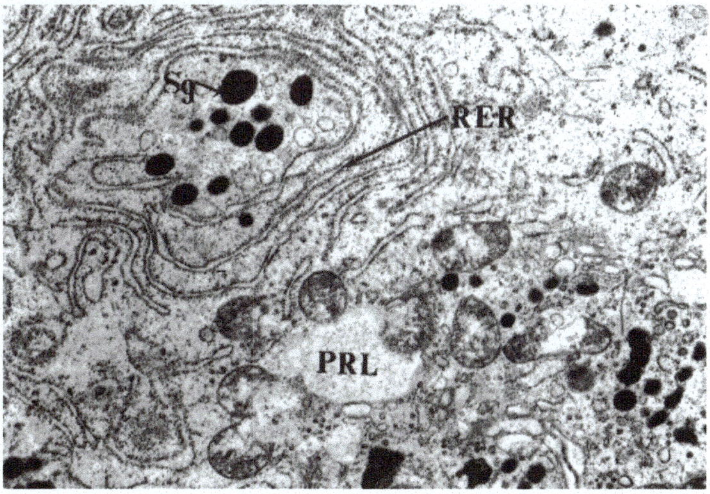

Fig. 1.52 PRL cell hypertrophy in R-06 treated female rat for 30 days marked increased with 'nebenkern' formulation. Note polymorphic secretory granules [X 13,000]

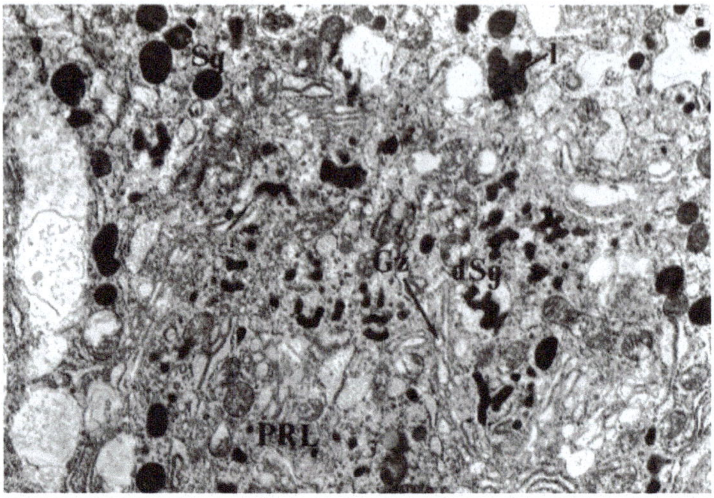

Fig. 1.53 PRL cell in EE + LG female rat for 30 days containing numerous polymorphic mature and immature secretory granules (Sg) within the Golgi zone [X 13,000]

such a predominantly euchromatic nucleus is a marker of metabolically active cell (Fawcett 1981). The stepped-up metabolic status of the cells under the tamoxifen citrate and oestradiol valerate regimens is also inferred on the basis of their nuclei with crenated margins and invaginations. So far as nuclear invaginations are concerned, Colonnier (1964) has expressed the view that it is an attempt to maintain the normal nuclear surface-to-volume ratio.

Our observations on the paired nucleoli in the PRL cell of progesterone regimen are compatible with that of Saeger (1992), who observed more than one nucleoli, frequently, in the stimulated PRL cells following the chronic administration of oestrogens to rats.

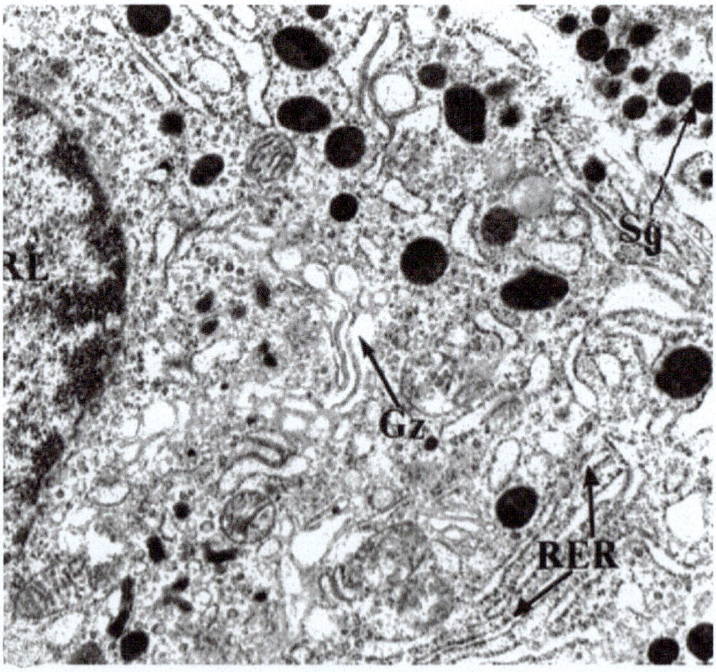

Fig. 1.54 EM of PRL cell after 30 days of ovral treatment. Note the dilated lamellar cisternae of RER. Enlarged Golgi zone and numerous secretory granules (Sg) [X 13,000]

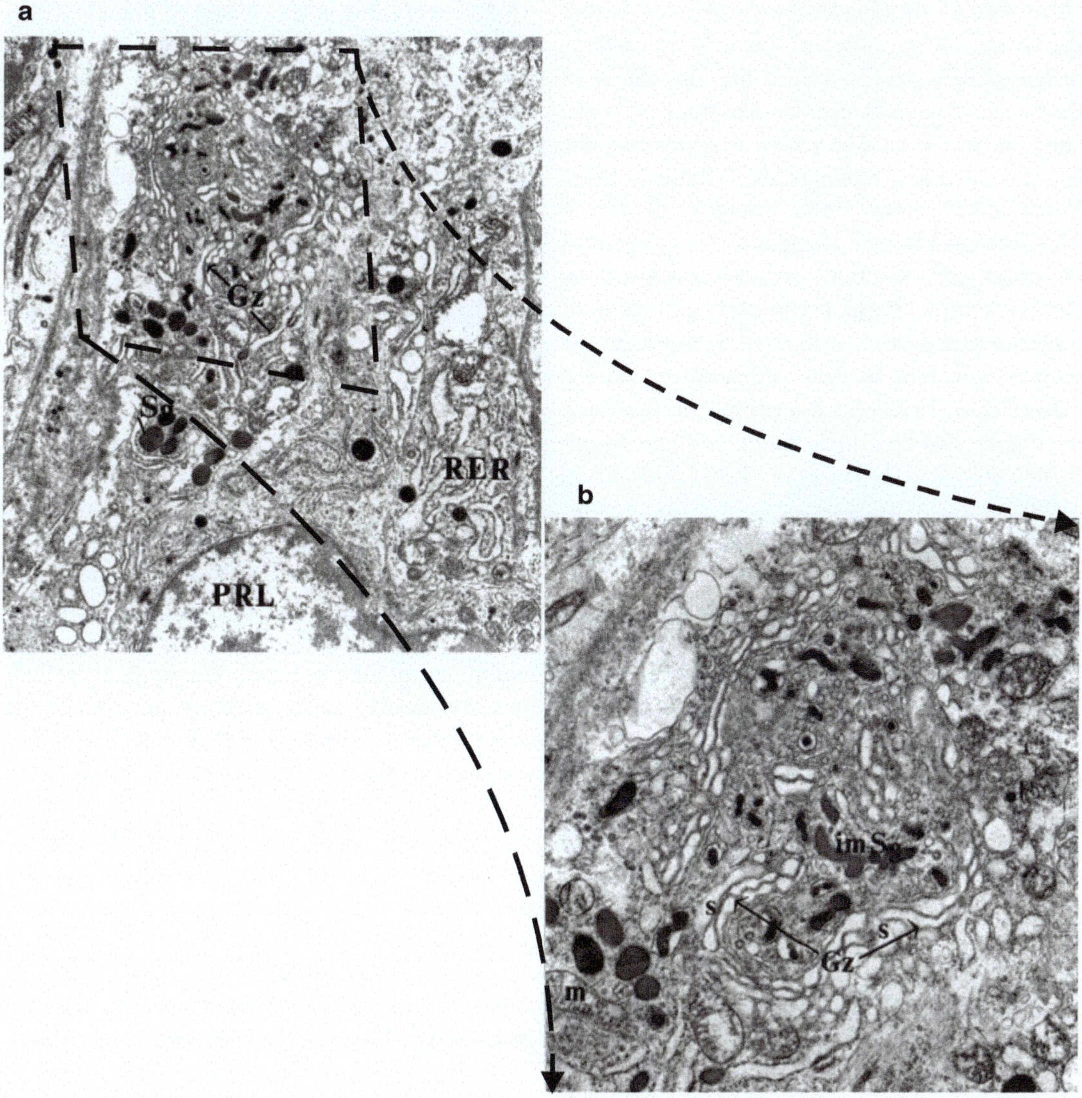

Fig. 1.55 (**a**) Anterior pituitary of male rat after DMPA + TE treatment for 120 days. The PRL cell show Golgi (GZ). Enlarged cisternae of RER and large secretory granules (Sg) [X 8500]. (**b**) Enlarged part of the Golgi zone as in the Golgi zone comprising dilated sacs and interspersed within the Golgi area are numerous immature secretory granules [X 16,500]

As can be seen from our observations, the dynamics of granulation is a measure of paramount importance in assessing the cellular activity under various influences. Barring the Norethisterone regimen, in all other drug treatments, the prolactin cell is sparsely granulated. However, it should be carefully noted that the sparse granulation of the PRL cell, as influenced by tamoxifen citrate, oestradiol valerate and levonorgestrel + ethinyl oestradiol combination regimen; is also coupled with frequent extrusion of granules in the former two cases, and misplaced exocytosis in the last regimen. These events require careful consideration.

Ordinarily sparse granulation, increased episodes of extrusion and a simultaneous occurrence of 'forming secretory granules' within the active Golgi zone of the cell as well as increased RER profiles reflect the 'discharge' of the secretory product and classical stage of 'packaging' of secretion, with a view to regranulate the cell (Farquhar and Wellings 1957; Rennels 1962; Smith and Farquhar 1966; Young et al. 1967). That these cytological changes are suggestive of increased cell function, is also endorsed by Dekker (1967). Particularly such extrusion of prolactin granules is considered to represent the morphological equivalent of prolactin release (Shiino et al. 1972). Such a cell picture has been invariably observed in lactating and oestrogen-treated animals (Hymer et al. 1961; Richardson 1979).

In case of our experimentally manipulated female rats, it is necessary to remember that the ultrastructure suggestive of sustained stimulation of PRL synthesis is exogenously induced either due to decreased dopamine inhibition or oestrogenic influence (Horvath and Kovacs 1988). Furthermore, it is relevant to note that oestrogen administration has been one of the most common methods of inducing pituitary tumours, which secrete prolactin (Clifton and Meyer 1956; Lloyd et al. 1973; Lloyd 1983). At this juncture, it must be stated that an overwhelming majority of prolactin-producing tumours belong to this sparsely granulated variety (McComb and Kovacs 1978; Horvath and Kovacs 1992). Our description of prolactin cells of tamoxifen citrate, oestradiol valerate and levonorgestrel + ethinyl oestradiol combination regimen are in agreement with the prescribed ultrastructural features for a sparsely granulated PRL cell adenoma, that have a striking appearance of hormonally active PRL cells (Stefaneanu and Kovacs 1991; Saeger 1992; Horvath and Kovacs 1992). It has been recorded that such sparsely granulated adenomatous PRL cells resemble non-tumorous PRL cells of lactating and oestrogen-treated rats (McComb and Kovacs 1978).

We have been able to observe a case of 'misplaced exocytosis' from the prolactin cell of a female rat treated with a combination regimen of levonorgestrel + ethinyl oestradiol. Undoubtedly this is also a case of granule extrusion, but unlike the 'orthotopic exocytosis', wherein the granule extrusion occurs at the basal portion of the cells facing the perivascular space; in misplaced exocytosis, the granule is extruded at the lateral cell surface, far from the basement membrane (Horvath and Kovacs 1974). In this case, instead of fusion and storage, the secretory granules are rapidly discharged in their immature stage, indicating that these adenomatous PRL cells are in a hyperactive secretory phase. Incidentally, the presence of misplaced exocytosis is a characteristic feature of sparsely granulated PRL cell adenoma (McComb and Kovacs 1978; McComb et al. 1984) and represents morphological manifestation of an abnormally high rate of hormone release.

As stated in our observations, a noteworthy feature of the PRL cell of tamoxifen-treated female rats is the presence of secretory granules within the RER cisternae. Young et al. (1967) have observed similar secretory granules in the ergastoplasmic cisternae of PRL cells during late pregnancy and lactation in rabbits. There is an autoradiographic evidence that the newly synthesized products pass into the RER cisternae, but are usually not visible as discrete granules (Nadler et al. 1964). Therefore, such sequestrated secretory granules may be the result of intense secretory activity of the cell, causing a precocious formation of visible secretory product before it reaches the Golgi region (Young et al. 1967). This phenomenon may also be viewed in the light of the observations made by Osamura and Watanabe (1986) and Osamura et al. (1962) that the prolactin can be directly secreted from the RER, without going through the Golgi or the secretory granule, as a matter of constitutive pathway.

The phenomenal augmentation of the RER in the PRL cells of almost all the drug regimens, and in particular of the oestradiol valerate treatment, reflects the activation of the synthetic machinery, since the secretory product is elaborated in RER (Amat and Boya 1973). The rapid rate of prolactin synthesis endorsed by extensive formations of concentric RER 'nebenkerns' in the oestrogen-stimulated PRL cells has been highlighted by many workers. Needless to say such accentuated RER is commonly seen in PRL cells of pregnancy and lactation (Hymer et al. 1961; Young et al. 1967; Saeger 1992) as well as hyperplastic and adenomatous PRL cells formed under oestrogenic stimulation (Hymer et al. 1961: Aumuller et al. 1978; McComb and Kovacs 1978; Horvath and Kovacs 1988; Lloyd 1991; Stefaneanu and Kovacs 1991). The sequestration of secretory granules in the nebenkem cores of Tamoxifen citrate, Norethisterone heptanoate and progesterone regimen have also been reported following Norethisterone treatment (Aumuller et al. 1978), and oestrogen treatment (Stefaneanu and Kovacs 1991; Horvath and Kovacs 1988). The papilliferous and vesicular profiles of RER formed as a result of intracisternal sequestration in the Norethisterone treatment is a mechanism by which excess RER is disposed of by the cell. The short solitary and long meandering confronting cisternae noticed in the PRL cells of the tamoxifen citrate and oestradiol valerate regimen are generally known to be occurring in the rapidly dividing cells progressing towards neoplasia. This is an important observation, since the mechanisms governing the development of oestrogen-induced pituitary tumours in rats, during progression from hyperplasia to neoplasia, are of applied interest and should be carefully investigated. Though as of today, pituitary adenomas are classified clinically according to the symptom complex, the importance of morphological clues such as the confronting cisternae should be emphasized in the accurate ultrastructural diagnosis of pituitary tumours.

The phenomenal increase and the accompanying bizarre formations of RER noticed in the PRL cell of oestradiol valerate treated rat illustrates the process of involution in a transitional stage, as a result of the toxic effect of the drug (Santolaya et al. 1979). As can be seen in this cell, the RER has appeared as dilated cisternae occupying extensive portions of the cytoplasm and displacing the other cell organelles. Yoshimura et al. (1977a, b) have interpreted such a cell picture as being indicative of an accelerated cell cycle.

The occurrence of an oligocilium seen at the end of the ergastoplasmic cisternal pair in the PRL cell of the levonorgestrel + ethinyl oestradiol-treated female rat is an interesting find of this study. Though there are scattered reports on the presence of cilia in the adenohypophysis, by light microscopists (Dawson 1937) as well as electron microscopists (Barnes 1961; Salazar 1963; Ziegler 1963; Kagayama 1965; Wheatley 1967; Dingemans 1969; Lawzewitsch et al. 1972), no unifying hypothesis regarding their significance has yet been proposed and their function still eludes us.

Some workers have assigned an embryological significance to these structures and interpret them to be remnants of the original ciliated ectoderm. Some others have opined that stimulation of centriolar reproduction without subsequent mitosis may lead to ciliary formation. In this context, it is significant to note that a negative correlation between mitosis and cilia production is also thought to occur in the adenohypophysis. Also noteworthy is the opinion of Munger (1958) who proposed a chemoreceptive function for cilia in endocrine organs.

Conspicuous enlargement has occurred in the Golgi zones of the cells following the administration of tamoxifen citrate, oestradiol valerate and levonorgestrel + ethinyl oestradiol in particular. Numerous reports are available to support such proliferative changes occurring in the Golgi zone following sustained stimulation of PRL cell. This is marked by strong or in some cases aggressive dilatation and subsequent distortion of the Golgi elements. Invariably, the cores of such hypertrophied Golgi fields show many forming secretory granules as well as aggregating prosecretory granules (Farquhar et al. 1978). The dense pleomorphic forming secretory granules are a potentially important marker in the identification of the PRL cell (Horvath and Kovacs 1988). Such drastic alterations in the morphology of the Golgi fields are not surprising when one considers that the Golgi complex is especially susceptible to the oestrogenic stimuli leading to endocrine hyperactivity (Aumuller et al. 1978; Lloyd 1991; Stefaneanu and Kovacs 1991; Saeger 1992). This contention is further supported by the grossly enlarged mitochondria that are in close strategic association with the Golgi zone. The augmentation of the lytic machinery as seen in the core of the Golgi zone of levonorgestrel + ethinyl oestradiol treated PRL cell, is in good agreement with the report of Aumuller et al. (1978) who observed similar cell picture following norethisterone administration.

Apparently, the heightened lysosomal activity in the PRL cell is a common feature of all the drug regimens but differs in intensity, being more pronounced in the tamoxifen, oestradiol valerate and levonorgestrel + ethinyl oestradiol regimens. In this context, it may be realized that the lysosomes control the rate of hormone production by digesting and diminishing some of the granules through 'crinophagy' thereby controlling the volume of the hormone to be secreted (Farquhar 1971) as a built-in check to prevent hyperprolactinemia.

The annular lipid inclusions with lucent centres seen in the PRL cell of female rats treated with oestradiol valerate and progesterone are also a manifestation of disposal of secretory granules by crinophagy (Stokreef et al. 1986). The prolactin granules are digested by a multistepped process that has resulted in the formation of such residual free lipid globules with electron lucent centres. A means for disposal or reuse of this criniphagy-related lipid in the PRL cell has not been clearly understood, but it is possible that the folliculostellate cells are involved in the metabolism of this residual matter (Stokreef et al. 1986). In the lactotroph of the levonorgestrel + ethinyl oestradiol regimen, the additional membrane-bound structure called 'GERL' with Golgi stack on the trans side was first reported by Novikoff (1964). It is designated as GERL since this structure is intimately linked to Golgi saccule (G), that it was a part of the RER and that it formed the lysosome (L).

Having discussed and analyzed the fine structure of the PRL cells as effected by the various regimens, it clearly stands out that in almost all the cases the dynamics of the hormone synthesis and secretion have been affected. Most, if not all the actions of steroid hormones are achieved by induction of gene expression (Brattin and Portanova 1977). A direct stimulating effect of oestradiol on prolactin gene transcription has been documented in cultured rat pituitary cells (Shull et al. 1987). Also, it has been reported that the oestrogen stimulation results in the accumulation of mRNA in the cytoplasm due to the elevated prolactin gene transcription rate (Maurer 1982). It has long been known that the oestrogens increase prolactin secretions both, in rat and in man (Ratner et al. 1963; MacLeod et al. 1969; Ehora et al. 1976) and stimulates the growth of the pituitary lactotrophs (Clifton and Meyer 1956; Farquhar et al. 1978).

Early phase of research was marked by an uncertainty as to whether the main action of oestrogens was at the level of pituitary or the hypothalamus. In either case, an interaction between the oestrogen and dopamine control of prolactin secretion seemed to be involved (Raymond et al. 1978; Cramer et al. 1979). As of today we know that the hypothalamic control of prolactin secretion is by inhibition and all the experimental evidences are consistent with the existence of the hypothalamic inhibitory factor, possibly dopamine (Donoso et al. 1974; MacLeod 1976). Stimulation of the PRL cell is mostly the effect of decreased dopamine inhibition (Von Werder 1988) and oestradiol has an integrated effect to prevent the dopaminergic inhibition of prolactin mRNA levels, prolactin synthesis, and prolactin secretion. The logical implication of the foregone account comfortably justifies the overwhelming response of the PRL cell to oestradiol valerate that has culminated in prolactinoma. A word of caution here is that most of the lesions induced by oestrogens are called as tumours and most often a distinction between hyperplasia, nodules and adenoma may not be possible (Stefaneanu and Kovacs 1991). Reportedly, oestradiol benzoate treatment also increases serum and pituitary prolactin levels (Nogami 1984). A significantly increased number of PRL cells and cytological criteria of increased secretory activity was seen following a high dose of oestradiol administration to female rats (Attia and Zayed 1989). Similarly, oestrogen-treated mice developed striking hyperprolactinemia and died prematurely (Walker et al. 1992).

Despite the acclaimed antioestrogenic activity of tamoxifen citrate, it may seem from our observations that the compound has elicited a cell picture suggestive of enhanced synthesis and secretion. Though prima facie this response may appear to be anomalous, one must consider that the pharmacology of tamoxifen is complex and is species as well as target site specific (Jordan 1995). Therefore, it is simplistic and inaccurate to call tamoxifen as an antioestrogenic agent (Paterson and Geggie 1993). It must be emphasized that tamoxifen can behave as an oestrogen agonist when endogenous oestrogen levels are low (Muylder and Neven 1993). This is supported by the observations of Bhatavdekar et al. (1992), that tamoxifen therapy in postmenopausal women resulted in hyperoestrogenemia followed by hyperprolactinemia. This is contradictory to the opinion of Quijada et al. (1980), that tamoxifen significantly increases the inhibitory effects of dopamine on prolactin secretion.

Thus, elevated prolactin levels may be caused by the oestrogenic potency of tamoxifen thereby blunting its inhibitory effects (Gibson et al. 1990). That in rats, tamoxifen is a partial agonist is known (Harper and Walpole 1967). Also, a point in consideration is that the isomerization of tamoxifen metabolites could decrease the net antioestrogenicity of this compound (Jordan 1995). Furthermore, the development of mutated oestrogen receptors, thereby altering the pharmacology of the antagonist to an agonist is not far from being true (Jordan 1995). Despite the antitumour activity of this compound, the ability of the pituitary to continue producing high levels of prolactin in circulation has also been reported (Turkington 1974; Jordan and Koerner 1976). Therefore, tamoxifen does not completely block oestrogen stimulated prolactin release (Jordan and Dowse 1976).

The enhanced synthetic function of the PRL cell of progesterone regimen is concordant with the demonstration of progesterone-induced hyperprolactinemia (Williams et al. 1994). It is also known that progesterone suppresses the pituitary gonadotrophins and promotes the development of breasts (Dollery 1991). Increased risk and induction of mammary carcinomas by progesterone have also been documented (Kordon et al. 1993; Misdorp et al. 1992). The long-term administration of progesterone, similar to the dosing protocol adopted in this study is known to have resulted in increased serum prolactin levels (Mishell et al. 1977). It is possible that some of this progesterone gets converted into oestrogen in vivo, before it acts on the pituitary. Reports are also available to suggest that progesterone decreases the hypothalamic Prolactin Inhibitory Factor content thereby increasing the pituitary prolactin secretion (Chen and Meites 1970).

The fine structural changes correlating the stepped-up synthesis in PRL cells of norethisterone heptanoate-treated animals should take into account the observations of Karim and El Mahgoub (1970), who commented that as a pure progestogen, this compound does not suppress milk production. Further, these authors have opined that protracted use of norethisterone heptanoate actually appears to promote and prolong lactation. Of paramount importance in understanding the stimulatory effect of this progesterone on the PRL cells as observed in this study, is the work of Aumuller et al. (1978). These workers have shown that norethisterone expresses a strong oestrogenic potency in rats and brings about an increase in the percentage of PRL cells. The fine structure of these PRL cells matches impressively with our observations. The combination regimen of levonorgestrel + ethinyl oestradiol has resulted in a cell picture suggestive of neuroendocrinoma. Our findings are concordant with the opinion of Jernstrom et al. (1992), who reported that though oestrogen–progestogen combination in standard contraceptives leads only to moderate and non-progressive stimulation of the pituitary activity in women, often excessive growth of prolactinoma can occur. Our results are also consistent with the reported stimulation of PRL cells in some experimental animals following long-term exogenous application of high doses of oestrogen and progestogen combination regimen. This is further supplemented by the observations of El Etreby and Gunzel (1974), which long-term treatment of men and women with high doses of such a combination drug schedule enhanced the development of PRL cell tumours. Probably, the oestrogenic fraction of combination oral pills evokes the typical oestrogenic response so far as the lactotrophs are concerned and augment the secretion of prolactin (Jernstrom et al. 1992). Several groups of investigators have reported that women using combination oral contraceptives have significantly higher levels of prolactin (Mishell et al. 1977). However, some investigators have shown conflicting results (Luciano et al. 1985; Gaspard et al. 1983).

Mammosomatotrophs (GH/PRL Cell)

Early phase of research involving categorization of anterior pituitary cells in man was marked by an uncertainty regarding the chemical distinction between prolactin and somatotropin (Dixon 1964). This uncertainty was supported by the observations of Couch et al. 1969, that administration of purified GRH resulted in granule extrusion from the somatotrophs as well as the lactotrophs of the rat. Though it has been confirmed histochemically, that these two hormones are secreted by two distinct types of acidophils in many mammalian species; histochemical criteria have often failed to distinguish between the two cell types leading to misinterpretations in other animals like the rat (Purves 1961; Rennels 1962) and bat (Herlant, 1964). This is also further validated by Hoeffler et al. 1985, who found through Reverse Hemolytic-Plaque-Assay, that the first PRL cells were derived from the GH cells. The recent study of Behringer et al. 1988, showing a severe reduction of PRL cells in transgenic mice with defects in GH-expressing cells; attests to this observation. Though election microscopy in conjunction with certain physiological experiments has accomplished this distinction in some species (Rennels 1962; Pasteels 1963a, b) there is an increasing body of evidence to believe that the individual cells do exit in which the growth hormone and prolactin are co-packaged.

Such bihormonal cells that secrete both GH and PRL are now known to be a common pituitary cell type in many mammalian species (Kineman et al. 1991). Rat, the experimental animal chosen by us is also known to have a large number of mammosomatotrophs in its adenohypophysis (Frawley et al. 1985). In our study, we have encountered mammosomatotrophs in the experimental animals of three drug regimens, viz., oestradiol valerate, norethisterone heptanoate and a combination regimen of levonorgestrel and ethinyl oestradiol.

It is obvious from the cell picture of the mammosomatotrophs of the oestradiol valerate regimen, that the cell is in an intensely activated state. This is evident from the dilated Golgi elements, precocious formation of visible secretory products within the RER cisternae and the occupancy of the cytoplasm around the Golgi zone by numerous enlarged mitochondria. The presence of numerous lysosomes and multivesicular bodies, is in all possibility, a prelude to an intracellular degradation of excessively synthesized secretory products. A similar picture suggestive of activity though of a lesser magnitude, has also emerged in the mammosomatotrophs of the combination regimen. However, in the mammosomatotroph of the norethisterone heptanoate regimen, the cell picture does not corroborate an active status. Such a differential response of the cell to the three different drug regimens can be understood with little difficulty when one takes into account the fact that oestradiol valerate and the combination drug contains oestrogenic principles. In the latter case, levonorgestrel negates a full-fledged response of the cell to ethinyl oestradiol, hence the activity seen is less intense. The third drug regimen of norethisterone is a case of exclusive progestational influence that cannot evoke an oestrogenic response.

It should be mentioned here that in a critical review of mammosomatotrophs, Arita 1993, has referred to several studies that show that most of the PRL cells arise from the somatotrophs via. The mammosomatotrophs. In the context of our observations, it is relevant to note that such transformations of the GH cells into prolactin-producing cells occur under oestrogen stimulation. This is correlated well with the increased signs of activity shown by the mammosomatotrophs under oestradiol valerate and combination regimen, both of which as stated earlier, contain oestrogenic compounds. Also, in both cases, elaboration of secretory granules by the Golgi zone is a good marker of their activated status. Since the newly synthesized PRL is preferentially released from oestrogen-stimulated cells, it is reasonable that the Golgi complex should be one of the earliest sites reflecting enhanced secretion. The emphasis laid on the prolactin component of the mammosomatotroph also finds support from the observations of Sasaki and Iwama 1989, who reported combined immunoreactivity only in a small area of the cell cytoplasm, whereas the cell was generally immunoreactive to the prolactin alone. The work of Asa et al. (1988) has shown that the mammosomatotrophs are the only source of prolactin in early gestation and also that this cell type is a progenitor of the PRL cell line.

Normally, mammosomatotrophs may function as stem cells during the development and impart functional plasticity to the adult pituitary, permitting bidirectional interconversion of GH and prolactin secretors. It is well known that the shift in the mono and bihormonal cells is triggered by fluctuations in the reproductive and annual cycles of many animals like rats (Ishibashi and Shiino 1989a; Porter et al. 1990); musk shrews (Ishibashi and Shiino 1989a) and bats (Ishibashi and Shiino 1989b). In this context, our observations imply that a shift in the hormonal milieu caused by exogenously administered oestrogenic compounds could disturb the normal dynamics of interconversion. This perception is rationalized when one takes into account the detection of bihormonal cells with co-packaged GH and prolactin in several non-tumorous human pituitaries (Bendayan 1982, 1984).

After treating the animals with DMPA (Progestin), R-06 (weak oestrogen) and Ovral [Combination of ethinyl oestradiol (oestrogen) and levonorgestrel (progesterone)], the PRL cells increase in number and show an abundance of organelles concerned with the synthesis and secretion in the cell.

After DMPA treatment, the RER membranes increase in number, the Golgi apparatus shows hypertrophy and there are a number of polymorphic secretory granules in the cell. These observations corroborate with the findings of Baker et al. 1973, who have reported the enlargement of PRL cells in rats with Norethindrone, another synthetic progestin and no changes in the PRL cells after medroxyprogesterone treatment. These morphological observations are in agreement with the findings of Mishell et al. 1977, Baker et al. 1973, and Wright et al. 1977, who were able to measure the increased serum prolactin levels after the treatment with MPA, norethindrone enanthate and progesterone to women, rats and sheep, respectively.

Oestrogenic drug, R-06 leads to hypertrophy and hyperplasia of PRL cells. The cell shows typical nebenkern formation of RER which was also observed by Horvath and Kovacs (1988), Stefaneanu and Kovacs (1991) and Saeger (1992). Also, increased cellular activity sometimes along with the tumour formations has been reported by El Etreby et al. (1972) and Aumuller et al. (1978) after oestradiol benzoate and norethisterone enanthate treatment in rats, and oestrogen treatment in rats by Farquhar and Rinehart (1954a, b). These morphological observations are in conformity with the increased serum prolactin as observed by Chen and Meites (1970), Meites et al. (1972) Ojeda and McCann (1974), Neumann et al. (1976), Andrews and Ojeda (1977), Gudelsky et al. (1981) and Pinilla et al. (1993) in rats. Andrews and Ojeda (1977) have reported no alternations in plasma prolactin after oestradiol benzoate treatment but an increase in the same following diethylstilbestrol and RU-2858 administration. These results indicate that dynamic changes occur in the RER of the acidophils with the 600 nm granules during oestrogen administration. Similar changes have been observed in lactating animals in the present study and further indicate that the lactogenic hormone is probably produced, stored and secreted by these cells.

The present results coincide with the previous findings stating that in rats after treatment with Norethisterone enanthate observed hypophyseal and mammary gland tumours via a stimulatory effect on the prolactin-producing cells due to the oestrogenic activity of this compound (Neumann and Schenck 1976; Von Berswordt-Wallrabe et al. 1977. The presented data supplement for the first time the biochemical and histological findings from ultrastructural analysis of the prolactin-producing cells after Norethisterone enanthate treatment. Norethisterone enanthate treatment caused a clear-cut augmentation of PRL cell number. These obviously newly formed PRL cells showed signs of structural alteration.

Our observations also corroborate with the findings of Hymer et al. 1961. Pantic and Genbaceu (1969, 1972); Schelin and Lundin 1964, Pasteels et al. 1972 and Shiino and Rennels (1976) as the PRL cells showed proliferative and hypertrophic nature after the administration of oestrogen in rats. Focal proliferation of rough ER in PRL cells after treatment with oestrogenic compounds results in the formation of concentric whorls consisting of both granular and agranular membranes with centrally located lysosomes or multivesicular bodies. From the foregoing discussion, it is suggested that the oestrogenic compounds are an effective stimulator of prolactin secretion and plays an important role in the morphological and developmental changes in the pituitary prolactin granules.

Administration of Ovral (combination of oestrogen and progesterone) leads to the dilation of RER cisternae, a hypertrophy of Golgi apparatus and an increase in numerous immature polymorphous secretory granules. The present observations are in agreement with those made by Goluboff and Ezrin (1969), Pellion et al. 1970, Herlant and Pasteels (1971), Hachmeister et al. (1972), Pasteels (1971) and El Etreby and Gunzel (1974) after a combination of oestrogen and progesterone treatment at light microscope level. The stimulated PRL cells indicative of active secretion result in increased prolactin levels in serum which were recorded by Wright et al. 1977, and Mishell et al. 1977, in sheep and women, respectively.

Beauvillain et al. 1977, have reported the existence of spherical granules in PRL cells of the guinea pig pituitary for the first time, some PRL cells with only small secretory granules (250-500 nm) in the porcine pituitary (Dacheux 1980) and one of the immunoreactive PRL cells to be identical with 'acidophils of the small granule type' (Nogami and Yoshimura 1980). Prolactin cells with small granules correspond to the 'acidophils of the small granules type' described by Yoshimura et al. 1974. These cells also resemble in fine structural properties, the Kurosumi and Oota (1968) LH gonadotrophs. The PRL cells of male adult rats are divided into four categories, and the polymorphic shape of the granules in not necessarily regarded as an absolute criterion for identification (Nogami 1984).

In male rats, we found great hyperplasia, hypertrophy and polymorphic shape of the secretory granules of PRL cells in the AP, as a result of treatment with Depo provera + testosterone enanthate. These observations are further supported by the elevated prolactin levels in serum, measured by using RIA after MPA + TE treatments in rats (Bellare 1994); and also reported by Herbert et al. 1977 and Mukku et al. 1981, in rats and monkey after TE treatments only.

It is obvious therefore that a combined regimen of MPA + TE shall have a synergistic effect by enhancing the synthetic and secretory activity of PRL cells.

Increase in prolactin levels can effect testosterone secretion either centrally by the effects on hypothalamus–pituitary axis or peripherally by acting directly on the testis (Segal et al. 1976). Increase in prolactin levels reduces the 5 alpha-reductase activity in peripheral tissues which results in a conversion ratio of dihydrotestosterone (DHT) to testosterone (Magrini et al. 1976; Bernini et al. 1983). This interference of conversion by testosterone to biologically active DHT caused by 5 alpha-reductase represents the modes of action of prolactin on the testis and/or on other androgen-sensitive target organs and presents yet another approach to study male infertility (Magrini et al. 1976). Prolactin is also known to inhibit LHRH release by influencing the hypothalamic aminergic functioning as a result, LH secretion is reduced and testosterone levels fall down (Pontiroli et al. 1980). Although the specific role of prolactin in the regulation of steroidogenic enzymes in the Leydig cell (Hafiez et al. 1972) is still unclear, there is evidence to show that prolactin is involved in male infertility (Segal et al. 1976; Sueldo et al. 1985).

1.3 Thyrotroph Cell

1.3.1 Introduction

Thyrotroph cell (TSH cell) is perhaps the least common and least characterized cell type, constituting less than 10% of pituitary cells. However, their number in the pituitary increases in the oestrous phase as reported for various chiropteran species (Herlant, 1964; Kobayashi 1966; Kobayashi and Herman 1966; Kurosumi and Oota 1968; Azzali 1971; Bhiwgade et al. 1989), rabbit (Salazar 1963) and dog (Gale 1972). Their ultrastructure in a wide variety of mammals has been investigated (Barnes 1962; Salazar 1963; Young et al. 1967; Murphy and James 1976; Horvath et al. 1977 and Bhiwgade et al. 1989). The human thyrotroph has been dealt with insufficient detail by Lawzewitsch et al. (1972) and Horvath and Kovacs (1988). The consolidation of the findings of all these workers show that though the TSH cells are variously shaped, in most species they are fairly large, angular, with varying amounts of slightly dilated RER and a prominent Golgi zone. An unfailing observation made by several workers in the cell pertains to its secretory granules which are of the smallest size among all the cells of the anterior pituitary (Surks and DeFesi 1977; Kiguchi 1978; Yoshimura et al. 1982; Cinti et al. 1985; Bhiwgade et al. 1989). The secretory granules often show a perigranular membrane, that in some animals may manifest into a perigranular halo (Murphy and James 1976). The granules are distributed throughout the cell, but when less in number these may be peripherally disposed (Bergland and Torack 1969; Horvath and Kovacs 1988; Saeger 1992).

Specific stimulation of the cell can be achieved by inducing hypofunction of the thyroid gland, this approach has been adopted through surgical and chemical means by some workers to distinguish the TSH cells from the other cells (Farquhar and Rinehart 1954b; Stratmann et al. 1972; Kiguchi 1978; Raza and Shakoori 1978; Horvath and Kovacs 1988; and Bhiwgade et al. 1989). Such stimulated hyperplastic TSH cells are designated as ‘thyroidectomy’ cells and they show a vacuolated cytoplasm due to a strong dilatation of RER and an enlarged Golgi zone (Kovacs and Horvath 1986; Horvath and Kovacs 1988; Stefaneanu and Kovacs 1991). Frequent presence of intracisternal granules was noticed by Ozawa, (1991), in such thyroidectomy cells. Ozawa (1991), combined immunoelectron microscopy and enzyme cytochemistry to trace the development of such thyroidectomized cells. Shiino et al. (1972), successfully induced the thyroidectomy cells by TRH treatment in a tissue culture of rat pituitary.

1.3.2 Light Microscopic Observations

Thyrotroph cells are basophilic mucoid cells. They stain red with Periodic Acid Schiff / Orange G, purple blue purple with Alcian Blue/ Periodic Acid Schiff /Orange G, violet with Aldehyde Fuschin /Carmoisine L/Orange G (Fig. 1.56) and blue purple with Aldehyde Thionin /Periodic Acid Schiff /Orange G (Fig. 1.57). They are thus Alcian Blue-positive, Aldehyde Thionin-Positive and Aldehyde Fuschin-positive. They are negative to Orange G in the combined histochemical techniques used. They are round or angular in shape with a diameter of 9–11 μ. They show distinct cell boundaries and the nuclei are round, centrally placed or eccentric with a diameter of 3–4 μ and contain clumped chromatin material with a small nucleolus. The cytoplasmic granules are fine. These cells are distributed throughout the gland, singly or in groups of two.

1. Action of thiourea solution in water—In the initial control animals, the thyrotrophs are moderately stained with Aldehyde Fuschin, Aldehyde Thionin and Alcian blue stains

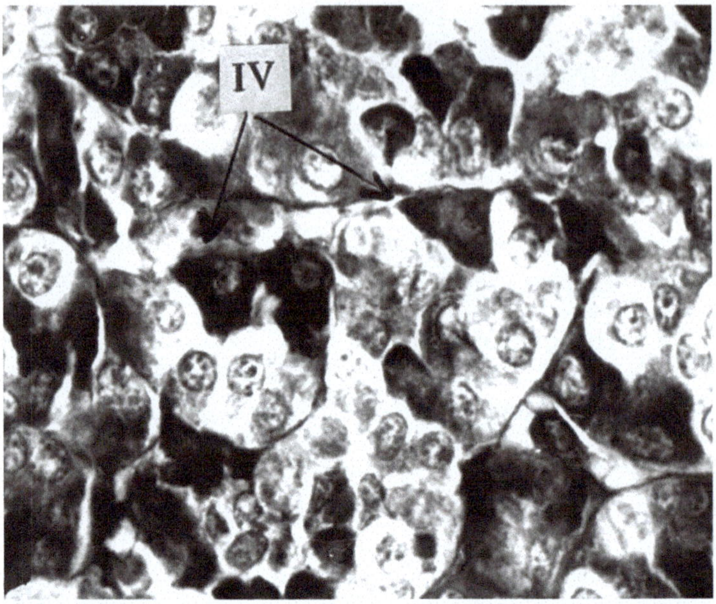

Fig. 1.56 Pars anterior of *Rousettus leschenaulti* oestrous female showing TSH cells stained with Aldehyde Fuchsin / Light Green /Orange G [X 500]. Unpublished light micrograph from Dr. Bhiwgade

(Figs. 1.58 and 1.59). After 9 days of thiourea treatment, the thyrotrophs are deeply stained, and increased in number and size (Figs. 1.60 and 1.61). After 15 and 18 days of thiourea treatment, a slight decrease in the staining intensity of thyrotrophs (Fig. 1.62a and b) is observed as compared to the 9 days treatment. The degranulation condition is gradually accentuated at the remaining intervals studied (18, 24, 30, 40, 65 and 90 days). The thyrotrophs are still lightly stained with Aldehyde Fuschin, Aldehyde Thionin and Alcian blue in the 30 days of treated animals (Fig. 1.62c, d, e, and f).

Degranulation commence by 30 days (Fig. 1.62c) and becomes gradually accentuated in 40, 65 and 75 days treated animals (Fig. 1.62d). Consequently, most of the thyrotrophs show vacuoles. By 90 days, the thyrotrophs show drastic changes. Most of the thyrotrophs are totally degranulated and become converted into

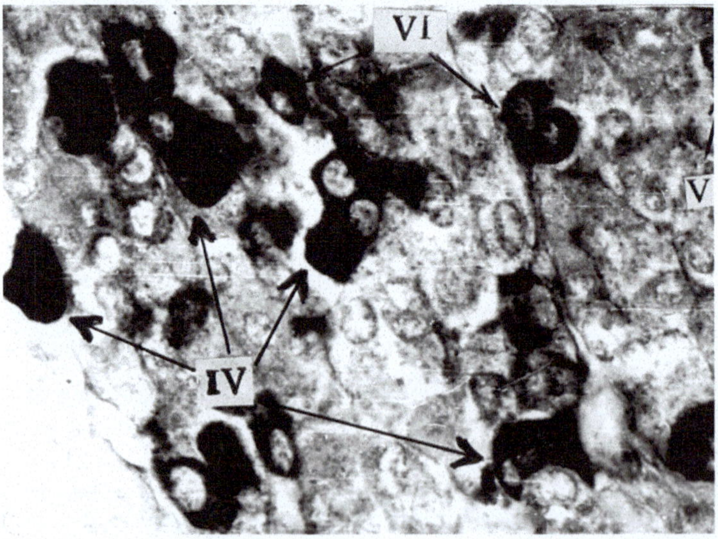

Fig. 1.57 Pars anterior of male bat, *Rousettus leschenaulti*, during active breeding period showing TSH cells (IV). Note the bluish-purple FSH cells (VI) and yellow-reddish LH cells (V) stained by Aldehyde Thionin /Periodic Acid Schiff/Orange G [X 500]

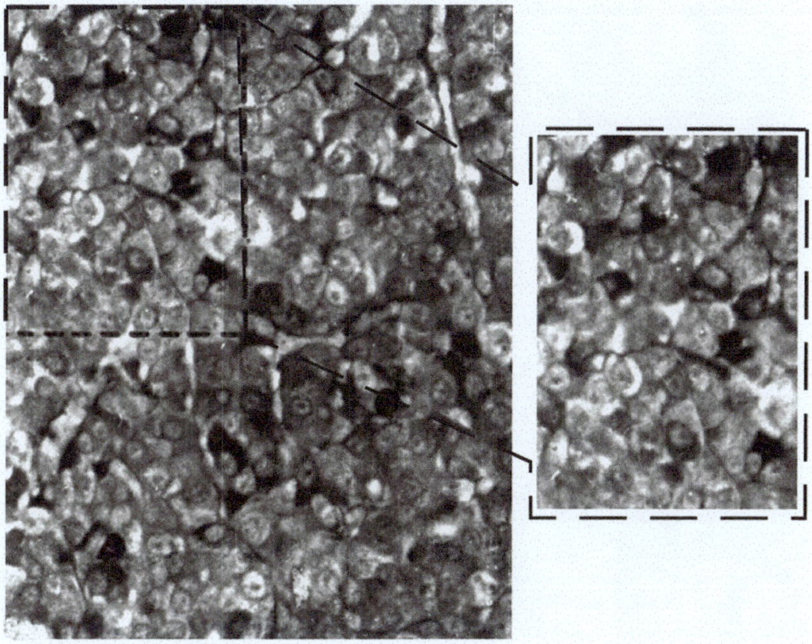

Fig. 1.58 Antero-median portion of the pars anterior of a control male squirrel showing the distribution of thyrotrophs. Alcian Blue / Aldehyde Fuschin/ Orange G [X 320]. Delineated area (inset) is enlarged in Fig. 1.4

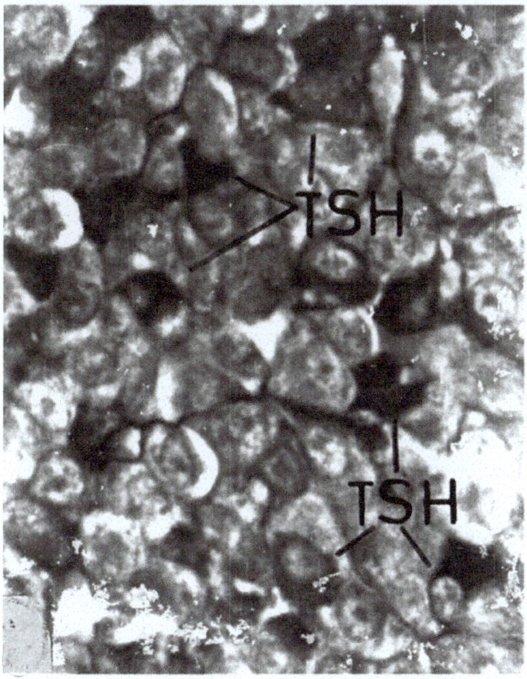

Fig. 1.59 Enlargement of the area delineated by the rectangle in Fig. 1.3. Note the angular thyrotrophs (arrows). [X 800]

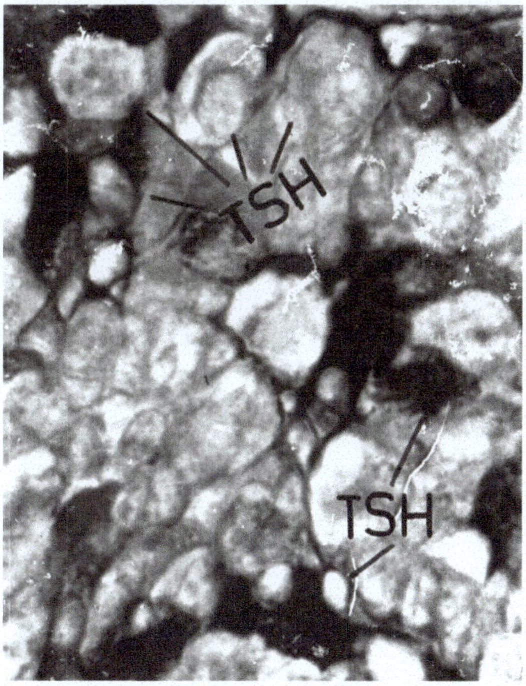

Fig. 1.60 Pars anterior of male squirrel (*Funambulus pennantii*) treated with thiourea for 9 days. Note the large darkly stained TSH cells. Aldehyde Thionin/Periodic Acid Schiff/Orange G [X 800]

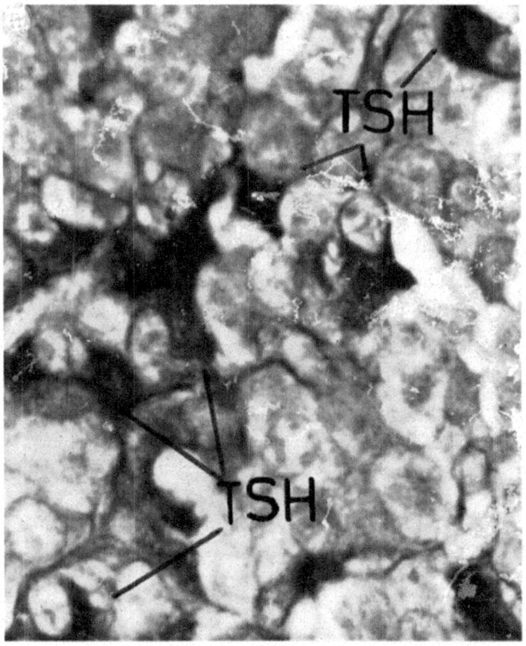

Fig. 1.61 Pars anterior of the male squirrel (*Funambulus pennantii*) treated with thiourea for 9 days. Note the angular TSH cells selectively stained with Alcian Blue/Aldehyde Fuschin/Giemsa [X 800]

chromophobes with large nuclei (Fig. 1.62e and f) which are significantly larger than those of the controls. On the other hand, the thyrotrophs of the control animals are brightly stained with Aldehyde Fuschin, Aldehyde thionin and Alcian blue.

2. Action of DL thyroxine—In the initial control animals, the thyrotrophs show a positive response to Periodic Acid Schiff, Alcian Blue, Aldehyde Thionin and stain violet with Aldehyde Fuschin (Fig. 1.63a). After 9 days of treatment with DL—thyroxin, the thyrotroph cells seem to be increased in number and staining intensity (Fig. 1.63b and c) as compared to those of the untreated controls. There is, however, a slight fall in the staining intensity and diminution in the numbers of Aldehyde Fuschin-positive cells after 15 days of thyroxine treatment (Fig. 1.63d). Eighteen days of thyroxine injections bring about a further reduction and degranulation of Aldehyde Fuschin-positive cells from the squirrel pituitary except a few cells showing a moderate staining (Fig. 1.63e and f). After 24 and 30 days of treatment, the thyrotrophs do not show any Periodic Acid Schiff, Alcian Blue, Aldehyde Thionin and Aldehyde Fuschin-positive material and are chromophobic in nature (Fig. 1.63g and h).

1.3.3 Electron Microscopic Observations

Normal thyrotrophs from adult female rats are elongated, polygonal or oval in shape. The rough endoplasmic reticulum is arranged in numerous parallel arrays of tubular cisternae. The Golgi apparatus is not prominent. Secretory granules are very small in size and scattered throughout the cytoplasm with a tendency to accumulate focally in a part of the cytoplasm; but they are not distributed in a row along the cytoplasmic membrane (Fig. 1.64a and b). Similar ultrastructural features are observed in the pituitary of bat, Cynopterus. The cytoplasm of the TSH cell is relatively electron transparent and devoid of fine structural details. The endoplasmic reticulum and the Golgi apparatus are minimally developed and a few endoplasmic reticulum membranes are studded with ribosomes (Figs. 1.65 and 1.66). There are either none lying free or only a few in the cytoplasm. The mitochondria are numerous, small, oval or elongated in shape with loss of cristae in some of them. The lysosomes are seen as, empty vesicles and vacuoles. The secretory granules are accumulated on one side of the cell.

The thyrotroph is the least common of the known cell types in the monkey pituitary gland. The TSH cells are elongated or roughly butterfly shaped (Figs. 1.67 and 1.68). The typical TSH cells are easy to recognize, they possess spherical or ovoid nuclei and a few parallel arrays of RER that may be slightly widened. In the more granulated cells, they are scattered throughout the cytoplasm. In the active, well-granulated cells, the secretory granules are spherical with somewhat varying densities.

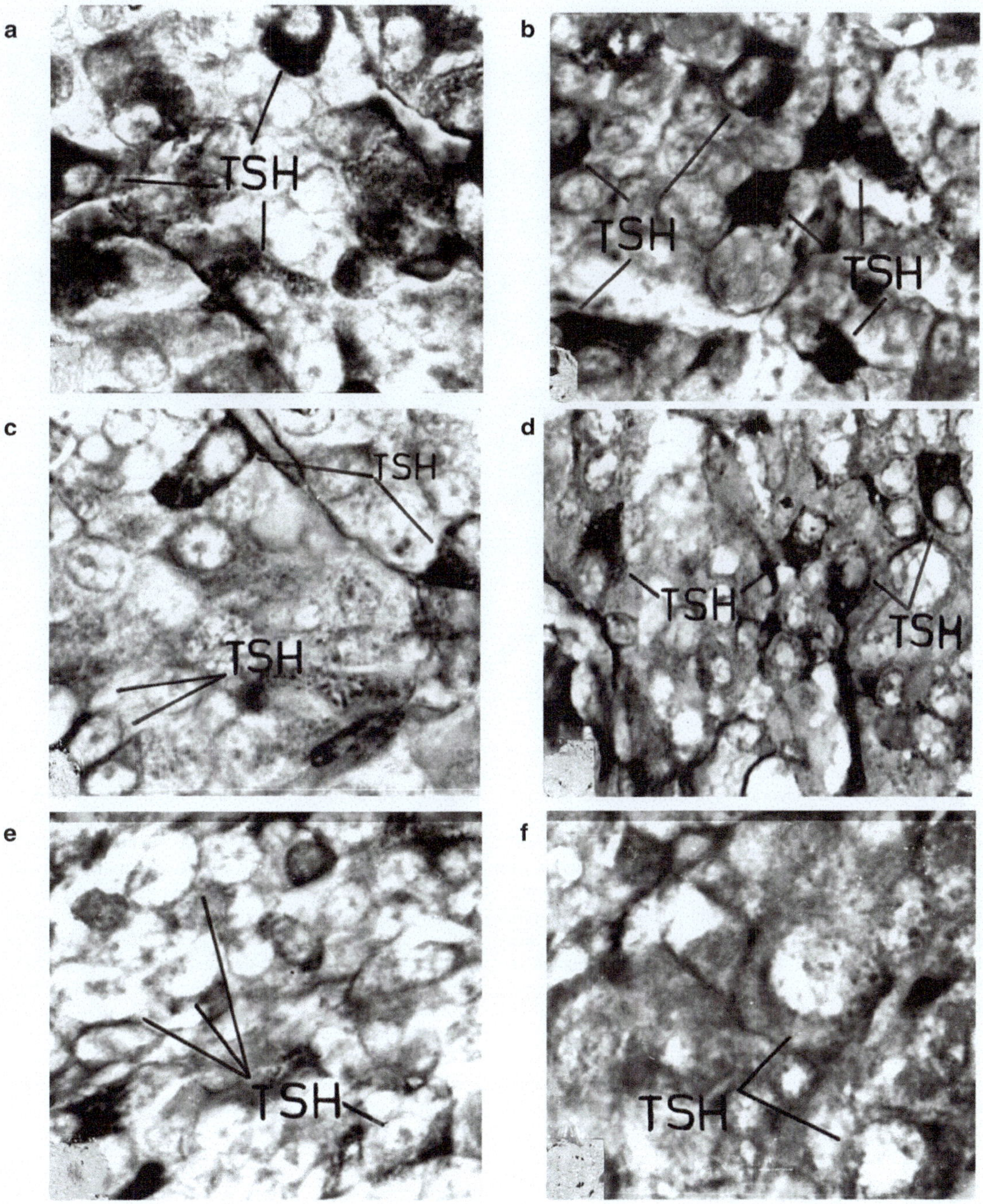

Fig. 1.62 Pars anterior of male squirrel (*Funambulus pennantii*) treated with thiourea for different days (**a**). For 15 days. Note TSH cells. Aldehyde Thionin/Periodic Acid Schiff/Orange G [X 800] for 18 days (**b**). Note angular thyrotrophs Alcian Blue/Aldehyde Fuschin/Giemsa [X 800] for 30 days (**c**). Note TSH cells Aldehyde Thionin/Periodic Acid Schiff/Orange G [X 800] for 40 days (**d**). Note vacuolization in thyrotroph cells. Herlant-II (Alcian Blue/Periodic Acid Schiff), (X 800) for 75 days (**e**). Note TSH cells decrease in staining intensity and increase in vacuolization Alcian Blue/Aldehyde Fuschin/Giemsa [X 800]. In the antero-median zone of the pars anterior of male squirrel for 90 days (**f**). Note chromophobic appearance of TSH cells with large nuclei, Herlant-II (Alcian Blue/Periodic Acid Schiff) [X 800]

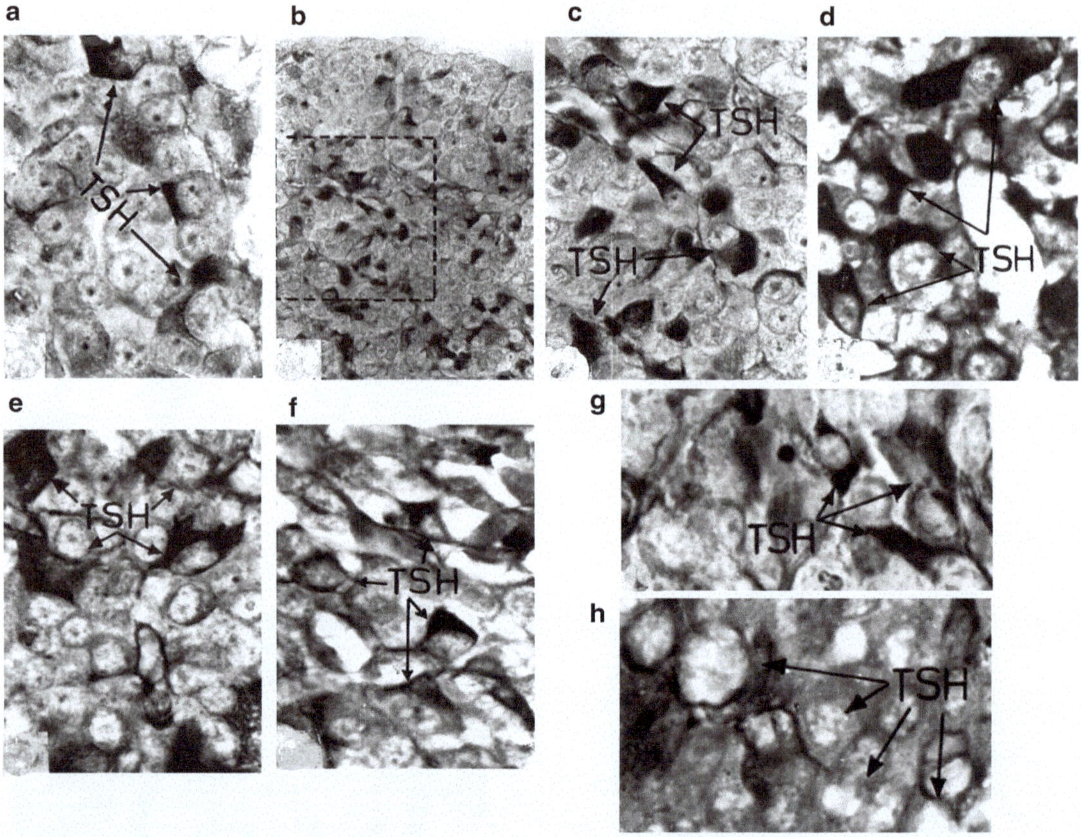

Fig. 1.63 (**a**)—Pars anterior of a control male squirrel (*Funambulus pennantii*) showing Aldehyde Fuschin-positive angular thyrotrophs (arrows), Aldehyde Fuschin/Light Green/Orange G, [X 800]. (**b**)—Pars anterior of male squirrel treated with DL-thyroxin for 9 days showing distribution of Aldehyde Fuschin-positive thyrotrophs [X 260]. (**c**)—Enlarged part of the area delineated by the rectangle in Fig. 1.81 showing darkly stained angular thyrotrophs (arrows) Aldehyde Fuschin/Light Green/Orange G [X 600]. (**d**)—Pars anterior of same animal as (arrows) in (**b**), but stained with Alcian Blue/Aldehyde Fuschin/Giemsa after bromine oxidation. Note selectively stained angular thyrotrophs (arrows) Alcian Blue/Aldehyde Fuschin/Giemsa [X 800]. (**e**)—Pars anterior of a male squirrel as in (**b**) but stained with aldehyde-thionine showing thionine-positive angular thyrotrophs (arrows), Aldehyde Thionin/Periodic Acid Schiff/Orange G [X 800]. (**f**)—Pars anterior of a male squirrel treated with DL-thyroxine for 15 days. Note TSH cells (arrows) Alcian Blue/Aldehyde Fuschin/Giemsa [X 800]. (**g**)—Pars anterior of a male squirrel treated with DL-thyroxine for 18 days to show further degranulation of AF-positive staining of angular thyrotrophs (arrows) Aldehyde Fuschin/Light Green/Orange G [X 800]. (**h**)—Antero-median zone of pars anterior of a male squirrel after 30 days of treatment with DL-thyroxine showing chromophobic nature of TSH cells (arrows). Herlant-II (Alcian Blue/Periodic Acid Schiff) [X 800]

1.3.4 TSH Cells in Bat Treated with Propylthiouracil

In bat, *Rousettus leschenaulti*, following propyl thiouracil treatment for 30 days, TSH cells show structural signs of hyperfunction as in hypothyroidism, similar to those seen in other species. Enlargement of the cell is due to progressive proliferation and dilation of RER, a marked expansion of Golgi apparatus along with numerous Golgi cisternae, dilated sacculi and hypertrophied mitochondria with distinct internal cristae. There is a decrease in the secretory granules and only the immature granules are seen (Figs. 1.69 and 1.70). Few lysosomes are

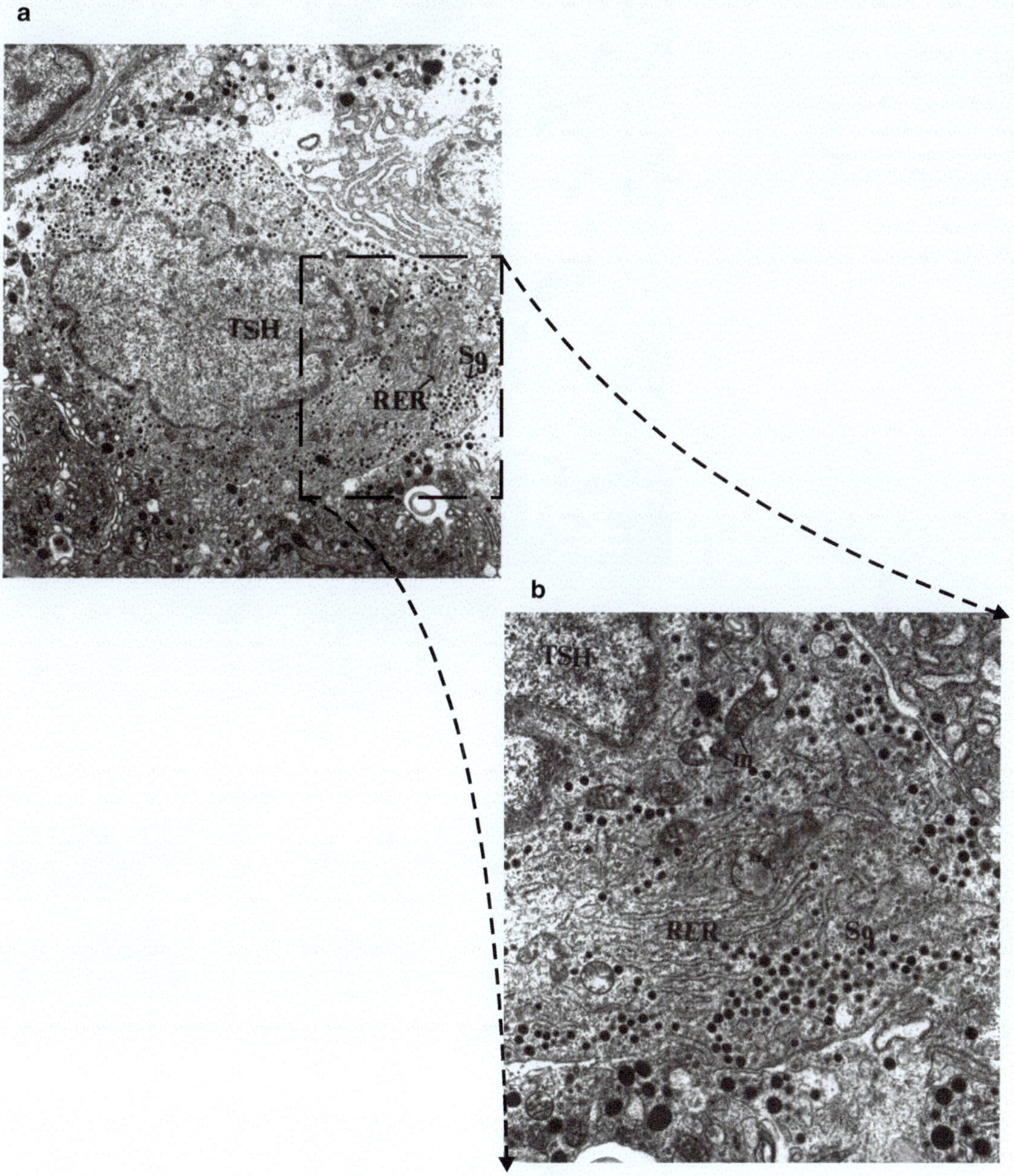

Fig. 1.64 (**a**) TSH cells of oestrous female rat. Note small sized secretory granules (Sg), scattered throughout the cytoplasm, few and short lamellar profiles of RER [X 6000] Unpublished electron micrograph from Dr. Bhiwgade and Dr. Mantri. (**b**) Enlarged portion of the area RER of the above cells (1). Note small- sized secretory granules (Sg) as well as mitochondria [X 8500] Unpublished electron micrograph from Dr. Bhiwgade and Dr. Mantri

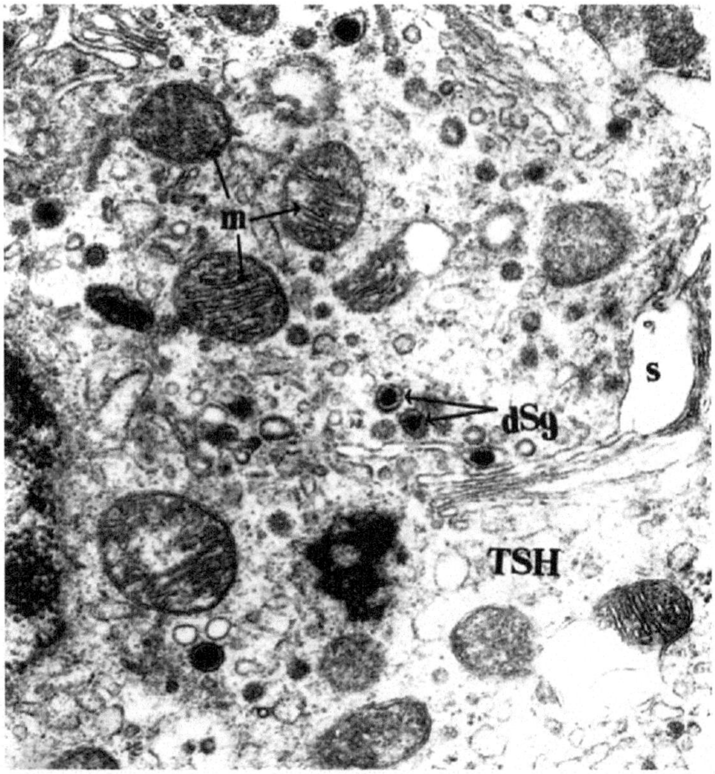

Fig. 1.65 Anterior pituitary of late pregnant bat, *Cynopterus sphinx*. Note TSH cell with conspicuous mitochondria (m), developing secretory granules (dsg) and dilated Golgi sacs [X 25,000]. Unpublished electron micrograph from Dr. Bhiwgade and Dr. Mantri

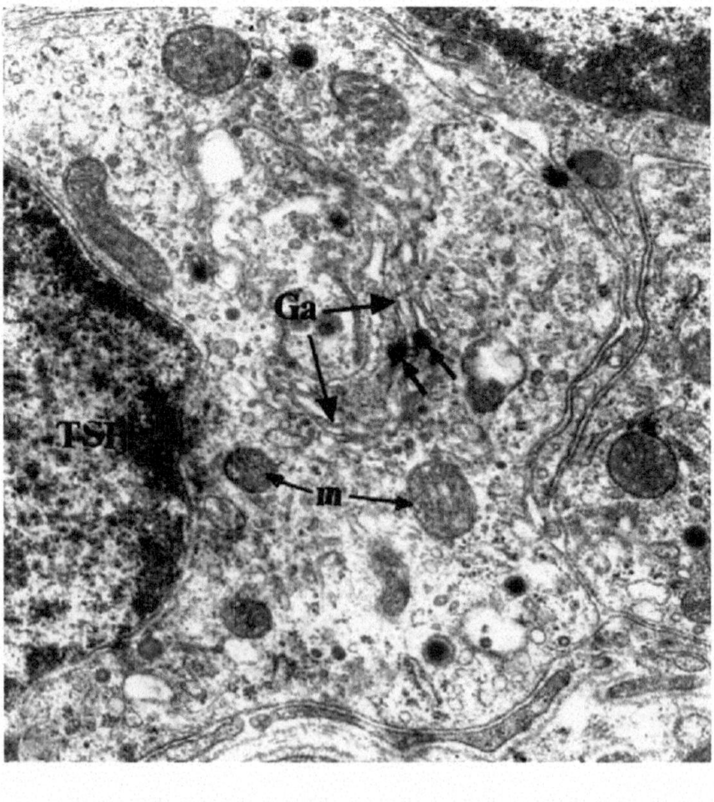

Fig. 1.66 Thyrotroph of late pregnant bat, *Cynopterus sphinx*, showing Golgi apparatus (Ga) within whose sacs are presently developing secretory granules (arrows) and numerous mitochondria (m) [X 16,000]. Unpublished electron micrograph from Dr. Bhiwgade and Dr. Mantri

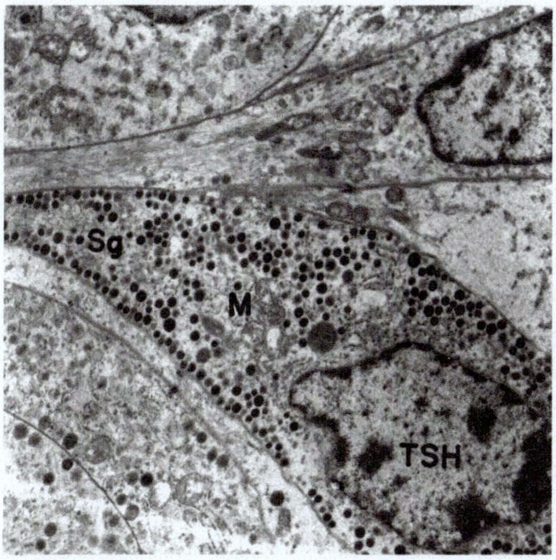

Fig. 1.67 A typical angular thyrotroph in male Bonnet monkey (Macaca radiata) showing minimum small secretory granules and few mitochondria (m) [X 8000] Unpublished electron micrograph from Dr. Bhiwgade and Dr. Thakur

also encountered and fragments of RER are scattered throughout the cytoplasm.

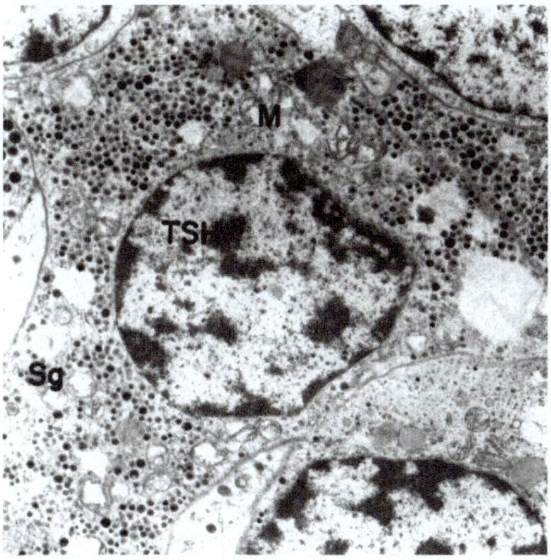

Fig. 1.68 TSH cells of male Bonnet monkey(Macaca radiata). Note the typical shape of the cell with minimum small secretory granules scattered throughout the cytoplasm [X 8000] Unpublished electron micrograph from Dr. Bhiwgade and Dr. Thakur

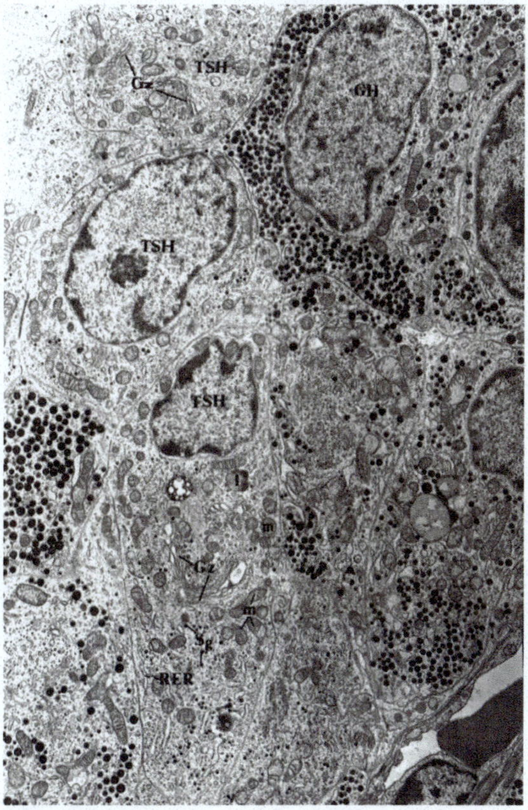

Fig. 1.69 Low-power EM of pituitary of male bat, *Rousettus leschenaulti* treated with propyl thiouracil (PTU). GH cells and TSH cells can be easily distinguished. Enlarged FSH cell is also seen with well developed Golgi zone (Gz), secretory granules (sg) and RER [X 8000]

1.3.5 TSH after Treatment with Oestradiol Valerate

This cell contains a nucleus with a circular or angular profile within which chromatin of uniform density is evenly distributed and a distinct marginated nucleolus (Fig. 1.69). Moderate to dense number of small and round secretory granules have been peripherally displaced within this cell. A few moderately dilated short, as well as numerous confronting cisternae are included in the cytoplasm. The Golgi zone is small but discrete and a few secretory vesicles arise from its 'trans' face, in the juxtanuclear portion. The lysosomal activity has apparently increased and the involved mitochondria have assumed 'U'- and

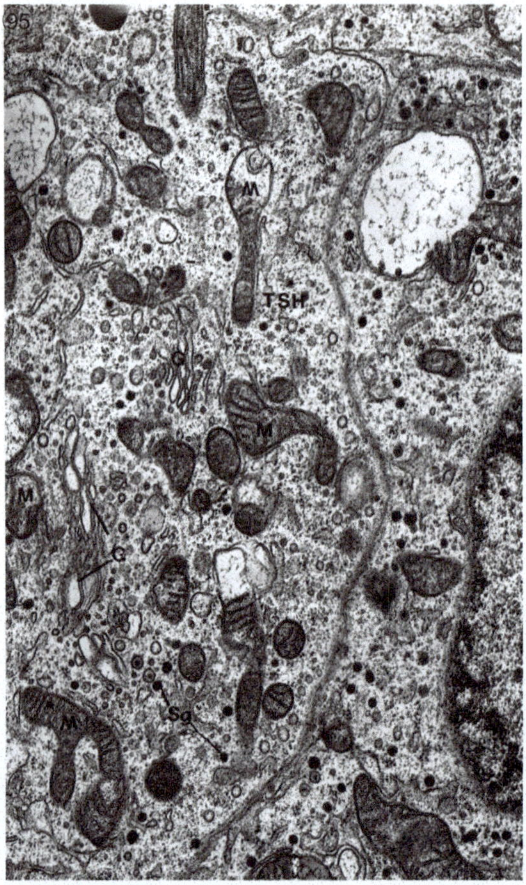

Fig. 1.70 TSH cell of male bat, *Rousettus leschenaulti*, after PTU treatment for 30 days. Note the hypertrophied elongated mitochondria (M), enlarged Golgi zone (G) and rER. Also, note few immature secretory granules (sg) [X 16,000]

'C'-shaped configurations (Figs. 1.70, and 1.71a, b).

TSH after Treatment with Norethisterone Heptanoate

Within the small angular TSH cell, the substantially large and oblong nucleus contains dense chromatin in clumps, besides a distinct reticular nucleolus. Noticeably, a distinct electron lucent space free of secretory granules has been created, whose proximal boundary is constituted by the nuclear membrane and the distal limit has been laid down by a single discontinuous strand of ribosomes studded endoplasmic reticular membrane. Many small, spherical mitochondria with dense matrix have been invested by the RER membrane. The peripheral cytoplasm is crowded with secretory granules of varying sizes and electron densities, amidst which a very few mitochondria occur. The continuity of the nuclear membrane with the RER membrane has been clearly established. The granular endoplasmic reticulum is represented by single strands of varying lengths that are encountered frequently in the peripheral cytoplasm (Fig. 1.72). Lysosomes have also appeared in the cytoplasm.

1.3.6 Discussion

Several authors have studied the action of various antithyroid drugs such as thiourea and thiouracil on the pituitary gland and observed hyperplasia and degranulation of certain basophil cells, which were later identified as thyrotrophs (Halmi 1950). It is usually agreed that antithyroid drugs such as thiourea and thiouracil impair the synthesis of thyroxine (Goldberg and Chaikoff 1951) and (D'angelo 1953). The resultant fall in the levels of circulating hormone induced an increase in the pituitary thyrotropin secretion which in turn causes hypertrophy and hyperplasia of the thyroid follicles. Degranulation of thyrotrophs followed by hypertrophy has been reported in different mammalian species following thiouracil administration (Serber 1961) and (Peters and Halmi, 1961), and after surgical and radio thyroidectomy.

Electron microscopic studies of Farquhar and Rinehart 1954a, b also revealed that the specific secretory granules decrease in the cells after thyroidectomy. In our studies, the basophils situated in the anteromedian zone and peripheral parts of the pars anterior in *Funambulus pennanti* are designated as thyrotrophs. They show signs of activation by 9 and 15 days and appear hypertrophied and become completely degranulated after prolonged treatment (40, 60 and 90 days). The disturbance in the thyroid hormone level is reflected in the cytology of the thyrotroph cells after thiourea treatment. The hypertrophied chromophobes of anteromedian

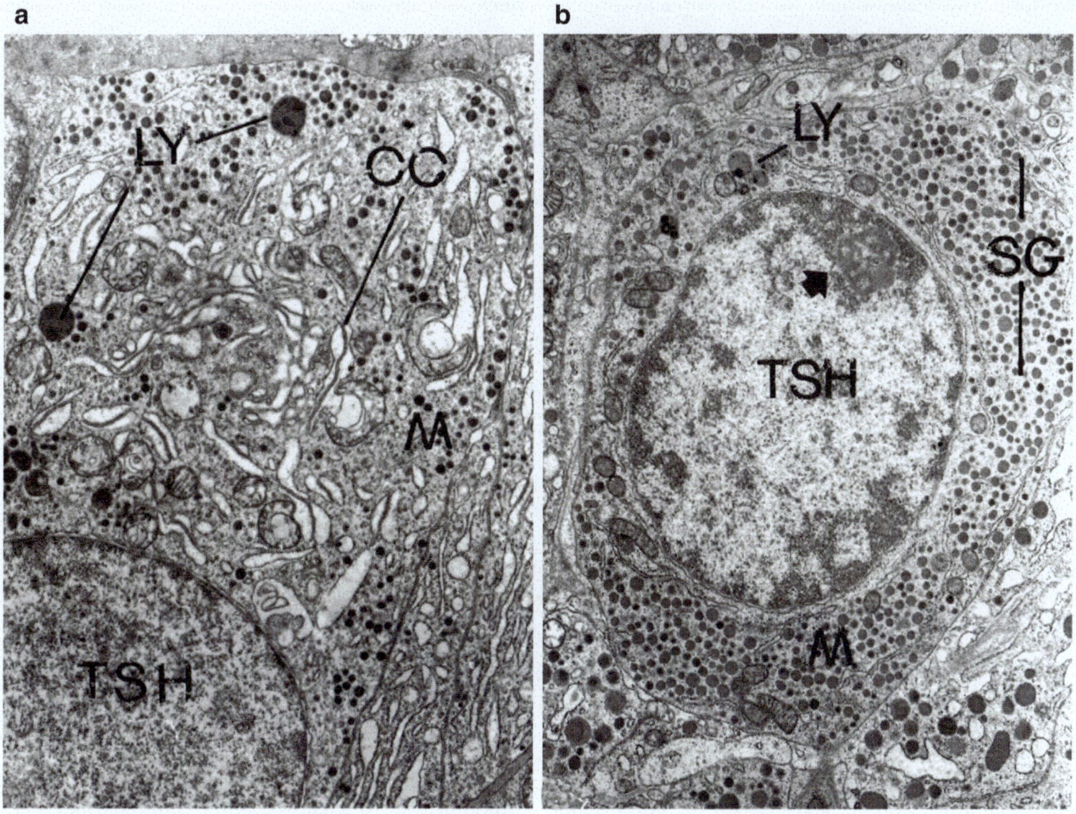

Fig. 1.71 (**a**) Electron micrographs of the thyrotrophs (TSH) of a female albino rat treated with Oestradiol valerate for 24 weeks. The cells harbour a large euchromatic nucleus with a distinct marginated nucleolus (bold arrow) (**b**). The small, round secretory granules (SG) are peripherally displaced. The cytoplasm also shows confronting cisternae (CC), Fig. 1.71a, involuted mitochondria (M), and lysosomes (LY) Fig. 1.71a. [X 10,000] Fig. 1.71b [X 8000] Unpublished electron micrograph from Dr. Bhiwgade and Dr. Borkar

zone and peripheral parts provide evidence that these are the sites of production of TSH since the hormone contained in them appears to have been released in order to stimulate the thyroid follicle.

Treatment of squirrels with DL-thyroxine led to regression of thyrotroph cells, presumably indicating an inhibition of the synthesis of TSH. Purves and Griesbach 3–4 described thyrotroph cells to be similar to those of control following thyroxine treatments after 2 weeks, but after 3 weeks of observation the regression of the thyrotrophic cells from the rat pituitary was observed. In our studies, in *Funambulus pennanti*, the thyrotroph cells remained as those of controls for up to 13 days, but prolonged treatment for 24 and 30 days caused total degranulation of the thyrotroph cells.

In our studies, the TSH cells are seen elongated or slightly butterfly shaped or triangular in shape. The cytoplasmic microfilaments have established contact with the basement membrane, the hypertrophied mitochondria dilation of RER is seen. Similar observations have been shown in other animals (Kurosumi and Fujita 1974; Childs et al. 1981a, b; Yoshimura et al. 1982; Kurosumi 1986). From the results of our studies, it is noted that the most pronounced effect of the PTU treatment on the thyrotrophs is the increase in number and size after chemical thyroidectomy. After 15 days of PTU treatment, the secretory granules decreased remarkably. Even so, the presence of

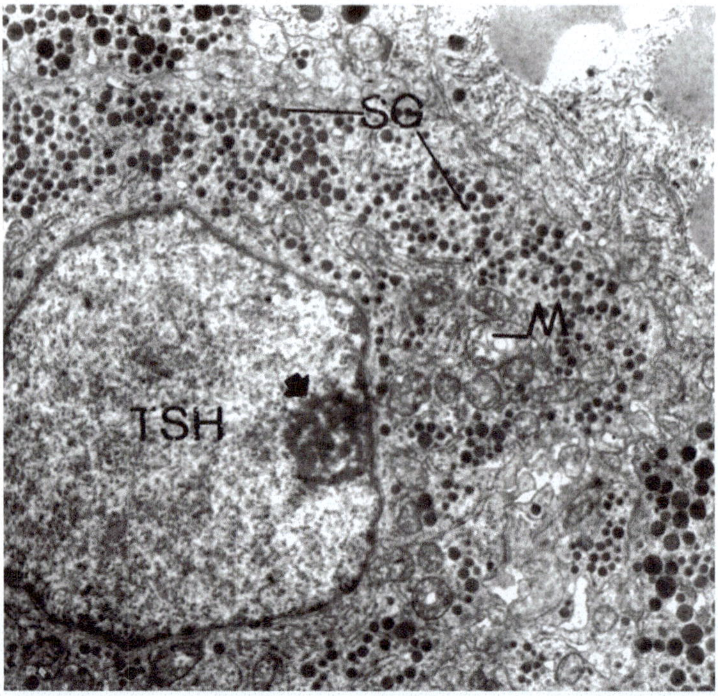

Fig. 1.72 Thyrotroph (TSH) of a female albino rat following norethisterone heptanoate administration for 24 weeks. The cell has an angular contour and a large nucleus with a distinct reticular nucleolus (bold arrow). Note the dense population of secretory granules (SG) and a very few mitochondria (M). Lysosomes have also appeared in the cytoplasm [X 8000]. Unpublished electron micrograph from Dr. Bhiwgade and Dr. Borkar

the remaining secretory granules is an indication of the synthetic activity of the cell and thus the complete inhibition of TSH secretion after PTU treatment cannot be concluded. Herlant (1964), Kobayashi (1966), Kobayashi and Herman (1966), Azzali (1971) and Bhiwgade et al. (1989) have confirmed the above ultrastructural observations in various species of bats. Salazar (1963), Gomez-Dumm and Echave-Llanos (1972) and Gale (1972) have also reported analogous observations on TSH cells in the female rabbit, male mouse and dog, respectively.

The response of TSH cells to thyroidectomy has been reported in the rat (Lever and Peterson 1960), rabbit (Salazar 1963), cow (Mikami 1970), mink (Murphy and James 1976) and bat (Bhiwgade et al. 1989). There is a hypertrophy of TSH cells after thyroidectomy, and cytoplasm contains few, small secretory granules. Many saccular dilations and the swollen RER is in the form of vacuoles studded with ribosome particles. The nuclear chromatin is dispersed and the nucleolus is prominent in most of the cells.

Raza and Shakoori (1978) performed chemical thyroidectomy by administering PTU treatment for 10 days to male albino rats. They have reported increased activity of thyrotrophs after PTU treatment, which collaborates with the observations of Goldberg and Chaikoff (1951) and Stratmann et al. (1972). Our studies have also confirmed this (Bhiwgade et al. 1989). However, after 10 days of PTU withdrawal, they observed a decrease in the synthetic and metabolic activity of the thyrotrophs, which ultimately lead to degenerative changes. These results are similar to those reported by Purves and Griesbach (1951a, b), Farquhar and Rinehart (1954a, b) and D'angelo (1953, 1967) after thyroidectomy.

1.4 Gonadotroph Cells (GTH Cells)

1.4.1 Introduction

The gonadotrophs comprise approximately 10–20% of the anterior pituitary cells. Their peculiar disposition, within the anterior pituitary, facing the vascular space has been noticed in several animals. The cellular origin of LH and FSH has been a contentious issue for a long time. The realization that these two hormones have a separate cellular origin, came on the basis of a non-parallel secretion of LH and FSH. Though the classical tinctorial methods did not fully resolve the problem of gonadotroph identity, the advent of immunocytochemistry marked an important phase in ascertaining the cellular sources of these two hormones. The fine structure variations between the two gonadotrophs in the pituitary of some animals prompted some workers to subtype them to FSH and LH cells (Barnes 1962; Gomez-Dumm and Echave-Llanos 1972; Lawzewitsch et al. 1972). Lawzewitsch et al. (1972) used nuclear morphology as a clue for subtyping the gonadotrophs. Soon it was realized that such variations do not consistently correlate with the synthesis or storage of either gonadotrophins. A co-localization of these two hormones in one and the same cell was demonstrated by Nakane 1970.

Early attempts at classifying the gonadotrophs were made on the basis of cytophysiological studies using conventional electron microscopic techniques. Inspired by the classic studies of Farquhar and Rinehart (1954a, b) and Barnes, (1962, 1963) on mouse pituitary, Kurosumi and Oota (1968) distinguished the FSH and LH-producing gonadotrophs in the rat. As said earlier, with the liberal use of immunocytochemical techniques, the identification of these separate gonadotroph cells came under a critical inquiry, since it was discovered that these two cell types contained both the gonadotrophic hormones.

In human (Pelletier et al. 1976) as well as in Ferret and porcine pituitaries (Beauvillian et al. 1975; Dacheux 1978) such bihormonal gonadotrophs containing both the hormones were found. However, the concentration of FSH and LH in such bihormonal cells varied from cell to cell and appeared to show species specificity. Extensive work has been done with respect to such bihormonal cells in the rat (Nakane 1970, 1975; Tougard et al. 1973, 1980; Moriarty 1975, 1976; Yoshimura et al. 1981; Herbert 1975; Bugnon et al. 1977; Kofler 1982; Inoue and Kurosumi 1984), in mouse (Baker and Gross 1978), in pig (Dacheux 1978) and in dog (El Etreby and Fath El Bab 1977).

Monohormonal gonadotrophs containing only FSH or LH were recorded in dog (El Etreby and Fath El Bab 1977), in rat (Fellmann et al. 1982), in human (Pelletier et al. 1976), in pig (Dacheux 1978, 1981) and in monkey (Girod et al. 1980). Exclusive studies by Childs and her coworkers have shown that the number of bihormonal gonadotrophs varied under various experimental conditions, thereby reflecting a certain fluidity of gonadotrophs. As of today, the most acceptable explanation is that the ultrastructural heterogeneity of the gonadotrophs may represent different stages of the secretory cycle of one single cell, that is capable of co-storing LH and FSH, and releasing them in different proportions depending on the physiological conditions (Tougard et al. 1980).

A reliable indicator of the identity of this cell was provided by some workers who demonstrated dramatic changes in the ultrastructure of this cell following surgical or chemical castration. Cells after such castration have been electron microscopically investigated by Foncin and LeBeau (1966); Kovacs & Horvath (1975) & Thakur (1991). The observed vacuolation in such cells is in fact due to the strongly dilated RER. Comparable observations have also been recorded by Costoff (1973); Kurosumi and Fujita (1974); Inoue and Kurosumi (1981); Childs et al. (1980, 1982) and Bhiwgade et al. (1989). The altered morphology of these cells following the administration of various drugs is reviewed by Saeger (1992). Following drug administration, these cells assume a bizarre appearance due to RER vacuolation (Bhiwgade et al. 1989; Saeger 1992; Raut 1993; Bellare 1994; Mantri 1994). A differential response of gonadotrophs to such experimental manipulations has also been used by some workers as the basis of subtyping (Farquhar and Rinehart 1954a, b; Serber 1961; Salazar 1963; Dekker 1967; Raza and Shakoori 1978; Bhiwgade et al. 1989). However, as of today, an overwhelming number of reports point convincingly to the bihormonal nature of gonadotrophs.

1.4.2 Gonadotroph Cells in Fish

The gonadotroph cells are globular and show variations in the abundance of secretory inclusions and rough endoplasmic reticulum. Most of these cells are sparsely granulated with secretory granules 300–400 nm in diameter and large globules 600–1300 nm in diameter. The variously dilated cisternae of rough endoplasmic reticulum occupy most of the cytoplasm, some of them are closely aligned to the cell membrane. Well-developed Golgi regions are common and mitochondria are abundant (Fig. 1.73).

On the other hand, the vesicles of the rough endoplasmic reticulum show extensive dilation. The number of granules and globules is few. However, these small granules are seen to be present close to the plasma membrane of the cell. Numerous rounded mitochondria are also present (Fig. 1.73).

During the prolonged active period, i.e. of the peak of spawning, the globular cells exhibit remarkable changes. The small granules in these cells are clearly diminishing in number due to degranulation, while the large globules appear to retain their number and become more electron dense. Since the cisternae of the rough endoplasmic reticulum are dilated extensively, they occupy most of the cytoplasm (Fig. 1.74).

The vesicles of rough endoplasmic reticulum are dilated extensively and fused with one another forming large vacuoles. The granular contents of the cells are thus expelled to the periphery of the cell, which indicates active exocytosis of the secretory granules. However, the large globules have remained unchanged (Figs. 1.75 and 1.76).

Two months after spawning, the vesicular cells are observed to be larger in size as compared to the earlier stages, with an enlarged nucleus placed at one pole (Fig. 1.77). In these cells, the rounded vesicular cisternae of the rough endoplasmic reticulum show an uneven form and some of them are distinctively large. The secretory granules are relatively plentiful, forming groups among the dilated vesicular cisternae, while the large globules are rare in these cells. The Golgi elements are comprised of minute vesicles and stacks of flattened lamellae are striking.

On the other hand, some of the globular cells are observed to become regranulated with small granules 150,350 nm in diameter, and a few large globules 400-1000 nm in diameter. Numerous small vesicular cisternae of the rough endoplasmic reticulum and the Golgi bodies are observed (Fig. 1.78).

There is a controversy regarding the number of gonadotrophic cells in the pituitary of teleosts. According to one school of thought, there is a single type of gonadotrophic cell, which has been reported in, *Cymatogaster aggregata*; *Anoptichthys jordan* (Mattheij 1970); *Poecilia reticulata* (Sage and Bromage 1970); *Gillichthys mirabilis* (Zambranok 1971); *Clarias batrachus* (Dixit 1970); *Heteropneustes fossilis* (Prasada Rao 1972; Baker et al. 1974); *Embiotoca lacksoni* (Lagios 1965); *Gasterosteus aculeatus*; *Phoxinus phoxinus* and *Pleuronectes flesus* (Benjamin 1975); *Cyprinus carpio* (Kurosumi et al. 1963); *Salmo gairdneri* (Peute et al. 1978) and *Pocilia latipinna* (Young and Ball 1981). In other species there are discordant reports. Although some investigators have identified only one gonadotropic cell type in Salmo salar (Ekengren et al. 1978b); *Carassius auratus* (Kaul and Vollrath 1974, Lam et al. 1976) and *Oncorhynchus nerka* (McKeown and Leatherland, 1973), other authors have described two types of cells in these species, respectively (Olivereau, 1976; Leatherland, 1972; Cook and van Overbeeke 1972 and Olivereau, 1962).

To date, various histochemical and cytological studies have repeatedly demonstrated the presence of two types of cells of possible gonadotrophic nature in the pituitary gland of a variety of teleost fishes, viz. *Anguilla anguilla*, (Olivereau and Herlant 1960, Olivereau 1961, 1967).

Mugil cephalus (Leray and Carlon 1963; Stahl 1963; Leray 1966); *Zoarces viviparous* (Oztan 1966); *Caecobarus geertsii* (Olivereau and Herlant 1954); *Rutilus rutilus* (Samuelson et al. 1968; Ekengren et al. 1978b) and *Salvelinus leucomaenls*, (Udea 1980).

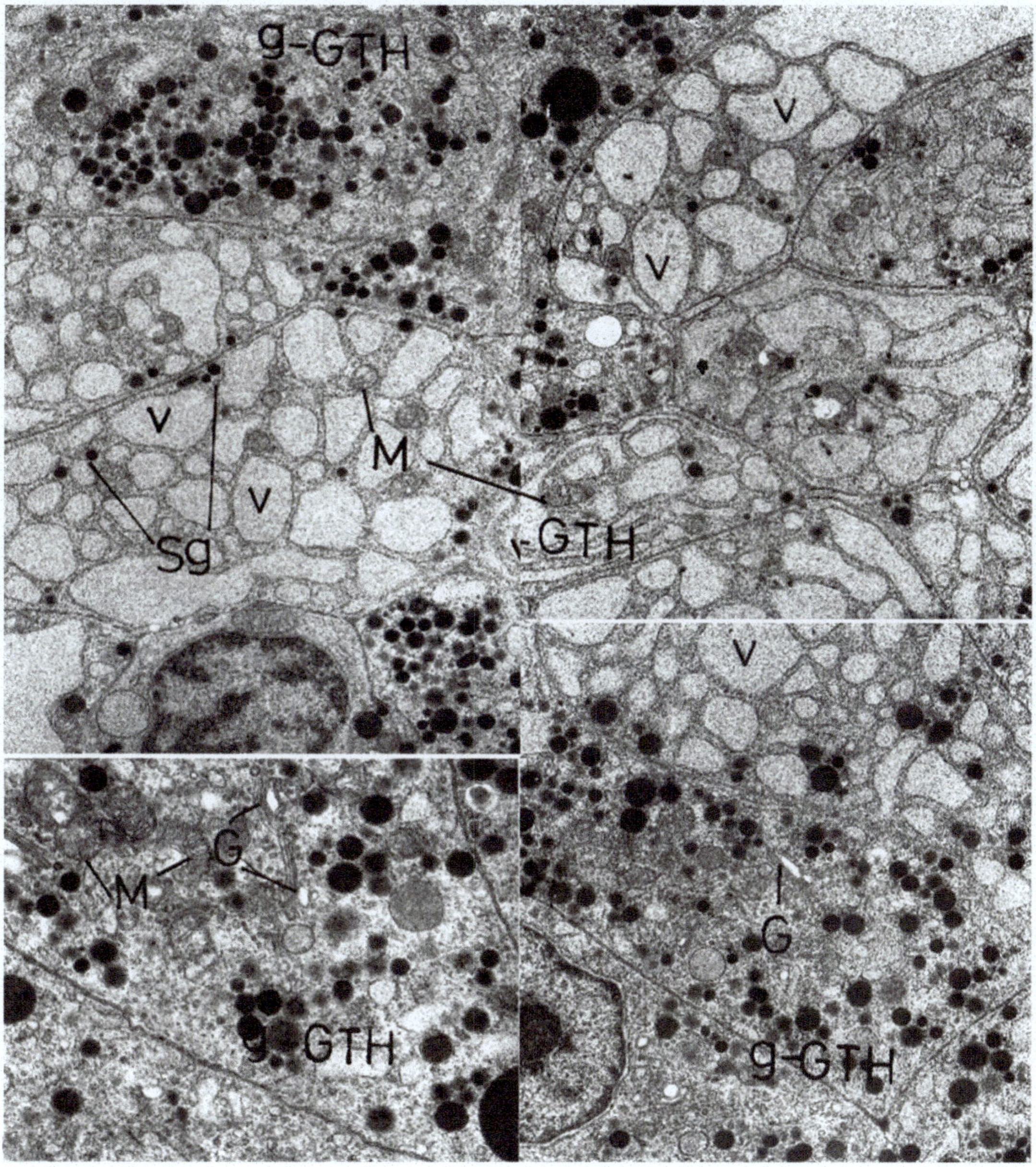

Fig. 1.73 Photomontage electron micrograph of vesicular cells (v-GTH) and globular cells (g-GTH) in the pituitaries of female *Johnius belengeri* sampled in November. (Stage III-Primary yolk globule stage). Note in the vesicular cells (v-GTH), the vesicles of rough endoplasmic reticulum (rER) are extensively dilated and the secretory granules are closely aligned to the cell membrane [X 9000]. Inset (left lower corner) Higher magnification of globular cell (g-GTH) showing alignment of secretory granules adjacent to plasma membrane (arrow), and coated pits (asterisks) on the plasma membrane, illustrating the exocytosis [X 16,000]. Unpublished electron micrograph from Dr. Bhiwgade and Dr. Mishra

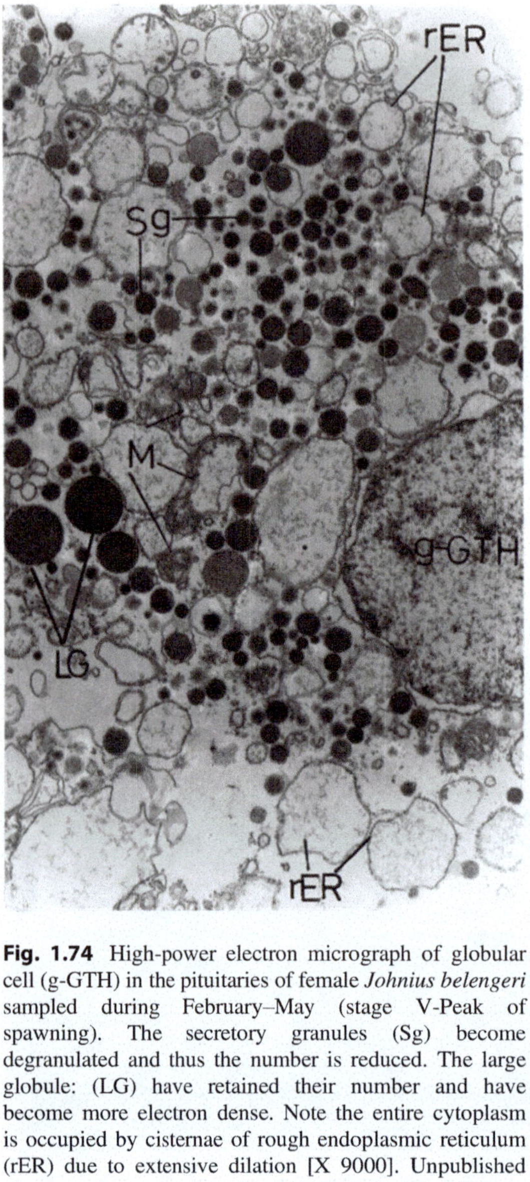

Fig. 1.74 High-power electron micrograph of globular cell (g-GTH) in the pituitaries of female *Johnius belengeri* sampled during February–May (stage V-Peak of spawning). The secretory granules (Sg) become degranulated and thus the number is reduced. The large globule: (LG) have retained their number and have become more electron dense. Note the entire cytoplasm is occupied by cisternae of rough endoplasmic reticulum (rER) due to extensive dilation [X 9000]. Unpublished electron micrograph from Dr. Bhiwgade and Dr. Mishra

Recent ultrastructural research have also indicated the duality of pituitary gonadotrophs in *Rutilus rutilus* (Bage et al. 1974; Ekengren et al. 1978a); *Gasterosteus aculeatus* (Slijkhuis 1978); *Oncorhynchus masou* (Ueda and Hirashima 1979) and *Salvelinus leucomaenis,* (Udea 1980). Similarly in *Johnius belengeri,* it is observed that there are two types of possible gonadotrophic cells in the adenohypophysis, which assume distinctive ultrastructural aspects and show different responses to various stages of gonadal development in natural conditions. Identical cell types have also been reported in *Misgurnus anguillicaudatus* (Ueda and Takahashi 1977); and *Salvelinus leucomaenis* (Udea 1980); *Oncorhynchus nerka* (Peute et al. 1978); *Oncorhynchus masou* (Ueda and Hirashima 1979). They are obviously different from possible somatotrophs and thyrotrophs in their ultrastructure, in the proximal pars distalis.

In *Johnius belengeri*, the centrodorsal gonadotrophs designated as globular cells have exhibited a remarkable, increases in size and number in correlation with the improving maturation of the gonads, during the reproductive cycle. Moreover, the cells eminently disclosed the depletion of secretory granules, extensive dilation of the rough endoplasmic reticulum and subsequent pycnotic changes of the nucleus following ovulation. Similar changes have been reported to occur in typical pituitary gonadotrophs of many other fishes at the time of spawning, especially those in *Carassius auratus*, after LHIH-induced ovulation (Kaul and Vollrath 1974 a and Lam et al. 1976) and after natural spawning in *Salmo salar* (Ekengren et al. 1978b) and *Salvelinus leucomaenis* (Udea 1980). Therefore, it seems reasonable to consider the globular cells of *Johnius belengeri* to be conventional gonadotrophs. In this way, the globular cells may secrete a gonadotrophin which promotes the ovulatory changes of the ovary in *Johnius belengeri*, as seen in many other teleosts (Fontaine and Olivereau 1975; Schreibman et al. 1973; Ueda and Takashashi 1980; Udea 1980; Peute et al. 1986).

Gonadotroph cells in garden lizard (*Calotes versicolor*)

In the active male pituitary, the FSH Gonadotroph cell shows large secretory granules. The abundant and variously dilated cisternae of rough endoplasmic reticulum occupy most of the cytoplasm. Numerous round mitochondria are abundant (Fig. 1.79).

In active females, the FSH cells show numerous secretory granules. The rough endoplasmic reticulum is seen throughout the cytoplasm. The

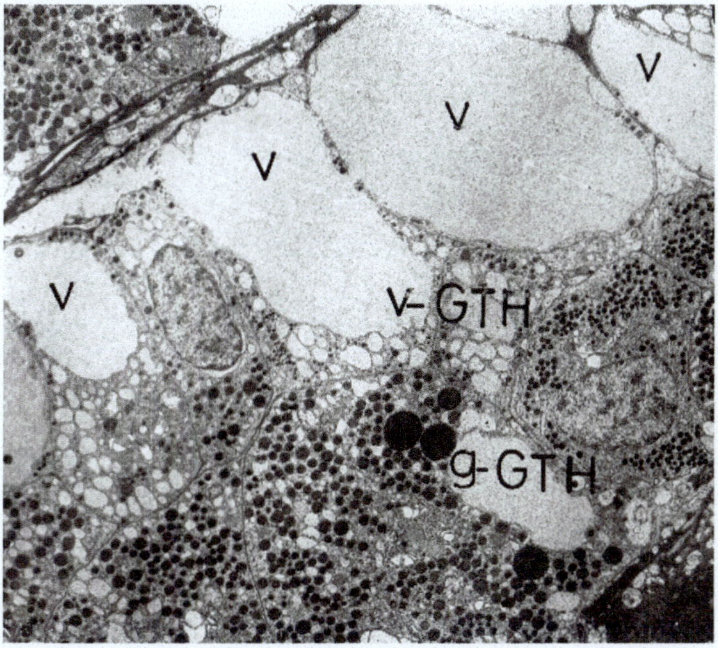

Fig. 1.75 Low-power electron micrograph of proximal pars distalis (PPD) from centro dorsal region of pituitary of female *Johnius belengeri* sampled during February–May (stage V-Peak of spawning). It illustrates the vesicular cells (v-GTH) and the globular cells (g-GTH). Note the extension and fusion of vesicles of rough endoplasmic reticulum which is resulted in the formation of big vacuoles (V) in the cytoplasm, and sometimes formation of a single large translucent vacuole. The granules are almost expelled to periphery of the cell [X 3000]. Unpublished electron micrograph from Dr. Bhiwgade and Dr. Mishra

Fig. 1.76 High-powered electron microphotograph magnified from Fig. 1.3 which illustrates the vesicular cell (v-GTH). It shows extensive dilation and fusion of vesicles of rough endoplasmic reticulum and pressing of secretory granules (Sg) to the periphery of the cell membrane [X 13,000]. Unpublished electron micrograph from Dr. Bhiwgade and Dr. Mishra

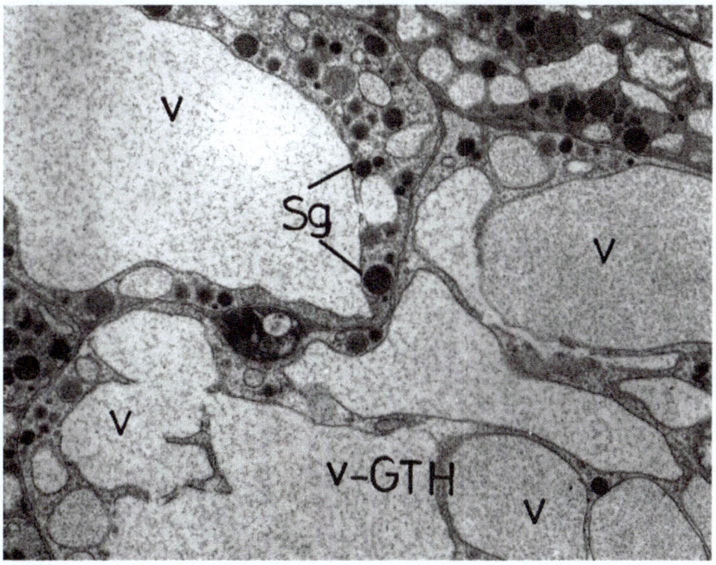

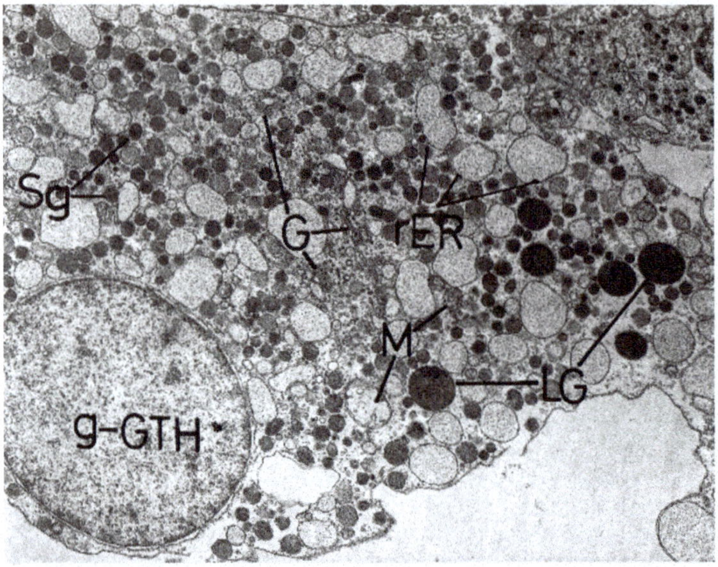

Fig. 1.77 Electron micrograph of globular cell (g-GTH) in the pituitary of the female *Johnius belengeri* sampled in July with an ovary containing only previtellogenic oocytes alongside few ripened oocytes. The Golgi bodies (G) are many, and are well developed with numerous granules around; small secretory granules (Sg) are numerous and few large globules (LG) with inconspicuous membranes and many mitochondria (M) indicating regranulation of the cell. Rough endoplasmic reticulum (rER) is vacuolar [X 9000]. Unpublished electron micrograph from Dr. Bhiwgade and Dr. Mishra

mitochondria are round to elongated and the Golgi zones is very well developed with numerous Golgi saccules (Fig. 1.80).

At ultrastructural level, the pituitary of active male Calotes shows very well-developed LH gonadotroph cells. The secretory granules are few in number. The variously dilated rough

Fig. 1.78 Magnified electron micrograph of Fig. 1.5 showing activated Golgi bodies (G) around which numerous granules are present. Note the rough endoplasmic reticulum (rER), small secretory granules (Sg), mitochondria (M) and large globules (LG) [X 25,000]. Unpublished electron micrograph from Dr. Bhiwgade and Dr. Mishra

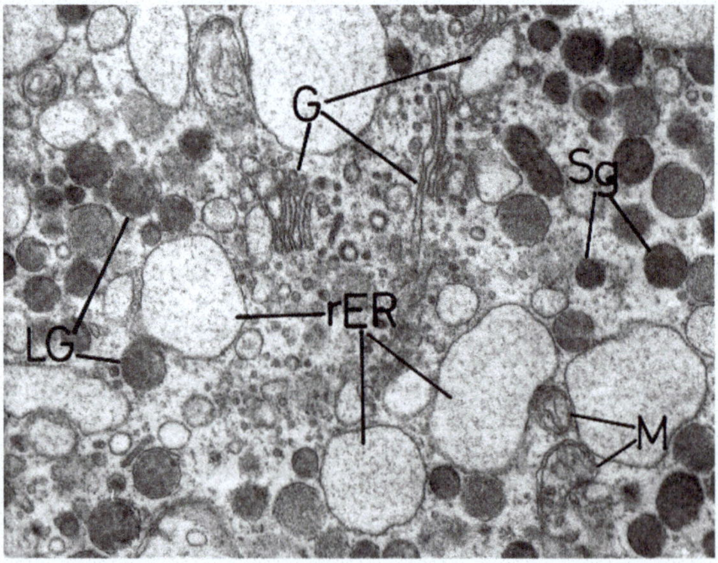

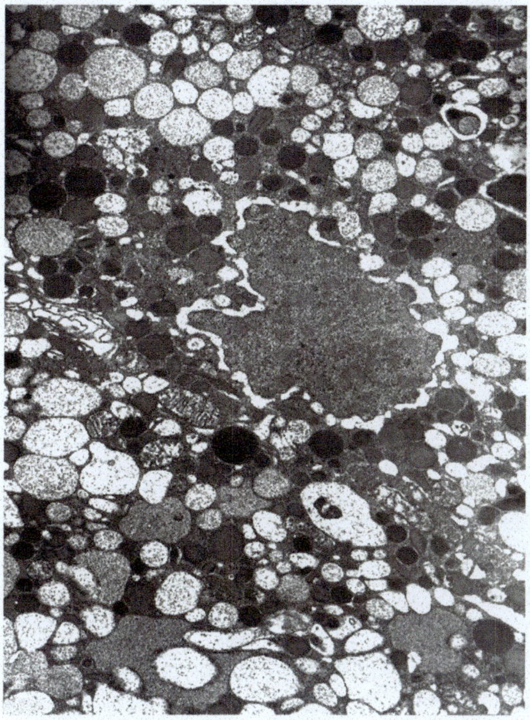

Fig. 1.79 The FSH gonadotroph of active male garden lizard (*Calotes versicolor*) is characterized by vesiculated cisternae of rER occupying an extensive portion of the cytoplasm occupying little room for discerning other cell organelles. However, round secretory granules are particularly concentrated in the perinuclear cytoplasm. Nucleus (N) presents a crenated envelope and is relatively small. There is a crowding of elongated and round mitochondria around the nucleus [X 8000]. Unpublished electron micrograph from Dr. Bhiwgade

cistern endoplasmic reticulum occupies most of the cytoplasm (Fig. 1.81).

The nucleus, round shaped mitochondria are observed along with a well-developed Golgi zone with variously dilated sacculi.

The LH cell in the active male during breeding season shows few secretory granules. They are round in shape and surrounded by variously dilated rough endoplasmic reticulum (Fig. 1.82).

The Golgi zone is well developed and round-shaped mitochondria with cisternae are also observed.

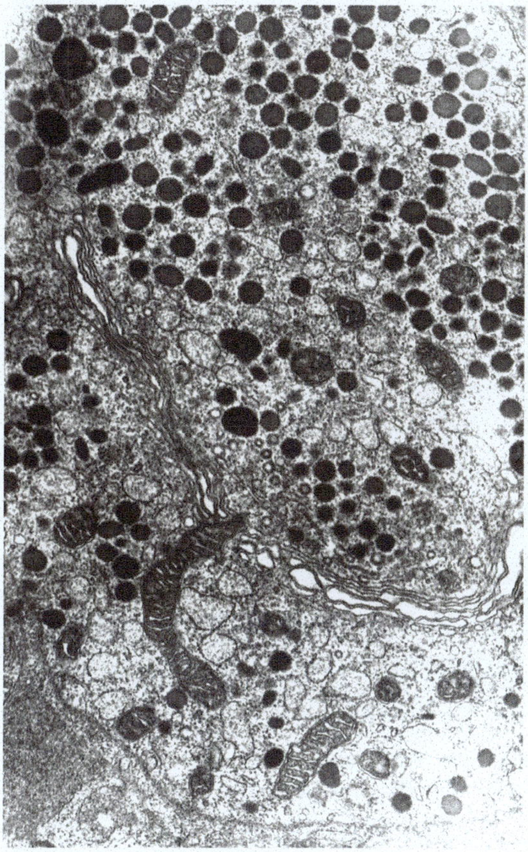

Fig. 1.80 A portion of FSH gonadotroph of an active female garden lizard (*Calotes versicolor*) shows a rich assortment of cell organelles. The conspicuous Golgi zone (GZ) shows a distinct parallel configuration closely associated with pro-secretory granules and elongated mitochondria (M) with Christmas stocking contours. Population of secretory granules is noticeably interspersed with rER moderately vesiculated [X 10,000]. Unpublished electron micrograph from Dr. Bhiwgade

1.4.3 Gonadotroph Cells in Mammals (FSH and LH): Light Microscopic Observations

FSH cells in the male bat, stained magenta or purple (Figs. 1.83 and 1.84) and LH cells stained yellow-red (Figs. 1.84 and 1.85), after combined histochemical techniques. The FSH cells showed a slightly greater affinity for AB (Fig. 1.86). They were also methyl-blue-positive (Fig. 1.85).

Gonadotroph cells after cyproterone acetate treatment: Following cyproterone acetate treatment for 60 days in male bats, the FSH

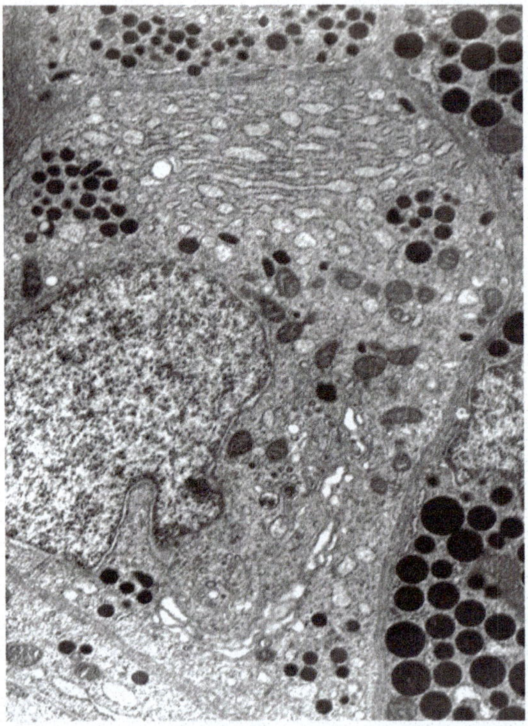

Fig. 1.81 The LH gonadotroph of the active male garden lizard (*Calotes versicolor*) shows a moderately sized invaginated nucleus (N), an rER showing dilated secretion-filled cisternae in parallel arrays; active Golgi zone (GZ) is seen in the juxtanuclear position. The mitochondrial association with GZ is striking, just as is the sparse presence of secretory granules in small, isolated clumps [X 10,000]. Unpublished electron micrograph from Dr. Bhiwgade

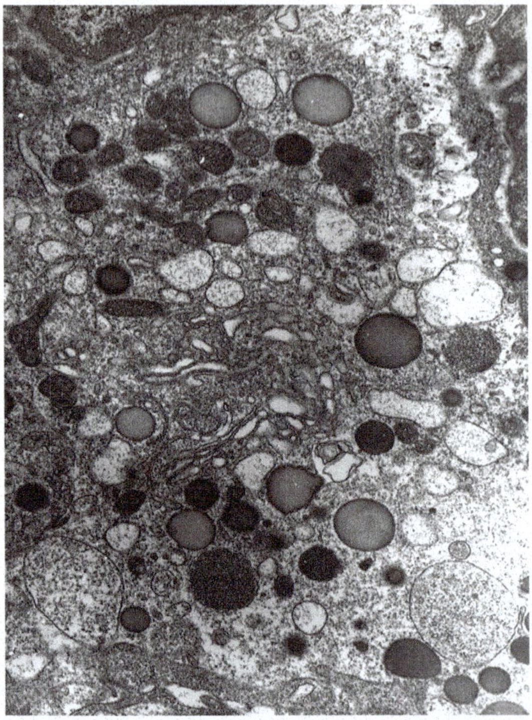

Fig. 1.82 High-power EM of LH Gonadotroph of active male Garden Lizard, *calotes versicolor*, showing a part of the cytoplasm having the Golgi zone (GZ). Note the initiation of Golgi saccule dilation, few yet significantly distended cisternae of rER are seen scattered around the GZ. The secretory granules are few and scattered. Lipid inclusions (L) with opaque core also surround the GZ [X 10,000]. Unpublished electron micrograph from Dr. Bhiwgade

gonadotroph cells (Fig. 1.88) are characterized by increased number and hypertrophy of mitochondria and Golgi apparatus as compared to control (Fig. 1.87). Secretory granules are scanty and tend to range along the cell membrane. These findings strongly suggest an increase in synthesis as well as the release of the hormone. The LH cells (Fig. 1.16) undergo nearly complete degranulation and the cisternae of endoplasmic reticulum become vacuolated. An increased number of mitochondria and the greatest hypertrophy of the Golgi apparatus are observed (Figs. 1.87 and 1.88).

Gonadotroph cells after gonadectomy: In the pituitary of castrated males, the FSH cells undergo partial degranulation. This occurs after 30 days following castration in male bats. The proportion of cells showing the changes seen in active secretion is markedly increased (Fig. 1.89). The LH cells undergo nearly complete degranulation indicating an increased cellular activity (Fig. 1.89).

Castration resulted in charges such as slight degranulation of both FSH and LH cells after 7 days of operation. Administration of testosterone ester restored the cytology of gonadotrophic cell types to normal (Fig. 1.90), while concomitant administration of testosterone ester and cyproterone acetate did not result in the restoration of post-castration changes in the anterior pituitary.

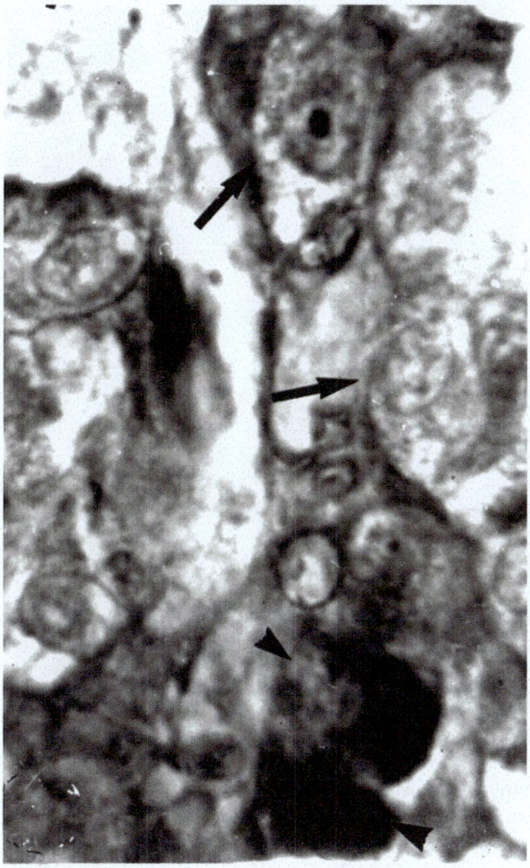

Fig. 1.83 Anterior pituitary of untreated male bat, Rousettus leschenaulti. Note FSH (arrows) and yellow red LH (arrow heads) are seen. Stained with Periodic Acid Schiff /Orange G [X 800]

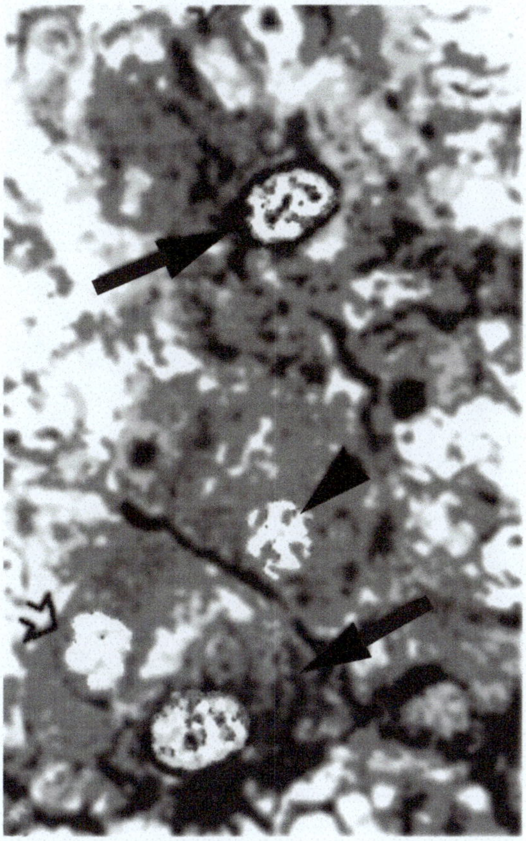

Fig. 1.84 Anterior pituitary of untreated male bat, Rousettus leschenaulti. Note purple stained FSH (Full arrows) yellow red LH (arrow head) and orange STH. Stained with Alcian Blue/Periodic Acid Schiff Orange G [X 800]

1.4.4 Electron Microscopic Observations

Though GTH cells can be separately observed in the anterior pituitary, there are many GTH cells, possessing ultrastructural characters of both FSH and LH in the same cell. These cells are bound to be bihormonal in nature and termed together as gonadotrophs which contain both FSH and LH.

The ultrastructural cytology of GTH cells does not have any difference with respect to sex of the animal and the changes exhibited by the induction of drugs/hormones are also similar in both sexes.

In electron microscopy, FSH/LH cells are seen as medium-sized cells harbouring spherical and eccentric euchromatic nuclei and abundant cytoplasm. The rough endoplasmic reticulum (RER) is well developed with short, often slightly dilated profiles. The Golgi apparatus is prominent with numerous sacculi and vesicles, and vesicles include many immature secretory granules. The mature secretory granules may vary considerably in size, structure and number.

There are many conditions in which gonadotroph cells are stimulated and hyperactive. In those states the gonadotrophs cells increase in number and size, showing densely arranged vacuoles and variably diminished secretory granules. Since these alterations are demonstrated after removal of gonads, these cells have been called castration cells or gonadectomy cells. Electron microscopically the vacuoles are identified as

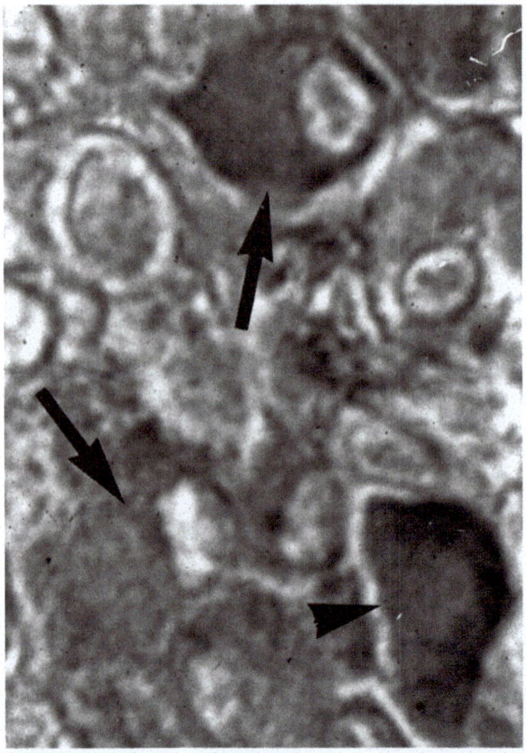

Fig. 1.85 Anterior pituitary of untreated male bat, *Rousettus leschenaulti*. Note methyl-blue-positive FSH (arrows) and yellow-red LH (arrow head) cells. Stained with Alcian Blue/Periodic Acid Schiff/Orange G [X 800]

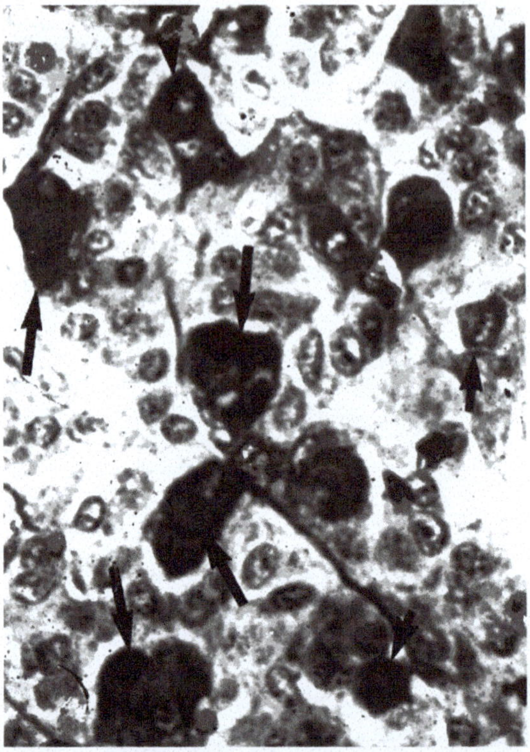

Fig. 1.86 Anterior pituitary of untreated male bat, *Rousettus leschenaulti*. Note FSH cells (full arrows) and LH cells (arrow head) and TSH cells (small arrows). Stained with Tetra chrom of Hertant-II Alcian Blue /Periodic Acid Schiff /Orange G [X 800]

strongly dilated rough endoplasmic reticulum. There seems to be a progression from smaller to larger vacuoles. The Golgi apparatus is enlarged and the number of secretory granules decreased. In animals, similar alterations can be observed after treatment with various drugs/hormones. Similarly, suppression of gonadotrophs can be expected in many states such as during inactive conditions and treatment with drugs/hormones since, in these conditions, a decrease in number and size of secretory granules has been reported.

(i) FSH/LH cells in Normal and Cyclical Animals

These cells are mostly polygonal in shape with an eccentrically placed nucleus, found near the vascular spaces. The cytoplasm contains round electron-dense medium-sized secretory granules, lamellar and poor-to-moderately developed RER, a typical juxtanuclear Golgi zone arranged in a circular configuration and a moderate number of mitochondria and lysosomal bodies.

The cell is polygonal in shape in male rat (Fig. 1.91), male bat, *Rousettus leschenaulti* (Fig. 1.92), male monkey (Fig. 1.93), female Suncus (Fig. 1.94), female bat, *Rhinopoma microphyllum* (Fig. 1.95) and in female monkey (Fig. 1.96). The granules are moderate in number in active male rat, male monkey, active female rat, whereas few granules are found in male bat, Rousettus and in female bat, *Rhinopoma microphyllum*.

The RER is poorly developed in male *Rousettus leschenaulti* and monkey, whereas lamellar profiles of RER are moderately seen in cells of male and female rats (Fig. 1.97) and female Suncus. A well-developed juxtanuclear Golgi zone is found in all aforementioned animals. In the male bat, Rousettus, the Golgi

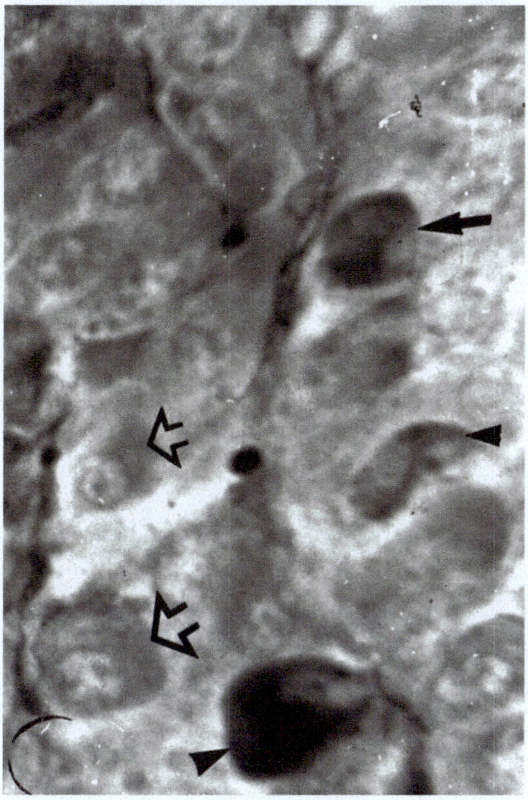

Fig. 1.87 Anterior pituitary of Control untreated male bat, Rousettus leschenaulti. Note FSH (full arrows) LH (arrow heads) and STH (hollow arrows) cells are seen. Stained with Alcian Blue/Periodic Acid Schiff/Orange G [X 600]

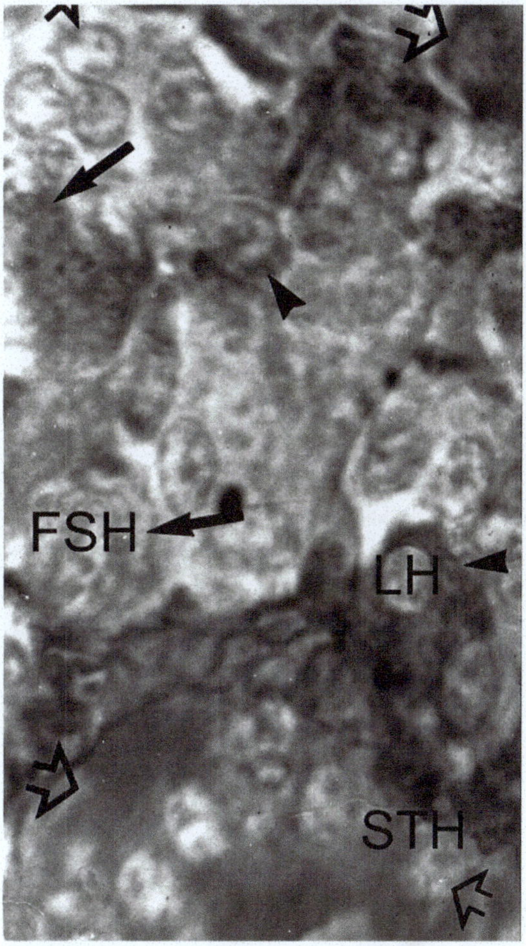

Fig. 1.88 Anterior pituitary of male bat, Rousettus leschenaulti, treated with CPA for 60 days. Note FSH (full arrow), LH (arrow heads) and STH (hollow arrows) cells are seen. Stained with Alcian Blue/Periodic Acid Schiff/Orange G [X 800]

apparatus has a circular arrangement of sacs and vesicles whereas in female bat, *Rhinopoma microphyllum,* four to five flattened sacs and many vesicles are present. The shape, size and number of mitochondria also differ from species to species.

Moderate number of granules are found in cells during early pregnancy in *Rhinopoma microphyllum.* The endoplasmic reticulum presents vesiculation in early stages of pregnancy in *Rhinopoma microphyllum* (Fig. 1.98). There is an increase in a number of mitochondria and the mitochondrial shape becomes irregular. During early pregnancy in Suncus (Fig. 1.99), the mitochondria become large, elongated with dense matrix and prominent cristae. Well-developed lysosomal bodies are also seen. The cisternae of RER become vacuolated and their membranes are provided with ribosomes.

The comparative ultrastructure of FSH cell during late pregnant stage in two species of bats. *Hipposideros fulvus* (Fig. 1.100) and *Megaderma lyra* (Figs. 1.101 and 1.102) reveal similarities rather than differences. However, some species-specific characters are also observed. During late pregnancy, the cell contains an irregular nucleus in both species of aforementioned bats. As the development of the embryo advances, there is a decline in the gonadotrophin synthesis corroborated by the presence of sparse

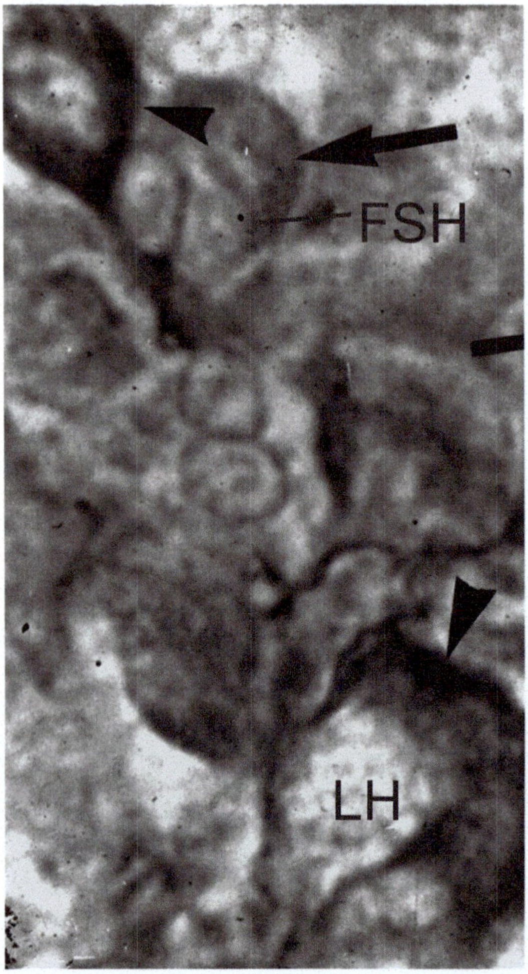

Fig. 1.89 Anterior pituitary of male bat, Rousettus leschenaulti, 30 days after castratiOn. Note FSH (arrow) and LH (arrow heads) cells. Stained with Alcian Blue / Periodic Acid Schiff /Orange G [X 800]

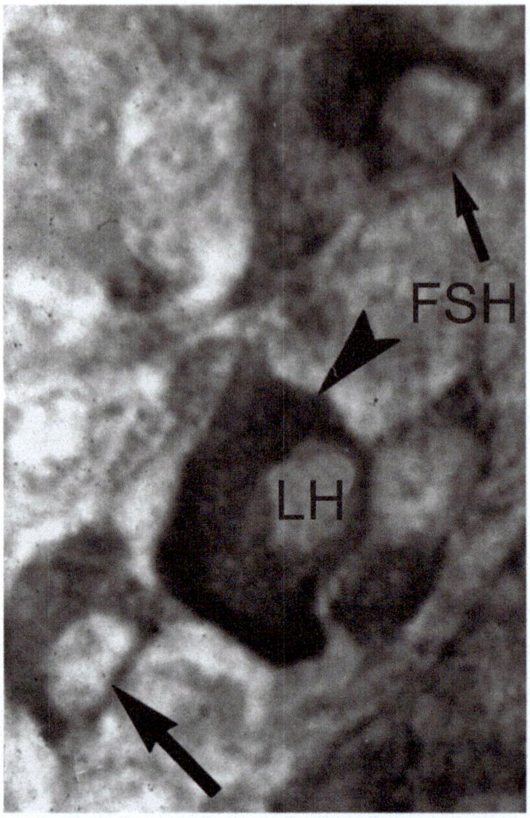

Fig. 1.90 Anterior pituitary of 7 days old castrated male bat, Rousettus leschenaulti, administered T.E. for 30 days. Note FSH (arrows), and LH (arrow head) cells. Stained with Alcian Blue / Periodic Acid Schiff /Orange G [X 800]

granulation in Hipposideros whereas a moderate granulation is found in Megaderma. Circular profiles of vesiculated RER are found in both the aforementioned species of bats. Mitochondria encountered in the cells of Hipposideros are either round or elongated in shape.

(ii) FSH cell in Anterior Pituitary of experimental animals

(a) Male bonnet monkey after castration

In male bonnet monkey after gonadectomy (Figs. 1.103 and 1.104a), the FSH cell becomes irregular in shape and undergoes complete degranulation and only a few immature secretory granules are present as compared to the cell of normal male monkey. Electron microscopically the vacuoles are identified as strongly dilated round-shaped RER occupying the entire cytoplasm. There seems to be a progression from smaller to larger vacuoles. The hypertrophied Golgi apparatus is juxtanuclear in position and appears circular with numerous dilated Golgi sacs and vesicles. The mitochondria also undergo hypertrophy with loss of cristae and are mostly found near the Golgi zone.

(b) Male rat after CPA treatment

In male rats, after CPA treatment (Fig. 1.104b) there is a decrease in granules as compared to the normal rats. The RER is hypertrophied and vacuolated typically in circular profiles. The number

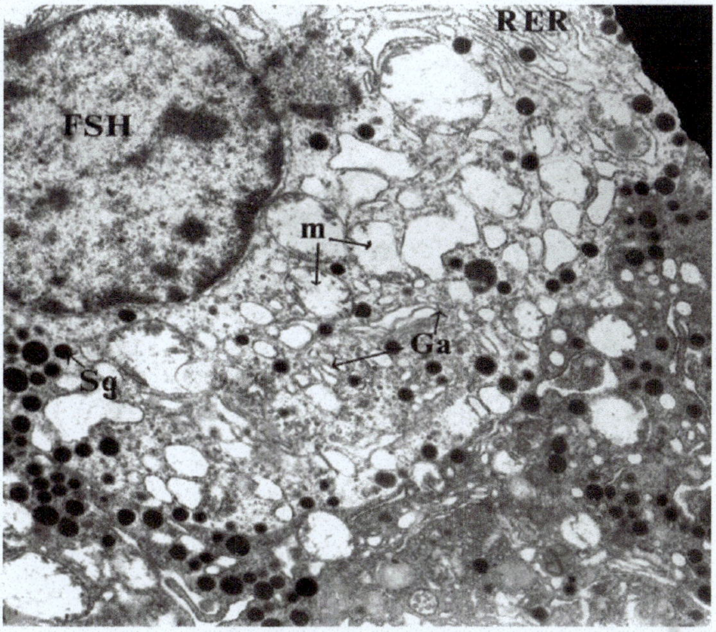

Fig. 1.91 FSH gonadotroph of the normal male rat. Note dilated lamellar cisternae of RER (RER), a prominent Golgi apparatus (Ga) comprising numerous sacs and vesicles and mitochondria with loss of internal matrix (m). Medium-sized secretory granules (Sg) are scattered throughout the cytoplasm [X 10,000]. Unpublished electron micrograph from Dr. Bhiwgade

of mitochondria decreases and those which are found are with inconspicuous matrix. There is an increase in lysosomal activity.

(c) Male rat after MPA + TE treatment

In the pituitary of male rats, treated with MPA + TE, the FSH cell shows a slight decrease in a number of granules (Fig. 1.105). The RER gets vesiculated and their membranes are

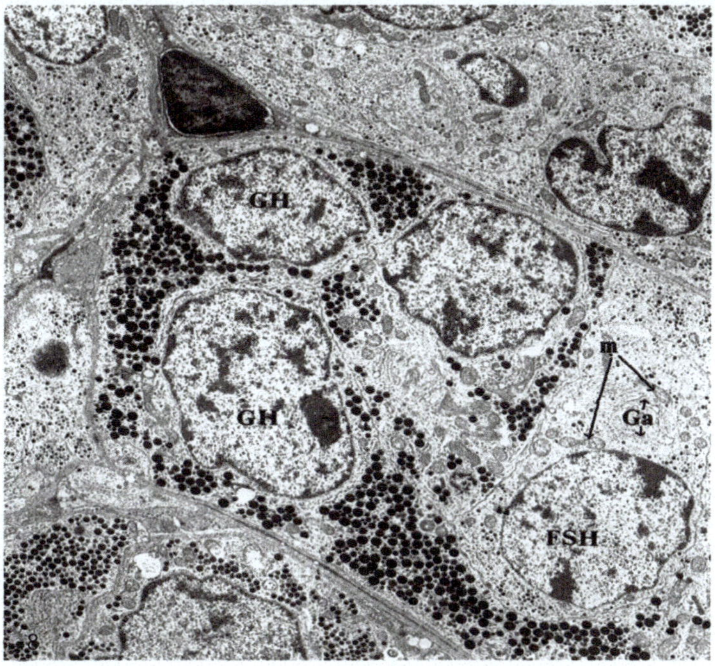

Fig. 1.92 Anterior pituitary of male bat, *Rousettus leschenaulti*. Population of cells showing GH cells and FSH gonadotrophs. Note the circular configuration of the Golgi apparatus (Ga) typically juxtaposed to the nucleus and a few large mitochondria (m) [X 6000]. Unpublished electron micrograph from Dr. Bhiwgade and Dr. Borkar

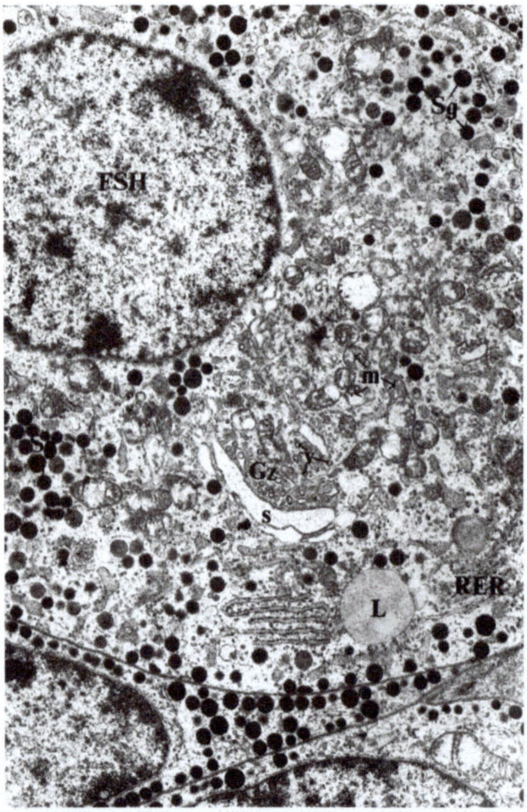

Fig. 1.93 FSH gonadotroph of active male bonnet monkey, *Macaca radiata*. Note lamellar profiles of RER (RER) at one side of the cell, numerous mitochondria (m) found in groups in the cytoplasm, a highly developed juxtanuclear Golgi zone (Gz) comprising dilated sacs (s), small vesicles, secretory granules of variable density (sg) and a few lipid droplets (L) are also seen [X 8000]. Unpublished electron micrograph from Dr. Bhiwgade and Dr. Thakur

provided with numerous ribosomes. The Golgi sacs get dilated indicating the hypertrophy of the organelle. The mitochondria increase in number, and show a dense matrix. Lysosomes in the form of dense bodies and multivesicular bodies are also found.

(d) Male rat after Gossypol treatment

The FSH cell (Fig. 1.106) demonstrates an increase in vesiculation of RER and appears like vacuoles, and a circular Golgi apparatus undergoes hypertrophy and is juxtanuclear in position. The number of immature secretory granules increases significantly and exocytosis is

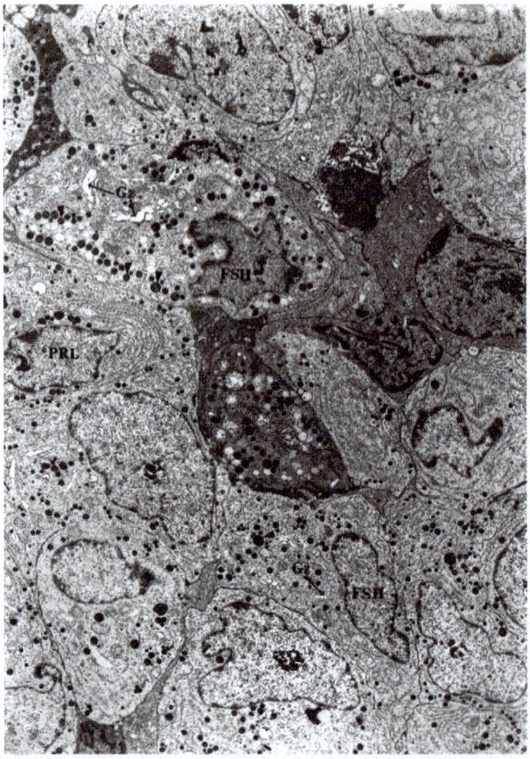

Fig. 1.94 Anterior pituitary of active female Suncus showing FSH and PRL cells. The FSH cells are easily distinguished because of the enlarged Golgi zone (Gz) near the nucleus. A few secretory granules are found in cell in the form of dark dense round dots (arrow heads) [X 3500]. Unpublished electron micrograph from Dr. Bhiwgade and Dr. Mantri

seen. There is a spectacular change seen in the mitochondria with disintegration along with the loss of cristae leading to involution of mitochondria.

(e) Female rat after DMPA treatment

DMPA administration in female rats brings about changes in FSH cell cytology to a greater extent (Fig. 1.107). The cell undergoes hypertrophy leading to the appearance of vesiculated RER and shows circular profiles as compared to the fragmented lamellar profiles of RER in normal rats. These circular profiles anastomose together to form a large vacuole. This vacuole displaces other cytoplasmic organelles and inclusions to one side of the cell to give it a signet ring appearance. Hypertrophy of the Golgi zone takes place

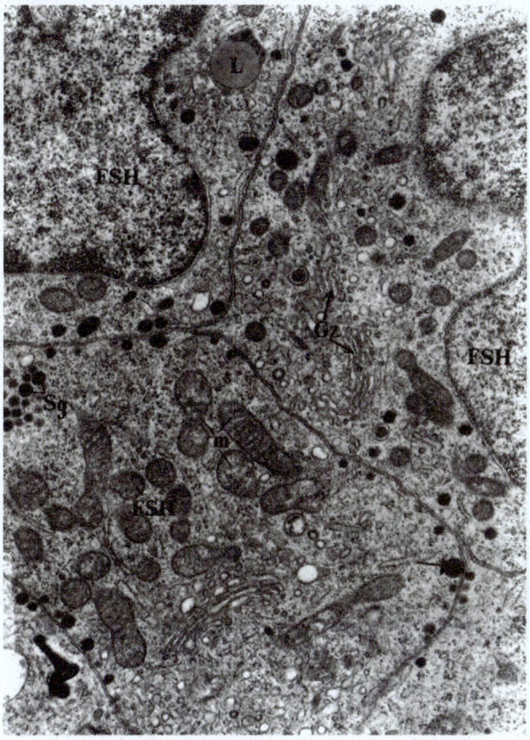

Fig. 1.95 A group of FSH cells found in oestrous female bat, *Rhinopoma microphyllum.* Note well-developed Golgi zones (Gz) with numerous flattened sacs and vesicles, enlarged distinct mitochondria (m) with prominent shapes, lipid droplets (L) and secretory granules (Sg) in the process of exocytosis (arrow) [X 16,500]. Unpublished electron micrograph from Dr. Bhiwgade and Dr. Mantri

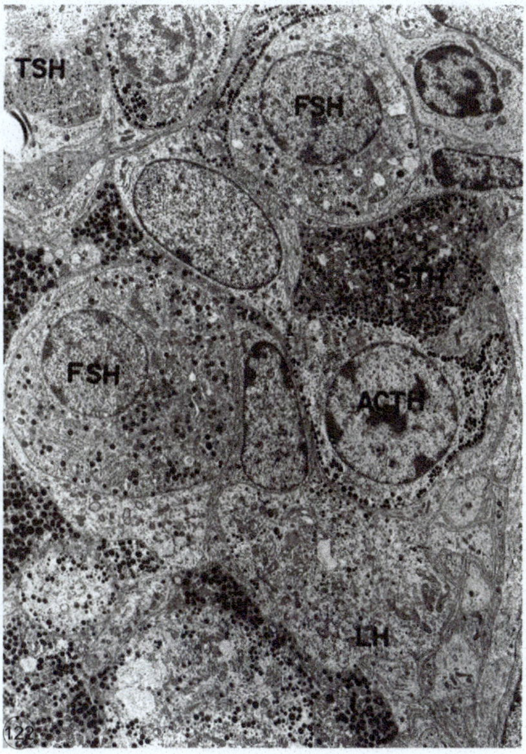

Fig. 1.96 Anterior pituitary of active female bonnet monkey *Macaca radiata*. A population of FSH, FS, ACTH and other cells are seen. The FSH gonadotroph is encircled by other cells and contains short lamellar profiles of RER, mitochondria are rod-shaped or elongated with well-formed cristae and secretory granules are of variable density [X 3500]. Unpublished electron micrograph from Dr. Bhiwgade and Dr. Thakur

and mitochondria with prominent mitochondrial matrix increase in number and get aggregated to one side of the Golgi zone. All these changes are due to blocking of the binding sites of the target hormone responsible for the feedback mechanism. The formation of secretory granules is continued along with its discharge.

(iii) LH cells in experimental animals

The LH cells are mostly polygonal to irregular in shape. The placement of nucleus is mostly eccentric and the nuclear shape is slightly irregular. The secretory granules are smaller than that of FSH cell granules, these are electron-dense and round in shape. The RER is mostly slightly vesiculated, but sometimes lamellar profiles of RER are also seen. The Golgi zone is not easily observed as compared to the Golgi zone of the FSH cells. Mitochondria from round to irregular shapes are observed. Lysosomal activity is less frequent.

(a) Male bonnet monkey after castration
The LH cells undergo hypertrophy following 60 days castration (Figs. 1.103 and 1.108). These cells respond more rapidly to castration than the FSH cells. The endoplasmic reticulum (RER) is markedly dilated and gives a foamy appearance to the cell. The mitochondria are hypertrophied with loss of cristae and the cell undergoes complete degranulation.
(b) Male rat after CPA treatment

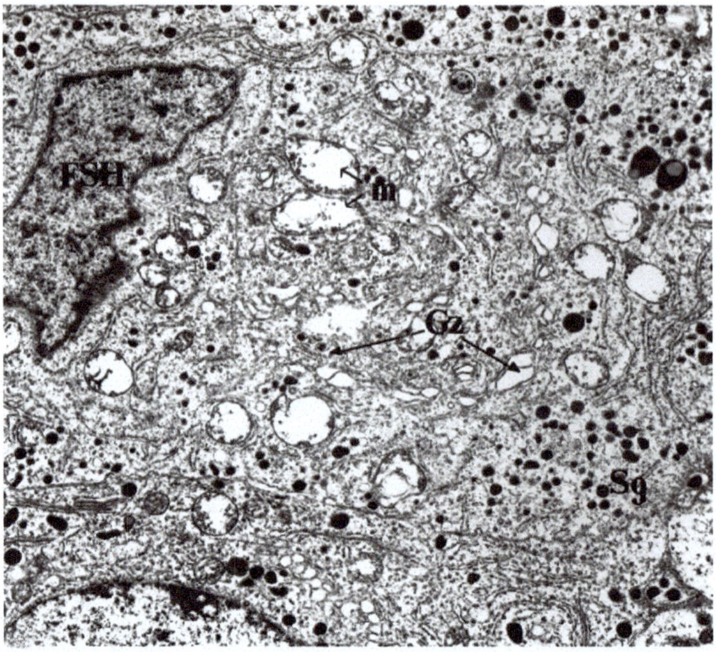

Fig. 1.97 FSH cell of normal female rat showing large swollen mitochondria (m) with loss of cristae, Golgi zone (Gz) and small secretory granules (Sg) scattered throughout the cytoplasm [X 8500]. Unpublished electron micrograph from Dr. Bhiwgade and Dr. Mantri

In the anterior pituitary (Fig. 1.104b), there is a slight decrease in the number of secretory granules, the mitochondria become round with prominent cristae and a few mitochondria show loss of cristae due to extensive enlargement.

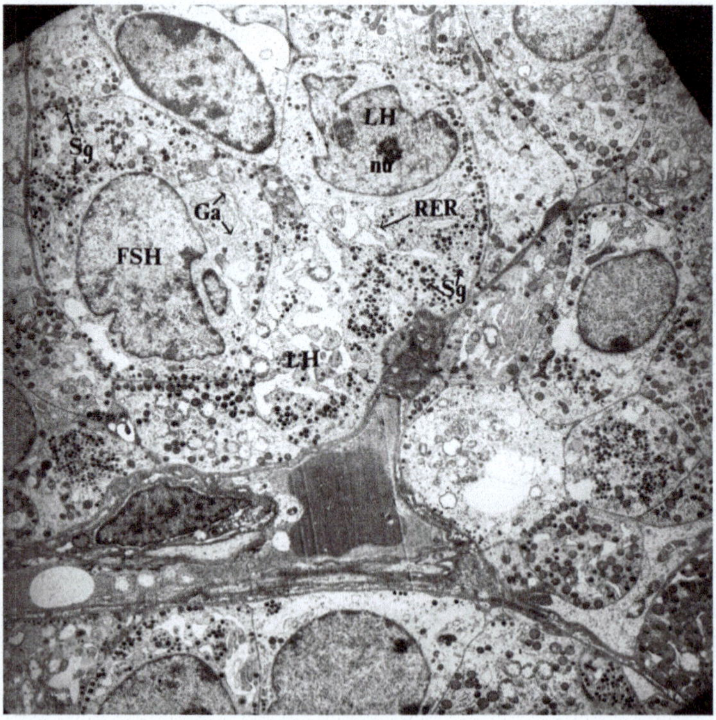

Fig. 1.98 Anterior pituitary cells of the early pregnant stage of female bat, *Rhinopoma microphyllum.* The FSH cell contains a typical juxtaposed Golgi apparatus (Ga) and round secretory granules (Sg) of variable density. Also, note the LH cells with vesiculated RER (RER) and small secretory granules (Sg) and nucleus with prominent nucleolus (nu) [X 3500]. Unpublished electron micrograph from Dr. Bhiwgade and Dr. Mantri

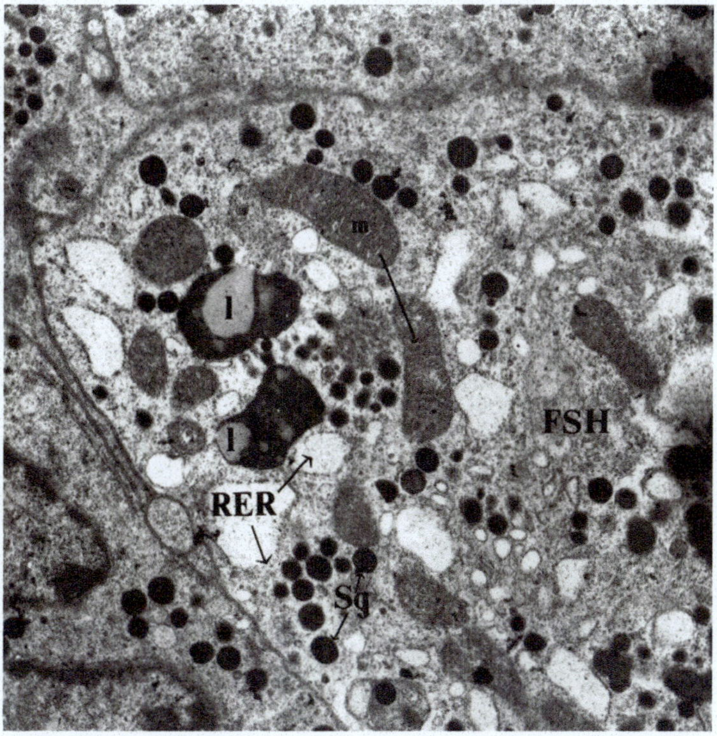

Fig. 1.99 FSH gonadotroph of female *Suncus murinus*. Note vesicular profiles of RER dark mitochondria (m), lysosomal (l) and secretory granules (Sg) scattered throughout the cytoplasm in groups [X 13,000]. Unpublished electron micrograph from Dr. Bhiwgade and Dr. Narayane

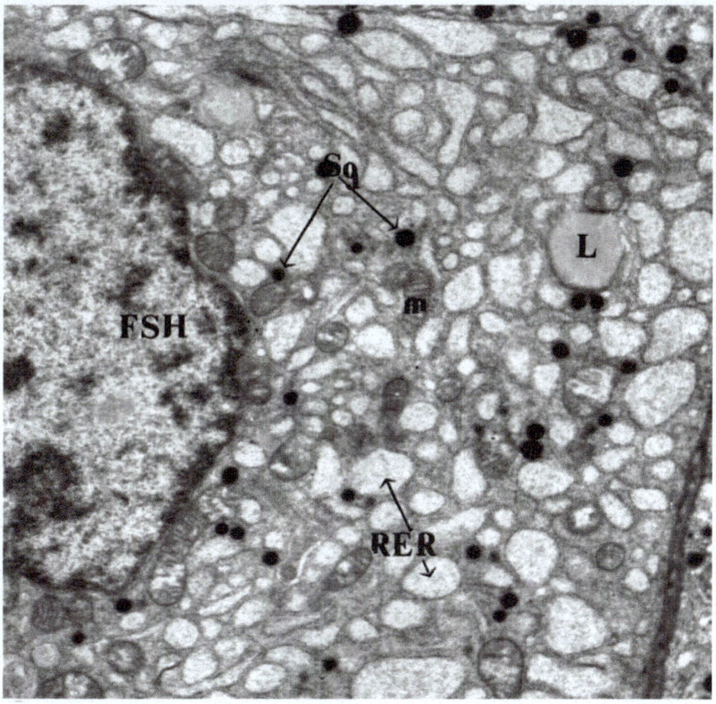

Fig. 1.100 FSH cell of late pregnant *Hipposideros fulvus*. The cell shows dilated vesicular RER (RER), round mitochondria (m), few secretory granules (Sg) and lipid droplets (L) [X 13,000]. Unpublished electron micrograph from Dr. Bhiwgade and Dr. Narayane

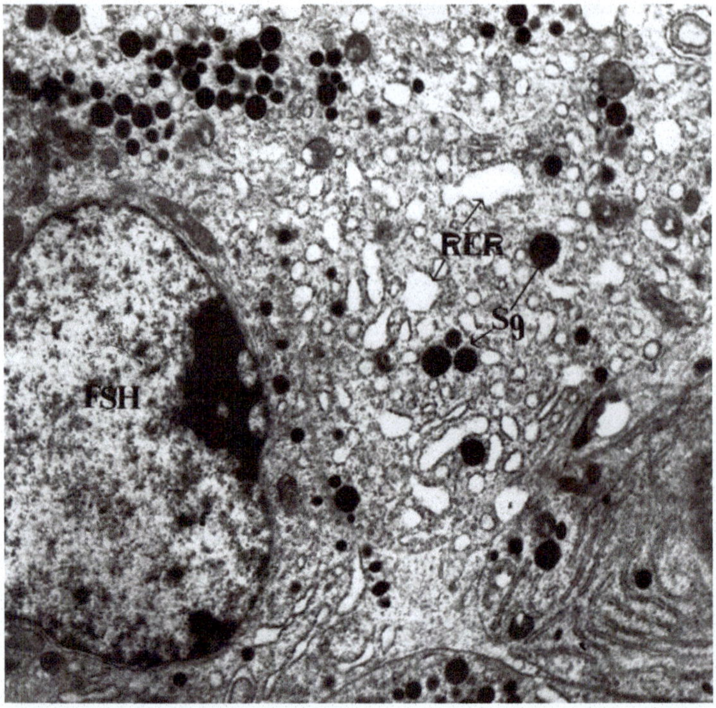

Fig. 1.101 FSH cell of late pregnant female bat, *Megaderma lyra.* Note typical small round vesicles of rough endoplasmic reticulum (RER) and secretory granules which are highly electron-dense (Sg) [X 10,000]. Unpublished electron micrograph from Dr. Bhiwgade and Dr. Mantri

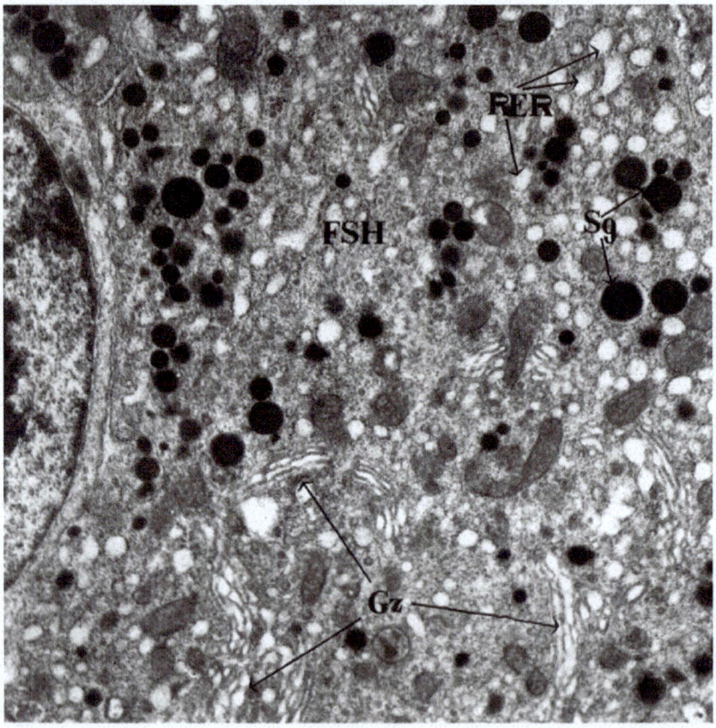

Fig. 1.102 FSH gonadotroph of late female bat, *Megaderma lyra.* Note scattered secretory granules (sg) of different sizes, typical round vesicular profiles of rough endoplasmic reticulum (RER) and well-developed extensive juxtanuclear Golgi zone (Gz) [X 13,000]. Unpublished electron micrograph from Dr. Bhiwgade and Dr. Narayane

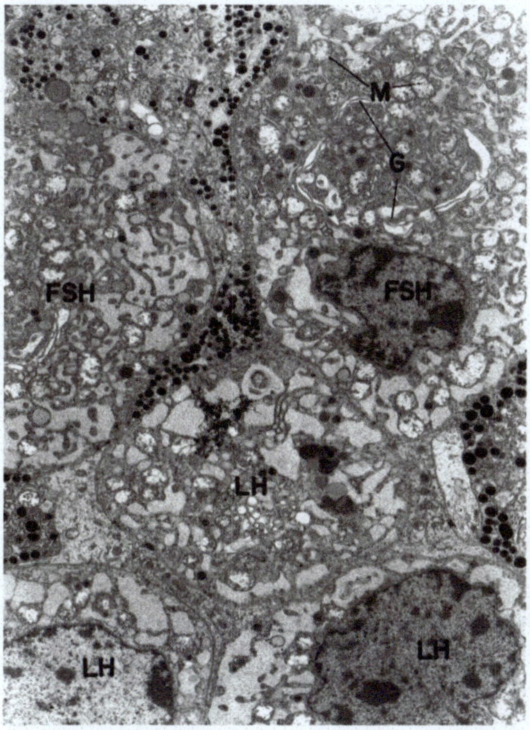

Fig. 1.103 Anterior pituitary of male bonnet monkey after 30 days of gonadectomy. Note FSH cell with vacuolated RER highly developed Golgi (G), minimum mitochondria (M) with loss of cristae. The LH cell shows complete degranulation show rER and mitochondria with loss of cristae [X 5020]. Unpublished electron micrograph from Dr. Bhiwgade, Dr. Thakur and Dr. Raut

(c) Male rat after MPA + TE treatment
There occurs a slight decrease in the number of secretory granules along with the irregularly vesiculated RER (Fig. 1.105). The Golgi apparatus also undergoes hypertrophy and there is a lysosomal degradation and some dense bodies are encountered.

(d) Male rat after Gossypol treatment
Administration of gossypol to male rats brings about the drastic changes in mitochondrial structure (Fig. 1.109). The mitochondria become extensively large and appear balloon shaped, and there is a complete loss of cristae. The cells undergo complete degranulation and granule extrusions may occur at the basal portion of the cell facing the perivascular space. The RER is arranged parallel and its membrane is provided with numerous ribosomes, and polyribosomes are also distributed throughout the cytoplasm. The Golgi apparatus is well-developed with dilated sacculi and vesicles. There is an increase in lysosomal activity.

(e) Male monkey after TE treatment

The LH cell undergoes complete degranulation after TE treatment. Only a few immature granules are seen in the Golgi area (Fig. 1.110). The RER becomes irregularly vesiculated giving the cell a bizarre appearance and similar changes are observed after castration. The cytoplasm is also characterized by the presence of hypertrophied Golgi apparatus, enlargement of mitochondria with loss of cristae and increase in lysosomal activity.

(f) Mature female rat after tamoxifen citrate treatment

The LH cell shows very few secretory granules, but an increased number of dense bodies, lysosomes, lipid droplets and degradation of secretory granules (Fig. 1.111). The vesiculated RER cisternae are fused together. There is a hypertrophy of Golgi apparatus along with numerous Golgi cisternae and sacculi.

(iv) Gonadotroph cells (GTH Cells) in female rats after treatment with female contraceptive drugs

a. Effect of tamoxifen citrate: The cell shape varies between round, ovoid and elongated. The consistently large and round nucleus is centrally located, euchromatic and frequently shows a distinct nucleolus. However, it has also assumed an irregular shape with a corrugated margin showing dense chromatin (Fig. 1.112a). The large nucleus does not favour expansive cytoplasm, hence the latter is restricted, thereby, resulting in an increased nucleo-cytoplasmic ratio. The electron density of the cytoplasm has increased. The granulation is scanty to moderate and the round secretory granules exist in variable sizes. As such, the RER is

a

b

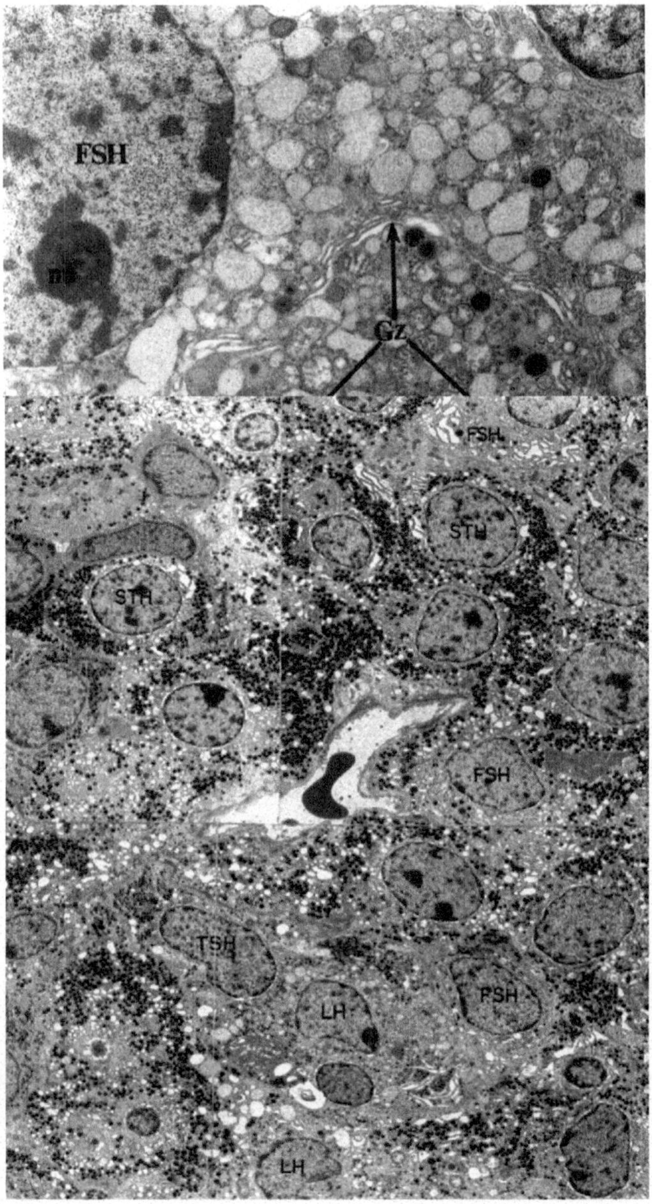

Fig. 1.104 (**a**) Anterior pituitary of male bonnet monkey after 30 days hemi castration to show the FSH cell. Note extreme vacuolation of rER in typical circular profiles (RER). Enlarged Golgi zone encloses mitochondria within and near Golgi zone, few secretory granules and a prominent nucleus [X 8500]. Unpublished electron micrographs from Dr. Bhiwgade and Dr. Thakur and Dr. Raut (**b**) Low-power electron micrograph of anterior pituitary of CPA treated male rat. Note various cell types namely STH, TSH, FSH and LH after CPA treatment for 60 days [X 3500]. Unpublished electron micrograph from Dr. Bhiwgade and Dr. Kadam

scanty, but has undergone dilatations or marked vesiculations, particularly in the polar cytoplasm. Extensive papuliferous and vesicular profiles sequestrated within the enormously enlarged cisternae are commonly encountered (Fig. 1.112a). Confronting cisterna that has undergone moderate dilatations are also present within the cytoplasm.

The Golgi zone is small and is not easily seen, however, some cells do show moderate to significantly stepped-up activity. This is endorsed by the hypertrophied Golgi elements that contain electron-lucent material with a dense periphery. A gradation exists in the size of vacuoles that arise from the Golgi. The round and rod-shaped mitochondria show numerous variations among the individual cells. In

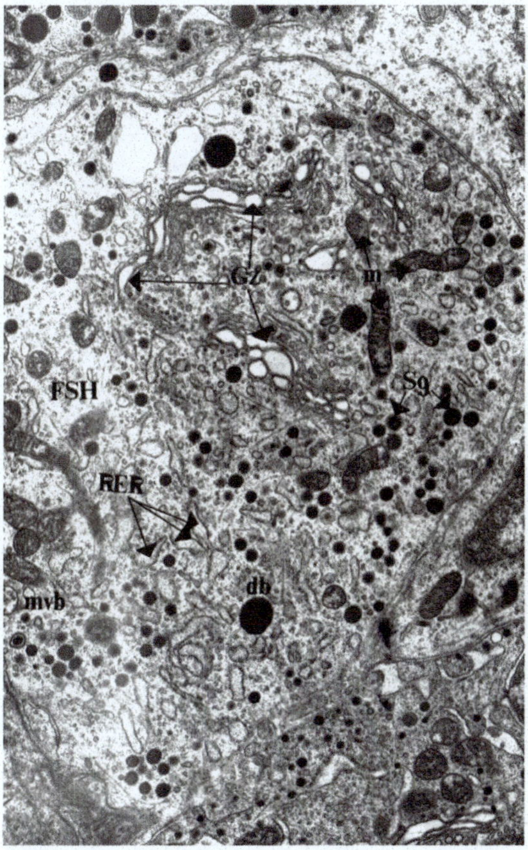

Fig. 1.105 Anterior pituitary of male rat after DMPA + TE treatment for 120 days. The FSH cells show vesicle of RER well-developed Golgi zone (GZ) with dilated sacs secretory granules scattered throughout the cytoplasm dense bodies (db), multivesicular bodies (mvb) and enlarged mitochondria (m) [X 10,000]. Unpublished electron micrograph from Dr. Bhiwgade and Dr. Borkar

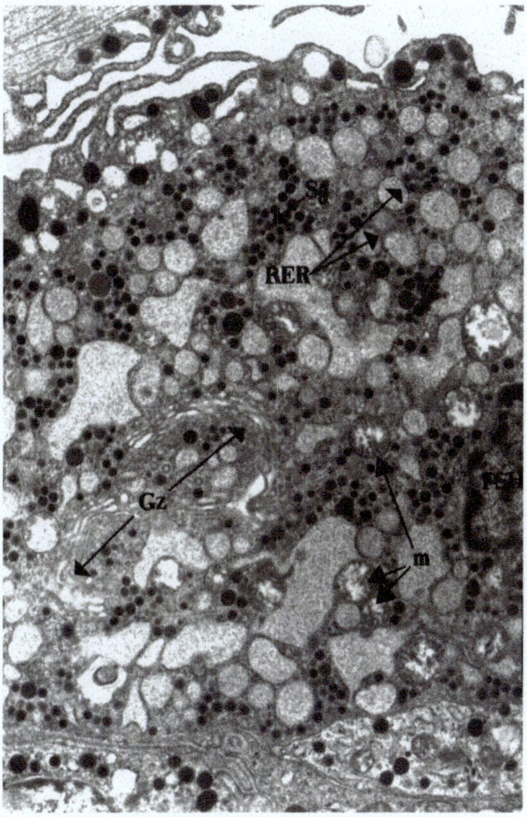

Fig. 1.106 Anterior pituitary of male rat after 70 days Gossypol treatment FSH cell shows vesicular RER in circular profile, Golgi zone highly developed comprising Golgi sacs are vesicular (Gz). Smaller mitochondria (m) with loss of cristae and secretory granules (Sg) are scattered throughout the cytoplasm [X 10,000]. Unpublished electron micrograph from Dr. Bhiwgade and Dr. Nair

some cells their number is sufficiently big. The enlargement of the nucleus has caused the displacement of the mitochondria to the peripheral cytoplasm. Mitochondria are also present near the maturing face of the Golgi zone (Fig. 1.112b). Enlargements and normal matricial densities of the mitochondria are common features, despite Tamoxifen treatment. Exceptionally, in some cells, a few mitochondria that are present have enlarged, but simultaneously show degenerative changes such as matricial loss, lucency, broken walls and loss of cristae (Fig. 1.113a). Highly increased lysosomal activity is arrested by the frequently occurring large single lysosomes and their aggregates (Fig. 1.113b).

b. Effect of oestradiol valerate: Following oestradiol valerate treatment, this cell harbours an inordinately large, circular nucleus, with an evenly smooth or crenated outline (Fig. 1.114a and d). Most of these nuclei have single or paired nucleoli (Figs. 1.114c and d). Clumps of dense chromatin are peripherally scattered in the generally euchromatic nucleus. In some cells, the nuclear pores are easily discernible (Fig. 1.114a). The relatively

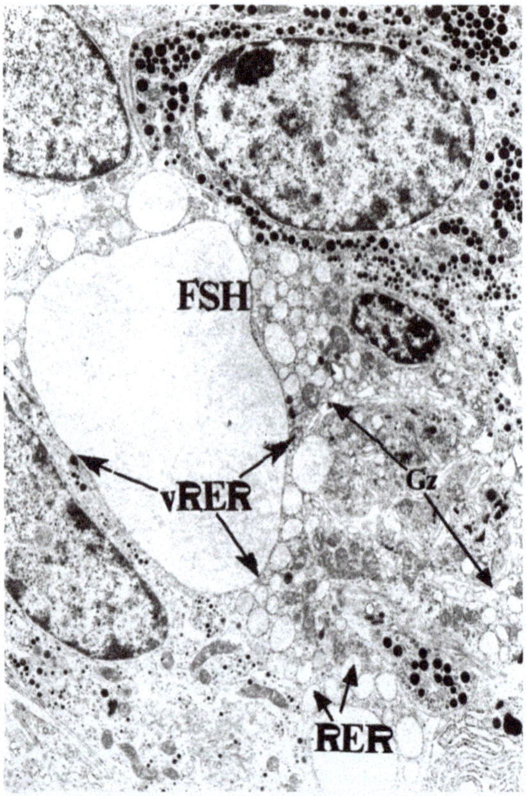

Fig. 1.107 FSH cell after 30 days DMPA treated female rat. A typical signet ring cell formation extreme of RER and Golgi zone (Gz) is highly developed [X 5000]. Unpublished electron micrograph from Dr. Bhiwgade and Dr. Borkar

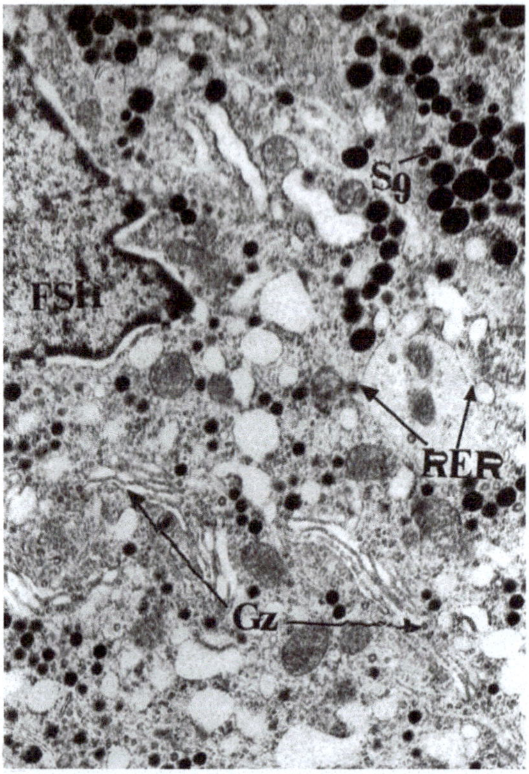

Fig. 1.108 Anterior pituitary of tamoxifen citrate-treated female rat for 30 days. FSH cell shows short vesiculated RER enlarged Golgi zone (Gz) and minimum secretory granules (Sg) [X 13,000]. Unpublished electron micrograph from Dr. Bhiwgade and Dr. Borkar

scanty cytoplasm results in an increase in the nucleo-cytoplasmic ratio. The tiny secretory granules are sparse and have been clumped in small patches within the cytoplasm or occur randomly scattered, more often within the marginal cytoplasm.

Vesicular profiles of RER have undergone extreme distention and occupy the cytoplasm extensively (Fig. 1.114c). Intercisternal sequestration has occurred resulting in the formation of a few papilliferous and many vesicular profiles of RER. In such cells, the cytoplasm has assumed a filigreed appearance. It is noteworthy that in some of these cells, the RER cisternae have undergone cystic dilatations, thereby imparting a vacuolated appearance to the cytoplasm (Fig. 1.114d). Amorphous secretion has accumulated in the vesiculated RER cisternae (Fig. 1.114b), endorsing the excessive synthetic activity of the cells. However, in some cells, the RER is represented by slightly dilated confronting cisternae that form 'dense lamina'.

The Golgi complex is not very well developed in the cells. The cell invariably contains a rich complement of mitochondria that have undergone enlargement, most of which are round, but some have a condensed configuration. Degenerative changes have occurred resulting in reduced matricial density, loss of matrix and cristae, and breaks in the wall. Some of the mitochondria have been myelinated and distorted beyond

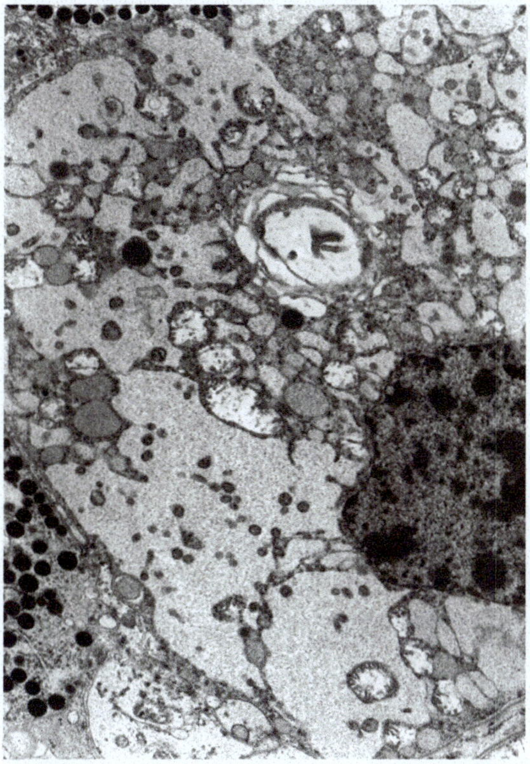

Fig. 1.109 Anterior pituitary EM of bonnet monkey after castration LH cell with a vesicular profile of RER containing amorphous material leading to vacuolation. Mitochondria (M) with loss of cristae [X 16,500]. Unpublished electron micrograph from Dr. Bhiwgade and Dr. Raut

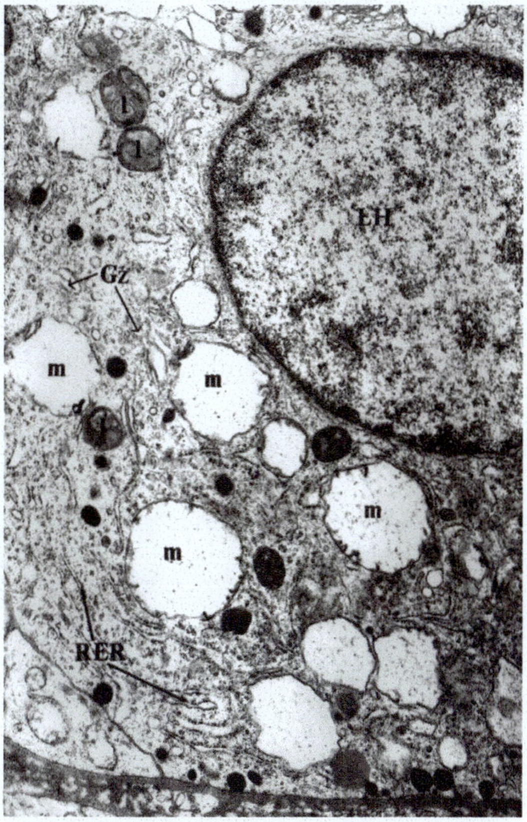

Fig. 1.110 Anterior pituitary of Gossypol treated male rat. Note LH cell with lamellar profiles of RER, lysosomes (l) extensively hypertrophied mitochondria (m) with loss of cristae and Golgi zone (Gz). [X 10,000]. Unpublished electron micrograph from Dr. Bhiwgade and Dr. Nair

recognition, as they lose their characteristic morphology. Some 'C'-shaped involuted mitochondria are also present (Fig. 1.114c). Lytic bodies and lipid residues are encountered regularly in the cells (Fig. 1.114d) following oestradiol valerate treatment (Fig. 1.114d).

c. Effect of progesterone: The cells are elongated and have undergone a drastic cytoplasmic shrinkage (Fig. 1.115). The euchromatic nucleus of the cell is relatively large with a perfectly spherical contour and contains a distinct nucleolus as well as heterochromatin clumps, along the nuclear margin. The cytoplasm has been rendered strikingly vacuolated, due to extreme vesiculation, caused by a defect in egress of the secretory material (Fig. 1.116a and c). Lysosomal activity has been significantly augmented (Figs. 1.116a). RER organization is suggestive of an advanced lesion, as the cisternae have undergone marked dilation as well as vesiculation, and the vesicles are filled with secretion. Ribosomes have been detached, thereby imparting a smooth texture to the endoplasmic cisternal membranes (Fig. 1.116c). A distinct irregularly circular, juxtanuclear Golgi zone in a state of hypertrophy is observed regularly. Numerous transport vesicles containing newly synthesized material have started budding off from the cisternal margins (Fig. 1.116b and c).
d. Effect of norethisterone: The gonadotrophs in female rats treated with norethisterone for 24 weeks show smoothened endoplasmic reticulum due to loss of ribosomes

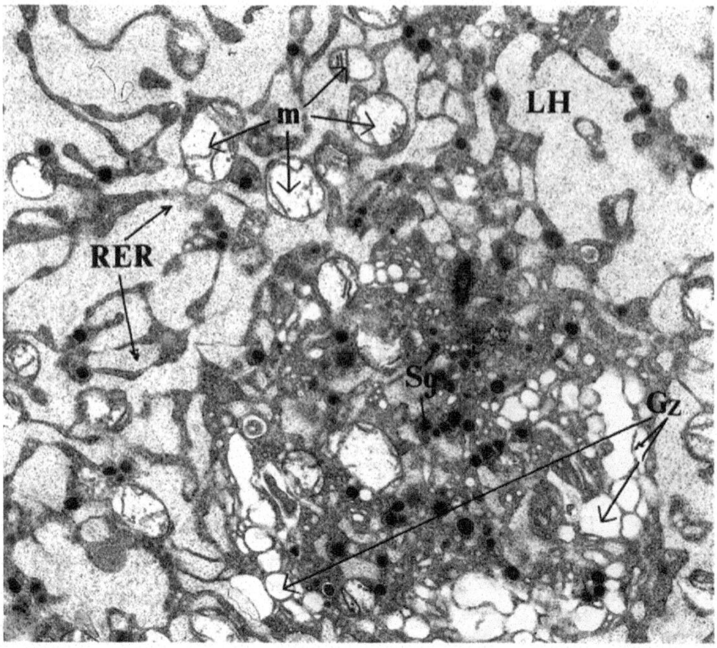

Fig. 1.111 LH gonadotroph in male monkey after 21 weeks of TE treatment. Note irregular reticulation of RER; immature secretory granule (Sg) near Golgi zone (Gz) and few mitochondria (m) with loss of cristae [X 8000]. Unpublished electron micrograph from Dr. Bhiwgade and Dr. Raut

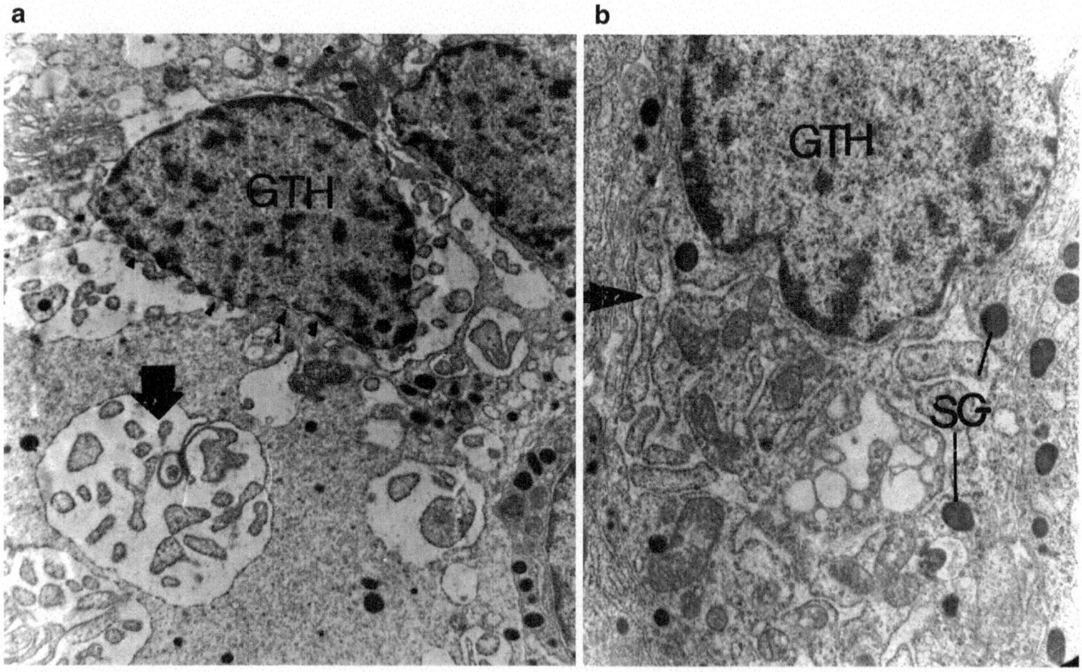

Fig. 1.112 (**a**, **b**) High-power electron micrograph of the gonadotrophs (GTH) of female rats treated with tamoxifen citrate for 24 weeks. Note the irregularly shaped nucleus with conjugated margin (arrows heads) and done chromatin clumps. The granulation is scanty. See extensive papilliferons and vesicular profile of RER (Bold arrow). (**a**) [X 8000], (**b**) [X 8000]. Unpublished electron micrographs from Dr. Bhiwgade and Dr. Borkar

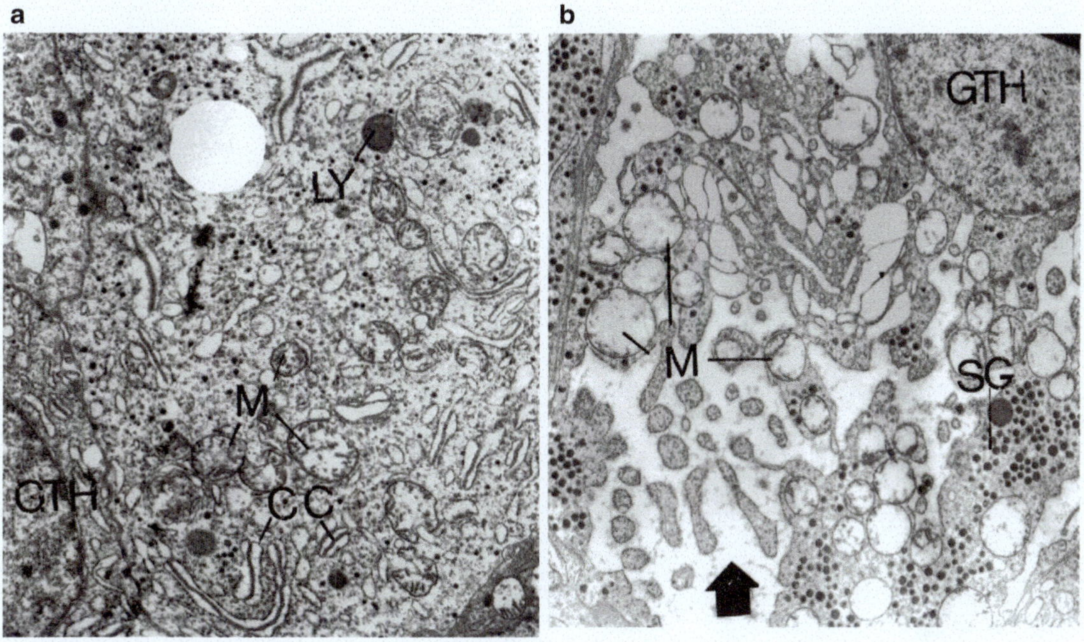

Fig. 1.113 Portion of the gonadotrophs of female albino rats treated with Tamoxifen citrate for 24 weeks (**a**) Cytoplasm is occupied by dilated confronting cisternae (CC) distorted and myelinated mitochondria (M) and lysosomes (LY) [X 6000] (**b**) The secretory granules (SG) are clumped in patches. See the extremely vacuolated / dilated reticulum (bold arrow). Also, note the enlarged mitochondria (M) [X 6000]. Unpublished electron micrographs from Dr. Bhiwgade and Dr. Borkar

(Fig. 1.117a). The rough endoplasmic reticulum shows vesicles and mitochondria are dilated (Fig. 1.117b). The lysosomal activity is also apparent indicating increased cellular activity.

e. Effect of levonorgestrel + ethinyl oestradiol: The cell is round or elongated in shape, has assumed a dull appearance with a low contrast cytoplasm and contains a large, spherical or indented, euchromatic nucleus occupying a central or basal position in the cell. In some cells, the nucleus has a distinct reticular nucleolus resting against the nuclear membrane. A very thin continuous streak of dense chromatin has appeared along the nuclear membrane. The cell has undergone a considerable degranulation. Small, round and immature secretory granules have been chiefly confined to the peripheral cytoplasm or are randomly distributed in clumps (Fig. 1.118a and b). The RER membranes in most of the cells have diminished and the cisternae of those present have been dilated, the dilatations being more pronounced in the cisternae lying towards the nucleus. Some other cells contain a good number of discontinuous cisternal fragments that are moderately distended with a medium-density material (Fig. 1.118b). Some dilated short confronting cisternae also occur in the cytoplasm (Fig. 1.118a). Polyribosomal rosettes have been formed particularly in the cytoplasm around the nucleus. A distinct juxtanuclear Golgi zone of horse-shoe shape, with well-developed and distended sacs and vacuoles reflect the initiation of advanced secretory activity (Fig. 1.118b). Several mitochondria have been closely associated with the Golgi zone which is complementary to the activated state of the latter. One or two mitochondria are occasionally contained within the core of the Golgi zone (Fig. 1.118b). Of the many mitochondria

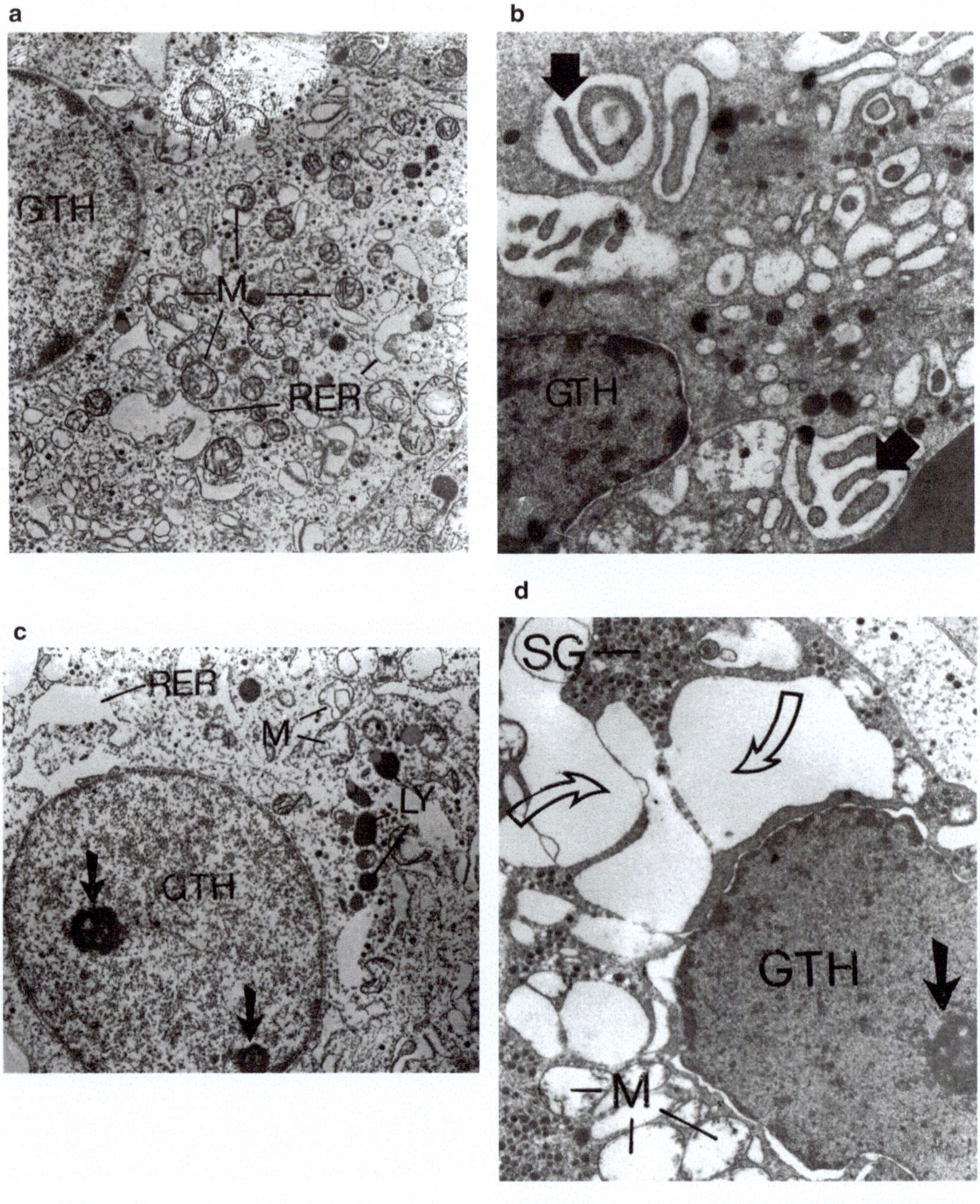

Fig. 1.114 (**a**) All the gonadotrophs of oestradiol valerate-treated female albino rats Fig. 1.114. Note the evenly smooth outline of the nucleus and the distinct pores (arrow heads). The ergastoplasmic cisternae have undergone dilatations (RER), the mitochondria (M) have been myelinated [X 6000]. Unpublished electron micrographs from Dr. Bhiwgade and Dr. Borkar (**b**). The amorphous secretion that has accumulated in the vesiculated cisternae of the rough endoplasmic reticulum (bold arrows) is suggestive of excessive synthetic activity [X 8000]. Unpublished electron micrographs from Dr. Bhiwgade and Dr. Borkar (**c**). Note the distinct reticular nucleolus (bold arrow) of gonadotroph (GTH). The cytoplasm bears a disorganized look due to the distention of the rough endoplasmic reticular cisternae (RER), distorted mitochondria (M) and lysosomal aggregates (LY) [X 6000]. Unpublished electron micrographs from Dr. Bhiwgade and Dr. Borkar. (**d**) See the inordinately large nucleus with a reticular nucleolus (bold arrow). The rough endoplasmic cisternae of this cell have undergone cystic dilatations (hollow arrows). The secretory granules (Sg) are displaced towards the periphery [X 6000]. Unpublished electron micrographs from Dr. Bhiwgade and Dr. Borkar

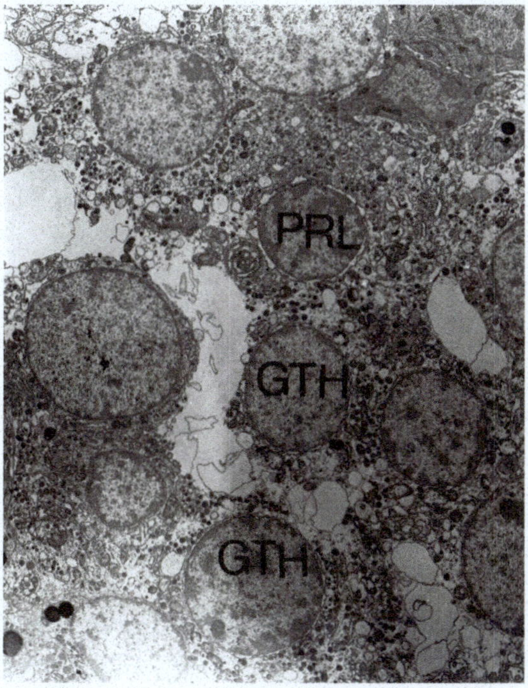

Fig. 1.115 A low-power electron micrograph of the anterior pituitary of the female albino rat was treated for 24 weeks with progesterone showing Prolactin cells (PRL) and gonadotrophs (GTH) [X 3500]. Unpublished electron micrographs from Dr. Bhiwgade and Dr. Borkar

contained in the cell, some are ring shaped, with central or eccentric perforation serving to support the increased mitochondrial–cytoplasmic exchange (Fig. 1.118b). The regressive changes shown by the mitochondria may be viewed in congruence with the distention of RER and increased cellular activity. Lysosomes have also appeared in the cytoplasm (Fig. 1.118b).

1.4.5 Discussion

Gonadotroph Cells

The presence of two different gonadotrophin hormones is well established and are measured from the blood serum separately. But whether two different types of cells are present in the pituitary, or a single cell produces both the hormones is still a confusion. Authors like Saeger (1992) state that 'Probably structural differences between the bihormonal and monohormonal gonadotrophs do not exist'. Although in our study we have confirmed ultrastructural differences between the FSH and LH cells in normal and experimental pituitaries. The two gonadotroph cells have been described separately by Herlant (1964), Nakane (1970), Raza and Shakoori (1978) in rats; Barnes (1962) and Gomez-Dumm and Echave-Llanos (1972) in mouse, Dekker (1967) in hamsters, Bhiwgade et al. (1989) in bats and Lawzewitsch et al. (1972) in humans.

FSH Cells

The shape of the FSH cell is variable according to the species and not very authentic evidence to characterize the cell shape, since in our study we have found polygonal to irregular shapes in male rat, bat and monkey and in female bat, Rhinopoma. Round to oval shape of FSH cells has been reported by Herlant (1964), Nakane (1970), Raza and Shakoori (1978) in rats, Barnes (1962) and Gomez-Dumm and Echave-Llanos (1972) in mouse; Dekker (1967) in hamsters, Bhiwgade et al. (1989) in bat and Lawzewitsch et al. (1972) in human. The numerical strength is also found to be variable depending upon the various phases of reproductive cycle, for instance, in active female rats, the granules are found to be moderate in number. The secretory granules are round, medium-sized and electron-dense in almost all the pituitary cells of the animals studied, similar observations have been noted in the other animals too. In active animals the RER is found to be lamellar whereas it is reported to be vesiculated in hamsters (Dekker 1967), but the Golgi zone is found to be juxtaposed to nucleus in most of the animals studied (Dekker 1967—in hamster; Bhiwgade et al. 1989—in bats and Lawzewitsch et al. 1972—in human).

During advance pregnancy, the RER appears to be vacuolated and a progressive decrease in granulation of FSH cells in Hipposideros, and Megaderma, which reflect decreased synthesis and output of FSH. The diminishing FSH output could be linked to increased levels of serum PRL which has an inhibitory influence on oestrogen and consequently on FSH. The hyperprolactinaemia has a direct as well as

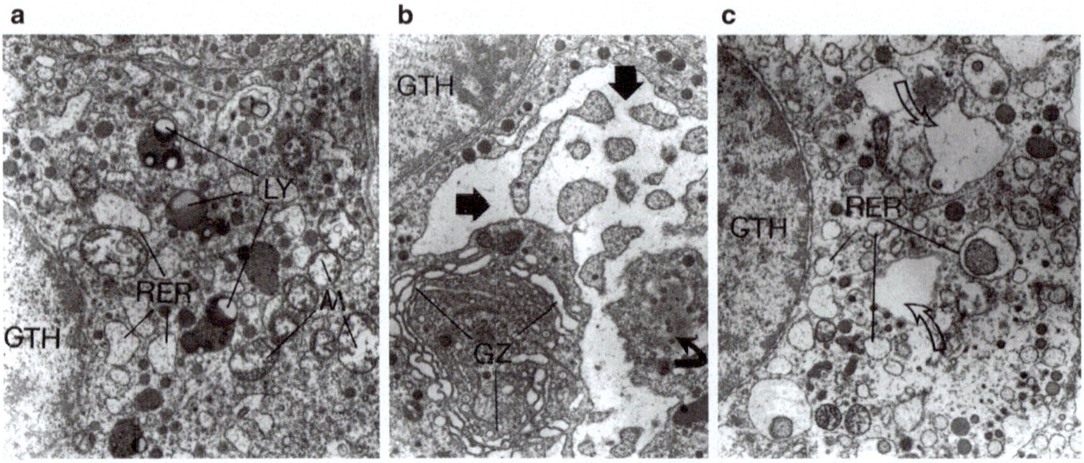

Fig. 1.116 (**a**, **b**, **c**) Show altered ultrastructure of the gonadotrophs (GTH) following 24 weeks of progesterone treatment. (**a**) Massive lysosomal aggregates (LY) and lipidic droplets have appeared in the cytoplasm. The mitochondria (M) show distortion of the cristae and loss of matrix [X 10,000]. Unpublished electron micrographs from Dr. Bhiwgade and Dr. Borkar. (**b**) The marked dilatations of rough endoplasmic reticular cisternae (bold arrows) are suggestive of an advanced lesion. See the transport vesicles budding off from the cisternal margins (curved arrow). The Golgi zone (GZ) has undergone hypertrophy [X 8000]. Unpublished electron micrographs from Dr. Bhiwgade and Dr. Borkar. (**c**) See the smooth texture of endoplasmic cisternal membrane (hollow arrow) due to the detachment of ribosomes [X 8000]. Unpublished electron micrographs from Dr. Bhiwgade and Dr. Borkar

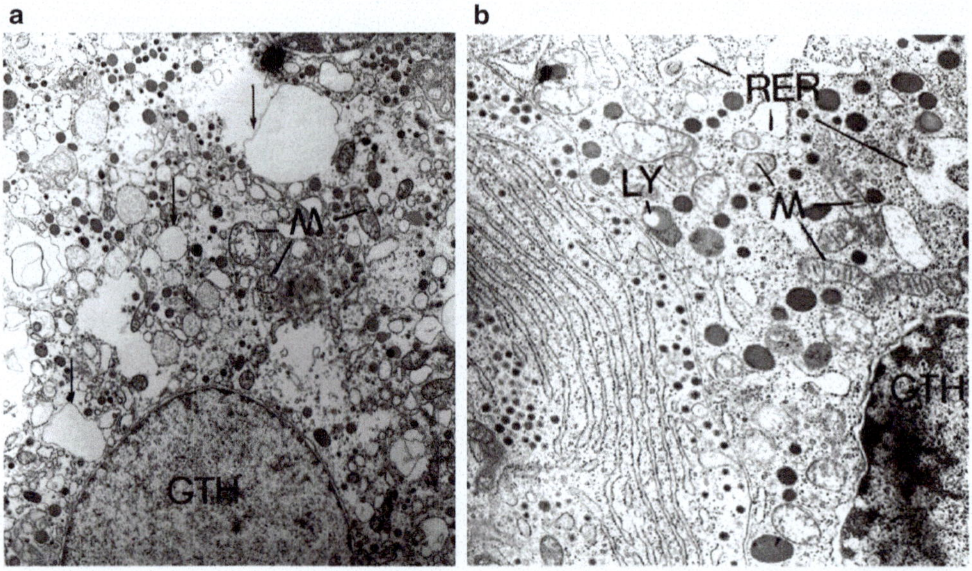

Fig. 1.117 (**a**, **b**) Gonadotrophs of female albino rats following 24 weeks of norethisterone treatment. Note the smooth nature of the endoplasmic reticulum due to the detachment of the ribosomes (arrows—Fig. 1.63e). The rough endoplasmic reticulum (RER) is generally vesiculated Mitochondria (M) are mildly dilated. Lysosomes (LY) have appeared. (**a**) [X 8000]. (**b**) [X 13,000]. Unpublished electron micrographs from Dr. Borkar

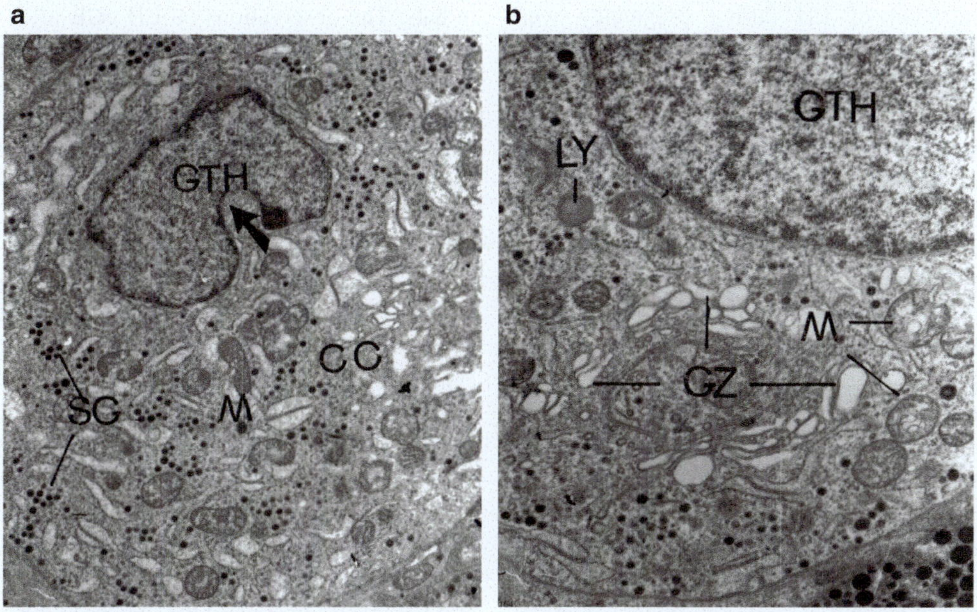

Fig. 1.118 (**a**) Gonadotrophs of female albino rats treated for 24 weeks with levonorgestrel + ethinyl oestradiol. The cell has an indented nucleus (bold arrow) and a considerably degranulated cytoplasm. The secretory granules (SG) are confined to the cell periphery. The dilated conforming cisternae (CC) as well as abundant mitochondria (M) are prevalent [X 10,000]. Unpublished electron micrographs from Dr. Bhiwgade and Dr. Borkar. (**b**) Observe the well-developed Golgi zone (GZ) with distended sacs. Also note the closely associated mitochondria (M). Lysosomes (LY) have appeared in the cytoplasm [X 13, 000]. Unpublished electron micrographs from Dr. Bhiwgade and Dr. Borkar

indirect influence on pituitary FSH release (McNeilly et al. 1978; Vasquez et al. 1980; Marchetti and Labrie 1982; Miyake et al. 1985).

LH Cells

Our study demonstrates that the LH cells are polygonal or irregular in shape with mostly lamellar rER, moderate number of secretory granules, and a Golgi complex which is sometimes discerned in male and female rats, monkey, bat and Suncus. Similar observations have been reported by earlier authors viz. Herlant (1964), Moriarty (1975) in rats, Barnes (1962) and Gomez-Dumm and Echave-Llanos (1972) in mouse; Shirasawa et al. (1985) in goat; Bhiwgade et al. (1989) in bats and Lawzewitsch et al. (1972) in human.

During early pregnancy, the LH cells show few secretory granules, which is indicative of increased secretory activity a supposition. This is further supported by the presence of vesiculated RER and numerous mitochondria with a prominent matrix in female bat, *Rhinopoma microphyllum*. Parlow et al. (1964) and Morishge et al. (1973) have reported an increase in serum LH levels during early pregnancy in gilts and mice, respectively. In our studies of late pregnant pituitaries of bats, Cynopterous and Taphozous, the accumulation of immature granules and the poorly developed RER are observed. Randman and Runner (1953) have stated that the decrease in LH levels during pregnancy is due to the inability of the animal to ovulate. It is well known that the prolactin level increases as the pregnancy advances. These elevated prolactin levels bring about a decreased oestrogen level which affects the positive feedback mechanism at the hypothalamic level.

The results from the pregnant rats indicated that the ovarian binding sites of LH developed earlier, whilst FSH did not appear to be induced. Although the binding sites for and the concentration of LH during pregnancy were inversely related, the pattern of increase of ovarian-specific binding was nearly identical to that of serum progesterone concentrations during pregnancy (Morishge et al. 1973), and was very similar to that of LH-stimulated adenylate cyclase activity in rat corpora lutea during pregnancy (Hunzicker Dunn and Birnbaumer 1975).

Castration and treatments with contraceptive drugs bring about drastic changes in the FSH and LH cell cytology leading to the transformation into castration cells. These effects are due to the direct effect of negative feedback on pituitary.

Castration

Our study demonstrates that the FSH cell undergoes degranulation following castration, hypertrophy of RER and Golgi apparatus and loss of mitochondrial cristae in bonnet monkey, whereas the LH cell undergoes complete degranulation, hypertrophy of mitochondria with loss of cristae and extreme dilation of RER giving the bizarre-shape appearance to the cell. Earlier authors like Farquhar and Rinehart (1954a, b) have stated that, the FSH cells response to castration prior to that of the LH cells in rats; these cells contain ovoid cytoplasmic vesicles of rough endoplasmic reticulum which coalesce with each other to form a large vacuole. El Etreby et al. (1972) have reported enlargement of both FSH and LH cells along with the vacuolation of cytoplasm, finally transforming into a signet-ring castration cell in pituitary of various animals. Murphy and James (1976) have shown the Type I cell undergoes partial degranulation, increase in electron density and vacuolation of the cytoplasm after castration in mink. Castration results in an increase in synthetic and secretory activity (Davis Davis and Meyer 1973) which ultimately results in increased serum gonadotrophin levels in rats (Hellbaum et al. 1961 and D'Angelo, 1966) in snow shoe hares (Davis Davis and Meyer 1973).

Cyproterone Acetate (CPA): Antiandrogen

After CPA treatment, the FSH and LH cells show extreme dilation of rER, hypertrophy of mitochondria and Golgi apparatus and decrease in granulation in rats and complete degranulation in bats. The observations on LH cells after CPA treatments are similar to those made by El Etreby et al. (1972), who observed changes in LH cells that are similar to castration. The morphological changes are in agreement with the findings of Neumann et al. (1970) and Neumann et al. (1976), who were able to measure an elevated level of serum FSH in rats. Cyproterone acetate causes a change in the secretory activity. Though we did not measure the levels of serum FSH and LH it is likely that they may not be altered significantly since much smaller doses of CPA were used than those used in the studies above. In an earlier study by Koch et al. (1976) and Morse et al., of RER, Golgi apparatus, loss of mitochondrial cristae and presence of few immature secretory granules were reported. These cellular changes are suggestive of a suppression of FSH and LH release in spite of the increased hormonal synthesis. Our observations also confirm the decreased serum FSH and LH levels recorded by Troen (1980), Rajalakshmi & Ramkrishnan (1989), Matsumoto (1990), Wallace et al. (1993), Raut (1993) and Bagatell et al. (1994) with different doses of testosterone, and its analogs for different duration of treatments in humans. Decreased serum LH levels have also been recorded by Peilion and Racodot et al. (1965); Giguere et al. (1981) in rats and Ramakrishnan et al. (1989) in Rhesus monkeys. This is possibly due to the decreased number of LH–RH receptors at the hypothalamic level because of the negative feedback effect of testosterone.

Depo-Provera (DMPA): Antifertility

Our results show that FSH and LH cells become vacuolated due to extreme dilation of the RER, which gives the cell a signet ring appearance, the Golgi apparatus gets extremely enlarged, the mitochondria with prominent matrix increase in size, and also there is absolute degranulation in DMPA treated female rat for 30 days. Our morphological observations confirm the results reported by Neumann et al. (1976) in rats, Jeppsson et al. (1982), Fraser and Susan (1983), Lobl et al. (1983), who reported a fall in FSH levels in the serum after MPA treatments with different doses and different durations of treatment in human. Gordon et al. (1970) have reported such a fall in FSH levels only in postmenopausal women after MPA treatments but LH levels fell in both pre and post-menopausal women. Another possible reason for the suppression of FSH seems to be hyperprolactinaemia caused by MPA treatments; as it is well known that hyperprolactinaemia causes inhibition of the release of FSH and LH from pituitary (Grandison et al. 1977; McNeilly et al. 1978; Frantz 1978; Winters and Louriaux 1978; Vasquez et al. 1980; Marchetti and Labrie 1982). Changes in LH levels because of hyperprolactinaemia resulting in a decrease in serum LH level were recorded by Aono et al. (1976) and Muralidhar et al. (1977). Mishell et al. (1977) observed no change in FSH levels after a short dose of MPA and with progesterone treatment, Roche and Ireland (1981).

Tamoxifen Citrate: (Anticancer)

After tamoxifen citrate treatment, in FSH cells there is a decrease in granules but an active synthesis of hormone is evident from the hypertrophied RER and Golgi apparatus. The LH cell cytoplasm shows an increased number of dense bodies, lysosomes, lipid droplets and degradation of secretory granules along with the enlargement of Golgi sacs and vesicles. This can be explained as due to antioestrogenic properties of tamoxifen which competes with oestrogens for the receptors in the hypothalamus, where the feedback inhibition of GnRH is inhibited which results in release of FSH from pituitary. Comhaire and Dhont (1975), Vermeulen and Comhaire (1978) have reported an increase in FSH and LH levels after tamoxifen treatment which support our present morphological observations on the ultrastructure of FSH and LH cells in the pituitary. Watson et al. (1975) have reported a decrease in pituitary LH concomitantly with an increase in serum LH after tamoxifen treatment in rats.

1.5 Corticotroph Cell (ACTH Cell)

1.5.1 Introduction

Though the exact frequency of ACTH cells is a matter of speculation, rough estimates state that these may represent approximately 15–25% of the pituitary cells. The ultrastructure of the cell, Probably, does not vary between the two sexes (Kovacs et al. 1981). So far as the rat is concerned, there is no agreement regarding the fine structure of the ACTH cell. Such lack of consensus was chiefly due to morphological variations encountered within the cell type of the same species (Costoff 1973; Moriarty 1973; Kurosumi and Fujita 1974; Girod 1977; Kurosumi 1986; Horvath and Kovacs 1988). The difficulty is further compounded by the immunoreactive interference of the other subunits of the ACTH molecule that rendered immunohistologic demonstration of these cells difficult (Celio et al. 1980; Saeger et al. 1990; Saeger 1992).

By ultrastructural examination of ACTH cells in a wide range of mammals, it is concluded that this cell has an ovoid, angular or stellate outline with an eccentric nucleus harbouring a marginated nucleolus (Kueosumi and Kobayashi 1966; Siperstein and Miller 1970, 1973; Weman 1974; Saeger 1977; Horvath and Kovacs 1988; and Bhiwgade et al. 1989). Haloed immature secretory granules often found in the Golgi areas of the cell, and the characteristic single-row alignment of mature secretory granules beneath the cell membrane are observed in rat (Kueosumi and Kobayashi 1966; Bowie et al. 1972) in mink (Weman 1974; Murphy and James 1976) and in bat (Bhiwgade et al. 1989). These are accepted as reliable fine structural features of this cell. Such characteristic features of corticotroph correspond to the immunostained fringe in these cells (Yoshimura and Nogami 1981).

A hyperfunctional state of the cell can be induced by stress, adrenalectomy or administration of ACTH-stimulating hormones and drugs (Girod et al. 1964; Kueosumi and Kobayashi 1966; Rennels and Shiino 1968; Siperstein and

Miller 1970; Cameron and Foster 1972; and Bhiwgade et al. 1989) to enable their identification. Such hyperactive 'adrenalectomy' cells increase in number and size. They have large nuclei, increased RER and Golgi area with occasionally lysosomes and increased granulation (Siperstein and Allison 1965; Costoff 1973).

1.5.2 Light Microscopic Observations

In the control animal, corticotroph cells are moderately stained with lead haematoxylin and possess fine granules (Fig. 1.119). After 9 days of metapyrone treatment, no significant change is observed in the nuclear size but a remarkable difference in the staining nature of the cytoplasm has been observed in the cells (Fig. 1.120a). There is a slight increase in the nuclear size on the 14th day after the commencement of metopirone injections. By 18th day, the nuclei in the experimental animals are larger than those of the control animals (Fig. 1.120b). The cytoplasm in the cells of the experimental animals is stained stronger than those of the control. After 30 days of metapyrone treatment, the nuclei are still larger than those of the control animal, a conspicuous hypertrophy and hyperplasia of the corticotroph cells are observed. The cytoplasm of these cells is brightly stained in experimental animals (Fig. 1.121) while that in the control animals is feebly stained.

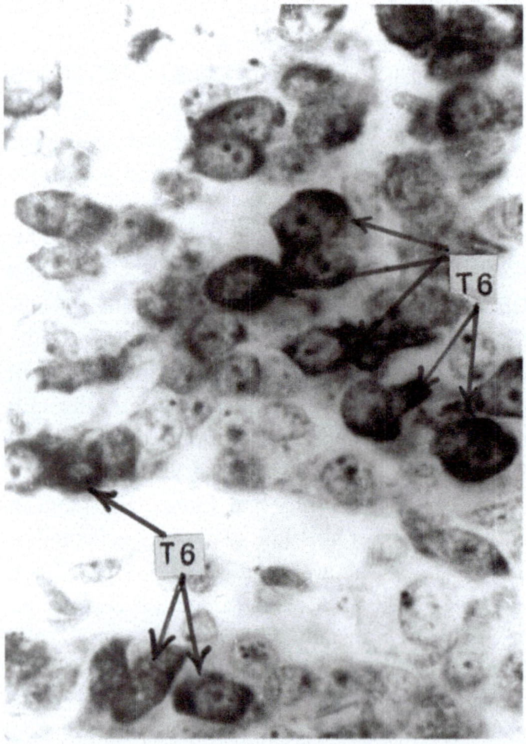

Fig. 1.119 Antero-median portion of the pars anterior of the control male squirrel (*Funambulus pennantii*) showing the lead haematoxylin-positive ACTH cells (T6). Staining: Lead haematoxylin [X 800]

1.5.3 Electron Microscopic Observations

The ATCH cells are generally stellate cells with their cytoplasmic processes extending in between two cells. Polygonal or oval cells are also observed. The nucleus is mostly round or elongated with smooth margins. The secretory granules are small in size but slightly larger than the granules of TSH cells. The density of granules is usually variable. The RER and Gz are discerned during the active state of cell.

The ACTH cell of the active male bat, *Tylonycteris pachypus*, is an elongated angular cell with a considerably large nucleus that mimics the shape of the cell (Fig. 1.122). The large size of the nucleus is an impediment to the cytoplasmic volume and is responsible for increasing the nucleo-cytoplasmic ratio. This cell has rather poor inclusion of electron-dense secretory granules, arranged in a single layer adjacent to the plasma membrane. The RER and the Golgi zone are poorly developed and non-discernible. Sometimes a very small Golgi zone is seen at one end of the cell. Very few mitochondria of round and elongated shapes are seen near the nucleus.

ACTH cell in *Funambulus pennanti* is elongated angular with electron-dense cytoplasm (Fig. 1.123). The nucleus is centrally located and small, with a condensed ovoid contour. Within the nucleus are seen distinct reticular, dense chromatin clumps. The cell is poorly granulated and the small, spherical, electron dense existent granules are closely opposed to

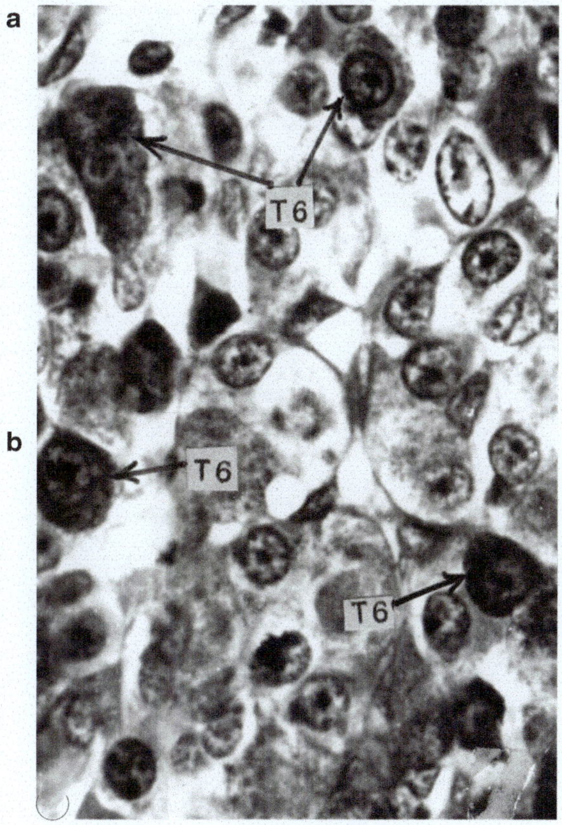

Fig. 1.120 (**a**) Antero-median portion of the pars anterior of the mature male squirrel (*Funambulus pennantii*) was treated with metopirone for 9 days. Note the lead haematoxylin-positive cells (T6). Staining: Lead haematoxylin [X 800]. (**b**) Antero-median portion of the pars anterior of the mature male squirrel (*Funambulus pennantii*) was treated with metopirone for 18 days showing deeply stained ACTH cells (T6). Staining: Lead haematoxylin [X 800]

the cell membrane in a single row. Few poorly developed confronting cisternae of the RER are seen scattered in the cytoplasm. An undeveloped Golgi zone is seen opposite one pole of the nucleus, variously shaped mitochondria are abundant in this cell.

A typical ACTH cell of a normal female (Fig. 1.124) and male bonnet monkey *Macaca radiata* (Figs. 1.125 and 1.126) is irregular or pear shape with elongated cytoplasmic processes. A striking accumulation of secretory granules, arranged in a single layer adjacent to the plasma membrane is observed. Both RER and Golgi zone are conspicuous and the mitochondria are large, round and some of them show loss of internal cristae. The electron opaque, large and rounded nucleus occupies the maximum portion of the cell. A few chromatin centres are observed in the nucleoplasm.

Folliculostellate Cell (FS Cell)

In course of the ultrastructural characterization of anterior pituitary cells, microscopists encountered, with a fair degree of regularity, a cell type devoid of secretory granules, in the pituitaries of many animals. It is for the same reason that these cells could not be recognized with tinctorial and histochemical reactions employed by the light microscopists. The agranular nature of the cell is highlighted by all the workers.

The characteristic feature of the folliculostellate cells (FS cells) is the absence of hormone-containing granules. The inactive FS cells are typically polygonal, angular or irregular in shape with large centrally placed nucleus. The RER membranes and Golgi zone are poorly developed in the cells. During cellular activity, these cells bear microvilli or cilia or their apical surfaces; these cells often show junctional complexes with the adjacent hormone-producing cells.

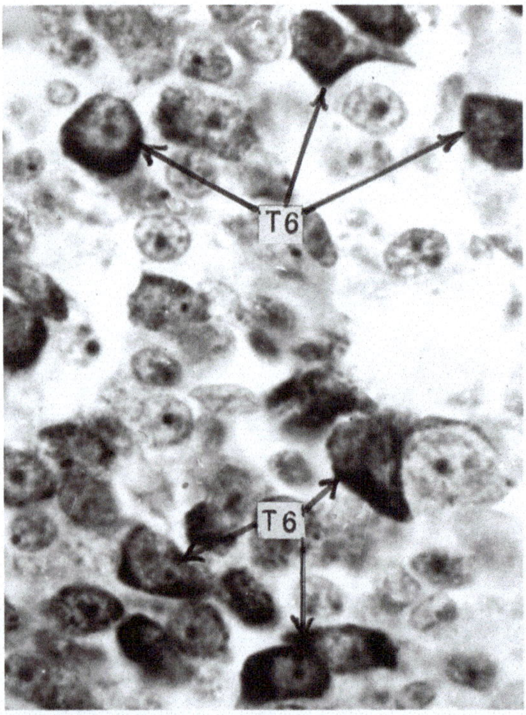

Fig. 1.121 Antero-median portion of the pars anterior of the mature male squirrel (*Funambulus pennantii*) was treated with metopirone for 30 days. Note the darkly stained ACTH cells (T6). Staining: Lead haematoxylin [X 800]

Dramatic changes are observed in the FS cells after MPA + TE treated male rat (Fig. 1.127), which are seen adjacent to the prolactin cells or the gonadotrophs. The cellular hypertrophy, presence of microvilli are the prominent features of hypertrophied FS cells. These structures increase the surface area of the FS cell within the newly formed system of extranuclear channels. These hypertrophied cells contain few homogenous mitochondria and small Golgi apparatus comprising of a few Golgi stacks and vesicles too. Hemidesmosomes, zona adherens-type junctions are also observed. Hemidesmosomes are found alone or close to zona adherens. The zona adherens are observed between FS cells or at a point where several granular and FS cells come together.

First described by Farquhar (1957), these cells have been a subject of controversy and speculation. These are often found fringing the borders of colloid-filled extracellular spaces. Based on animal studies several investigators have now concluded that these follicular or stellate (folliculostellate) cells are a separate cell type of the pituitary, that forms an interconnecting meshwork throughout the gland (Kurosumi 1986; Cardell 1969; Schechter 1969; Costoff 1973; Girod 1976).

Descriptions of the FS cells have been given for a wide variety of species like rats (Surks and DeFesi 1977; Cinti et al. 1985; and Stokreef et al. 1986), dog (Kagayama 1965; Gale 1972), rabbit (Young et al. 1967), mouse (Gomez-Dumm and Echave-Llanos 1972), mink (Weman 1974), goat and sheep (Shimada 1992), bat (Mantri 1994) and humans (Horvath and Kovacs 1988; Bergland and Torack 1969). Some common observations across the species line are the presence of characteristic intracellular lipid droplets, lysosomes (Dingemans and Feltkamp 1972) and underdeveloped cytoplasmic organelles. The nucleus is irregular with a prominent nucleolus. These cells are invariably seen with their microvilli projecting into the lumen of the colloid-filled extravascular spaces that they line. Junctional complexes are commonly seen adjoining such cells.

1.5.4 Discussion

Very little information is available on corticotrophs of bats. The problem has been further accentuated by the variations observed within the cell type, which can be mainly attributed to inherent differences in the effects of the various primary fixatives used in electron microscopy. In our study, this cell type was found to be scarce in the pituitaries of male and lactating individuals of *Tylonycteris pachypus*. Such a low frequency of occurrence has also been reported in normal bat pituitaries (Bhiwgade et al. 1989). The cell shape has been observed to

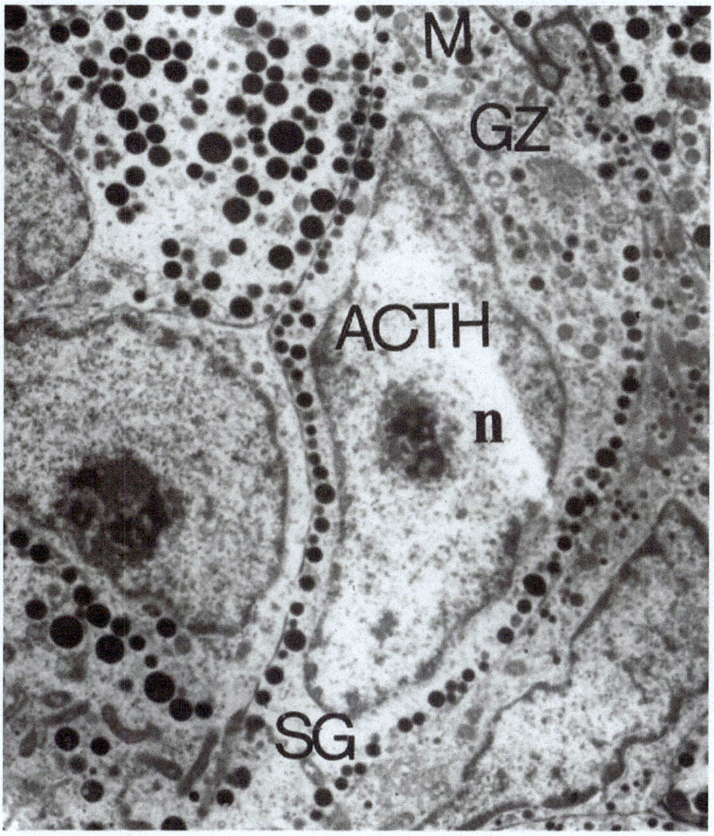

Fig. 1.122 A corticotroph (ACTH) of an active male *Tylonycteris pachypus* showing an angular profile, a large nucleus with a distinct nucleolus (n) and characteristically aligned secretory granules (SG) beneath the plasma membrane. The Golgi zone (GZ) is poorly developed and a few mitochondria (M) are seen [X 5000]

be angular in active males as well as lactating females of this bat, though the cell shape is reportedly variable and does not form any valid criteria for identification.

In rats (Kueosumi and Kobayashi 1966; Siperstein and Miller 1970; Bowie et al. 1972; Cinti et al. 1985; Horvath and Kovacs 1988), and men (Lawzewitsch et al. 1972) polyhedral, stellate or elongated ACTH cells have been reported. The scanty granulation of the ACTH cell and the nuclear shape mimicking the cell contour have also been reported in rat (Horvath and Kovacs 1988).

The peculiar alignment of the secretory granules in a single row along the cell membrane of the cell as seen in the male as well as in lactating bats (*Tylonycteris pachypus*) and the early pregnant squirrel has been similarly observed in the corticotrophs of rats (Kueosumi and Kobayashi 1966; Bowie et al. 1972; Siperstein and Miller 1970, 1973), as well as in mink (Weman 1974; Murphy and James 1976). Such a characteristic peripheral disposition of secretory granules in a corticotroph corresponds well to the immunostained fringe in these cells. (Yoshimura and Nogami 1981).

The corticotroph of the lactating female *Tylonycteris pachypus*; shows signs of stepped-up activity as endorsed by the enlarged nucleus, an electron-dense cytoplasm and an elaborate active Golgi zone; with visible condensation of secretion. Such variations in the Golgi apparatus of the ACTH cells, as influenced by functional conditions have been reported in the human pituitary (Lawzewitsch et al. 1972).

The poorly developed RER, Golgi zone and a relatively poor inclusion of mitochondria are seen in the corticotrophs of the male and lactating

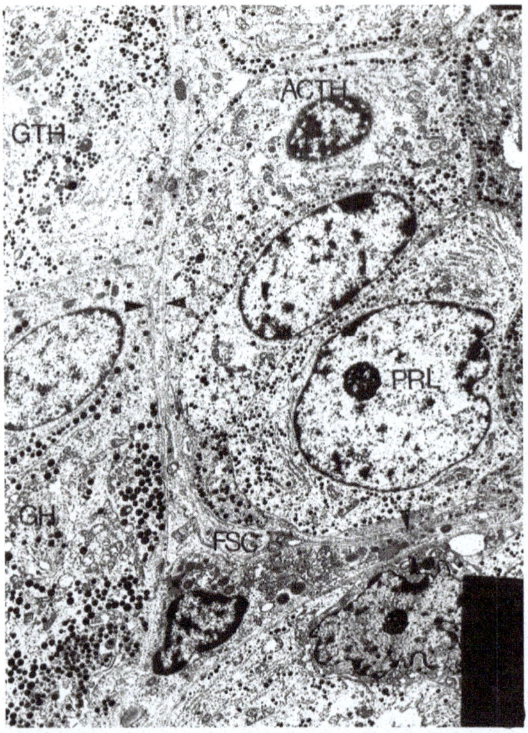

Fig. 1.123 A low-power electron micrograph of the anterior pituitary of the female *Funambulus pannanti* during early pregnancy shows an assortment of cells. The corticotroph (ACTH) is elongated and angular and has a small, condensed, and ovoid nucleus. The secretory granules are opposed to the cell membrane in a single row. Note the characteristic triradiate folliculostellate cell (FS) whose thin sheet-like cytoplasmic processes (paired arrow heads) pass between the granulated cells. The cytoplasm of this cell shows abundant lipid droplets and a few dense bodies [X 5000]

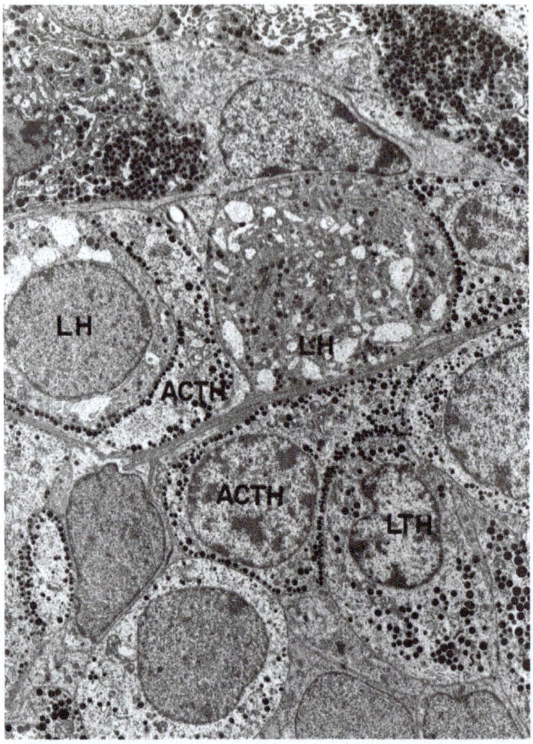

Fig. 1.124 Low-power electron micrograph of anterior pituitary of active female bonnet monkey. LH, ACTH and LTH cells can be easily seen [X 7000]

individuals of *Tylonycteris pachypus* and of the Indian fruit bat, *Rousettus leschenaulti*. In the corticotroph, the mitochondria are plenty probably in anticipation of a functional change.

Our studies demonstrate that the ACTH cells of the pituitary gland of males and females are irregular or pear shaped with elongated cytoplasmic processes. The RER and Golgi apparatus are poorly developed and mitochondria are large. The nucleus occupies a large portion of the cell. The secretory granules are typically arranged in a single line beneath the cell membrane and also distributed throughout the cytoplasm.

There are controversies regarding the shape of the ACTH cells. Kurosumi and Kobayashi (1966); Sipertien and Miller (1969, 1970); Bowie et al. (1972); Cinti et al. (1985) and Horvath and Kovacs (1988) in rats; Bhiwgade et al. (1989) in bats and Lawzewitsch et al. (1972) in human have reported polyhedral, elongated or stellate cells whereas Shirasawa et al. (1985) reported polygonal or oval ACTH cells in goat. Today, it is generally accepted that the normal corticotrophs in rat pituitary are elongated, angular or stellate cells (Costoff 1973; Moriarty 1973; Kurosumi and Fujita 1974; Girod 1976, 1977; Kurosumi 1986; and Horvath and Kovacs 1988). The arrangement of secretory

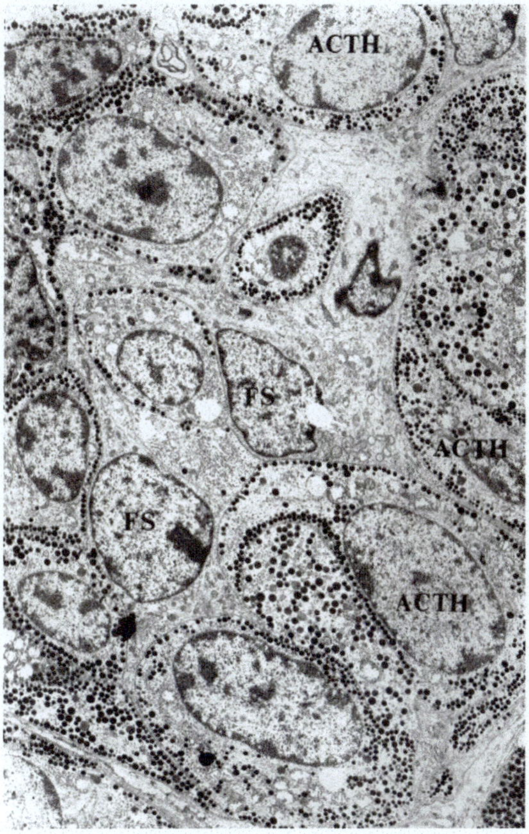

Fig. 1.125 Low-power electron micrograph of AP of male bonnet monkey showing a population of FS and ACTH cells. Note a typical pear shape of ACTH cells. The secretory granules are aligned to the margins of the cell [X 3500]. Unpublished electron micrograph from Dr. Bhiwgade and Dr. Thakur

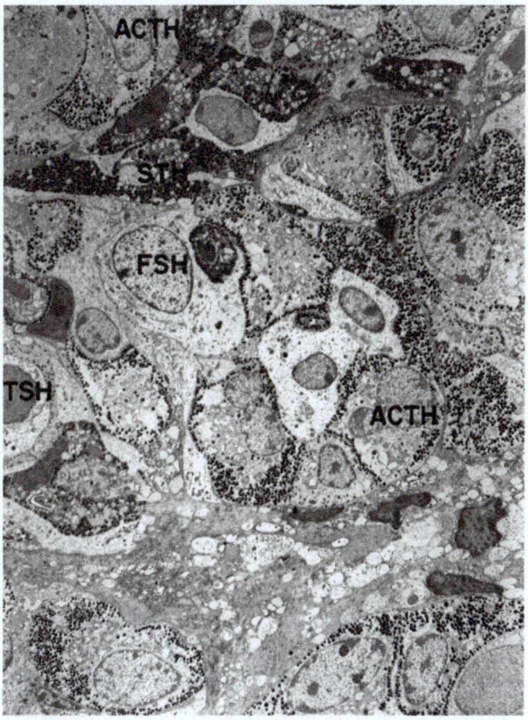

Fig. 1.126 Low-power electron micrograph of pars anterior of normal female bonnet monkey showing the distribution of STH, LTH, TSH, FSH, LH and ACTH [X 3500]. Unpublished electron micrograph from Dr. Bhiwgade and Dr. Mantri

granules as observed in our study is in conformity with the observations of Kurosumi and Kobayashi (1966), Bowie et al. (1972) in rats and Weman (1974) and Murphy and James (1976) in minks.

Our results after surgical adrenalectomy and treatment with metapyrone show the ACTH cells with slight degranulation, extensive enlargement of mitochondria with loss of cristae and presence of short lamellar cisternae of RER. A similar observation has been reported by Bhiwgade et al. (1989) after adrenalectomy and metapyrone treatments in bat, Rousettus. These changes are similar to those observed following adrenalectomy in the ACTH cells of other species (Herlant 1963; Siperstein and Allison 1965; Kurosumi and Kobayashi, 1966; Yamada and Yamashita 1967; Siperstein and Miller 1970; Murphy and James 1976; Horvath and Kovacs 1988) and after metapyrone treatments (Girod et al. 1964; Cameron and Foster 1972).

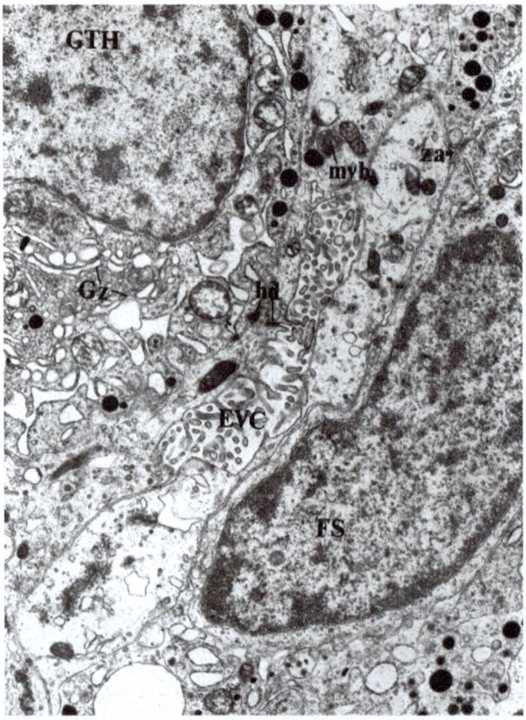

Fig. 1.127 Anterior pituitary of male rat treated with DMPA + TE for 120 days. Junctional complex in between FS cell and a gonadotroph (GTH) forming an extravascular channel (EVC). Hemidesmosomes (hd) and zona adherens junctions (Za) are also observed. Multivesicular bodies which are the remains of Golgi apparatus, forming lysosomes are also observed near the EVC. The gonadotroph contains a very well-developed Golgi zone (Gz) [X 8000]. Unpublished electron micrograph from Dr. Bhiwgade, Dr. Avari and Dr. Borkar

References

Amat P, Boya J (1973) Ultrastructural observations on the genesis and extrusion of secretory granules in the adenohypophysis. Acta Anat 86:44–52

Amenomori Y, Chen CL, Meites J (1979) Serum prolactin levels in rats during different Reproductive states. Endocrinology 86:506–510

Andrews WW, Ojeda SR (1977) On the feedback actions of estrogen of gonadotropin and prolactin release in infantile female rats. Endocrinology 101:1517–1523

Aono T, Miyake A, Shioii KT, Onishi T, Kurachi K (1976) Impaired LH release following exogenous estrogen administration in patients with amenorrheagalactorrhea syndrome. J Clin Endocrinol Metab 42:696–702

Arita J (1993) Analysis of the secretion from single anterior pituitary cells by cell immunoblot assay. Endocr J 40(1):1–15

Asa SL, Horvath E, Kovacs K, Ezrin C (1984) Cytology of the normal pituitary and pituitary tumours. Pituitary Tumours. Eds. D. Odell and D. H. Nelson., Futura Publishing Company Inc., Mount Kisco, New York, In, pp 71–143

Asa SL, Kovacs K, Horvath E, Losinski N, Laszlo F, Domokos I, Halliday W (1988) Human fetal adenohypophysis Electron microscopic and ultrastructural immunocytochemical analysis. Neuro Endocrinology 48:423–434

Attia MA, Zayed I (1989) Thirteen weeks subcutaneous treatment with high dose of natural sex hormones in rats with special reference to their effect on the pituitary gonadal axis. I Oestradiol. DTW Dtsch Tierarz H Wochenschr 9:438–445

Aumuller G, Wagner R, Graf KJ (1978) Fine structure of rat prolactin cells after treatment with a long acting depot contraceptive. Acta Endocrinol 89:251–262

Azzali G (1971) Cytologia adenoipotisaria dei Chirotteri con particolare riguardo alle cellule FSH, LH, ACTH e LTH. Ateneo Parmense Sezione I. Acta Biomed 42: 169–229

Bagatell CJ, Dahl KD, Bremner WJ (1994) The direct pituitary effect of testosterone to inhibit gonadotropin secretion in men is partially mediated by aromatization to estradiol. J Androl 15:15–21

Bage C, Ekengern B, Fernholm B, Fridberg G (1974) The pituitary gland of roach, Leuciscus rutilus. II. The proximal pars distalis and its innervations. Acta Zool (Stockholm) 55:191–204

Bailey P, Davidoff LM (1925) Concerning the microscopic structure of the hypophysis cerebri in acromegaly (Based on the tissues removed at the operations from thirty five patients). Am J Pathol 1:185–207

Baker BL, Eeskin TA, August LN (1973) Direct action of synthetic procrestins on the hypophysis. Endocrinology 92:965

Baker BI, Gross DS (1978) Cytology and distribution of secretory cell types in mouse hypophysis as demonstrated with immunocytochemistry. Am J Anat 153:193–215

Baker BI, Keshavanath P, Sundararaj BI (1974) The cell types in the adenohypopysis of the Indian catfish, Heteropneustes fossilis (Bloch). Cell Tissue Res 148: 347–358

Barnes BG (1962) Electron microscope studies on the secretory cytology of the mouse anterior pituitary. Endocrinology 71:618–628

Barnes BG (1961) Ciliated secretory cells in the pars distalis of the mouse hypophysis. J Ultrastruct Res 5: 453–467

Barnes BG (1963) Fine structure of the mouse adenohypophysis in various physiological states. In: Cytologic de l'adenohypophyse, Colloques international du centre de la recherche scientifique, vol 128, pp 91–110

Beauvillain JC, Mazzuca M, Dubois MP (1977) The prolactin and the growth hormone producing cells of the guinea pig pituitary. Cell Tissue Res 184:343–358

Beauvillian JC, Tramu G, Dubois MP (1975) Characterization by different techniques of adrenocorticotropin and gonadotropin producing cells in Larot pituitary (Eliomys quercinus). A superimposition technique and an immunocytochemical technique. Cell Tissue Res 158:301–317

Behringer RR, Mathews LS, Palmiter RD, Brinster RL (1988) Dwarf mice produced by genetic ablation of growth hormone expressing cells. Genes Dev 2:453–461

Bellare AY (1994) Effects of medroxyprogesterone acetate and testosterone enanthate on hormonal levels, clinical biochemistry, and histopathology in the male white rat. Ph.D. Thesis. University of Bombay

Bendayan M (1982) Double immunocytochemical labeling applying the protein A gold technique. J Histochem Cytochem 30:81–85

Bendayan M (1984) Protein A–Gold electron microscopic immunocytochemistry: methods, applications and limitations. J Electron Micros Tech 1:243–270

Benjamin M (1975) The adenohypophysis of the flounder, Pleuronectes flessus, and the minnow, Phoxinus phoxinus. Cell Tissue Res 157:391–409

Bergland RM, Torack RM (1969) An ultrastructural study of follicular cells in the human anterior pituitary. Am J Pathol 57:273–297

Bernini GP et al (1983) Effects of sulpiride induced hyperprolactinemia on testosterone secretion and metabolism before and after h c G in normal men. J Endocrinology Invest 6:287

Bhatavdekar JM, Shah NG, Patel DD, Karelia NH, Trivedi SN, Vora HH, Ghosh N, Giri DD, Balar DB (1992) Endocrine function in premenopausal and postmenopausal advanced breast cancer patients treated with CMF or tamoxifen. Neoplasma (Czechoslovakia) 39(2):123–127

Bhiwgade DA, Akolkar VV, Menon SN, Manekar AP, Senad DG (1989) Ultrastructural and functional characteristics of anterior pituitary cells in the Indian fruit bat, Rousettus leschenaulti (Desmarest). Acta Anat 135:129–141

Bhiwgade DA, Borkar MR, Avari KM (2001) Fine structure of prolactin cell of female albino rat as affected by some antifertility drugs. A comparative electron microscopic study I. J Expt Bio 39:119–131

Boockfor FR, Frawley LS (1987) Functional variations among prolactin cells from different pituitary regions. Endocrinology 120:874–879

Borkar MR (1998) Ultrastructural studies: identification of Anterior pituitary cells in some mammals. Ph.D. thesis. University of Mumbai

Bowie EP, Williams G, Shiino M, Renneles E (1972) The corticotroph of the rat adenohypophysis: a comparative study. Am J Anat 138:499–520

Brattin WJ, Portanova R (1977) Corticosterone induction of pituitary hyporesponsiveness to CRF: divergent effects of puromycin and cycloheximide. Mol Cell Endocrinol 7:221–231

Bugnon C, Fellmenn D, Lenys D, Bloch B (1977) Etude cyto-immunologique des cellules gonadotropes et des cellules thyreotropes de l'adenohypophyse du rat. C R Soc Biol:907–913

Cameron E, Foster CL (1972) Ultrastructural changes in the adenohypophysis of metapyrone treated rabbits. J Endocrinol 52:343–347

Cardell RR (1969) The ultrastructure of stellate cells in the pars distalis of the salamander pituitary gland. Am J Anat 126:429–456

Celio MR, Pasi A, Burgisser E, Buetti G, Hollt V, Gramsch C (1980) Proopiocortin fragments in the normal human adult pituitary distribution and ultrastructural characterization of immunoreactive cells. Acta Endocrinol 95:27–40

Chen CL, Meites J (1970) Effects of estrogen and progesterone on serum and pituitary prolactin levels in ovariectomized rats. Endocrinology 86:503–505

Childs GV, Ellison DG, Garner LL (1980) An immunocytochemists view on gonadotrophin storage in the adult male rat. Cytochemical and morphological heterogeneity in serially sectioned gonadotropes. Am J Anat 158:397–409

Childs GV, Ellison DG, Lorenzen JR, Collins TJ, Schwartz NB (1982) Immunocytochemical studies on gonadotropin storage in developing castration cells. Endocrinology 111:1318–1328

Childs GV, Ellison D, Foster L, Ramaley J (1981a) Postnatal maturation of gonadotrophs in the male rat pituitary. Endocrinology 109:1683–1692

Childs G, Yang H, Tobin R, Wilber JF, Kubek M (1981b) Effects of thyroidectomy, propylthiouracil, and thyroxine on the pituitary content and immunocytochemical staining of thyrotropin (TSH) and thyrotropin releasing hormone (TRH). J Histochem Cytochem 29(3): 537–363

Cinti S, Sbarbati A, Marelli M, Osculati F (1985) An ultrastructural morphometric analysis of the adenohypophysis of the lactating rats. Anat Rec 212:381–390

Clifton KH, Meyer RK (1956) Mechanism of anterior pituitary tumor induction by estrogen. Anat Rec 125: 65–81

Colonnier M (1964) On the nature of intranuclear rods. J Biophys Biochem Cytol 25:646–7653

Comhaire F, Dhont M (1975) Influence of current modes of treatment in male infertility on the hypothalamo-pituitary testicular function. In: Schellen T (ed) Releasing factors and gonadotropic hormones in male and female sterility. European Press Mediicon, Ghent, p 117132

Cook H, van Overbeeke AP (1972) Ultrastructure of pituitary gland (pars distalis) in sokeye salmon (Oncorhynchus nerka) during gonad maturation. Z. zellforsch Mikrosk. Antaeus 130:388–350

Costoff A (1973) Ultrastructure of rat adenohypophysis: correlation with function. Academic Press, New York, pp 1–207

Couch EF, Arimura A, Schaly AV, Saito M, Swano S (1969) Electron microscope studies of somatotrophs of rat pituitary after injection of purified growth hormone releasing factor (GRF). Endocrinology 85:1084–1091

Cramer O, Parker C, Porter J (1979) Estrogen inhibition of dopamine release into hypophysial portal blood. Endocrinology 104:419–422

Crosby WM, Block MF (1980) Physiology of pregnancy–obstetrical practice, edited by Salvio Aladjem, Publisher C. V. Mosby & Company

Dacheux F (1980) Ultrastructural immunocytochemical localization of prolactin and growth hormone in the porcine pituitary. Cell Tissue Res 207:277–286

Dacheux F, Dubois MP (1976) Ultrastructural localisation of prolactin, growth hormone and luteinising hormone by immunocytochemical techniques in the bovine pituitary. Cell Tissue Res 174:245–260

Dacheux F (1978) Ultrastructural localization of gonadotropin releasing hormone in the porcine pituitary using the immunoperoxidase technique. Cell Tissue Res 158: 301–317

Dacheux F (1981) Proportions of FSH/LH cells, LH cells and FSH cells in the porcine anterior pituitary. IRCS Med Sci 9:952–953

D'angelo SA (1953) Cytophysiologic aspects of thyrotrophic hormone secretion in the goitrous guinea pig. Endocrinology 52:331–337

D'angelo SA (1966) A comparative study of TSH and FSH secretion in rat and guinea pig: effects of gonadectomy and goiterogens. Endocrinology 78:1230–1237

D'angelo SA (1967) Pituitary thyroid inter-relations in maternal fetal neonatal guinea pig. Endocrinology 81: 132–138

Davis GJ, Meyer RK (1973) FSH and LH of the pituitary gland of gonadectomized snow shoe hares. Endocrinology 92(1):340–344

Dawson AB (1937) The relationship of the epithelial components of the pituitary gland of the rabbit and cat. Anat Rec 69:471–485

Dekker A (1967) Pituitary basophils of the Syrian hamster: an electron microscopic investigation. Anat Rec 158: 351–368

Dingemans KP, Feltkamp CA (1972) Nongranulated cells in the mouse adenohypophysis. Z Zellforsch 124:387–405

Dingemans KP (1969) The relation between cilia and mitosis in the mouse adenohypophysis. J Cell Biol 43:361–365

Dixit VP (1970) Neurosecretion and feedback mechanism in Clarias batrachus Linn. Ovariectomy and administration of exogenous sex hormones. Cellule 68:213–221

Dixon HBF (1964) In: Thimman KV, Astwood EB (eds) The hormones, vol 5. New York, Acad Press, p 54

Dollery C (1991) Therapeutic Drugs. Churchill Livingstone

Donoso AO, Banzan AM, Barcaglioni JC (1974) Further evidence on the direct action of L-dopa on prolactin release. Neuroendocrinology 15:236–239

Echave-Llanos JM, Gomez-Dumm CLA, Nessi AC (1971) Ultrastructure of STH cells of the pars distalis of the hypophysis of hepatectomized mice. Z Zellforsch 113: 29–38

Ehora Y, Siler T, Yen S (1976) Effects of large doses of estrogen on prolactin and growth hormone release. Am J Obstet Gynecol 125:455–458

Ekengren B, Lindahl K, Fridberg G (1978a) Immunocytology and innervations of gonadotropic cell in teleost fish Rutilus rutilus. Acta Zool (Stockholm) 59:125–133

Ekengren B, Peute J, Fridberg G (1978b) Gonadotropic cells in the Atlantic salmon, Salmo salar: an experimental immunocytological–electron microscopical study. Cell Tissue Res 191:187–203

El Etreby MF, Fath El Bab MR (1977) Effect of cyproterone acetate on cells of the pars distalis of the adenohypophysis in the beagle bitch. Cell Tissue Res 183:177–189

El Etreby MF, Gunzel P (1974) Sex hormones–Effects on prolactin cells in rat, dog, monkey and man. Acta Endocrinon Suppl:3–15

El Etreby MF, Richter KD, Gunzel P (1972) Histological and histochemical differentiation of the glandular cells of the pituitary in various experimental animals. Toxicology 14:270–281

Ezrin C, Swanson HE, Humpherey JC, Dawson JW, Wilson WD (1958) Delta cell in acute and chronic illness. J Clin Endocrinol Metab 18:917

Farquhar MG, Rinehart JF (1954a) Electron microscopic studies of the anterior pituitary gland of castrate rats. Endocrinology 54:516–541

Farquhar MG, Rinehart JF (1954b) Cytologic alterations in the anterior pituitary gland following thyroidectomy: an electron microscope study. Endocrinology 55:857–876

Farquhar MG, Wellings SR (1957) Electron microscopic evidence suggesting secretory granule formation within Golgi apparatus. J Biophys Biochem Cytol 3: 319–322

Farquhar MG (1957) "Corticotrophs" of the rat adenohypophysis as revealed by electron microscopy. Anat Rec 127:291

Farquhar MG (1971) Processing of secretory products by cells of the anterior pituitary gland. Mem Soc Endocr 19:79–124

Farquhar M, Reid JA, Daniell LW (1978) Intracellular transport and packaging of prolactin: a quantitative electron microscope autoradiographic study of mammotrophs dissociated from rat pituitaries. Endocrinology 102:296–311

Fawcett DW (1981) The cell Philadelphia and. W. B. Saunders, London

Fellmann D, Bresson JL, Clavequin MC, Bugnon C (1982) Quantitative immunocytochemical studies on the gonadotrophs isolated from the pituitary of the male rat. Cell Tissue Res 224:137–144

Fernandez-Moran H, Luft R (1949) Submicroscopic cytoplasmic granules in the anterior lobe cells of rat hypophysis as revealed by electron microscopy. Acta Endocrinol 2:199–121

Foncin JF, LeBeau J (1966) Cellules de castration et cellules FSH dans l'hypophyse humaine vues au microscope electronique. J Microsc Paris 5:523–526

Fontaine M, Olivereau M (1975) Some aspects of the organization and evolution of the vertebrate pituitary. Am Zool 15(Suppl. 1):61–79

Frantz AG (1978) Prolactin physiology in medicine. N Engl J Med 298:201

Fraser IS, Susan H (1983) Depot medroxyprogesterone acetate–Long acting steroid. Contraception

Frawley LS, Boockfor FR, Hoeffler JP (1985) Identification by plaque assays of a pituitary cell type that secretes both growth hormone and prolactin. Endocrinology 116:734–737

Gale TF (1972) An electron microscopic study of the dog adenohypophysis. Z Anat Entwickl Gesch 137:188–199

Gaspard UJ, Romus MA, Gillain D, Duvivier J, Demey-Ponsart E, Franchimont P (1983) Plasma hormone levels in women receiving oral contraceptives containing ethinyloestradiol plus levonorgestrel or desogestrel. Contraception 27:577–590

Gemelli A (1907) Les processus de la secretion de l'hypophyse des mammiferes. Arch Ital Biol 47:185

Gibson DFC, Gottardis MM, Jordan VC (1990) Sensitivity and insensitivity of breast cancer to tamoxifen. J Steroid Biochem Mol Biol 37(6):765–770

Giguere V, Lefebvre F-A, Labrie F (1981) Androgens decrease LHRH binding sites in rats anterior pituitary cells in culture. Endocrinology 108:l

Girod C (1977) Apport de l' immunohistochimie a l'etude cytologique de l' adenohypophyse. Bull Assoc Anat 62:417–603

Girod C, Dubois P, Cure M (1964) Identification experimentale des cellules corticotropes antehypophysaires chez le hamster dore (mesocricetus auratus waterh). Observations all microscope optique et au microscope electronique. C.R. Acad sci Paris 259:1229–1232

Girod C (1976) Etat actuel des connaissances sur la description morphologique et la signification fonctionelle des cellules antehypophysaires. Lyon Med 236:323–357

Girod C, Dubois MP, Trouillas J (1980) Mise en evidence de cellules gonadotrops de l'adenohypophyse du singe Macacus irus. Etude en immunofluorescence a l'aide d'anticorps anti-B FSH humain et anti-B LH ovine. C.R. Soc Biol 174:303–313

Goldberg RC, Chaikoff IL (1951) On the nature of the hypertrophied pituitary gland induced in the mouse by I131 injection and the mechanism of its development. Endocrinology 48:1–5

Goluboff and Ezrin (1969) J Clin Endocrinol:29

Gomez-Dumm CLA, Echave-Llanos JM (1972) Further studies on the ultrastructure of the pars distalis of the male mouse hypophysis. Acta Anat 82:254–266

Gordon G, Southern LA, Tochimoto S, Olivo J, Altman K, Rand J, Lemberger L (1970) Effect of medroxyprogesterone acetate (Provera) on the metabolism of biological activity of testosterone. J Clin Endocrinol 30:499

Grandison L, Hodson C, Cnen HT, Advis J, Simpking J, Meites J (1977) Inhibition by prolactin of postcastation rise in LH. Neuroendocrinology 23:312

Gudelsky GA, Dale D, John N, Porter C (1981) Role of estrogen in dopaminergic control of prolactin secretion. Endocrinology 108:440

Hachmeister U, Fahlbusch R, Werder KV (1972) Ultrastructural identity of pituitary adenoma cells in Forbes-Albright syndrome and of adenohypophyseal pregnancy cells. Acta endo (Kbh) Supp 159(42)

Hafiez AA, Lloyd CW, Bartle A (1972) The role of prolactin in stimulation of androgen production by the testis of hypophysectomized rats. J Endocrinol 52: 327–832

Halmi NS (1950) Two types of basophils in the anterior pituitary of the rat and their respective cytophysiological significance. Endocrinology 47: 289–299

Harper MJK, Walpole AL (1967) A new derivative of triphenylethylene: effect on implantation and mode of action in rats. J Reprod Fertil 13:101–119

Hellbaum AA, Mcarthur LG, Campbell PJ, Finerty JC (1961) The physiological fractionation of pituitary gonadotrophic factors correlated with cytological changes. Endocrinology 68:144–153

Herbert DC, Cisneros PL, Rennels EG (1977) Morphological changes in prolactin cells of male rats after testosterone administration. Endocrinology 100:487–495

Herbert DC (1975) Localization of antisera to LH-B and FSH-B in the rat pituitary gland. Am J Anat 144:378–385

Herlant M (1963) Apport de la microscopie electronique a l'etude du lobe anterieur de l'hypophyse. In: Benoit J, Dalage C (eds) Cytologie de l'adenohypophyse. pp. 73-90. Editions du Centre National de la Recherche Scientifique, Paris

Herlant M (1964a) The cells of the adenohypophysis and their functional significance. Int Rev Cytol 17:299–382

Herlant M, Pasteels JL (1967) Histology of human anterior pituitary. In: Bajusz E, Jasmin G (eds) Methods and achievements in experimental pathology, vol. III, p. 250. Karger, Basel

Herlant M, Pasteels JL (1971) Les cellules productrices de prolactine, la regulation de leur secretion et leur role dans les adenomas secretant la prolactine. In: Les adenomes hypophysaires secretants, endocrinopathies et immunologic, p. 21. Masson et cie, Paris

Herlant M, Racadot S (1957) Lc lobe anterieur del'hypophyse de la chatte au cours de la gestation et al lactation. Arch Biol 58:217–248

Herlant M (1949) Study of pituitary body with the PAS reaction. Nature (London) 164:703–704

Herlant M (1950) Methodes nouvelles applicables a l'etude histologique du lobe anterieur de l'hypophyses. Ann Dent Endocr 6:644–657

Hoeffler JP, Bookfor FR, Frawley LS (1985) Ontogeny of prolactin cells in neonatal rats: initial prolactin secretor also release growth hormone. Endocrinology 117:187–195

Horvath E, Kovacs K (1988) Fine structural cytology of the adenohypophysis in rat and man. J Elect Micro Tech 8:401–432

Horvath E, Kovacs K (1974) Misplaced exocytosis: distinct ultrastructural feature in some pituitary adenomas. Arch Pathol 99:221–224

Horvath E, Kovacs K (1992) Ultrastructural diagnosis of human pituitary adenomas. Microsc Res Tech 20:107–135

Horvath E, Kovacs K, Penz G, Ezrin C (1974) Origin, possible function and fate of "Follicular cells" in the anterior lobe of human pituitary. An electron microscopic study. Am J Pathol 77:199–212

Horvath E, Kovacs K, Ezrin C (1977) Junctional contact between lactotrophs and gonadotrophs in the rat pituitary. IRCS Med Sci 5:511

Hsu SM, Raine L, Fanger H (1981) The use of antiavidin antibody and avidin-biotin-peroxidase complex in immunoperoxidase techniques. Am J Clin Pathol 75: 816–821

Hunzicker Dunn M, Birnbaumer L (1975) Physiologic aspects of adenylate cyclase in follicles and corpora lutea of rats. In: Program of the 57th annual Meetings on the Endocrine Society, vol P. 105, Abstr. No. 110, New York

Hymer WC, McShan WH, Christiansen RG (1961) Electron microscopic studies of the anterior pituitary glands from lactating and estrogen treated rats. Endocrinology 69:81–90

Inoue K, Kurosumi K (1981) Mode of proliferation of gonadotrophic cells of the anterior pituitary after castration. Immunocytochemical and autoradiographic studies. Arch Histol Jpn 44:71–85

Inoue K, Kurosumi K (1984) Ultrastructural immunocytochemical localization of LH and FSH in the pituitary of the untreated male rat. Cell Tissue Res 235:77–83

Ishibashi T, Shiino M (1989a) Co-localization pattern of growth hormone (GH) and prolactin (PRL) in the anterior pituitary cells of the female rat and female musk shrew. Anat Rec 223:185–193

Ishibashi T, Shiino M (1989b) Subcellular localisation of prolactin in the anterior pituitary cells of the female Japanese house bat, Pipistrellus abranius. Endocrinology 124:r056

Jeppsson S, Gershogen S, Elof DB, Johansson and Ranneik G. (1982) Plasma levels of medroxyprogesterone acetate (MPA), hormone binding globulin gonadal steroid, gonadotropins and prolactin in women during long term use of depo-mpa (depoprovera) as contraceptive agent. Acta Endocrinol 1982-99:339–343

Jernstrom, H., Knutsson, M., Taskila, P. & Olsson, H. (1992). Plasma prolactin in relation to menstrual cycle phase, oral contraceptive use, arousal time and smoking habits.

Jordan VC, Dowse LJ (1976) Tamoxifen as an antitumour. agent: effect on estrogen binding. J Endocrinol 68: 297–303

Jordan VC, Koerner S (1976) Tamoxifen as an antitumour agent: role of oestradiol and prolactin. J Endocrinol 68: 305–311

Jordan VC (1995) Tamoxifen: toxicities and drug resistance during treatment and prevention of breast cancer. Annu Rev Pharmacol Toxicol 35:195–211

Kagayama M (1965) The follicular cell in the pars distalis of the dog pituitary gland. An electron microscopic study. Endocrinology 77:1053–1060

Karim M, El Mahgoub S (1970) Injectable steroids as a method of contraception. Am Shams Med J 21(5):543

Kaul S, Vollrath L (1974) The goldfish pituitary I. Cytology Cell tissue Res 154:201–230

Kiguchi Y (1978) The process of development of thyroidectomy cells from the so called thyrotrophs. Endocrinol Japon 25:75–86

Kineman RD, Faought WJ, Frawley LS (1991) Mammosomatotropes are abundant in bovine pituitaries: influence of gonadal status. Endocrinology 128:2220–2233

Kobayashi Y (1966) The effect of hibernation upon the fine structure of the bat adenohypophysis. Fed Proc Fed Am Socs Exp Biol 25:536

Kobayashi Y, Herman L (1966) Effect of hibernation upon the fine structure of the bat adenohypophysis. Proc. 6th Int. Congr Electron Microscopy, Koyoto, 2:537-538.

Koch UJ, Lorenz F, Danehl K, Ericsson R, Hasan SH, Keyserlingk DV, Lubke K, Mehring M, Rommler A, Schwartz U, Hammerstein J (1976) Continuous oral low-dosage cyproterone acetate for fertility regulation in male. A trend analysis in 15 volunteers. Contracep 14:2

Kofler R (1982) Immunofluorescence localization of glycoprotein hormones in the rat anterior pituitary gland using monoclonal antibodies and polyclonal antisera. J Histochem Cytochem 30:645–649

Kordon EC, Molinolo AA, Pasqualini CD, Charreau EH, Pazas P, Lanari C (1993) Progesterone induction of mammary carcinomas in BALB/C female mice. Correlation between progestin dependence and morphology. Breast Cancer Res Treat (Netherlands) 28(1):29–39

Kovacs K, Horvath E (1975) Gonadotrophs following removal of ovaries–fine structural study of human pituitary glands. Endocrinologie 66:1–8

Kovacs K, Horvath E (1986) Tumors of the pituitary gland. Atlas of tumor pathology. Second series fasicle, vol 21. Armed Forces Institute of Pathology, Washington, pp 1–269

Kovacs K, Horvath E, Ryan N (1981) Immunocytology of the human pituitary. In: DeLellis RA (ed) Diagnostics

Immunohistochemistry. Masson Publ, New York, pp 17–35

Kovacs K, Horvath E, Higgins HP, Hudson AR, Gezina I, Ryan N (1977) Case histories in human medicine:15

Kueosumi K, Kobayashi Y (1966) Corticotrophs in the anterior pituitary glands of normal and adrenalectomized rats as revealed by electron microscopy. Endocrinology 78:745–758

Kurosumi K (1986) Cell classification of the rat anterior pituitary by means of immunoelectron microscopy. J Clin Elec micro 19:299–319

Kurosumi K, Oota Y (1968) Electron microscopy of two types of gonadotrophs in the anterior pituitary glands of persistent estrous and diestrous rats. Z Zellforsch 85: 34–46

Kurosumi K, Fujita H (1974) Functional morphology of endocrine glands. An atlas of electron micrographs. Igaku, Shoin Ltd., Tokyo, pp 1–100

Kurosumi K (1991) Ultrastructural immunocytochemistry of the adenohypophysis in rat: a review. J Electron Micros Tech 19:42–56

Kurosumi K, Kobayashi Y, Watanabe A (1963) Light and electron microscope studies on the anterior pituitary (ubergangsteil of Stendell) of carp (Cyprinus carpio). Archives of Histology, Japan 23:489–515

Lagios MD (1965) Seasonal changes in the cytology of the adenohypophysis, testes and ovaries of the black surfperch Embiotoca lacksoni, a viviparous percomorph fish. Gen Comp Endocrinol 5:207–221

Lam TJ, Pandey S, Nagahama T, Hoar WS (1976) Effect of synthetic luteinizing hormone–releasing hormone (LH–RH) on ovulation and pituitary cytology of the goldfish Carassiue auratus. Can J Zool 54:816–824

Lawzewitsch IV, Dickmann GH, Amezua L, Pardal C (1972) Cytological and ultrastructural characterization of the human pituitary. Acta Anat 81:286–316

Leong DA, Lau SK, Sinha YN, Kaiser DL, Thorner MO (1985) Enumeration of lactotropes and somatotropes among male and female pituitary cells in culture: evidence in favour of mammosomatotrope subpopulation in the rat. Endocrinology 116:1371–1378

Leray C (1966) Apport de la method de precipitation selective par lacide trichloacetique dans letude histophyiologique de lhypophyse dun Teleosteen: Mugil cephalus L. Compt Rend Soc Biol 160:582–585

Leray C, Carlon N (1963) Sur la presence dune dualite parmi les cellules cyaophiles de ladenohypophyse de Mugil cephalus L. (Teleosteen Mugilidae). Comp Rend Soc Biol 157:572–575

Lever JD, Peterson RR (1960) Cellular identities in the pars distalis of the rat pituitary. Trans N Y Acad Sci 22: 504–508

Lloyd HM, Meares JD, Jacobi J (1973) Early effects of stilbestrol on the growth hormone and prolactin secretion and on the pituitary mitotic activity in the male rat. J Endocrinol 58:227–231

Lloyd RV (1983) Estrogen induced hyperplasia and neoplasia in the rat anterior pituitary gland. An immunohistochemical study. Am J Pathol 113:198–206

Lloyd RV (1991) Ultrastructure of spontaneous and transplanted pituitary tumours in laboratory animals. J Electron Micro Tech 19:64–79

Lobl TJ, Musto NA, Gunsalus GL, Bardine WC (1983) Medroxyprogesterone acetate has opposite effects on the androgen binding protein concentrations in serum and epididymis. Biol Reprod 29:697–712

Luciano AA, Sherman BM, Chapler FK, Hauser KS, Wallace RB (1985) Hyperprolactinemia and contraception: a prospective study. Obstet Gynecol 56:506–510

MacLeod RM (1976) Regulation of prolactin secretion. In: Martini L, Ganong WF (eds) Frontiers in neuroendocrinology, vol 15. Raven Press, New York, pp 236–239

MacLeod RM, Abad A, Edison LL (1969) In vivo effects of sex hormones on the in vitro synthesis of prolactin and growth hormone in normal and pituitary tumor bearing rats. Endocrinology 84:1475–1483

Magrini G, Ebiner JR, Burckhardt P, Felber JP (1976) Study on the relationship between plasma prolactin levels and androgen metabolism in man. J Clin Endocrinol Metab 43:944

Mantri UU (1994) Functional classification of cell types of the anterior pituitary gland accomplished by electron microscopy. Ph. D. Thesis.,. University of Bombay

Mantri UU, Borkar MR, Bhiwgade DA, Muni M (1995) Fine structure of lactotrophs during late pregnancy in some Indian bats. Bioved 6(1):85–90

Marchetti B, Labrie F (1982) Prolactin inhibits pituitary luteinizing hormone releasing hormone receptors in the rat. Endocrinology III:1209

Matsumoto A (1990) Effects of testosterone administration in normal men: safety and efficacy of high dosage testosterone and parallel dose–Dependent suppression of luteinizing hormone fo1lic1e-stimulating hormone and sperm production. J Clin Endocrinol Metab 70:280

Mattheij JAM (1970) The gonadotropic cells in the adenohypophysis of the Mexican cave fish, Anoptichthys jordani. Z Zelloforsch Mikrosk Ant 105:91–106

Maurer RA (1982) Relationship between estradiol, ergocryptine and thyroid hormone: effects on prolactin synthesis and prolactin messenger ribonucleic acid levels. Endocrinology 110:1515

McComb DJ, Kovacs K (1978) Ultrastructural morphometry of sparsely granulated prolactin cell adenomas of the human pituitary. Acta Endocrinol 89:521–529

McComb DJ, Kovacs K, Beri J, Zak F (1984) Pituitary adenomas in old Sprague-Dawley rats: a histologic, ultrastructural and immunocytochemical study. J Natl Cancer Inst 73:1143–1166

McComb DJ, Ryan N, Horvath E, Kovacs K, Nagy E, Bercz I, Domokos I, Laszlo FA (1981) Five different adenomas derived from the rat adenohypophysis Immunocytochemical and ultrastructural study. J Natl Cancer Inst 66:1103–1111

McKeown BA, Leatherland JF (1973) Fine structure of adenohypophysis in immature Sockeye salmon, *Oncorhynchns nerku. Z. Zellforsch* 140:459–471

McNeilly AS, Sharpe RM, Davidson DW, Fraser HM (1978) Inhibition of gonadotropin secretion by induced hyperprolactinaemia in the male rat. J Endocrinol 79: 59–68

Meites J, Lu KH, Wuttke W, Welsch CW, Nagasawa H, Quadri SK (1972) Res Prog Horm res 128:471

Merchant-Larois H, Mena F (1982) Evidence that extracellular glycoconjugates interact with prolactin granules during exocytosis in lactating rat adenohypophysis. J Ultrastruct Res 80:53–61

Mikami SL (1970) Light and electron microscopic investigation of six types of glandular cells of the bovine adenohypophysis. Z Sellforsch Mikrosk Anat 105: 457–482

Misdorp W, Romijin A, Hart AA (1992) Over de betekenis van ovariectomie en progestativa voor het onstaan van het mammacarinoom by de kat. Tijdschr. Diergeneeskd (Netherlands) 117(l):2–4

Mishell DR, Kletzky OA, Brener PF, Roy S, Nicoloff J (1977) The effect of contraceptive steroids on hypothalamic pituitary axis. Am J Osbtet Gynecol 128:60–69

Miyake A, Terakawa N, Tasaka K, Shimizu I, Ohtsuka S, Lee J, Aono T (1985) Prolactin inhibits oestradiol induced luteinizing hormone release at the pituitary level. Acta Endocrinol 109:204–207

Moriarty GC (1973) Adenohypophysis: ultrastructural cytochemistry. A review. J Histochem Cytochem 21: 855–892

Moriarty GC (1975) Electron microscopic immunocytochemical studies of rat pituitary gonadotrophs: a sex difference in morphology and cytochemistry of LH cells. Endocrinology 97:1215–1225

Moriarty GC (1976) Immunocytochemistry of the pituitary glycoprotein hormones. J Histochem Cytochem 24: 846–863

Morishge WK, Pepe GJ, Rotchild I (1973) Serum luteinizing hormone, prolactin and progesterone levels during pregnancy in the rat. Endocrinology 92:1527–1530

Mukku VR, Murthy GSRC, Srinath BR, Ramasharma K, Kotagi SG, Moudgal NR (1981) Regulation of testosterone rhythmicity by gonadotropins In bonnet monkey (macaca radiata). Biol Reprod 24:814–819

Munger BL (1958) A light and electron microscopic study of cellular differentiation in the pancreatic islets of the mouse. Am J Anat 103:275

Muralidhar K, Manekjee R, Moudgal NR (1977) Inhibition of in vivo pituitary release of luteinizing hormone in lactating rats by exogenous prolactin. Endocrinology 100:1137–1142

Murphy BD, James DA (1976) Cells of the adenohypophysis of the mink (Mustella vison) identified by immunohistochemical and functional criteria. Acta Anat 94: 184–203

Murr SM, Eric BG, Geschwind II (1974) Plasma luteinizing hormone and prolacting during pregnancy in mouse. Endocrinology 94:112

Muylder XD, Neven P (1993) Tamoxifen and potential adverse effects. The Cancer J 6(3):111–115

Nadler NJ, Young BA, Leblond CP, Mitmaker B (1964) Elaboration of thyroglobulin in the thyroid follicle. Endocrinology 74:333–354

Nakane PK (1970) Classification of anterior pituitary cell types with immunoenzyme histochemistry. J Histochem Cytochem 18:9–20

Nakane PK (1975) Identification of anterior pituitary cells by immuno electron microscopy. In: Tixier-Vidal A, Farquhar MG (eds) The anterior pituitary. Academic Press, New York

Neumann F, Schenck B (1976) New anti-androgens and their mode of actions. J Reprod Fertil 24:129–145

Neumann F, Diallo FA, Hasan SH, Schenck B, Traore J (1976) The influence of pharmaceutical compounds on male fertility. Andrologia 8(3):203–235

Neumann F, Von Berswordt-Wallrabe R, Elger W, Steinbeck H, Hahn JD, Knamer M (1970) Aspects of androgen-dependant events as studied by antiandrogens. Recent Prog Horm Res 26:337–410

Nikitowich-Winer MB, Atkin J, Maley BE (1987) Colocalization of prolactin and growth hormone within specific adenohypophysial cells in male, female and lactating female rats. Endocrinology 121:625–630

Nogami H, Yoshimura F (1980) Prolactin immunoreactivity of acidophil of the small Granule type. Cell Tissue Res 211:1–4

Nogami H (1984) Fine structural heterogeneity and morphologic changes in rat pituitary prolactin cells after estrogen and testosterone treatment. Cell Tissue Res 237:195–202

Novikoff AB (1964) GERL, its form and function in neurons of rat spinal ganglia. Biol Bull 127:358

Ojeda SR, McCann SM (1974) Development of dopaminergic and estrogenic control of rat prolactin release in the female rat. Endocrinology 95:1499

Olivereau M (1961) Maturation sexuelle de lAnuille male en eau douce. Compt Rend 252:3660–3662

Olivereau M (1967) Observations sur lhypophyse de lAnguilla female en particular lors de la maturation sexuelle. Z F Zellforsch U mikr Anat 80:286–306

Olivereau M, Herlant M (1960) Etude de lhypophyse de lAnguille male au cours de la reproduction. Compt Rend Soc Biol 154:706

Olivereau M, Herlant M (1954) Etude histologique de lhypophyse de Caecobarbus geertsii. Blgr Acad Roy Belg, Cl Sci 40:50–57

Osamura RY, Watanabe K (1986) In vivo and in vitro studies on the effects of bromocyptine and colchicine in estrogen induced rat prolactin secreting tumours. In: Yoshimura F, Gorbman A (eds) The pars distalis of the pituitary gland. Excerpta Medica, Tokyo, pp 413–425

Osamura RY, Kamatsu N, Izumi S (1962) Ultrastructural localization of prolactin in the rat anterior pituitary glands by preembedding peroxidase labelled antibody method: observations in normal, castrated or estrogen stimulated specimen. J Histochem Cytochem 30:919–925

Ozawa H (1991) Changing ultrastructure of the thyrotrophs in the rat pituitary after thyroidectomy as studied by immuno electron microscopy and enzyme cytochemistry. Cell Tissue Res 263:405–412

Oztan N (1966) The fine structure of adenohypophysis of Zoarces viviparous L. Z Zellforsch Mikroskop Anat 69:699–718

Paget GE, Eccleston E (1960) Simultaneous demonstration of the thyrotroph, gonadotroph and acidophil cells in the anterior hypophysis. Stain Technol 35:119

Paiz C, Hennigar GR (1970) Electron microscopy and histochemical correlation of human anterior pituitary cells. Am J Pathol 59:43–73

Pantic V, Genbaceu O (1969) Ultrastructure of pituitary lactotropic cells of oestrogen treated male rats. Z Zellforsch 95:280–289

Pantic V, Genbaceu O (1972) Z Zellforsch 126:41

Parlow AF, Anderson LL, Melampy Robert M (1964) Pituitary follicle–Stimulating hormone and luteinizing hormone concentration in relation to reproductive stages of the pig. Endocrinology 75:365–376

Parry DM, McMillen IC, Willcox DL (1978) Immunocytochemical localisation of prolactin and growth hormone in the ovine pituitary. A morphological and quantitative study. Cell Tissue Res 190:203–240

Parry DM, McMillen IC, Robinson JS, Thorburn GD (1979) Immunocytochemical localisation of prolactin and growth hormone in the perinatal sheep pituitary. A morphological and quantitative study. Cell Tissue Res 197:501–514

Parson JA, Erlandsen SL (1974) Ultrastructural immunocytochemical localisation of prolactin in rat anterior pituitary by use of the unlabelled antibody enzyme method. J Histochem Cytochem 22:340–352

Pasteels JL (1971) Morphology of prolactin secretion–ciba foundation Symposium on lactogenic hormones, p. 241

Pasteels JL, Gausset P, Danguy A, Ectors F (1972) Prolactin and carcinogenesis. In: Boyns AR, Griffiths K (eds) Immunofluorescent studies on prolactin and pituitary. Alpha Omega Alpha Publishing, Cardiff, Wales U. K

Pasteels JL (1963a) Recherhes morphologiques et experimentales sur la secretion de prolactine. Arch Biol Liege 74:439–453

Pasteels JL (1963b) Etude experimentale des cellules responsables de la secretion de prolactine chez les mammiferes. In: Cytology of Adenohypophysis., Eds. J. Benoit and C DaLage Colloq. Internat Centre Nat Rech Sci, 128: 137-148.

Pasteels JL, Eotors F, Danguy A, Robyn C, L'Hermite M, Dujardin . (1973) Histological immunofluorescent and electron microscopic identification of prolactin producing cells in the human pituitary. IV til International Congress of Endocrinology, Excerpta Medica, International congress, Series no. 273, pp. 616-621.

Paterson AHG, Geggie PHS (1993) Can tamoxifen prevent breast cancer? Can Med Assoc J 148(2):141–144

Pearse AGE (1948) Cytochemistry of the gonadotropic hormones. Nature 162:651

Pearse AGE (1949) The cytochemical demonstration of the gonadotropic hormone in the human anterior hypophysis. J Path Bact 61:195–202

Pelletier G, Leclerc R, Labrie F (1976) Identification of gonadotrophic cells in the human pituitary by immunoperoxidase technique. Mol Cell Endocrinol 6: 123–128

Pelletier G, Robert F, Hardy J (1978) Identification of human anterior pituitary cells by immunoelectron microscopy. J Clin Endocrinol Metab 46:534–542

Pellion FJ, Viola-Porcile E, Olivier L, Racodot J (1970) Ann Endocr (Paris) 31:259

Peters BH, Halmi NS (1961) Changes in the thyroidectomy cells of the rat pituitary under the influence of large doses of triiodothyronine; Endocrinology 68 (5):844–849

Peute J, Goos HJT, de Bruyn MGA, van Oordt PGWJ (1978) Gonadotropic cells of the rainbow trout pituitary during the annual cycle. Ultrastructure and hormone content. Ann Biol Bioch Biophys 18:905–910

Peute J, Zandbergen M, Goos HJT, de Leeuw R, Pinkas R, Viveen WJAR, van Oordt GPGWJ (1986) Pituitary gonadotrophin contents and ultrastructure of the gonadotrophs in the African catfish Clarias gariepinus during the annual cycle in a natural habitat. Can J Zool 64:1718–1726

Phifer RF, Midgley AR, Spicer SS (1973) Immunohistologic and histologic evidence that follicle stimulating hormone and luteinizing hormone are present in the same cell type in the human pars distalis. J Clin Endocrinol Metab 36:125–141

Pinilla L, Trimino E, Garnelo P, Bellido AR, Gaytan F, Aguilar E (1993) Changes in pituitary secretion during the early postnatal period and anovulationary syndrome induced by neonatal estrogen or androgen in rats. J Repro Fertil 97:13–20

Pontiroli et al (1980) Interaction of dopaminergic and antiserotoninergic drugs in the control of prolactin and LH release in normal women. Acta Endocrinol 93:271

Poole MC, Kornegay WD (1982) Cellular distribution within the rat adenohypophysis. A morphometric study. Anat Rec 204:45–53

Porter TE, Hill JB, Wiles CD, Frawley LS (1990) Is the mammosomatotrope a transitional cell for the functional interconversion of the growth hormone and prolactin secreting cells? Suggestive evidence from virgin, gestating and lactating rats. Endocrinology 127:2789–2794

Prasada Rao PD (1972) Identification of gonadotrophin– and thyrotrophin–secreting cells in the hypophysis of catfish, Heteropnuses fossilis (Bloch). Z mikrosk Anat Forsch 86:21–38

Purves HD, Griesbach WE (1951b) Specific staining of the thyrotrophic cells of the rat pituitary by the gomori stain. Endocrinology 49:427–428

Purves HD, Griesbach WE (1951a) The site of thyrotrophin and gonadotrophin production in rat pituitary studied by Mcmanus-Hotchkiss staining of glycoprotein. Endocrinology 534:244–263

Purves HD (1966) Cytology of adenohypophysis–the pituitary gland, vol I, p 147

Purves HD (1961) Morphology of the hypophysis related to its function. In: Young WC, Corner GW (eds) Sex and internal secretions, 3rd edn. Williams & Wilkins, Baltimore. Maryland, pp 161–239

Quijada M, Timmermans HAT, Lamberts SWJ (1980) Tamoxifen suppresses both the growth of prolactin-secreting pituitary tumors and normal prolactin synthesis in the rat. J Endocrinol 86:106–119

Racodot J, Olivier L, Porcile E, Droz B (1965) Appareil de Golgi et origin des grains de secretiopn dans les cellules adenohypophysaires chez le rat E'tude radiographique en microscopic electronique apres injection de leucine tritiee. C.r.hebd. Seanc Acad Sci (Paris) 261:2972–2974

Ramakrishnan PR, Kaur J, Rajalaxmi M (1989) Effect of dihydroxytestosterone on testicular and accessory gland function in male rhesus monkey. Contraception 40:1

Randman and Runner (1953) Demonstration of storage and release of gonadotrophin by the anterior pituitary of the mouse during gestation. Endocrinology 53:367–379

Ratner A, Talwalker PK, Meites J (1963) Effect of oestrogen administration in vivo on prolactin release by rat pituitary in vitro. Proc Soc Exp Biol Mech 112: 12–15

Raut PD (1993) Endocrinological approach to male fertility in Bonnet monkey, Macaca radiata: testosterone enanthate. Ph. D. Thesis,. University of Bombay

Raymond V, Beaulieu M, Labrie F (1978) Potent anti-dopaminergic activity of estradiol at the pituitary level on prolactin release. Science 200:1173–1175

Raza TA, Shakoori AR (1978) Cytologic alterations in the adenohypophysis of rat after chemical thyroidectomy: an electron microscopic study of TSH cells. Pak J Zool 10(2):139–162

Rennels EG, Shiino M (1968) Ultrastructural manifestations of pituitary release of ACTH in the rat. Arch Anat (Strasbourg) 51:575–590

Rennels EG (1962) An electron microscope study of pituitary autograft cells in the rat. Endocrinology 71:713–722

Richardson BA (1979) The anterior pituitary and reproduction in bats. J Reprod Fertil 56:379–389

Rinehart JF, Farquhar MG (1953) Electron microscopic studies of the anterior pituitary gland. J Histochem Cytochem 1:93–113

Roche JF, Ireland JJ (1981) The differential effect of progesterone on concentration of luteinizing hormone and follicle stimulating hormone in heifers. Endocrinology 108:568

Ruitenberg EJ, BerKvens JM, Van Nesselrooij JHJ (1974) A histological study of the adenohypophysis of the rat: evaluation of some specific staining methods. Z Versuchstierk 16:247251

Saeger W (1977) Die Hypophysentumoren cytologische und ultrastructurelle klassifikation, pathogenese, endokrine funktionen und terexperiment. Veroff Pathol 107:1–240

Saeger W (1992) Effect of drugs on pituitary ultrastructure. Microsc Res Tech 20:162–176

Saeger W, Ludecke DK, Geisler F (1990) The anterior lobe in Cushing's disease/syndrome. In: Ludecke DK, Chrousos G, Tolis G (eds) ACTH, Cushings syndrome and other hypercortisolemic states, vol 5, Raven Press, New York, pp 147–156

Saeger W, Rubenach-Gertz K, Caselitz J, Ludecke DK (1987) Electron microscopical morphometry of GH producing pituitary adenomas in comparison with normal GH cells. Virchows Arch A 411:467–472

Sage M, Bromage NR (1970) The activity of pituitary cells of the teleost Poecillia during the gestation cycle and the control of gonadotropic cells. Gen Comp Endocrinol 14:127–136

Salazar H (1963) The pars distalis of the female rabbit hypophysis: an electron microscope study. Anat Rec 147:467–497

Samuelson B, Fernholm B, Fridberg G (1968) Light microscopic studies on the nucleus lateralis tuberis and pituitary of the roach, Leuciscus rutilus, with reference to nucleus pituitary relationship. Acta Zool 49: 141–153

Santolaya RC, Ciocca D, Maneschi E (1979) Effects of pimozide on the ultrastructure of the pars distalis in the rat. Cell Tissue Res 199:483–492

Sar M, Meites J (1967) Proc Soc Exp Biol Med 125:1018

Sasaki F, Iwama Y (1989) Two types of mammosomatotropes in mouse adenohypophysis. Cell Tissue Res 256:645–648

Schechter J (1969) The ultrastructure of the stellate cell in the rabbit pars distalis. Am J Anat 126:477488

Scheithauer BW, Horvath E, Kovacs K, Laws ER, Randall RV, Ryan N (1986) Plurihormonal pituitary adenomas. Semin Diagn Pathol 3:69–82

Schelin U, Lundin PM (1964) Endocrinology 75:893

Schreibman MP, Leatherland JF, Mckeown BA (1973) Functional morphology of the teleost pituitary gland. Am Zool 13:719–742

Segal S, Polishuk WZ, Ben-David M (1976) Hyperprolactinemic male infertility. Fertil Steril 27: 1425

Serber BJ (1961) Large nuclear inclusions in the pituitary gland basophils of golden hamster. Anat Rec 139:345–355

Shiino M, Rennels EG (1976) Recovery of rat prolactin cells following cessation of estrogen treatment. Anat Rec 185:31–48

Shiino M, Williams G, Rennels EG (1972) Ultrastructural observation of pituitary release of prolactin in the rat by suckling stimulus. Endocrinology 90:176–187

Shimada T (1992) Immunohistochemical localization of keratin in bull, goat and sheep anterior pituitary glands. Cell Tissue Res 267:251–260

Shirasawa N, Kihara N, Yoshimura F (1985) Fine structural and immunohistochemical studies of goat adenohypophysial cells. Cell Tissue Res 240:315–321

Shull JD, Walent JH, Gorski J (1987) Estradiol stimulates prolactin gene transcription in primary cultures of rat anterior pituitary cells. J Steroid Biochem 26:451–456

Siperstein ER, Miller KJ (1970) Further cytophysio1ogic evidence for the identity of the cells that produce adrenocorticotrophic hormone. Endocrinology 86: 451–486

Siperstein ER, Allison VF (1965) Fine structure of the cells responsible for secretion of adrenocorticotrophin in the adrenalectomized rat. Endocrinology 76:70–79

Siperstein ER, Miller KJ (1973) Hypertrophy of the ACTH producing cell following adrenalectomy: a quantitative electron microscopic study. Endocrinology 93:1257–1268

Slijkhuis H (1978) Ultrastructural evidence for two types of gonadotropic cells in the pituitary gland of the male three–spined stickleback, Gasterosteus aculeatus. Gen Comp Endocrinol 36:639–641

Smith RE, Farquhar MG (1966) Lysosome function in the regulation of the secretory process in cells of the anterior pituitary gland. J Cell Biol 31:319–347

Stahl A (1963) Cytophysiologie de ladenohypophyse des poissons (Specialement en relation avec la function gonadotrope). In: Benoit J, Da Lage C (eds) Cytologic de ladenohypophyse. CNRS, Paris, pp 331–344

Stefaneanu L, Kovacs K (1991) Effects of drugs on pituitary fine structure in laboratory animals. J Electron Micro Tech 19:80–89

Sternberger LA, Hardy PH Jr, Cuculis HH, Meyer HG (1970) The unlabelled antibody enzyme method of immunohistochemistry. Preparation and properties of soluble antigen-antibody complex (horse radish peroxidase-anti horse radish peroxidase) and its use in the identification of spirochetes. J Histochem Cytochem 18:315–333

Stokreef JC, Reifel CW, Shin SH (1986) A possible phagocytic role for folliculo-stellate cells of anterior pituitary following estrogen withdrawal. Cell Tissue Res 243:255–261

Stratmann IE, Ezrin C, Seller EA, Simon GT (1972) The origin of thyroidectomy cells as revealed by high resolution radioautography. Endocrinology 90:728–734

Sueldo CE, Berger T, Kletzky O, Marrs RP (1985) Seminal prolactin concentration and sperm reproductive capacity. Fertil Steril 43:632

Surks MI, DeFesi CR (1977) Anterior pituitary of euthyroid and hypothyroid rats. Endocrinology 101:946–958

Thakur VM (1991) Endocrinology of reproduction in bonnet monkey, Macaca radiata. (Geoffroy). Ph. D. Thesis,. University of Bombay

Thorn W (1901) Untersuchungen uber die normale und pathologische hypophysis cerebri. Arch Micr Anat 57: 632–652

Tougard C, Kerdelhue B, Tixier-Vidal A, Jutisz M (1973) Light and electron microscopic localization of binding sites against ovine luteinizing hormone and its two sub units in rat adenohypophysis using peroxidase labelled antibody technique. J Cell Biol 58:503–521

Tougard C, Picart R, Tixier-Vidal A (1980) Immunocytochemical localization of glycoprotein hormones in the rat anterior pituitary. A light and electron microscope study using anti sera against rat B sub units. J Histochem Cytochem 28:101–114

Troen P (1980) Physiology and pharmacology of testosterone. Department of medicine, Montefiore hospital, University of Pittsburgh, school of medicine, Pittsburgh, Pennsylvania

Turkington RV (1974) Prolactin receptors in mammary carcinoma cells. Cancer Res 34:758–763

Udea H (1980) Changes of two types of pituitary gonadotrophs in white-spotted char, Salvelinus leaucomatenis, during gonadal development. Bull Fac Fish Hokkaidc Univ 28:106–117

Ueda H, Hirashima T (1979) On two different types of pituitary gonadotrophs in the pituitary gland of Oncorhynchus masou. Annot Zool Jpn 52:114–124

Ueda H, Takashashi H (1980) Response of two different types of pituitary gonadotrophs of loach, Misgurnus anguillicaudatus, to gonadectomy and to exogenous sex steroids. Gen Comp Endocrinol 40:463–472

Ueda H, Takahashi H (1977) Promotion of ovarian maturation accompanied with ovulation and changes of pituitary gonadotrophs after ovulation in loach, Misgurnus angullicaudatus, treated with clomiphene citrate. Bull Fac Fish Hokkaido Univ 28:106–117

Vasquez JM, Ellegood JO, Nazian SJ, Mahesh VB (1980) Effect of hyperprolactinemia on pituitary sensitivity to luteinizing hormone–releasing hormone following manipulation of sex steroid. Fertil Steril 33:5

Vermeulen A, Comhaire F (1978) Hormonal effects of an antiestrogen, tamoxifen, in normal and oligospermic men. Fertil Steril 29:320–327

Von Berswordt-Wallrabe R, Mehring M, Graf KJ, Beier S, Elger W (1977) In: Garfatmi S, Berendes HW (eds) Pharmacology of steroid contraceptive drugs. Raven Press, New York, p 163

Von Werder K (1988) Physiologie und pathophysiologic der regulation der prolactinsekretion. In: Jurgens O (ed) Hypcrprolactinemic prolactinome. Physiologie-Klinik-Therapie. Springer, p 326

Walker SE, McMurray RW, Besch-Williford CL, Keisler DH (1992) Premature death with bladder outlet obstruction and hyperprolactinemia in New Zealand black x New Zealand white mice treated with ethinylestradiol and 17B Estradiol. Arthritis Rheum 35(11):1387–1392

Wallace EM, Gow SM, Wu FCW (1993) Comparison between testosterone enanthate induced azoospermia and oligospermia in a male contraceptive study I: Plasma luteinizing hormone, follicle stimulating hormone, testosterone, estradiol and inhibin concentration. J Clin Endocrinol Metab 77:290293

Watson J, Anderson FB, Alam M, O'grady JE, Heald PJ (1975) Plasma hormones and pituitary luteinizing hormone in the rat during the early stages of pregnancy and after post-coital treatment with tamoxifen. J Endocrinol 65:7–17

Weman B (1974) Fine structure of the pars distalis of the pituitary gland in the female mink, mustela vison. Acta Zool 55:119–136

Wheatley DN (1967) Cells with two cilia in the rat adenohypophysis. J Anat 105:351

Williams RF, Gordon K, Fung H, Kolm P, Hodgen GD (1994) Hypothalamo–pituitary effects of RU 486: inhibition of progesterone induced hyperprolactinemia. Hum Reprod 9(suppl. 1):63–68

Winters SJ, Louriaux DL (1978) Suppression of plasma luteinizing hormone by prolactin In the male rat. Endocrinology 102:864

Wright P. J., Jenkin G., Heap R. B., and Walters D. E. (1977): Pituitary responsiveness to LH-RH and TRH and the effects of progesterone or progesterone and estradiol treatment in anestrous sheep

Yamada K, Yamashita K (1967) An electron microscopic study on the possible site of production of ACTH in the anterior pituitary of mice. Z Zellforsch 80:29–41

Yoshimura F, Soji T, Kumagai J, Yokoyoma M (1977a) Secretory cycle of the pituitary basophils and its morphological evidence. Endo Jon 24:185–202

Yoshimura F, Nogami H, Yashiro T (1982) Fine structural criteria for pituitary thyrotrophs in immature and mature rats. The anat Rec 204:255–263

Yoshimura F, Soji T, Takasaki Y, Kiguchi Y (1974) Pituitary acidophils with small or medium sized granules alone in normal and adrenalectomized rats with special reference to possible ACTH secretion. Endo Jpn 21: 297–316

Yoshimura F, Nogami H (1981) Fine structural criteria for identifying rat corticotrophs. Cell Tissue Res 219:221–228

Yoshimura F, Haramiya K, Yachi H, Soji T, Yokoyama M (1973a) Degranulated acidophils as a possible original source of' thyroidectomy cells' in the rat hypophysis. Endocrinol Jpn 20:181–198

Yoshimura F, Harumiya K, Soji T, Yokoyama M, Kumagai T (1973b) Possible reversion of pituitary 'thyroidectomy cells' into their original acidophils in rat. Endocrinol Jpn 20:249–262

Yoshimura F, Nogami H, Shirasawa N, Yashiro T (1981) A whole range of fine structural criteria for immunohistochemically identified LH cells in rat. Cell Tissue Res 217:1–10

Yoshimura F, Sato S, Soji T, Yokoyama M (1977b) Development and differentiation of rat pituitary follicular cells under normal and some experimental conditions with special reference to an interpretation of renewal cell system. Endocrinol Jpn 24:435–449

Young BA, Chaplin RE (1975) Some observations of the ultrastructure of the adenohypophysis of certain cervidae. J Zool (Lond) 175:493–508

Young BA, Foster CL, Cameron E (1967) Ultrastructural changes in the adenohypophysis of the pregnant and lactating rabbits. J Endocrinol 39:437–443

Young G, Ball JN (1981) Ultrastructural changes in the adenohypophysis during the ovarian cycle of the viviparous teleost, Poecilia latipinna. Gen Comp Endocrinol:39–59

Zambranok D (1971) The nucleus lateralis tuberis system of the gobiid fish Gillichthys mirabilis. III. Functional modifications of neurons and gonadotropic cells. Gen Cop Endocrinol 17:164–182

Ziegler B (1963) Light und elektronemikroscopische untersuchungen an pars intermedia und neurohypophyse der ratte zur frage der bezienhungen zwischen pars intermedia und hinterlappen der hypophys. Z Zellforsch Mikrosk Anat 59:486–491

Electron Microscopic Observations of Testis in Normal and Experimental Rat and Bonnet Monkey (*Macaca radiata*)

2

2.1 Introduction

The spiralling rise of the population is a global concern. No nation, however, advanced or undeveloped, can afford to be complacent regarding the hazards of this uncontrolled human population growth. All Demographers realize that the goal of controlling human fertility can be achieved through in-depth understanding of the basic cellular and molecular events in male and female reproduction. Besides the various classical methods of male contraception; the use of condoms, interrupted coitus and vasectomy, currently people are more interested in developing a new endocrinological method, in the form of hormonal contraceptives, to induce reversible azoospermia in healthy and fertile men. In principle, the endocrinological approaches have the potential to interrupt the male reproductive system in a more natural fashion to effectively produce a non-fertile but otherwise functionally competent state. In this context, the most appropriate form would be the long-acting injectable contraceptive, which obviously provides a prolonged protection and is easy to administer.

Hormone-induced sterility can be achieved by intramuscular injections of hormones is, by interfering with the hypothalamic- pituitary-testicular interaction, so as to arrest the normal spermatogenesis and sperm transport through the epididymis by creating androgen deprivation which otherwise is essential for spermatogenesis and normal accessory organs function.

In the research for a long-acting injectable regimen, various steroidal combinations have been studied, such as the use of androgen alone (Means 1975), Progestogen alone or in combination with testosterone. In the latter case, Depot Medroxyprogesterone Acetate (Depo provera, DMPA), a clinically important progestin and Testosterone enanthate (TE), a Testosterone ester has been widely studied as a probable 'Male contraceptive' by various investigators (Flickinger 1977a, b; Brenner et al. 1970; Alvarez-Sanchez et al. 1977; Faundes et al. 1981; Hedman et al. 1988; Pangkahila 1991; Avari and Bhiwgade 1992a, b; WHO 1992).

The Drug Depot Medroxyprogesterone acetate, commonly known as depo provera, is a synthetic hormone with the chemical formula; 17-acetoxy-6 methyl-Preg-4-ene-3, 20-dione. It belongs to the class of C-21, 17-acetoxyprogesterone and has a close structural similarity to natural progesterone. DMPA is the ester of Progesterone and esterification of this compound delays its absorption and metabolism, making it suitable for long-acting intramuscular injection.

Microcrystals of DMPA have a very slow solubility in body fluid and this provides prolonged release from the surface of the crystals at the deposition site. As there is usually some free MPA at the crystal surface, an initial high release

D. A. Bhiwgade, S. Menon, *Ultrastructural Investigations on the Pituitary-Gonadal Axis*,
https://doi.org/10.1007/978-981-99-3276-4_2

occurs into the surrounding tissues and the blood stream. The levels in plasma fluctuate from day to day and fall steadily over the next few months. The rate of all these depends on absorption, metabolism, protein binding and excretion. MPA is rapidly metabolized with a metabolite clearance rate of 1.668 l/24 h and has a plasma half-life of about 4–5 h. Levels of MPA decline more slowly and could be detected even 200 days or more after a single injection. These characteristics contribute to a very high contraceptive efficiency.

DMPA has little or no effect on body metabolism such as coagulation, fibrinolytic factors, platelet functions, carbohydrate and lipid metabolism, and liver, renal and thyroid functions.

1. Use of DMPA can lead to a decrease in HDL cholesterol and thus increase the risk of cardiovascular disease. However, this effect is less in the case of DMPA users as compared to other synthetic progestin. DMPA does not seem to have any influence on fasting triglyceride levels or on the composition of phospholipids.
2. Carbohydrate metabolism gets affected by synthetic progestogens. DMPA raised fasting glucose and insulin levels and caused an increased response of both glucose and insulin to a glucose load in comparison with pretreatment levels. Several other studies failed to show these changes (Amatayakul et al. 1979; Tankeyoon et al. 1976; Back et al. 1977).
3. No deterioration in hepatic, renal vitamin, hemostatic or thyroid function or in standard haematological parameters were found in short- and long-term DMPA users (Amatayakul et al. 1979).

In developing an endocrinological method to suppress temporary male fertility by using synthetic progestin, a combination of depot medroxyprogesterone acetate and testosterone enanthate and its gradual restitution is adopted. This approach relied upon the complete withdrawal of serum and intratesticular testosterone mediated via the in vivo suppression of Leydig cell function. In the adult testis, testosterone is a fundamentally important hormone for the quantitative maintenance of spermatogenesis because without an adequate supply of testosterone, spermatogenesis fails completely. DMPA possesses both antiandrogenic, antigonadotrophic properties as well as sperm-suppressing properties. The treatment of males of various animals species and men with synthetic progestogen, Depo provera (DMPA) causes arrest of spermatogenesis in a dose-dependent manner (Heller and Clermont 1964) and lowers testosterone level (Rivarols et al. 1968). Being an anti-androgen, the accompanying loss of libido, potency and decrease in accessory organs weight could be overcome by simultaneous administration of testosterone. The antispermatogenic action of DMPA is threefolded through a direct action on testis and epididymis as well as the endocrine pathway; (1) Spermatogenesis is endocrinologically regulated by pituitary gonadotrophin; Luteinizing hormone (LH) influences gamete production via high intratesticular testosterone (T) concentration resulting from stimulation of Leydig cells, while follicle stimulating hormone (FSH) expresses its effects through Sertoli cells. DMPA being an antigonadotrophic compound suppresses the secretion of both these hormones, LH and FSH from pituitary and consequently arrests spermatogenesis (Faundes et al. 1981). (2) The direct and local effects of DMPA on the testis includes suppression of testosterone biosynthesis in the Leydig cells via regulation of multiple enzymatic activity of 17-â hydroxysteroid dehydrogenase and? 5,3 â -hydroxysteroid dehydrogenases involves in testosterone synthesis (Barbieri and Ryan 1980). The inhibition of steroid dehydrogenase by synthetic progestin is probably due to direct binding of synthetic progestin to the active site of the dehydrogenase. (3) MPA also increases the metabolic clearance rate of testosterone within the liver by inducing hepatic testosterone A ring reductase activity as observed in rat (Altman et al. 1972).

Despite the considerable amount of data available regarding the effect of DMPA and TE on the reproductive organs and hormone levels of rats and human beings, the precise mechanism of action is not known at the electron microscopic level. Therefore, our study attempts to find out the effects of DMPA + TE on ultrastructure of the reproductive organs in rat and bonnet monkey.

2.2 Observations

1. Electron microscopic observations in DMPA + TE treated Rat

Ultrastructural observations of the testis of control rat show the seminiferous tubules enclosed by one or more layer of myoid cells constituting the myoid layer of the lamina propria. The supporting cells and the spermatogenic cells lined by a complex stratified epithelium are two major categories of cells seen. Single kind of supporting cell—the Sertoli cell is present whereas the spermatogenic cells include several morphologically distinguishable types of spermatogonia, primary spermatocytes, secondary spermatocytes and the spermatidvs (Figs. 2.1 and 2.2).

The Sertoli cell is seen resting on the basal lamina showing upward extensions through the full thickness of the epithelium. The cytoplasm contains numerous slender elongated mitochondria oriented parallel to the longitudinal axis of the cell. The other cellular organelles like the smooth surfaced tubules, membrane limited lysosomes and occasional lipid droplets are also present. The junctional complex on the boundary separating the Sertoli cell is prominently visible (Figs. 2.1 and 2.2).

Both A and B types of spermatogonia with their characteristic nuclear and cytoplasmic inclusions are observed. Well-developed spermatocytes with their characteristic circular nucleus are also visible (Fig. 2.1). Further down the lumen a number of spermatids are present (Fig. 2.3a, b, and c). Spermatids possess a spherical nucleus with pale staining finely granular cytoplasm which is less electron dense. Spermatids at the periphery are lined by the mitochondria. The first signs of differentiation and the appearance of a few small granules within

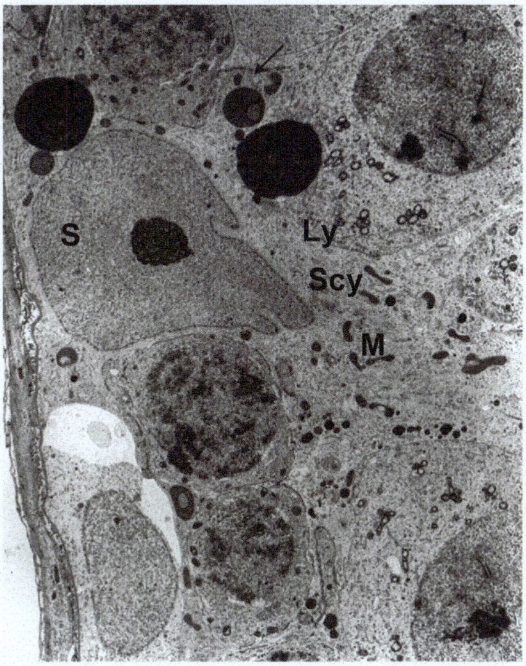

Fig. 2.1 Electron micrograph of the seminiferous tubule of the 17 weeks control rat, showing the seminiferous epithelium and the Sertoli cell cytoplasm (Scy). Note the clearly demarked junctional complex between the Sertoli cell and the neighboring germinal elements (arrows). Clearly visible is the spermatogonium Type A (Sg) having an ellipsoid nucleus with fine chromatin granules. The Sertoli cell cytoplasm contains numerous, slender elongated mitochondria (m), and a few membrane limited lysosomes (Ly) [X 5000]. Unpublished electron micrograph from Dr. Bhiwgade and Dr. Belare

the Golgi apparatus are visible. The maturation phase or phase four of the spermatids with coarse nucleoplasm is seen. Disappearance of acrosome granules and its contribution to the formation of the acrosomal cap is seen. The nucleus is flattened and pyriform. The cytoplasm of the maturing spermatid is well lined by mitochondria along the peripheral border. Tiny secretory tubules, chromatid body and endoplasmic reticulum are seen. Various stages of spermatozoa cut lengthwise are noticed.

The seminiferous tubules are enclosed by one or more layers of myoid cells constituting the myoid layer of the lamina propria. The germinal layer consists of chiefly the Sertoli cells resting on the basal lamina and the spermatogonia—Type A and Type B, both of which are clearly visible

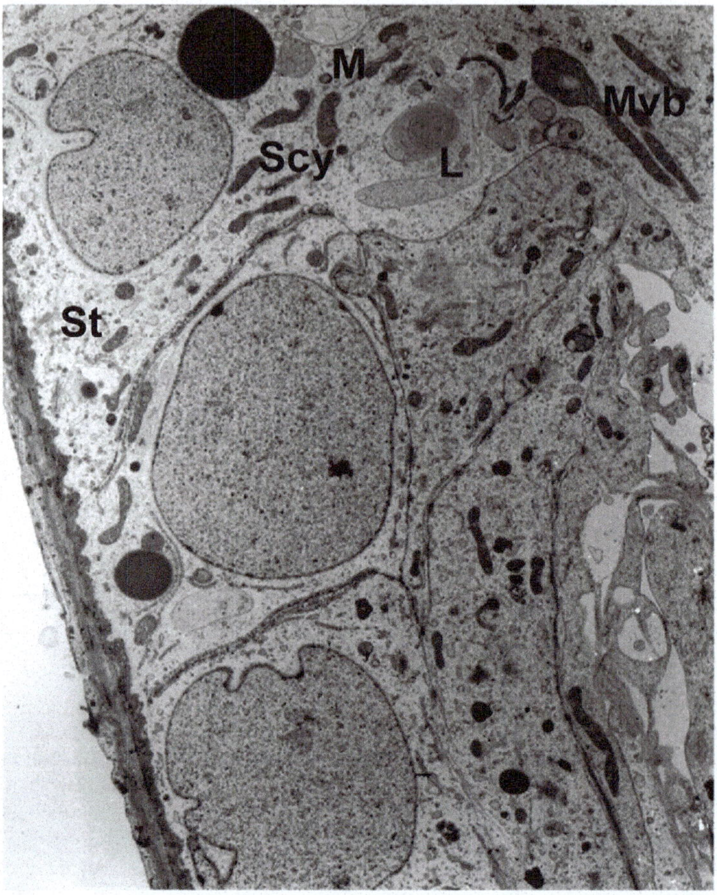

Fig. 2.2 The electron micrograph of the seminiferous tubules of the 17 weeks control rat shows the Sertoli cell cytoplasm (Scy) as we progress further down the lumen. All the normal cellular organelles are seen. Slender elongated mitochondria (m), lipid droplets (L), smooth surfaced tubules, multivesicular bodies (mvb etc. are seen. Few membrane limited lysosomes are also seen [X 13,000]. Unpublished electron micrograph from Dr. Bhiwgade and Dr. Belare

(Figs. 2.4 and 2.5). The characteristic infolding of the Sertoli cell nucleus at the surface with the homogenous nucleoplasm is seen (Fig. 2.5). The Type A spermatogonium with its oval-shaped electron dense nucleus firmly resting on the lamina propria, separated by a continuous tight junctional complex is clearly seen (Fig. 2.4). The Type B spermatogonium with its spherical nucleus and its uniform chromatin content also resting on the lamina propria is clearly visible (Fig. 2.5). The initial cytoplasmic region en route towards the lumen reveals the Sertoli cell cytoplasm composed of numerous elongated mitochondria, lipid droplets of varying size and density, endoplasmic network, scattered multivesicular bodies and occasional membrane limited lysosomes. The onset of vacuolation is prominent in some places. An occasional degenerating germinal element probably a spermatocyte is seen. With its lysed nuclear membrane, darkened but diminished chromatin content and an increased lysosomal content. The germinal layer clearly remains unaffected compared to the initial mid-luminal region.

Further down towards the lumen the formation of spermatids is clearly visible. Formation of cap phase spermatids as the acrosomal cap or granule is well established with the acrosomal membrane (Fig. 2.6a). The formation of the manchettes and the appearance of the chromatid body is visible at the opposite end of the acrosomal granule (Fig. 2.6b and c).

It is from this stage onwards that the onset of degeneration seems to have set in. There is a break in the continuity of the acrosomal membrane of some spermatids (Fig. 2.7a, b). The cytoplasm surrounding these degenerating cap phases also shows many degenerative changes

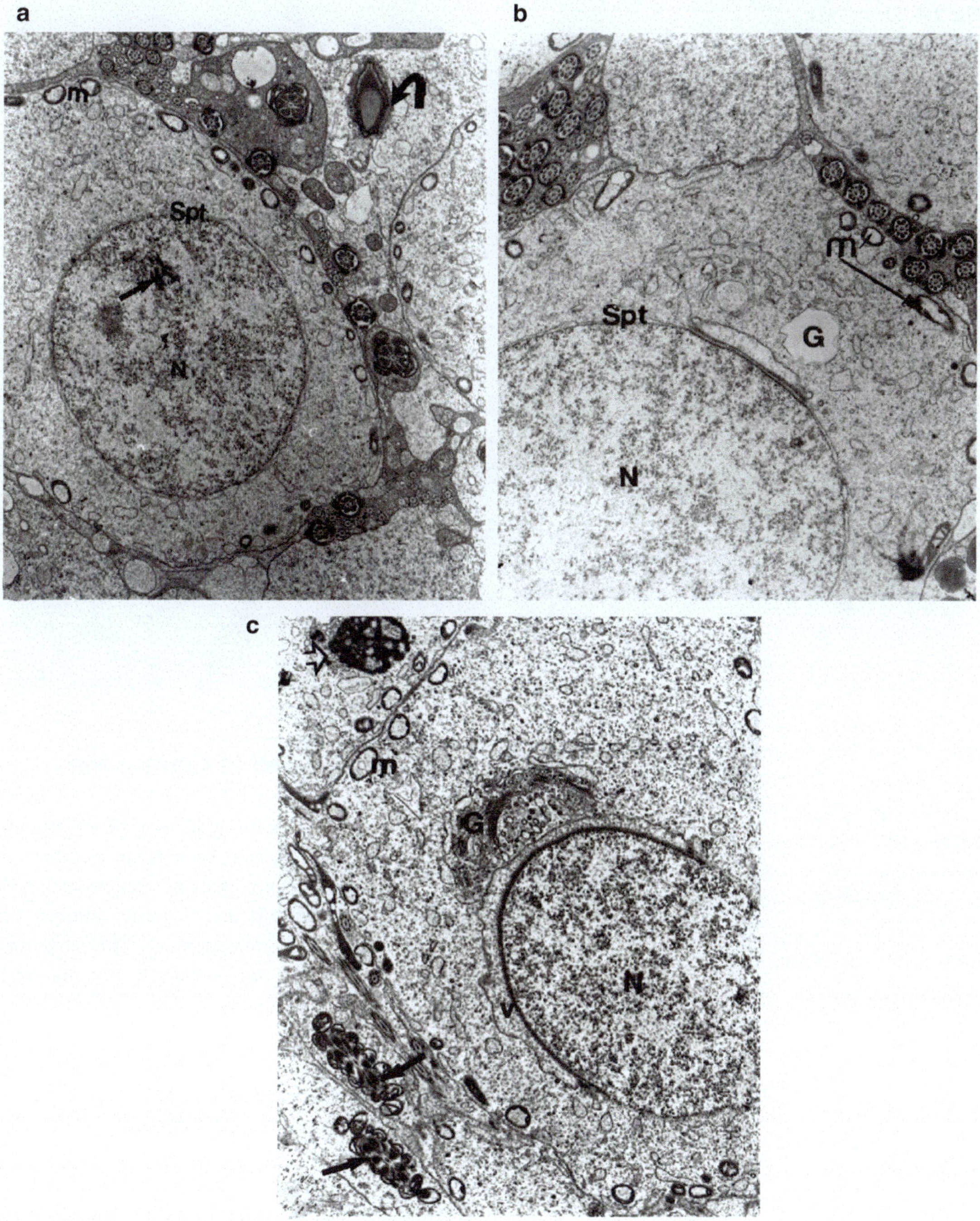

Fig. 2.3 Electron micrograph of the seminiferous tubule of the 17-week control rat showing the spermatid (Spt) indicating the first sign of differentiation in the nucleus (N) (**a**). The chromatin starts aggregating (arrow). Note the periphery of the spermatid is lined by mitochondria (m) and steps 17, 18 and 19 spermatids are visible from their cross-section between spermatids. The longitudinal section of the step 16 spermatid is clearly visible (curved arrow) (**b**) shows the area half way towards the lumen. The cap phase spermatid (Spt) formation has commenced and a perfect picture of the transverse section of spermatozoa through various places is seen indicating normal spermatogenesis. Spermatid nucleus (N), Golgi body (G) and mitochondria (m) are seen (**c**) shows the cap phase spermatid with the limiting membrane of the acrosome vesicle having increased its adherence to the nuclear envelope, forming a thin fold that spreads over the pole of the nucleus covering the entire periphery of the cell. Acrosomic vesicle (V), chromatid body (hollow arrows), Golgi body (G). Also visible are transverse sections of Stage 16/17 spermatids (bold arrows) [X 5000] [X 10,000] [X 7000]. Unpublished electron micrograph from Dr. Bhiwgade and Dr. Belare

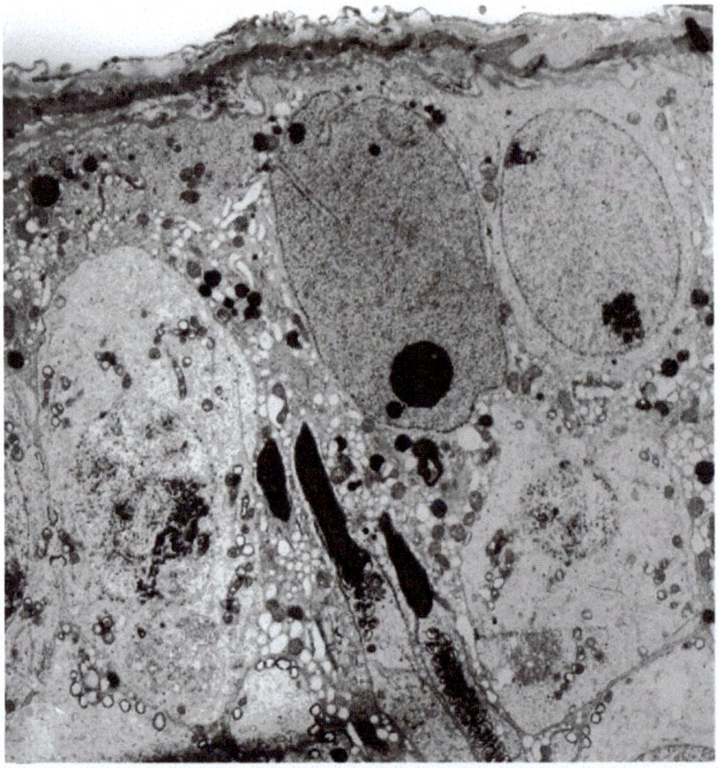

Fig. 2.4 Low-power electron micrograph of seminiferous tubule of a rat after MPA + TP treatment for 8 weeks spermatogonia, spermatocytes and Sertoli cells are recognized [X 4000]. Unpublished electron micrograph from Dr. Bhiwgade and Dr. Belare

(Fig. 2.6a). This is characterized by a predominance of numerous lipid droplets of various sizes capable of displacing various other cellular organelles like endoplasmic reticulum and mitochondria (Fig. 2.8). This increased lipid accumulation, lysosomal and vesicular activity all point towards the 'fatty degeneration' of the Sertoli cell cytoplasm (Fig. 2.6c). Some germinal stages as early as the Golgi phase spermatids have also undergone degeneration (Fig. 2.8). The degeneration could be made out due to the non-uniform continuity of the nuclear membrane as many breaks or gaps are distinctly visible in the nucleolemma. An advanced degeneration of the spermatid is observed with a total disorientation of the cell and its cytoplasmic organelles, total lysis of the nuclear membrane and total degeneration of the cytoplasm (Figs. 2.7b and 2.8).

2.3 Leydig Cell in Control Rat

The Leydig cell shows evidence of enhanced secretory activity as seen by a large number of lipid droplets, dense bodies, rough-surfaced endoplasmic reticulum and a large number of mitochondria with prominent cristae. Endocytosis and coated vesicles are frequently encountered (Fig. 2.9a, b, c, and d).

2.4 Leydig Cell after DMPA + TE Treatment (17 Weeks)

The ultrastructural observations of the DMPA + TE treated testis show the Leydig cell located in the angular interstices between convoluted seminiferous tubules. The Leydig cell nucleus shows loosened chromatin content (Fig. 2.10a, b). The cytoplasm shows a highly elaborate agranular reticulum appearing like a filigree. Hypertrophied

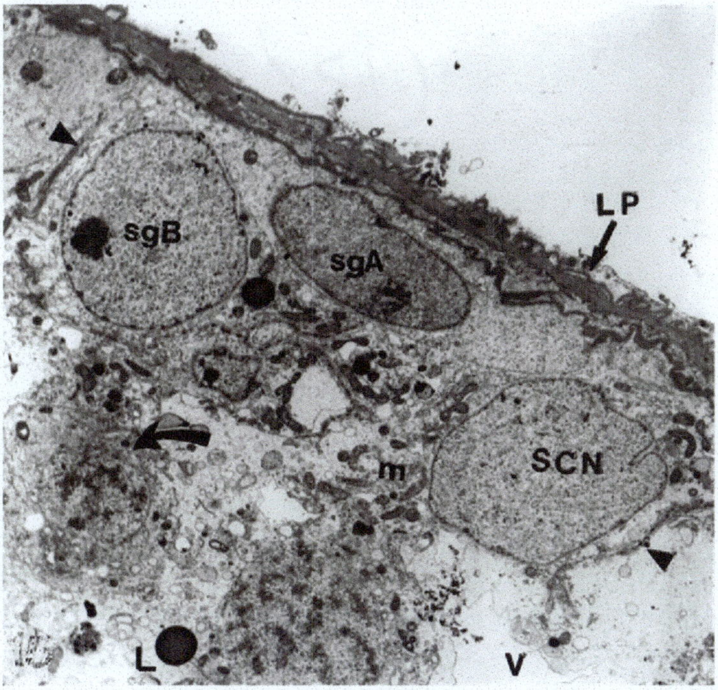

Fig. 2.5 The electron micrograph of the testis of the rat treated with DMPA + TE for 17 weeks. The germinal layer and the initial region are clearly visible. The type A spermatogonium (sgA), with its characteristics oval shape and electron-dense nature is firmly placed by a continuous, tight, junctional complex (arrow heads). It is flanked on one side by type-B spermatogonium (sgB), spherical in shape and with uniform chromatin content. On the other side lies a typical indented nucleus of the Sertoli cell (SCN). All are separated by junctional complexes. Note the dominance of rod-shaped mitochondria in the region (m), with an occasional scattered lipid droplet (L). Towards the lumen are visible degenerating germinal elements probably spermatocytes (curved arrow) with their lysed nuclear membranes, darkened but diminished chromatin content and the onset of vacuolation (V). [X 3500]. Unpublished electron micrograph from Dr. Bhiwgade and Dr. Belare

mitochondria, a large number of mitochondrial structures; spherical intramatrical bodies are also seen. The lysosomal system ranged from many membrane-limited lysosomes to spherical primary lysosomes. Numerous small vesicles are present suggesting an active endocytic activity (Fig. 2.10b). The disorientation of the Leydig cell cytoplasm could be due to the result of the injected regimen and hence suggests a poor site of Testosterone synthesis.

2. Electron microscopic observations CPA + TE Treated rat

Antiandrogens like Cyproterone Acetate (CPA) have received a lot of attention by reproductive endocrinologists all over the world as a potential male contraceptive. The mechanism of action of this compound and clinical and experimental studies have been reviewed on several occasions by Neumann and his associates. CPA is known to inhibit spermatogenesis in all species in a dose-dependent manner (Hamada et al. 1963; Junkmann and Neumann 1964; Ott 1968; Ott and HofTet 1968; Neumann 1972, 1973; Bhiwgade et al. 1989).

The selective inhibition of post-testicular process of spermatozoal maturation in the epididymis by CPA has been extensively studied by Prasad and co-workers (Prasad et al. 1970).

The changes, in the fine structure of the testis and epididymis and the accessory sex organs, associated with CPA treatment have helped to elucidate the mechanism of action of this

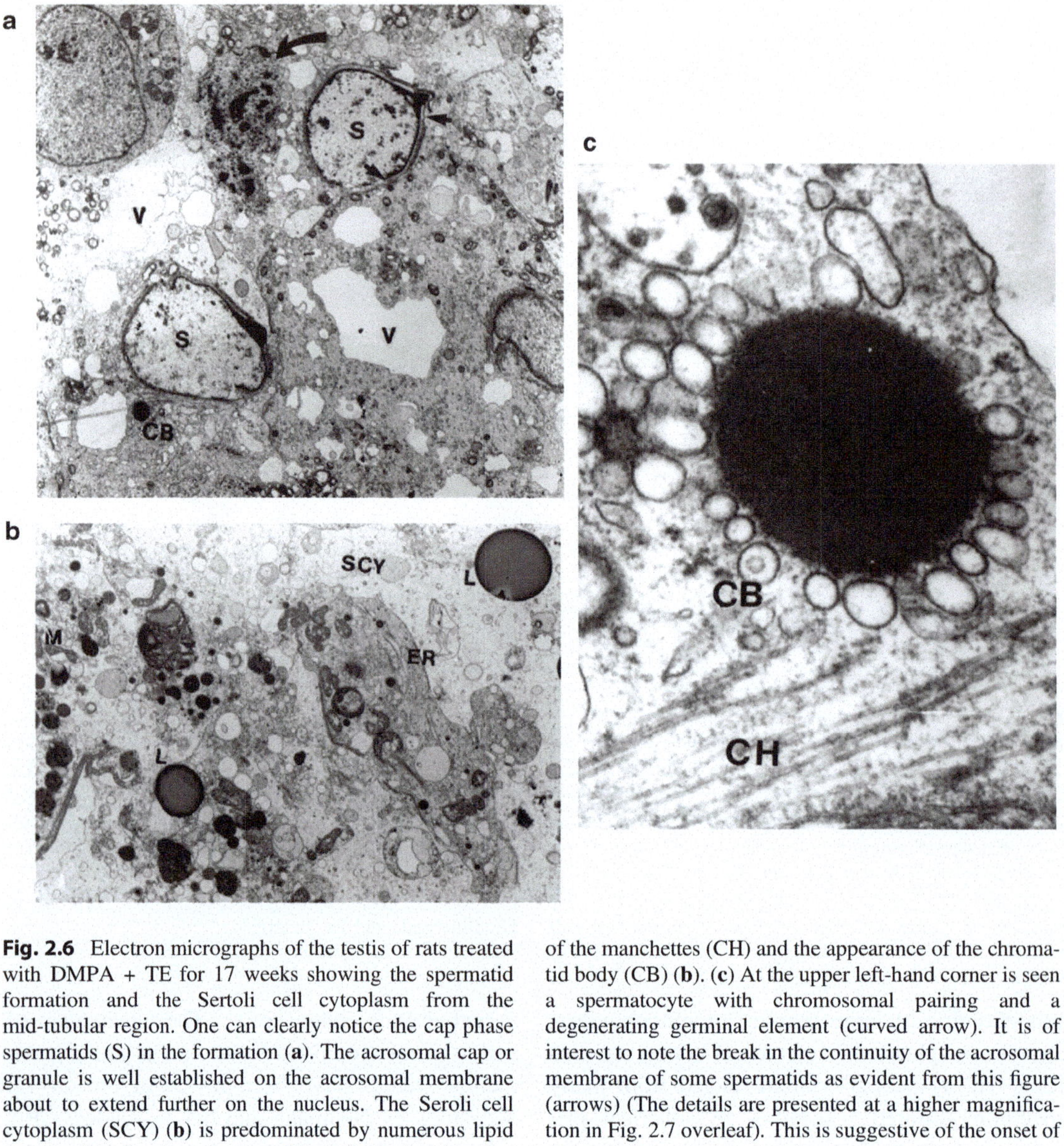

Fig. 2.6 Electron micrographs of the testis of rats treated with DMPA + TE for 17 weeks showing the spermatid formation and the Sertoli cell cytoplasm from the mid-tubular region. One can clearly notice the cap phase spermatids (S) in the formation (**a**). The acrosomal cap or granule is well established on the acrosomal membrane about to extend further on the nucleus. The Seroli cell cytoplasm (SCY) (**b**) is predominated by numerous lipid droplets (L) of various sizes capable of displacing the other cellular organelles like endoplasmic reticulum (ER), mitochondria (M) and lysosomes. Increased lipid accumulation, lysosomal and vesicular activity point towards 'fatty degeneration' of the Sertoli cell cytoplasm. At the opposite end of the acrosomal membrane is the formation of the manchettes (CH) and the appearance of the chromatid body (CB) (**b**). (**c**) At the upper left-hand corner is seen a spermatocyte with chromosomal pairing and a degenerating germinal element (curved arrow). It is of interest to note the break in the continuity of the acrosomal membrane of some spermatids as evident from this figure (arrows) (The details are presented at a higher magnification in Fig. 2.7 overleaf). This is suggestive of the onset of degeneration, as the cytoplasm surrounding them also reveals many degenerative changes and an onset of vacuolation is also visible (V) [X 3500] [X 35,000] [X 5000]. Unpublished electron micrographs from Dr. Bhiwgade and Dr. Belare

antiandrogen on various androgen target organs (Loving and Flickinger 1976; Flickinger and Loving 1976; Bhiwgade et al. 1990).

CPA being a potent anti-androgen with progestational activities, it has been known to cause several undesirable side effects like the loss of libido (Neumann and Schenck 1976; Neumann 1977) and the development of gynecomastia in men. To abolish these untoward side effects, it was suggested that a long-acting ester of

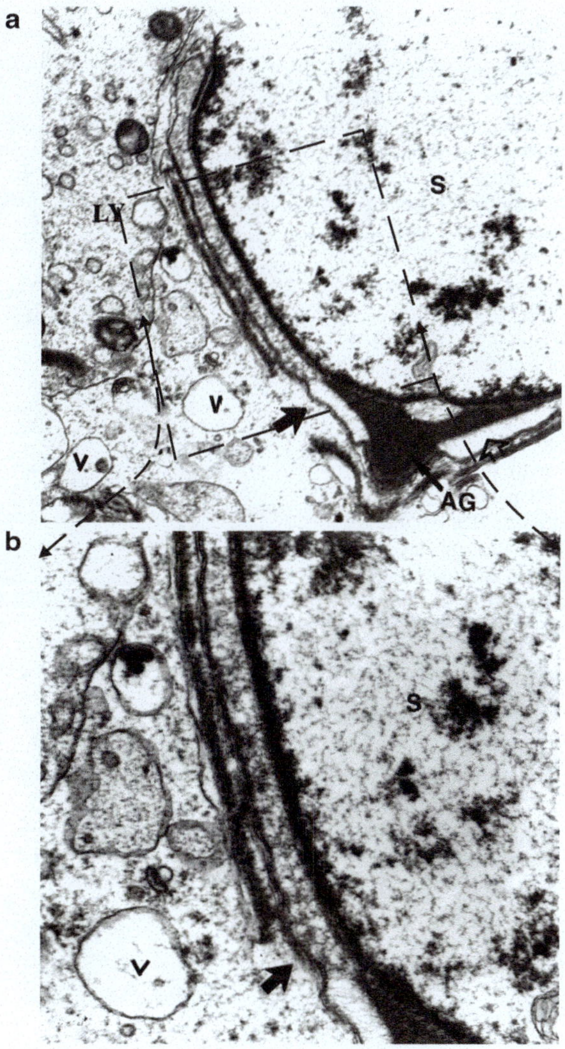

Fig. 2.7 (**a**, **b**) High-power electron micrographs of the cap phase spermatids (S) of the 17 weeks DMPA + TE treated rat, showing the commencement of degeneration, as there is a break in the continuity of the acrosomal membrane (bold arrow). Note the formation of the acrosomal granule (AG) and the acrosomal vesicle. Also, note the altered cytoplasmic contents predominated by increased vesicles (V) and lysosomes (Ly). [X 15000] marked area in (**a**) is enlarged in (**b**) [X35000]. Unpublished electron micrographs from Dr. Bhiwgade and Dr. Belare

testosterone should be administered in conjunction with the CPA treatment (Wang and Yeung 1980; Lohiya and Sharma 1983; Bhiwgade et al. 1990). Moreover, the possible synergistic effect of testosterone with CPA was expected to lead to further suppression of male fertility. Inspite of this, scanty literature is available on the combined effect of CPA and testosterone esters on the histology of the male reproductive tract and on spermatogenesis.

Our study, hence, looked at the synergistic effects of CPA and testosterone enanthate (TE) combination treatment on the fine structural changes in the various male reproductive organs.

2.5 Sertoli Cell After CPA + TE Treatment (60 Days)

Ultrastructurally, the Sertoli cell cytoplasm contains a large accumulation of dense bodies and lysosomes filled with heterogenous material (Figs. 2.11 and 2.12). The cytoplasm also contains a few myelin figures indicating the onset of degenerative changes (Fig. 2.12).

Slight vacuolation is noticed in the basal region of the cytoplasm of the spermatogonial cells (Fig. 2.11). Spermatogenesis has proceeded upto the spermatid stage even in the presence of the drug as is indicated by the occurrence of

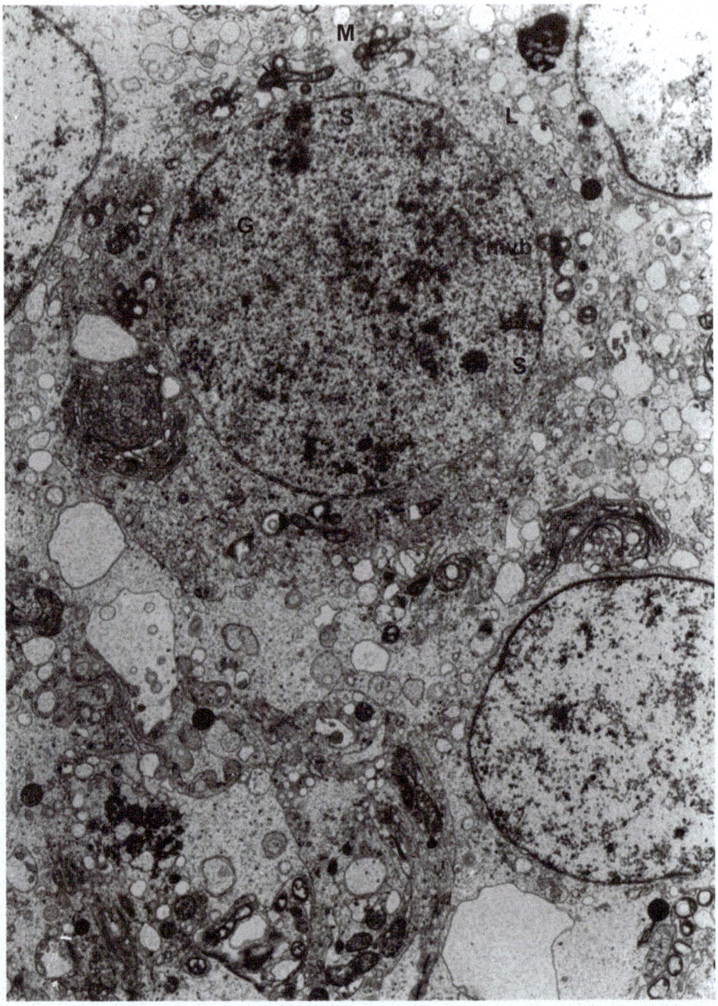

Fig. 2.8 Electron micrograph of the Golgi phase spermatid (S) of 17 weeks DMPA + TE treated rat. The onset of nuclear degeneration could be noticed due to its non-uniform continuity as many 'breaks' of 'gaps' (arrows) are distinctly visible. Note the formation of the Golgi complex (G) at one end, and the presence of a chromatid body at the opposite end. The cytoplasmic inclusions seen are multivesicular bodies (mvb), mitochondria (M) and occasional lipid droplets (L). All the organelles appeared to be in a state of hypertrophy [X 3500]. Unpublished electron micrographs from Dr. Bhiwgade and Dr. Belare

synaptonemal complexes in the nuclei of the primary spermatocytes (Fig. 2.13).

Cap-phase spermatids, though present, have got breaks in the continuity of the outer membrane of the acrosomal vesicle (Fig. 2.14). The Golgi complex shows vacuolization in many of the spermatids (Fig. 2.15). In most severely affected spermatids, the nuclear membrane begins to show degenerative changes. The nucleolemma is disintegrating and there are breaks in it. Such changes are visible not only in the cap phase spermatids but also in the acrosomal phase spermatids (Fig. 2.16).

The spermatids which are in the later stage of spermiogenesis show deformative changes which include a vacuolization at the acrosomal level. The shape of the acrosome in the 18/19 spermatids is distorted and the posterior region of the nucleus does not contain the characteristic condensed chromatin material (Fig. 2.17). Sections through spermatozoa at the middle-piece or tail region, however, appear normal exhibiting the 9 + 2 doublets of the axial filament, the outer coarse fibers and the mitochondrial helix of the middle piece (Fig. 2.18).

2.6 Discussion

Our study, the acrosomal membrane of the mature phase spermatids shows lysis at various places and has undergone degenerations following

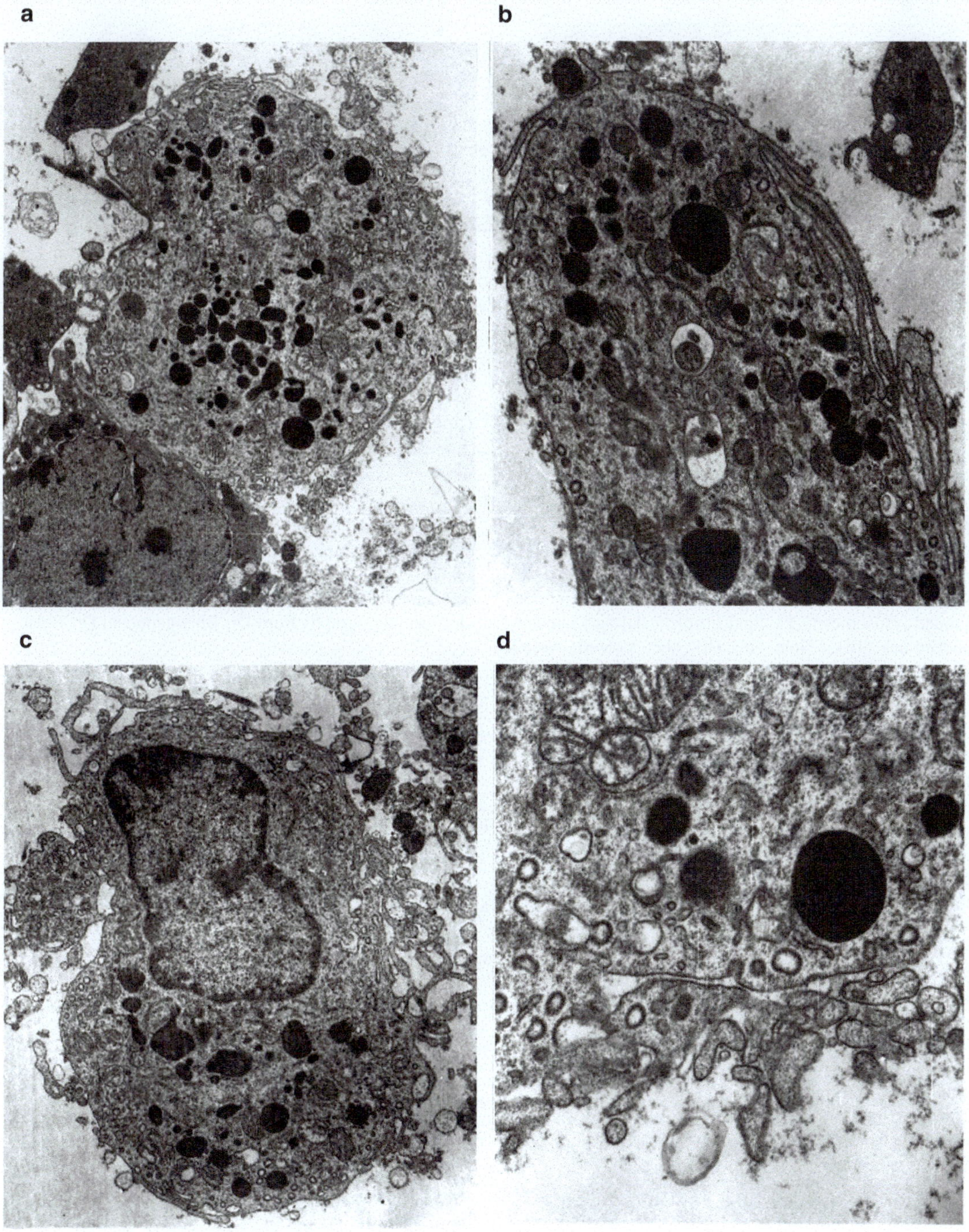

Fig. 2.9 (**a**, **b**, **c**, **d**) Electron Micrographs of Leydig cell of control rat. Note numerous pinocytotic invaginations, vesicles of varying density at the luminal surface, coated vesicles, lipid droplets, dense bodies, mitochondria and rough-surfaced endoplasmic reticulum and Lysosomes [X 8000] [X 8000] [X 8000]. Unpublished electron micrographs from Dr. Bhiwgade and Dr. Belare

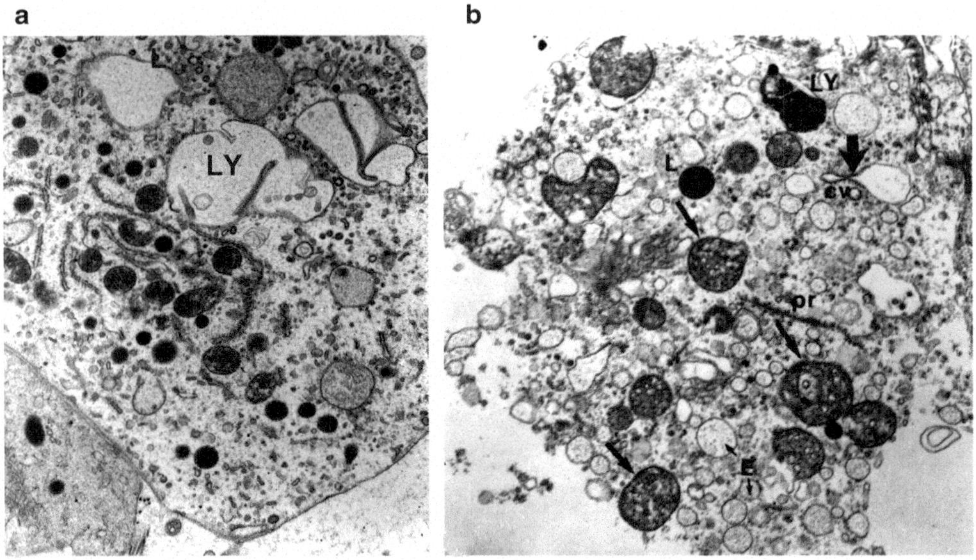

Fig. 2.10 (**a**) The electron micrographs of Leydig cells of the rat treated with DMPA + TE for 17 weeks. The peripheral cytoplasm appears like a filigree due to an immense number of intramatrical bodies, membrane-limited lysosomes (LY), lipid droplets (L) and uncoated endosomes (E). Note in (**b**) the coated vesicle (cv) delivering its secretion by membrane-fusion to an uncoated endosome (bold arrow), suggestive of endocytosis. The reduced mitochondrial activity reveals hampered cellular functioning [X 10,000] [X 15,000]. Unpublished electron micrographs from Dr. Bhiwgade and Dr. Belare

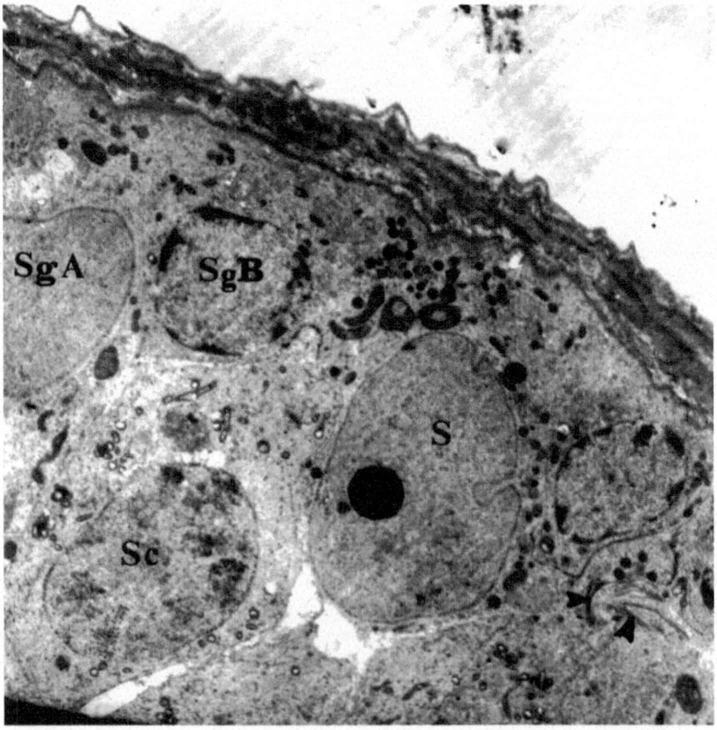

Fig. 2.11 Low-power electron micrograph of 60 days CPA + TE treated rat testis showing the various germs cells—the Sertoli cell (S), the type A (SgB) spermatogonia and the spermatocytes (Sc). The Sertoli cell cytoplasm is filled with several dense bodies and lysosomes and even a few myelin fibres (arrowheads) indicating the onset of degenerative changes [X 3600]. Unpublished electron micrographs from Dr. Bhiwgade and Dr. Kadam

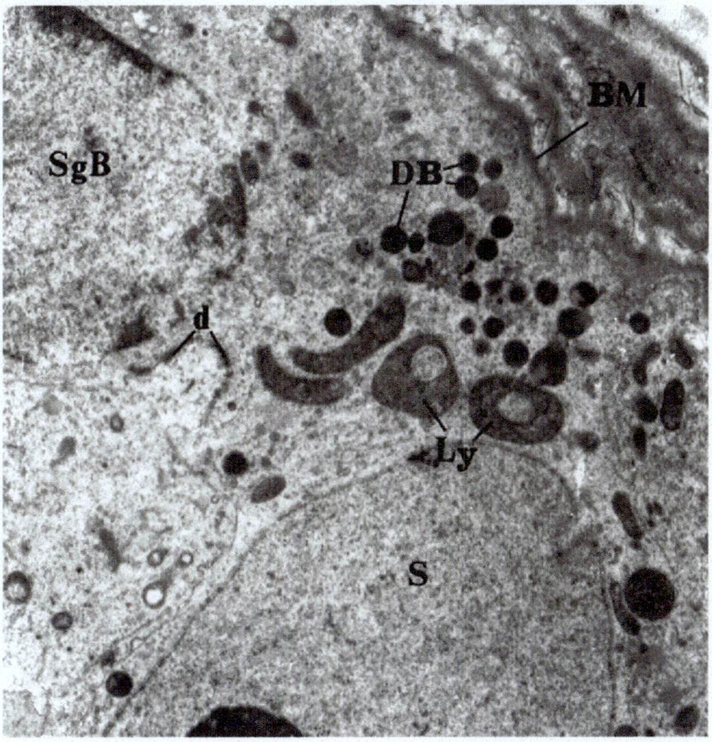

Fig. 2.12 Higher magnification of the Sertoli cell cytoplasm of 60 days CPA + TE treated rat testis, showing several dense bodies (DB) and lysosomes (Ly). Also seen are the Sertoli cell nucleus (S), type B spermatogonium (SgB), the basement membrane on which they rest (BM) and the demosomes (d) implying cell-to-cell interaction between the various cell types [X 8500]. Unpublished electron micrographs from Dr. Bhiwgade and Dr. Kadam

DMPA + TE treatment. Similar observations have been recorded in rats after DMPA + TE treatments by Flickinger (1977a) and Avari and Bhiwgade (1992a, b). Decreased levels of testosterone accompanied with the lysis of the acrosomal membrane due to the anti-gonadotrophic action of DMPA + TE combination regimen points towards a possibly reduced level of hyaluronidase activity (Avari and Bhiwgade 1992a, b) as testosterone bears a direct relation with hyaluronidase (Males and Turkington 1971a, b).

The ultrastructural changes show evidence of a possible impaired energy generation as far as the late stages of spermatogenesis are concerned. The late spermatids undergo ultrastructural changes such ashypertrophy and/or ballooning of the mitochondria with loss of cristae and lysis. However, Avari and Bhiwgade (1992a, b) have reported an increase in the activity of Lactate dehydrogenase (LDH) and Sorbitol dehydrogenase (SDH) following DMPA + TE regimen and suggested it might be due to the prolonged administration of the androgen. Long-term administration of androgen results in an increase in testicular enzymes since many mitochondrial enzymes are known to be androgen dependent (Mills and Means 1972). LDH and SDH play an important role as a shuttle system between the cytoplasm and the mitochondria and serve to generate energy (Mills and Means 1972; Blanco 1980).

At the ultrastructural level, the cytoplasm of the Sertoli cell revealed degenerative changes like vesiculated endoplasmic reticulum, dumbell and ovoid mitochondria, accumulation of large lipid droplets and an increase in the number of lysosomes. The Sertoli cells are supposed to be influenced by FSH (Fritz 1978; Means 1975), and suppressed release of FSH from the pituitary caused a hampered functioning of the Sertoli cell due to additive effects of DMPA + TE, which were evident from the ultrastructure observations of pituitary gonadotrophs.

The ultrastructural studies after DMPA + TE treatment revealed degenerating germinal elements and changes in the organelle content en route towards the lumen, in the Sertoli cell indicating a probable reduction in â-glucuronidase activity (Avari and Bhiwgade

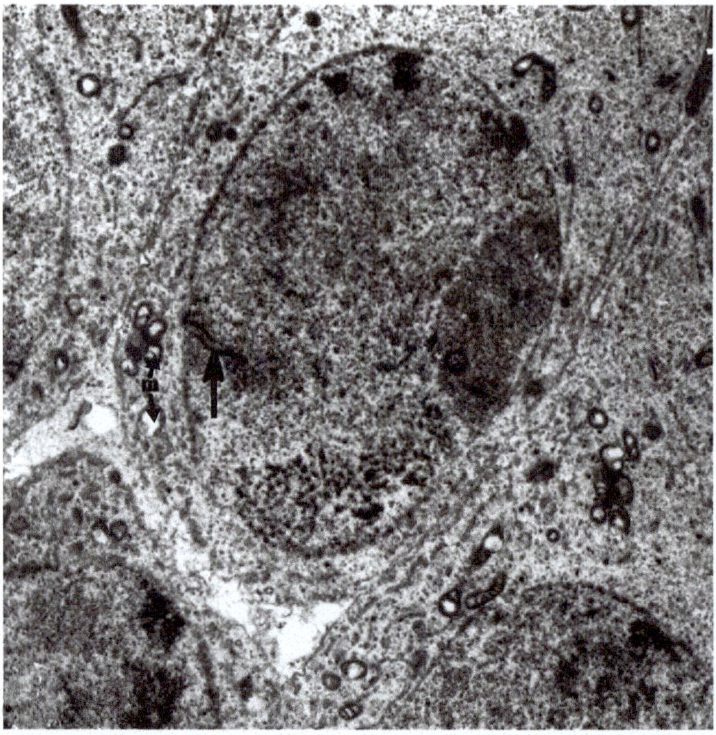

Fig. 2.13 High-power electron micrograph of 60 days CPA + TE treated rat testis. The spermatocyte (Sc) shows synaptonemal complexes (arrowhead) in its nucleus suggestive of normal cell division until this stage. The mitochondria lie scattered in the cytoplasm (m) [X8,500]. Unpublished electron micrographs from Dr. Bhiwgade and Dr. Kadam

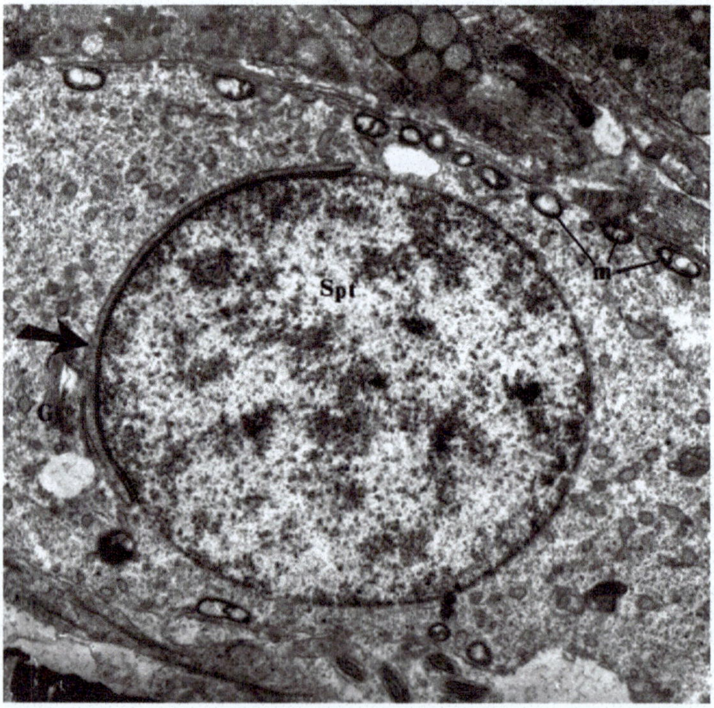

Fig. 2.14 High-power electron micrograph of 60 days CPA + TE treated rat testis. The cap-shape spermatids (Spt) have got breaks in the continuity of the outer membrane of the acrosomal vesicle (big arrow) and the Golgi complex (G) shows vacuolization. Also seen are the peripherally arranged mitochondria (m). [X 8500]. Unpublished electron micrographs from Dr. Bhiwgade and Dr. Kadam

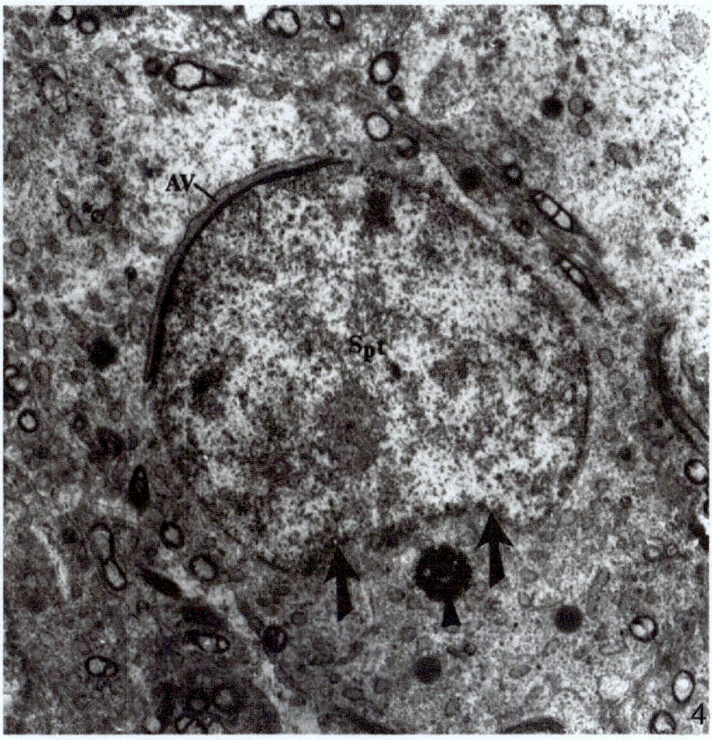

Fig. 2.15 High-power electron micrograph of 60 days CPA + TE treated rat testis, showing a severely affected (Stp) with a disintegrating nuclear membrane (big arrows). Note the acrosomal vesicle (AV) and the chromatoid body arrowhead). [X 8, 500]. Unpublished electron micrographs from Dr. Bhiwgade

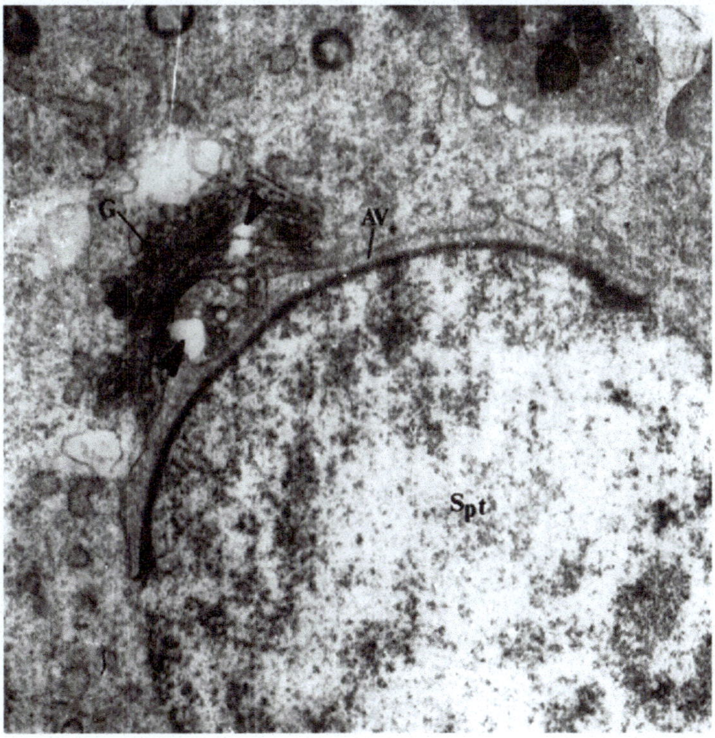

Fig. 2.16 Electron micrograph of 60 days CPA + TE treated rat testis at higher magnification showing a spermatid Stp) with acrosomal vesicle (AV) spreading over the nucleus and the development of the acrosomal granule from the Golgi (G). Note the vacuolization in the Golgi complex (arrowheads). [X 13,500]. Unpublished electron micrographs from Dr. Bhiwgade

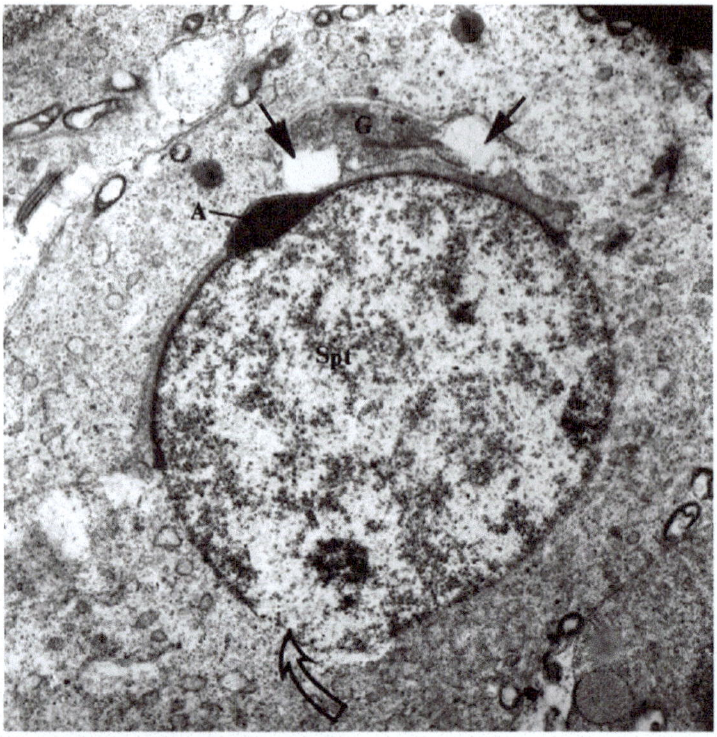

Fig. 2.17 Electron micrograph of 60 days CPA + TE treated rat testis showing a spermatid (Spt) with the acrosome (A) developing at one pole of the nucleus. Note the vacuolization (arrowheads) in the Golgi region (G). The nucleolemma has started disintegrating (curved arrow) due to the treatment [X 8500]. Unpublished electron micrographs from Dr. Bhiwgade

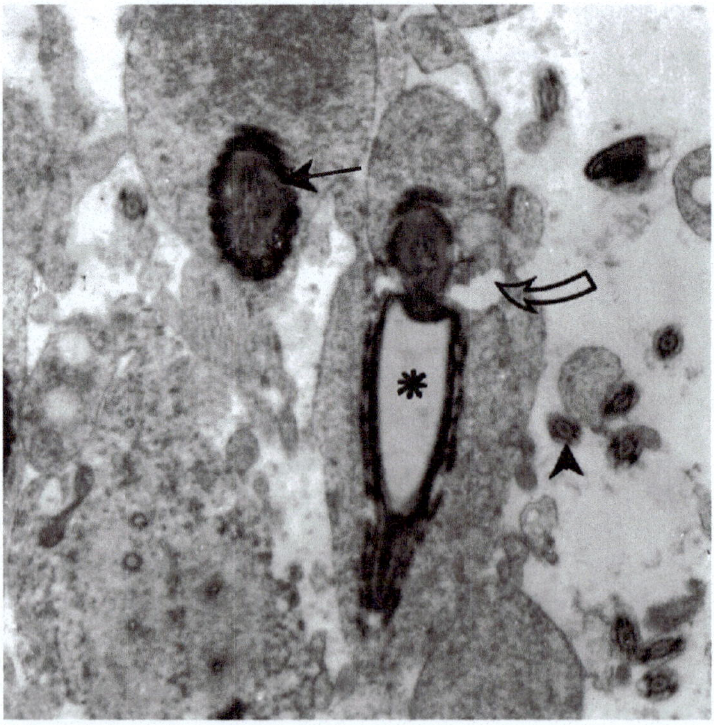

Fig. 2.18 High-power electron micrograph of rat testis following 60 days CPA + TE treatment illustrating a longitudinally cut spermatozoan. The acrosome is distorted and there is vacuolization in it (curved arrows). The nucleus does not contain the characteristic condensed chromatin (asterisk). Sections through the mid-piece and tail regions of the spermatozoa exhibit no abnormality (arrowheads) [X 13,500]. Unpublished electron micrographs from Dr. Bhiwgade

1992a, b) as it is a lysosomal enzyme and a marker for the earlier stages of spermatogenesis and is detected mainly in the Sertoli cells and spermatogonia (Males and Turkington 1971a, b).

In the present study, ultrastructural observations of the Leydig cell showed hypertrophied mitochondria and Golgi apparatus, a disoriented agranular endoplasmic reticulum and a well-developed lysosomal system. These changes revealed hampered functioning of the Leydig cell due to diminished LH secretion as evident from the ultrastructural observation of the pituitary resulting in further suppression of testosterone levels. The suppressed release of LH may have been caused due to the additive effect of DMPA + TE. Since LH acts directly on the Leydig cell which in turn is responsible for testosterone biosynthesis, suppressed testosterone levels are recorded which thus confirms the anti-gonadotrophic and anti-androgenic effect of DMPA + TE. Our results are in good agreement with the findings of the earlier workers (Faundes et al. 1981; Friedl et al. 1985; Hedman et al. 1988; Pangkahila 1991) who were able to detect reduced FSH, LH and testosterone levels in the serum of men after combination regimen of DMPA + TE.

Dorfman (1970) defined anti-androgens as 'substances which prevent androgens from expressing their activity at target sites. The inhibitory effect of these substances, therefore, should be differentiated from compounds which decrease the synthesis and/or release of hypothalmic (releasing) factors, from anterior pituitary hormones (gonadotropins, particularly luteinizing hormone) and from material which acts directly on the gonads to inhibit biosynthesis and/or secretion of androgens'. Cyproterone acetate (CPA) fulfills these criteria and it influences all those organs and organ systems that are functionally or morphologically androgen dependant.

The effects of CPA on various aspects of male fertility are well documented. It inhibits spermatogenesis in all species in a dose-dependant manner (Hamada et al. 1963; Junkmann and Neumann 1964; Neumann 1972, 1973; Ott 1968; Ott and HofTet 1968). Since Whalen and Luttge (1969) first proposed the idea that CPA might be suitable for male contraception, several trials have been conducted and in rats and pigs, the treatment with this steroid led to sterility without loss of libido (Zucker 1966; Beach and Westbrook 1968; Bloch and Davidson 1971; Whalen and Edwards 1969; Whalen and Luttge 1969). However, in certain species, such as the dog, rabbit, boar and man, CPA inhibits libido (Schmidtke and Schmidtke 1968; Horst and Bader 1969; Raspe 1972; Morse et al. 1973; Horn 1974; Neumann and Schenck 1976; Jeffcoate et al. 1980); and especially at high doses, the effect is more obvious.

Sterility during CPA treatment is due to the inhibition of spermatogenesis (Junkmann and Neumann 1964; Neumann et al. 1970). However, it is not possible to use CPA as a contraceptive on the basis of its spermatogenic inhibitory effect because in experimental investigations, much larger doses are needed to inhibit spermatogenesis than the secretions of the accessory sex glands (Steinbeck et al. 1971).

A considerable amount of success has been achieved by using low doses of CPA which did not affect spermatogenesis but altered the functioning of the epididymis and accessory sex organs (Prasad et al. 1970; Nag and Ghosh 1979). But, these results are also controversial and there are many reports where micro doses or low doses failed to induce sterility or suppress the accessory sex glands (Schenck et al. 1975).

In our study, an androgenic ester has been concomitantly administered with the CPA treatment to abolish the feminizing effects of the steroids and also to perceive, if the androgen exerts a synergistic effect of suppression of spermatogenesis.

After the treatment ultrastructures of various cell types in the seminiferous tubules have undergone considerable change. The Sertoli cell shows an accumulation of a large number of dense bodies and lysosomes. The formation of many myelin figures within the Sertoli cell cytoplasm also implies that these are all drug-induced changes. The 6-glucuronidase is a lysosomal enzyme marker for the earlier stages of spermatogenesis and is detected only in the Sertoli cells and spermatogonia (Males and Turkington

1971a, b). Significant reduction in the level of this enzyme also indicates an impaired Sertoli cell and the spermatogonial function evident from the vacuolation of the latter (Bhiwgade et al. 1990; Avari and Bhiwgade 1992a, b).

The impaired Sertoli cell function leads to a decreased production of the androgen-binding protein, ABP, which is transported via the rete testis fluid to the epididymis (Hansson et al. 1973; Schenck and Neumann 1978). Since the Sertoli cells are primary androgen target cells and contain receptors (Mulder et al. 1975; Sar et al. 1975; Sanborn et al. 1977; Tindall et al. 1977), they are also the primary target cells for anti-androgens (Schenck and Neumann 1978). A direct effect of CPA on the germ cells has not been recorded and, though, there are nuclear binding sites on these cells for CPA (Sanborn et al. 1975; Sanborn and Steinberger 1975; Tsai et al. 1977), cytoplasmic receptors have not been demonstrated (Grootegoed et al. 1977).

It is obvious that though the spermatogenic cycle appears to have progressed normally, almost at every step there are physiological alterations due to the CPA treatment. This leads to abnormally developed spermatozoa, which is a very common occurrence in animal models and humans after CPA administration (Koch et al. 1974, 1976; Lorenz et al. 1974; Hammerstein 1974; Roy et al. 1976; Moltz et al. 1978; Fogh et al. 1979; Wang and Yeung 1980). Testosterone supplementation also does not prevent the development of spermatozoa with abnormal morphology in the langur monkey (Lohiya and Sharma 1983).

One of the main objectives of externally administering small doses of a testosterone ester in our study, has been to maintain the normal functioning of the accessory sex glands. In our observations, however, the cells of the seminal vesicles as well as the prostate are completely atrophied and shrunken in size. The secretory apparatus of the cells of these two organs has become totally defunctory. The rough endoplasmic reticulum is reduced to only a few scattered cisternae and it no longer forms parallel arrays. The Golgi apparatus is also very small and very few secretory vacuoles are present. The significant reduction in the weights of these two accessory organs, in our studies, is symbolic of their atrophic state. The prostate is more severely affected than the seminal vesicles and has not recovered fully even after 6 weeks.

The inhibitory effect of CPA on the accessory sex glands is widely known. Since the secretory activity and proliferation of the prostate and seminal vesicles is androgen dependent, a potent antiandrogen like CPA causes regression of these glands in most of the mammals like rats, mice, hamsters, gerbils and guinea pigs. CPA causes atrophy of the prostate and seminal vesicles in intact animals as well as in castrated and substituted with androgens (Steinbeck et al. 1971; Loving and Flickinger 1976; Back et al. 1977; Kalla and Bhasin 1977; Purvis et al. 1978; Schenck and Neumann 1978; Umapathy and Rai 1980). The seminal vesicles and prostate are more androgen-sensitive than the epididymis but this does not corroborate to the findings of Prasad et al. (1970, 1972, 1973). In dogs, which are more sensitive to CPA than rodents, after only 32 days of treatment with 10 mg/kg daily i.m., marked signs of regression in the prostate have been observed (Neumann 1977).

The mechanism of action of CPA involves the ability to directly block the uptake of labelled androgens in the prostatic cell nucleus (Geller et al. 1969). CPA also inhibits the transport of the labelled testosterone metabolite dihydrotestosterone (DHT) from the prostatic cytoplasm to the nucleus (Fang and Liao 1969; Stern and Eisenfeld 1969; Tveter and Aakvaag 1969; Walsh and Korenman 1970; Tymoczko and Liao 1976) and blocks the binding of DHT to the nuclear receptor protein. This results in a decrease in the weight of the prostate and its DNA and RNA contents (Mertelsmann et al. 1968; Neumann 1977; Neumann 1994). This in turn leads to a decline in the synthesis of secretory proteins and, consequently, to the changes that are seen in the cytoplasmic organelles responsible for secretory functions.

The drastic changes in the endoplasmic reticulum, Golgi apparatus and secretory vacuoles would be expected to result in a greatly decreased output of secretory material, and in men treated with CPA, the volume of the ejaculate is indeed reduced (Morse et al. 1973).

In our study, it was hoped that the exogenous testosterone administration would maintain the normal functioning of the accessory sex glands. However, this has not proven to be the case and the testosterone levels were insufficient and low to prevent the alternations in the morphology and physiology of the accessory sex glands. Higher doses of androgens are required to overcome the side effects of CPA treatment. Nevertheless, we have been successful in abolishing the feminizing effects and the effects on libido of CPA. Many other parameters also need to be studied before any conclusions can be drawn as to the efficacy of this combination regimen.

Electron microscopic observations on testis of DMPA + TE treated (45 days) Bonnet monkey, *Macaca radiata*:

The testis of the control Bonnet monkey is composed of a large number of seminiferous tubules which are bounded by a thick fibromuscular layer, the tunica propria and a non-cellular basal lamina. The tunica propria or the peritubular tissue is a complex structure and consists of fibroblast cells with alternately arranged collagen fibres (Figs. 2.19 and 2.20). The fibroblast cells are elongated with tapering ends. The nucleus is elongated with clumps of chromatin material attached to the nuclear membrane. These cells are connective tissue fibroblasts. The innermost layer of tunica propria adhering to the basal lamina consists of flattened myoid cells which exhibit smooth muscle characteristics followed by a layer of collagen fibres. A typical non-cellular basal lamina lies next to the tunica propria. Its inner portion consists of very fine interlacing filaments arranged in parallel lamellae and numerous knob-shaped micro infoldings protruding towards the germ cells (Figs. 2.21, 2.22a, b).

The seminiferous tubule of testis in the control Bonnet monkey is made up of stratified epithelium of about 5–8 layers of cells having different morphologies. The tubule contains the non-germinal cell population, the Sertoli cells and the germ cells in various maturational phases viz. spermatogonia, primary and secondary spermatocytes, spermatids and spermatozoa. The seminiferous epithelium is structurally partitioned by the Sertoli cells into basal and adluminal compartment via the specialized tight junctions between the Sertoli cells (Fig. 2.23). The spermatogonial cells reside in the basal compartment and are found towards the outer margin of the seminiferous tubule resting over the basal lamina. The primary spermatocytes in different stages of cell division are characterized by the chromosomes in different stages of coiling processes within their nuclei. The Spermatocytes at the commencement of their development move into the adluminal compartment. The spermatids are also found in different stages of development as early and late spermatids forming a spermatozoa.

The Sertoli cells in the control monkey are tall columnar cells, attached to the basal lamina and extending radially from basal lamina to the lumen of the tubules. Laterally, the Sertoli cell gives out numerous cytoplasmic extensions forming adluminal compartments, which surround each germ cell, except the spermatogonia (Figs. 2.20 and 2.23). A large lobed nucleus located in the basal portion of cell with an electron lucent nucleolus and uniform chromatin (Fig. 2.22a). The basal Sertoli cell cytoplasm contains rough endoplasmic reticulum, numerous variable mitochondria and shows a characteristic accumulation of large Zebra striped lipid droplets (Fig. 2.22b). In apical cytoplasm, the mitochondria are elongated or dumble shape. Free ribosomes, lysosomes, Golgi membranes, SER and coated vesicles are noted, particularly towards the apex of the cell. Many microtubules and filaments are also observed running parallel to the cell in the basal and apical parts of the cell (Fig. 2.26).

The specialized tight junctional complexes are found between the adjacent Sertoli cells, which provide the anatomical basis of the blood–testis barrier and restrict the entry of large molecular weight particles into the seminiferous epithelium, effectively creating basal and adluminal compartments. The tight junctions create a unique intratubular environment necessary for the process of spermatogenesis (Fig. 2.23).

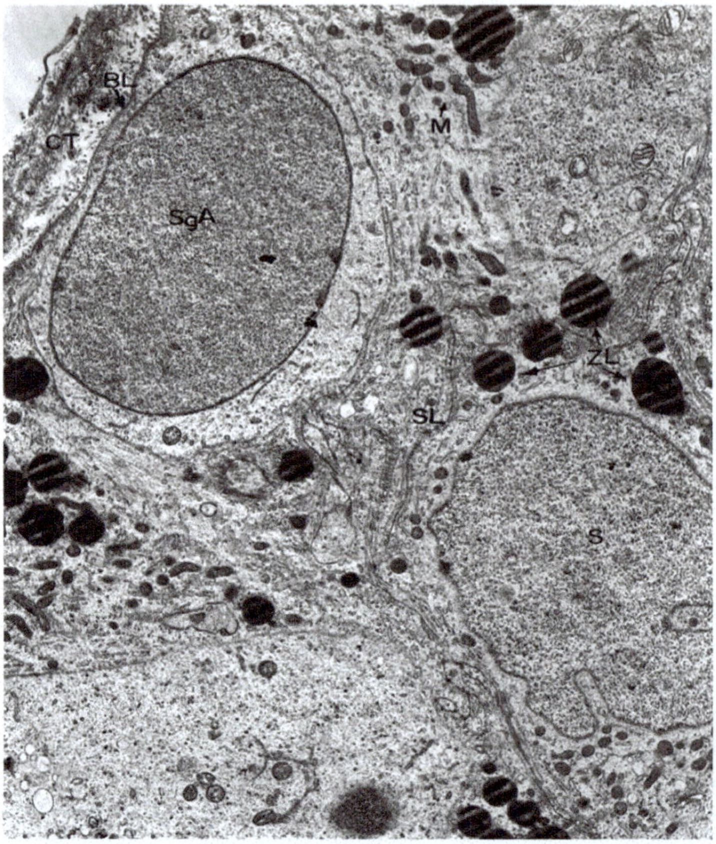

Fig. 2.19 High-power electron micrograph of control Bonnet monkey seminiferous epithelium. Note A type spermatogonium (SgA) resting on basal lamina (BL), connective tissue (CT). Also, note Seroli cell (S) with Sertoli cell junction (SL), zebra lipid droplet (ZL) and numerous mitochondria (M) [X 5000]. Unpublished electron micrographs from Dr. Bhiwgade and Dr. Bansal

2.7 Effect of DMPA+TE Treatment on Sertoli Cells

The seminiferous tubules after DMPA + TE treatment are composed of the Sertoli cells and the germinal cells but the germinal cells show some degenerative changes, deranged and lost normal cyclicity of spermatogenesis at the later stages of development (Fig. 2.24 and 2.25).

Following DMPA + TE treatment, a number of Sertoli cells are observed in seminiferous tubules which appear normal in cytoarchitecture at the basal side, however, they show regressive changes at the apical end (Fig. 2.20). Lipid droplets have been increased significantly, especially in the basal cytoplasm (Fig. 2.25). These are so large in treated animal as to indent the nucleus (Fig. 2.27) and appears to displace other cell organelles. The Golgi cisternae, rough endoplasmic reticulum and mitochondria of different shapes and sizes are reduced after treatment although there is an enlargement of smooth endoplasmic reticulum within Sertoli cells which have undergone hypertrophy and are found in the form of extensive vesiculation after DMPA + TE treatment (Fig. 2.27). Large number of lysosomes like structure filled with either floculent material or pleomorphic contents (the cellular debris) are found within the Sertoli cell cytoplasm (especially in apical cytoplasm) in close vicinity of degenerating gametes after treatment which suggest the phagocytosis or digestion of these structures within the Sertoli cells (Fig. 2.29). The junctional complexes are well marked between the neighbouring Sertoli cells with normal ultrastructure even after DMPA + TE treatment. These are composed of cisternae of rER lying parallel to the plasma membrane and bundles of the actin filaments interposed between the ER and the cell membrane of the two neighbouring cells (Fig. 2.26 and 2.27).

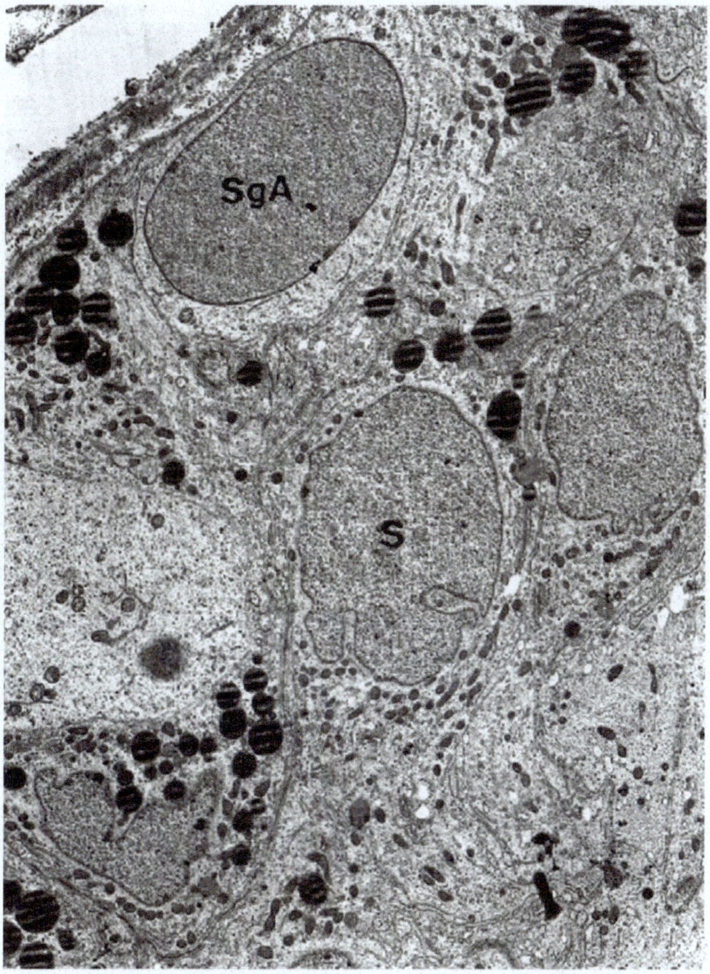

Fig. 2.20 Electron micrograph of control monkey seminiferous epithelium from control Bonnet monkey. Note a type Aspermatogonia (SgA) with spherical nucleus, Sertoli cell (S) cytoplasm with folded nucleus, zebra lipid droplets and numerous mitochondria [X 3500]. Unpublished electron micrographs from Dr. Bhiwgade and Dr. Bansal

2.8 Effect of DMPA+TE Treatment on Spermatogenesis

The germ cell proliferation and maturation within the seminiferous tubules is a complex yet orderly series of cytological transformation in which spermatogonial stem cell gives rise to a cohort of spermatozoa. In control monkey two types of spermatogonia A and B were observed under electron microscope, resting over the basal lamina. A type spermatogonia are oval cells with a large nucleus containing uniform chromatin surrounded by little cytoplasm with scanty membraneous cell organelle such as rough endoplasmic reticulum. Golgi complex, multivesciular bodies and mitochondria (Fig. 2.19).

B type spermatogonia are large, oval in shape containing a large oval nucleus with clear chromatin and many lobed nucleolus and a prominent sex vesicle. The cytoplasm is less dense containing numerous mitochondria with well-developed cristae, polyribosomes and lysosomes (Fig. 2.21). Both types of spermatogonia, 45 days after DMPA + TE treatment reveal no morphological alterations as compared to control except the mitochondria have undergone hypertrophy. Both types of spermatogonia are attached to the

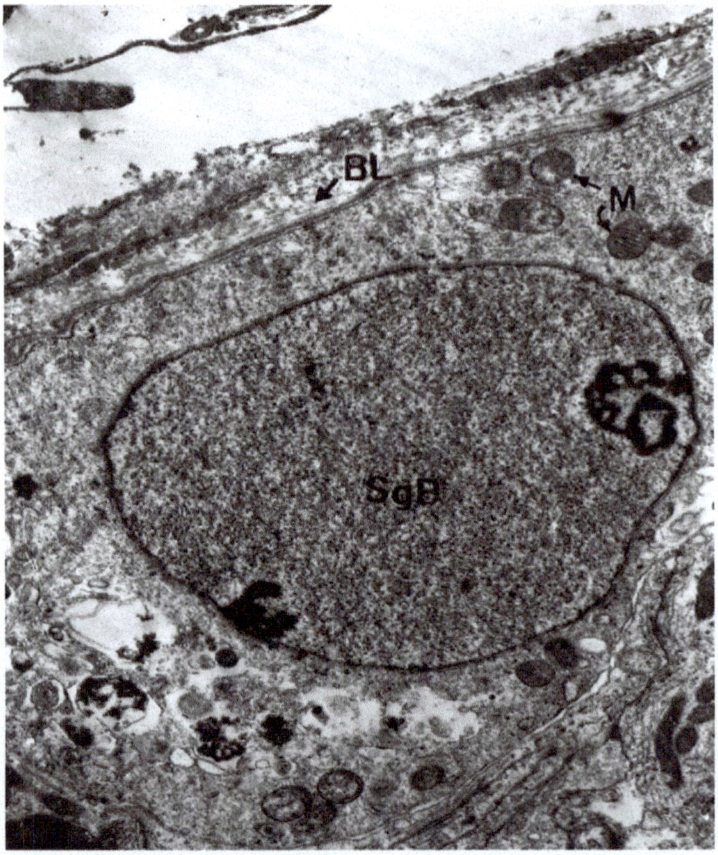

Fig. 2.21 Electron micrograph of seminiferous epithelium from control Bonnet monkey. Note B-type spermatogonia (SgB) with oval nucleus resting on basal lamina (BL). Cytoplasm with few mitochondria and multivesicular bodies [X 8000]. Unpublished electron micrographs from Dr. Bhiwgade and Dr. Bansal

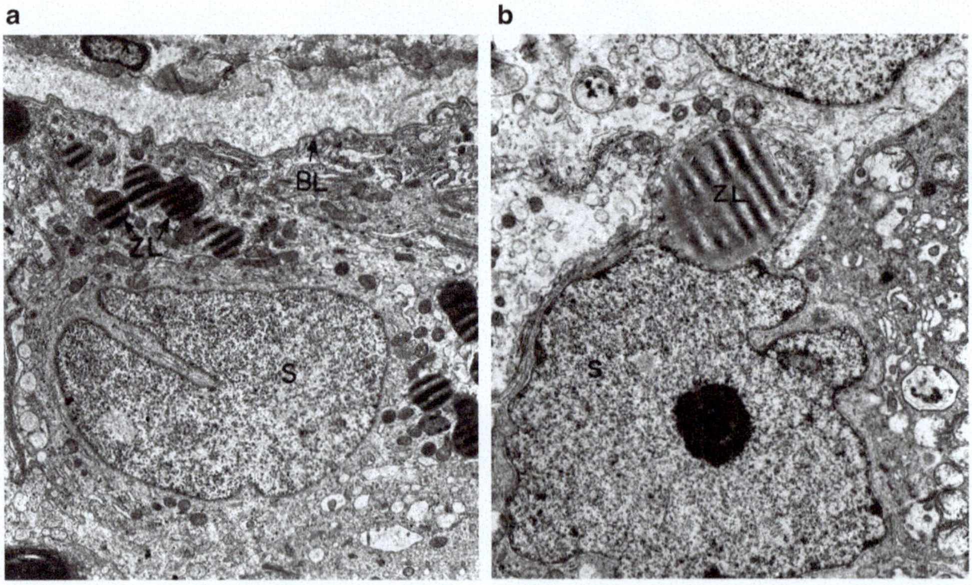

Fig. 2.22 (**a**, **b**) Electron micrographs of seminiferous epithelium from control Bonnet monkey. Note Sertoli cell (S) with a characteristic folded nucleus and resting on basal lamina (BL). Cytoplasm shows numerous zebra lipid droplets (ZL), Golgi complex and mitochondria [X 5000] [X 6000]. Unpublished electron micrographs from Dr. Bhiwgade and Dr. Bansal

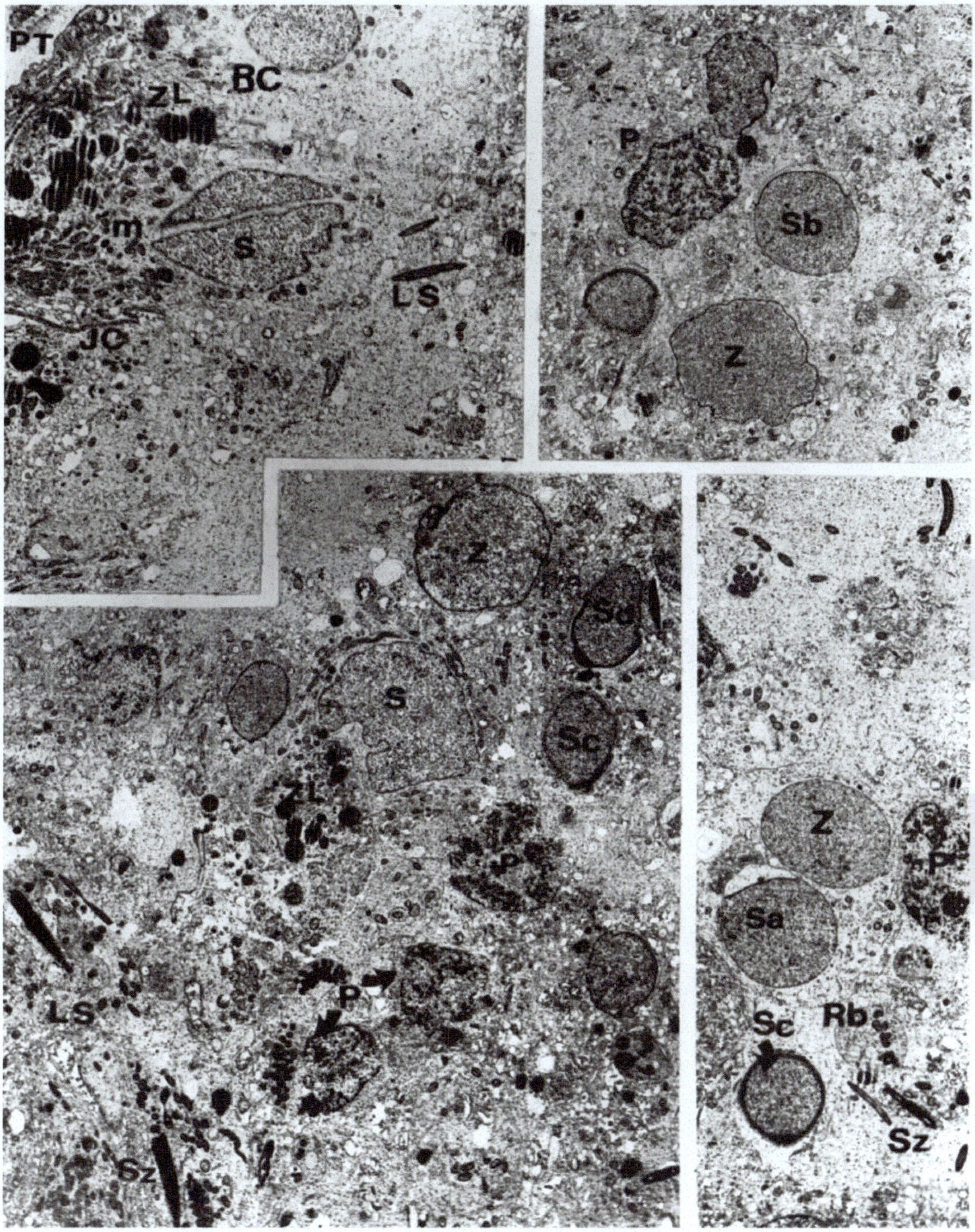

Fig. 2.23 Photomontage: Low-power electron micrograph of testis of control male Bonnet monkey showing seminiferous epithelium. Note peritubular tissues (PT), Sertoli cells (S), and Primary spermatocytes in different stages of morphogenesis (Sa, Sb, Sc, Sd). The Sertoli cells exhibit intercellular junctional complexes (JC), Basal compartment (BC) and cytoplasm containing nucleus, variable mitochondria (m) and Zebra striped lipid inclusions (ZL) [X 3000]. Unpublished electron micrographs from Dr. Bhiwgade and Dr. Bansal

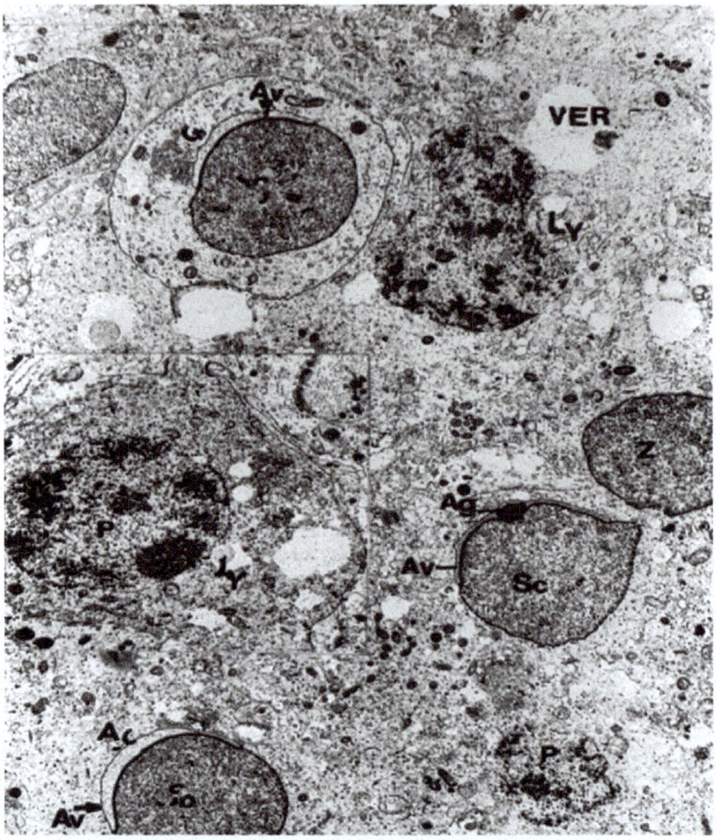

Fig. 2.24 Electron micrograph of seminiferous tubule of Bonnet monkey treated with DMPA + TE for 45 days showing normal appearing Zygotene spermatocytes (P) with degenerating nuclear membrane, associated lysosome (LY) and enlarged vesiculated endoplasmic reticulum (vER). Note the formation of Golgi phase spermatids (Sa), acrosomal and acrosomal cap organelle visible are mitochondria (m) and Golgi apparatus (G). Note Acrosome (AC), Acrosomal granule (Ag) and Acrosomal Vesicle (Av) [X 3500]. Unpublished electron micrographs from Dr. Bhiwgade and Dr. Bansal

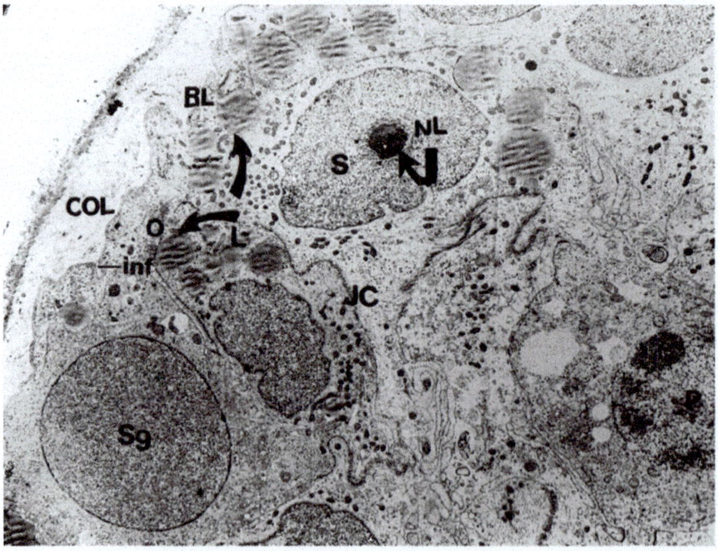

Fig. 2.25 Electron micrograph of basal aspect of the seminiferous epithelium in a DMPA + TE treated Bonnet monkey illustrating Sertoli cells nucleus (S) with Nucleoli (NL), and their specialized intercellular junctional complexes (JC). Note the presence of accumulated lipid droplets (L) in the basal compartment. The spermatogonia (Sg) and the pachytene primary spermatocyte (P) (without intact nuclear membrane) are visible. The peritubular tissue in the form of myoid cells (My), collagen fibers (COL) and Basal Lamina (BL) is well indicated [X 6000] Unpublished electron micrographs from Dr. Bhiwgade and Dr. Bansal

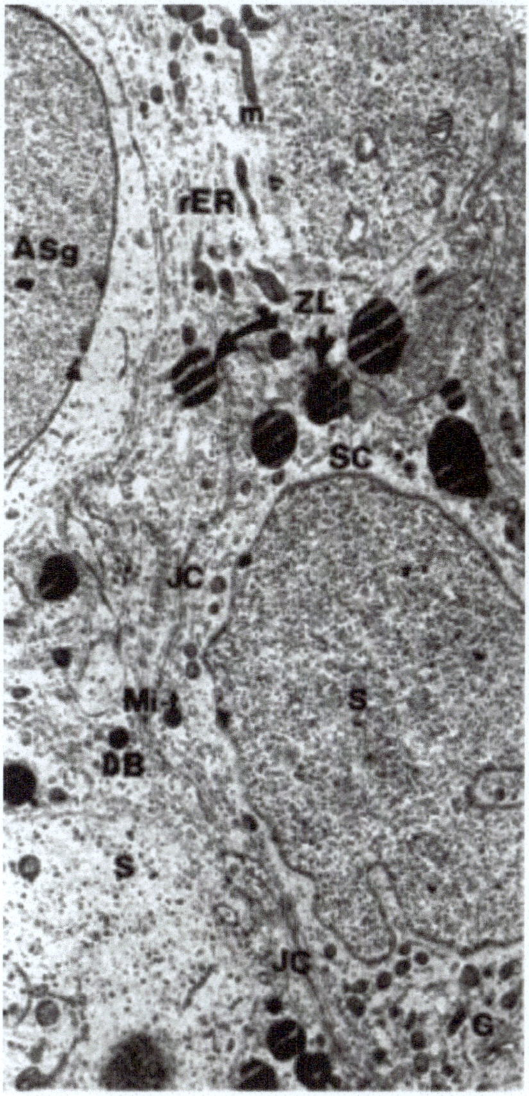

Fig. 2.26 High-power electron micrograph of seminiferous epithelium from control Bonnet monkey illustrating type A spermatogonia (A Sg) with large nucleus containing uniform chromatin and scanty cytoplasmic organelle. Note the Sertoli cell (S) with folded nucleus, Mitochondria (M), dense bodies (DB), rough endoplasmic reticulum (rER) and Zebra lipid droplets (ZL) between the neighboring cells [X 5000]. Unpublished electron micrographs from Dr. Bhiwgade and Dr. Bansal

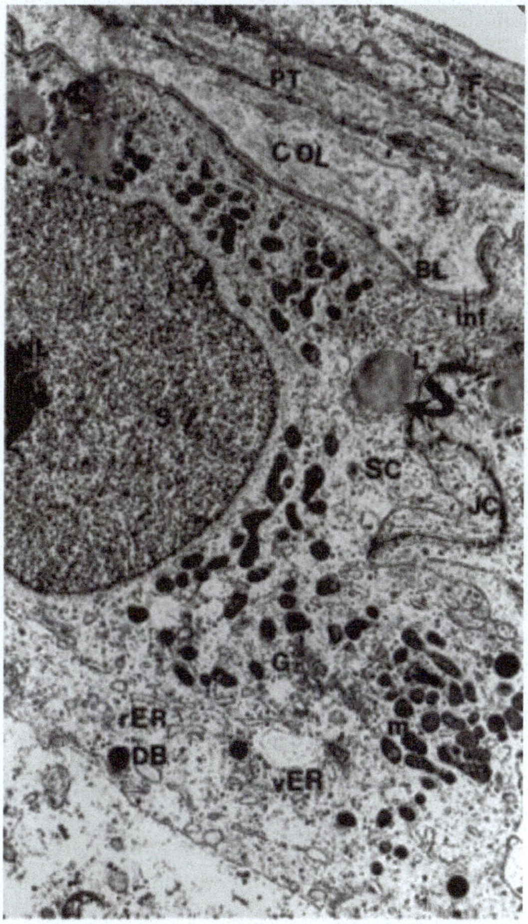

Fig. 2.27 High-power electron micrograph of treated monkey seminiferous epithelium from treated Bonnet monkey illustrating peritubular tissue (PT) with a thick layer of collagen fibre (COL), reduced fibroblast (F), Myoid cells (My) and much folded (inf) basal lamina (BL). Note the basal portion of Sertoli cells containing nucleolus (NL), variable mitochondria (m), reduced Golgi (G), accumulated large lipid droplets (L), Dense bodies (DB) and rough and vesiculated type endoplasmic reticulum (rER, vER) [X 6500]. Unpublished electron micrographs from Dr. Bhiwgade and Dr. Bansal

basal lamina as in the case of control one. The cytoplasm of B type of spermatogonia contains reduced endoplasmic reticulum, Golgi complex and lysosome-like structures seems to be engulfing mitochondria after DMPA + TE treatment (Figs. 2.20 and 2.25).

B type spermatogonia produce spermatocytes as a result of mitotic division which are classified as primary and secondary spermatocytes. Morphological distinction between the various types of spermatocytes is achieved via recognition of their position within the seminiferous epithelium, the arrangement of chromatin within the nucleus and the size of the cells. Primary spermatocytes are the largest cells in the seminiferous tubules and are round in shape. During Leptotene stage thin threads of chromatin are visible, Zygotene is being marked by the presence of synaptonemal complexes formed by the union of homologous chromosomes. The pachytene stage is identified by the short and thick chromosomes. The pachytene stage is the most frequently seen stage due to the most prolonged part of the meiosis (Fig. 2.23). The spermatocytes, after DMPA + TE treatment, are observed in Zygotene and pachytene stages with almost normal cytoarchitecture (Figs. 2.24 and 2.25). The Zygotene spermatocyte has a large rounded nucleus with groups of paired chromosomes in the nucleus. In treated animals, the nuclear membrane in pachytene spermatocytes is not intact which is an indication that the process of degeneration has been started. The surrounding cytoplasm contains hypertrophied mitochondria, degenerating Golgi complexes, free ribosomes and numerous vacuoles (Fig. 2.24).

Development of early spermatids into spermatozoa constitutes the process of spermiogenesis. The early formed spermatids are the spherical cell and during spermiogenesis, the nucleus progressively becomes condensed to form the sperm head which is connected viz. neck to a flagellum (Fig. 2.23). Many of these maturational stages are also observed in DMPA + TE treated seminiferous tubules. Some of them have undergone degeneration and even death as indicated by their altered cytoarchitecture. In control monkeys, the Golgi phase spermatids, Sa, Stage 5 (Fig. 2.28) is characterized by a large nucleus with uniform chromatin. The Golgi vesicles are grouped at one side of the nucleus, forming acrosomal vesicle containing a prominent core, the acrosomal granule and is filled with light floculent material. A dense layer is seen along the contact zone of acrosomal vesicle. The Mitochondria have undergone hypertrophy with plate-like cristae. Rough endoplasmic reticulum, many small coated vesicles are also found in surrounding cytoplasm. A prominent Golgi complex with several vesicles and granules is desernible very close to the acrosomal vesicle (Fig. 2.28).

The Golgi phase spermatid does not exhibit any alteration in structure following DMPA + TE treatment except the cytoplasmic membraneous organelles are scanty and show an atrophy viz. Rough endoplasmic reticulum and Golgi complex (Fig. 2.29).

In control monkey, next stage, the Cap phase spermatid, Stage 6 marked as Sb in Fig. 2.30, is characterized by a slightly elongated nucleus with an acrosomal vesicle covering half of the nucleus to form a head cap. The denser layer is also seen at the contact zone in between acrosomal vesicle and nucleus, formed by the distribution of dense acrosomal material. Nucleus has the uniform chromatin. A centriole is found to be attached to the nucleus, opposite to the acrosomal cap in the form of basal body of future flagellum. The cytoplasm contains rough endoplasmic reticulum, hypertrophied mitochondria and lysosomal bodies.

The cap phase spermatid after DMPA + TE treatment (Fig. 2.31) has the normal appearing nucleus with an acrosomal vesicle and granule and a denser layer at the contact zone, but degenerative changes are observed in the surrounding cytoplasm. An atrophied Golgi complex lies juxtanuclear position, with few cisternae along with a few secretory vesicles and vacuoles. Atrophied mitochondria with collapsed cristae, coated vesicles, lysosomes, vesiculated rough endoplasmic reticulum and free ribosomal rosettae are visible in the cytoplasm. The important structures observed are two large vesiculations filled with light material, resembling lysosomes, which lies very close to the cap phase spermatid in the surrounding Sertoli cell cytoplasm, probably for the digestion of degenerating gametes (Fig. 2.24).

In the next stage, the Acrosomal cap phase spermatids, Stage 7 Sc (Fig. 2.32a) of the control monkey, the nucleus shows further elongation of acrosomal cap and acrosomal granule which covers almost half of the nucleus. Chromatin becomes more dense and uniformly distributed within the nucleus. A thick band of microtubules, the manchetae are observed at both sides of the

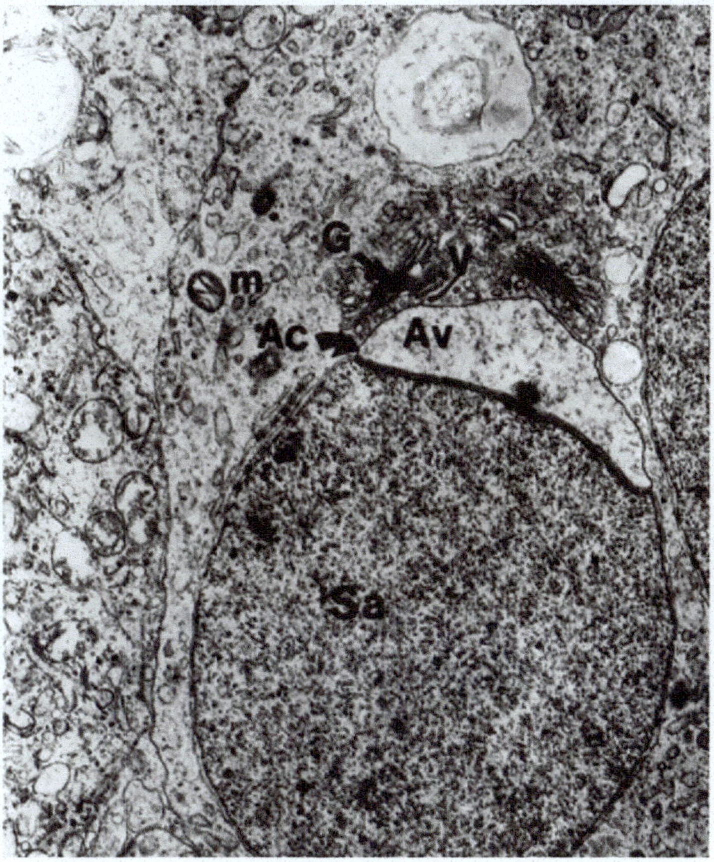

Fig. 2.28 Electron micrograph of Golgi phase spermatid (Sa) Stage 5, from a control Bonnet monkey. The nucleus is slightly elongated with a forming acrosomal visible (AV) and acrosomal granule (Ag) attached to one pole of nucleus. These structures are formed by the contribution of Golgi complex (G). Note the presence of vaculated mitochondria (m) and vesiculated endoplasmic reticulum (vER) arranged peripherally [X 10,000]. Unpublished electron micrographs from Dr. Bhiwgade and Dr. Bansal

remainder nucleus. The cytoplasm along with cell organelle has shifted to the lower side of the nucleus. The cytoplasm contains a darkly stained body, composed of anastomosing strands of packed filaments, the chromatid body of obscure significance.

Following the DMPA + TE treatment the acrosomal cap phase spermatid with the normal appearing nucleus, and the degenerating surrounding cytoplasm has been identified. These degenerative changes seen are atrophied mitochondria, accumulation of lysosomes with cellular debris, dense bodies, atrophied Golgi complex and vesiculated rER (Fig. 2.32b).

The acrosomal head cap phase Sd, (Stage 9/10) Spermatid in control bonnet monkey (Fig. 2.33), is characterized by the elongation of cell and nucleus with condensation of nuclear material. The nucleus becomes eccentric, with the displacement of spermatid cytoplasm to one side of the nucleus. The anterior protruded part of the nucleus is covered with an acrosomal cap having an acrosomal granule, spread over two-third of the nucleus in the form of acrosomal vesicle. The acrosome and acrosomal vesicle consists of an outer and inner acrosomal membrane containing electron-dense acrosomal material. The bundles of microtubules are seen which limits the cell laterally and possibly acting as a framework for shifting of cytoplasmic contents. The flagellum is seen emerging from lower side of the nucleus. A large number of hypertrophied mitochondria, rough endoplasmic reticulum, many lysosomes and chromatid body are observed in the surrounded cytoplasm. The acrosomal head cap phase spermatids are not observed in the testis of DMPA and TE treated bonnet monkey.

In testis of control monkey, the late spermatid (Stage 15) at the later stages of maturation is seen

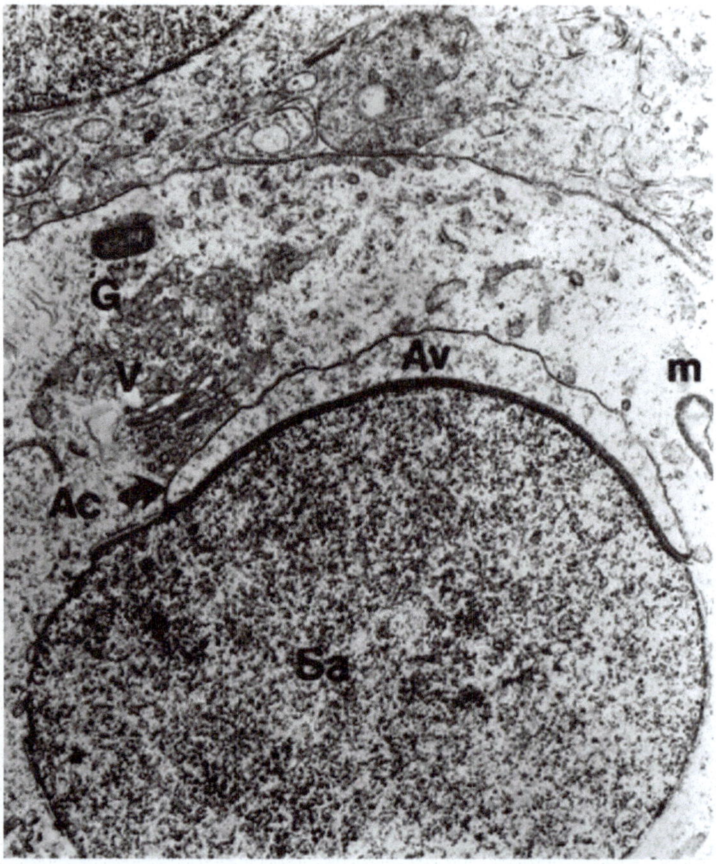

Fig. 2.29 The Golgi phase spermatid (Sa) from a Bonnet monkey treated with provera and testosterone enanthate for 45 days. This cell does not exhibit any alteration following treatment. It displays a reounded nucleus, large Golgi apparatus (G) with vacuoles (V) and the acrosomal vesicle (Av) covering the front pole of the nucleus forming acrosome (AC). The cytoplasmic organellae are scanty with few mitochondria and vER shows atrophy [X 10,000]. Unpublished electron micrographs from Dr. Bhiwgade and Dr. Bansal

(Fig. 2.34). There is complete elongation of nucleus as well as stretching of the acrosomal head cap over the head end. The nucleoplasm within the head is very dense (Figs. 2.34 and 2.35). The acrosome of sperm is a cap like with an inner acrosomal membrane adherent to the nuclear envelope and is continuous at the posterior margin of the cap with the outer acrosomal membrane. The two membrane run parallely, enclosing a cavity filled with enzyme-rich homogenous material. During Late Spermatid, a conspicuous thickening of the cap, i.e. the apical segment, the main portion of the acrosome, the principal segment and the caudal region with slight condensation of contents, i.e. equatorial segment are clearly visible. The distal centriole of maturing spermatid is attached to the equatorial region of nucleus and emerges as flagellum. The cytoplasm shifts back as a cytoplasmic projection around the initial path of the tail, converging aggregation of mitochondria to form the middle piece of spermatozoa. Many residual cytoplasmic bodies are also seen (Fig. 2.34). This stage has not been observed in the testis of bonnet monkey following DMPA + TE treatment.

The residual cytoplasm contains Golgi complex, lipid droplets, mitochondria, lysosomes and spherical masses. The residual cytoplasm undergoes degeneration and is phagocytosed within the apical Sertoli cell cytoplasm (Fig. 2.34). Although late spermatids are totally absent in DMPA + TE treated testis but some degenerating late spermatids can be observed at some places within Sertoli cell cytoplasm. One late spermatid with an elongated dense nucleus surrounded by dense cytoplasm with cell organelles is observed in the treated testis (Fig. 2.25). Many degenerating residual cytoplasmic bodies bounded by a limiting membrane containing hypertrophied mitochondria, dense

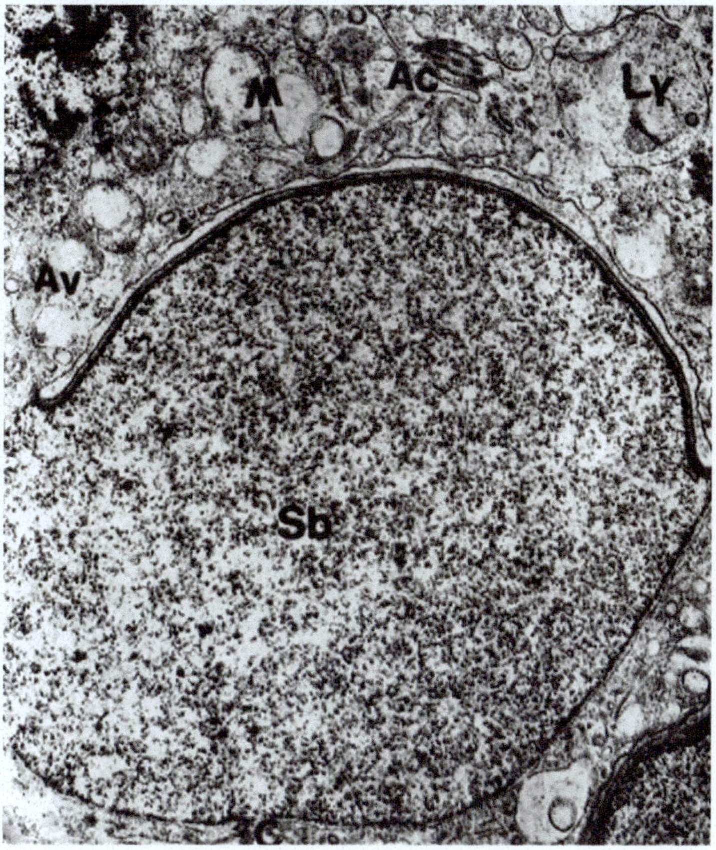

Fig. 2.30 Electron micrograph of the cap phase spermatid (Sb) stage 6, from the control monkey's testis. The nucleus is slightly elongated and the front pole of the nucleus is covered by the acrosomal cap (Ac) formed by Acrosomal Vesicle (Av) covering half of the nucleus. A dense layer is also seen at the contact zone. Note the vesiculated endoplasmic reticulum (vER) and few lysosomal bodies LY) in the peripheral cytoplasm [X 13,000]. Unpublished electron micrographs from Dr. Bhiwgade and Dr. Bansal

bodies and other cell organelles are also observed after DMPA + TE treatment (Fig. 2.25).

2.9 Discussion

The administration of a single dose of 120 mg of DMPA and 125 mg of TE through parenteral route has reduced the weight of the testis after 45 days of treatment, as compared to the normal bonnet monkey. Following treatment the gross morphological appearance of the testis is the same, however, they were a bit smaller than that of the control one. The reduction in the weight and size of the testis of DMPA treated monkey points to the reduced androgen level in the testis as well as in blood and reduced androgen binding proteins (ABP) in the testis. The male sex organs are highly dependent on androgenic hormones to maintain their normal structure and functions (Sugimura et al. 1986) and are very sensitive to the level of circulating androgen. The drop in the testis weight of male bonnet monkey, after 45 days of treatment is the resultant effect of diminished gonadotrophs, FSH and LH which otherwise stimulates the testicular seminiferous tubules and interstitial cells, respectively, to promote the spermatogenesis. LH profoundly influences the Leydig cell function and Sertoli cells are influenced by FSH. The additive effect of DMPA + TE have also played an important role in suppression of testicular weight and spermatogenesis (Heller et al. 1950a, 1970; Flickinger 1977a, b; Avari and Bhiwgade 1992a, b). Scarcity of late germ cell also helps to account for the small size and reduced weight of treated testis.

In the present study, following DMPA + TE treatment, the light microscopic results show a marked decrease in testicular cytoarchitecture parameters in the tubular and intertubular

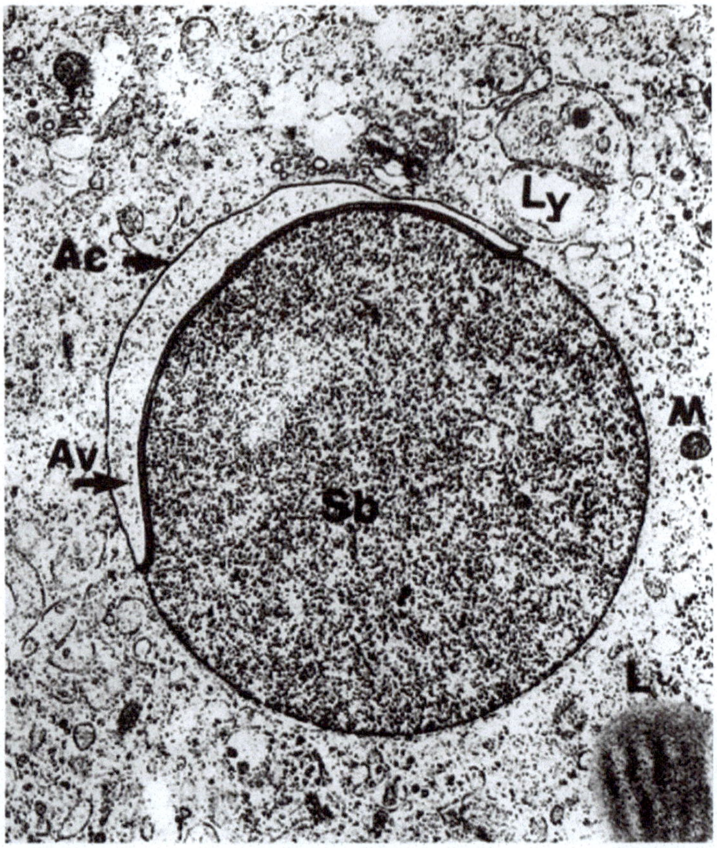

Fig. 2.31 A cap phase spermatid (Sb) Stage 6 form the Bonnet monkey treated with DMPA + TE for 45 days. In treated animal cap phase spermatids retained the normal appearance and exhibits a round nucleus, reduced Golgi apparatus (G), the characteristic acrosomal cap (AC) with Acrosomal vesicle (AV) and few lysosomes, mitochondria and lipid droplet in the surrounding cytoplasm [X 10,000]. Unpublished electron micrographs from Dr. Bhiwgade and Dr. Bansal

compartments, and some changes in the Sertoli cells. The Seminiferous tubules show a marked shrinkage with some of the later stages of spermatogenesis not observed due to the androgen deficiency. The epithelial cells lost their rigidity, regressed and the secretory activity is reduced as the central lumen is devoid of any secretion. A significant reduction in the interstitial volume, total Leydig cells and macrophage are also observed.

The electron microscopic study of the seminiferous tubules epithelium in the testis of DMPA + TE treated animals shows only early stages of germ cells and Sertoli cells due to arrest of spermatogenesis at late spermatid stage.

The ultrastructure of the lamina propria is also disturbed in case of DMPA + TE treated bonnet monkey as compared to the control one as a result of arrest of spermatogenesis. The lamina propria grows thick following the treatment caused by increasing deposition of extracellular matrix between the cellular components. The fibroblast and myoid cells are arranged in a non-continuous layer surrounding the seminiferous tubules. It has been reported that various forms of hypospermatogenesis is accompanied by different forms of thickening of lamina propria. In human beings, the cells of the lamina propria produced increased levels of extracellular matrix under different pathological conditions. An intact lamina propria is of paramount importance for the maintenance of spermatogenesis.

DMPA + TE treatment of the male bonnet monkey impairs the reproductive performance of males as indicated by stage-dependent arrest of spermatogenesis. Electron microscopic study suggests that the arrest of spermatogenesis is visible in Stage VII tubules only and primary pachytene spermatocytes, cap phase spermatids, Step 7 and Step 19 spermatids exhibit pycnotic changes in the treated testis. The tubules contains mainly spermatogonia, spermatocytes and early

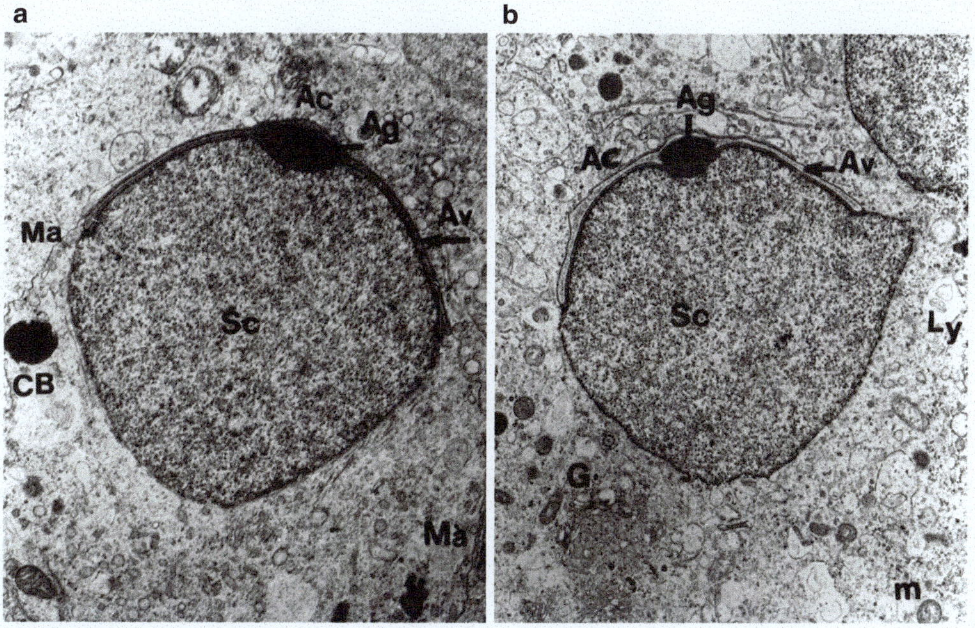

Fig. 2.32 (**a**, **b**) Electron micrograph of Acrosomal cap phase (Sc) Stage 7 spermatid of the control and treated monkey. The nucleus is elongated and the front pole of the nucleus is covered b the acrosomal cap (Ac) containing Acrosomal granule (Ag). The substance of granule spreads laterally to occupy the entire interior of the acrosomal cap. The nucleus shows fine chromatin. From the ends of the cap start the tubules of the manchatae (Ma). A chromatoid body (CB) is also visible. Note the atrophied mitochondria (m), accumulation of lysosomes (Ly), few dense bodies and atrophied Golgi complex in the surrounding sytoplasm [X 10,000]. Unpublished electron micrographs from Dr. Bhiwgade and Dr. Bansal

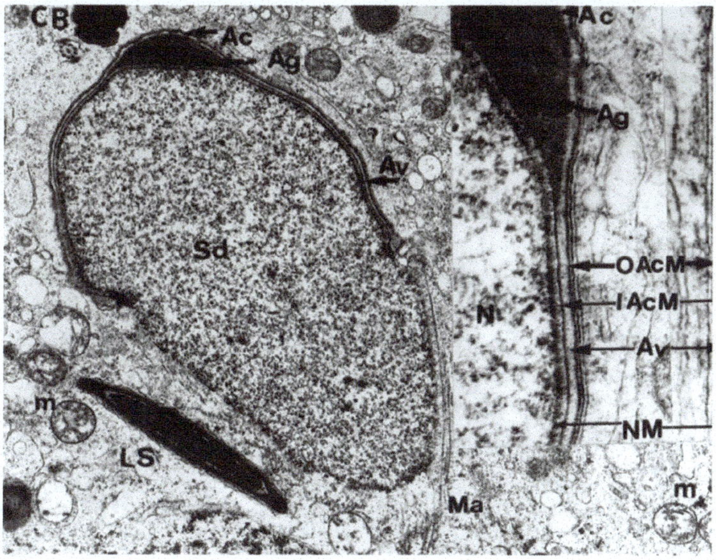

Fig. 2.33 Electron micrograph of control monkey showing elongated head cap phase spermatid (Sd) Stage 9/10. Note the presence of Acrosome (AC) with anterior Acrosomal granule (Ag) and lateral Acrosomal vesicle (AV). Note the outer and inner acrosomal membrane (OACM + IACM) enclosing acrosomal cesicle (NM). Also note the presence of manchatae (Ma), chromatoid body (CB) close to the spermatid. Many hypertrophied mitochondria (m) multivesicular bodies (mVB) and one late spermatid (LS) are also visible in the surrounding cytoplasm [X 10,000] [X 50,000]. Unpublished electron micrographs from Dr. Bhiwgade and Dr. Bansal

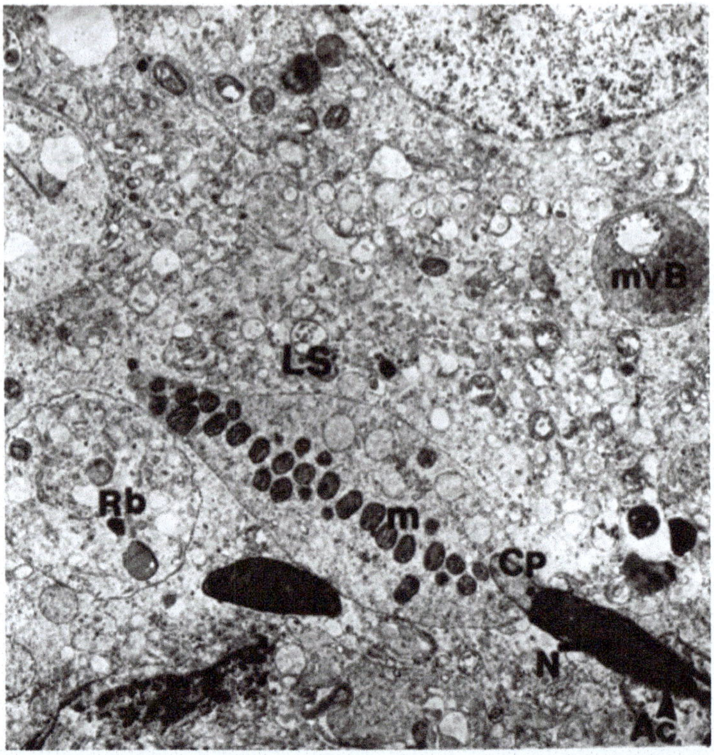

Fig. 2.34 Electron micrograph of late spermatid (Stage 15) illustrating elongated nucleus (N), with dense chromatin and covered with an Acrosome (Ac). Note the cytoplasmic projection containing converging aggregation of mitochondria to form the middle piece. A small connecting piece (CP) with distal centriole is visible between head piece and tail piece. Note the presence of Residual body (Rb) and multivesicular body (mvB) [X 7000]. Unpublished electron micrographs from Dr. Bhiwgade and Dr. Bansal

spermatids at Golgi and cap phase stage (Kerr 1991), however, the later stages of spermatids and spermatozoa are missing due to arrest of spermatogenesis at the later stages of germ cells as an effect of drug. Flickinger (1977a, b) had noticed similar changes in DMPA + TE treated rat testis. Similar results are also observed in hypophysectomized rats and hamsters (Russel and Clermont 1977; Ghosh et al. 1992), and CPA treated rats and Rhesus monkey testis (Prasad et al. 1970; Jagdeep Kaur et al. 1992). These deleterious effects could be a result of inhibition of androgens at the receptor level (Dym 1977; Flickinger 1977a, b; Russell et al. 1981). Pachytene primary spermatocytes and Steps 7 and 19 spermatids at Stage VII are the most sensitive to testosterone withdrawal following DMPA treatment as confirmed in rats testes after treatment with, a Leydig cells cytotoxic drug (Kerr et al. 1993). One important conclusion can also be drawn from our study is that the synthetic progestogen is not directly cytotoxic to the seminiferous epithelium as androgen supplementation in sufficient dose or withdrawal of DMPA treatment can maintain the normal spermatogenesis with no disruptive effects on the Sertoli cells.

A further indication of androgen-dependent seminiferous epithelial disruption following DMPA + TE treatment is the appearance of large basally situated vacuoles. The development of vacuoles occurred chiefly in Stage VII at or above the position of Sertoli cells nuclei and in association with the appearance of the cavities around the pycnotic pachytene spermatocytes or rounded spermatids, similar to the vacuoles described in the same stage in hypophysectomized rats, and seasonal breeders (Russel and Clermont 1977) and treated rats (Kerr et al. 1993; Avari and Bhiwgade 1992a, b). It is still not known whether the vacuole formation occurs as a direct consequence of germ cell necrosis or a non-specific response of Sertoli cells to androgen withdrawal. Earlier reports have suggested that the vacuole formation is a morphological indicator of Sertoli cell damage in hypophysectomized rats. In the present

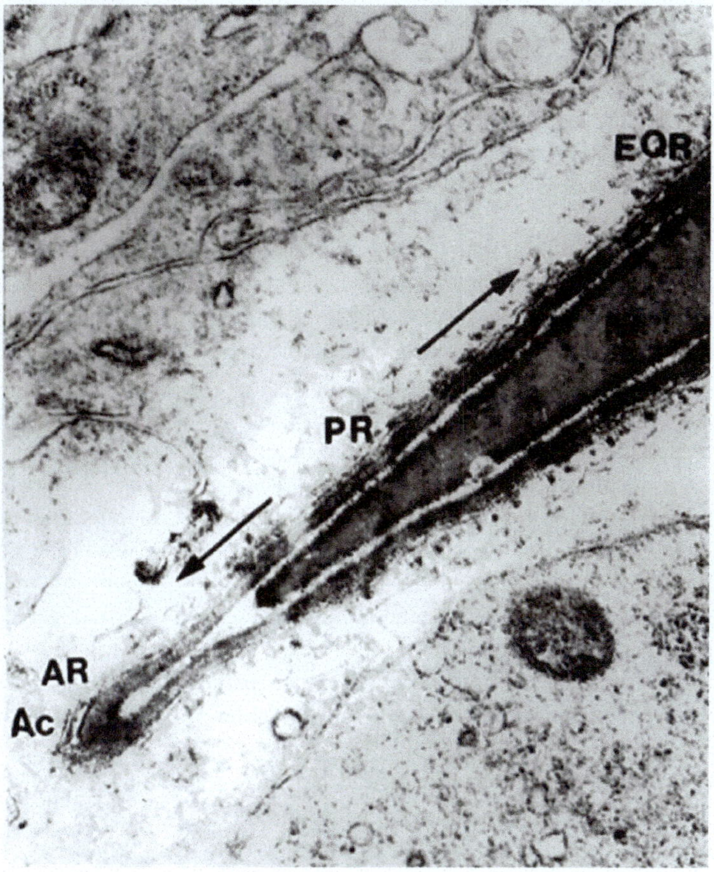

Fig. 2.35 High-power electron micrograph of control testis showing sperm acrosome. Note three distinct regions viz. Apical region (AR) covered with acrosome (Ac), Principal region (PR) and equatorial region (EQR). Note the condensation of chromatin material and two distinct acrosomal membrane [X 30,000]

study, the presence of large vacuoles within the Sertoli cells and occasionally extracellular, under the ultrastructure analysis are lysosomal in nature for the phagocytosis of degenerating germ cells.

Sertoli cells and Leydig cells are the other key cells that are known to respond to endocrine-induced changes due to the use of contraceptives because these cells are very sensitive to the gonadotroph hormone deprivation produced by the use of DMPA + TE (Flickinger 1977a, b); similar to the hypophysectomized hamster and rats (Ghosh et al. 1992) and testicular disorder in human being. It has been reported that DMPA + TE at normal dose (200 mg DMPA and 200 mg TE) induced intratesticular testosterone deficiency and impairs Sertoli cell functions, prevents normal condensation of sperm head leading to defective sperms maturation and failure to subsequent expression of sperm function.

The Sertoli cells, which contain FSH and androgen receptors and are thought to be a mediator of FSH and testosterone (Sanborn et al. 1975) action exhibits few changes following DMPA + TE treatment when the levels of this hormone are already decreased; maximally due to the effect of progestogen (Alvarez-Sanchez et al. 1977; Pangkahila 1991). The Sertoli cell nucleus and other cell organelles do not change much, especially in the basal region although the cell surface in the apical region decreased significantly. The cell organelles viz. mitochondria, ER and Golgi apparatus show slight atrophy and the only organelle that increased is the lysosome; due to an expected increase in germ cell degeneration. It is not clearly known whether these changes are the primary effect of hormonal deprivation or are secondary due to the loss of germinal cell contact subsequent to germ cell degeneration. Since FSH is known to influence Sertoli cell shape in culture to increase its surface area, the lack of FSH may be responsible for these changes and decrease in stimulation of adenyl cyclase, cyclic AMP, protein Kinase and protein biosynthesis resulting in lowered ABP production (Bardin et al. 1981).. These factors might play a role in testicular changes.

Accumulation of lipid droplets of considerable large size following treatment is the noteworthy change that could be responsible for displacing the otherwise normal cytoarchitecture of the Sertoli cells. Similar ultrastructural changes have also been reported earlier in DMPA and TE treated rats (Flickinger 1977a, b) and in hamsters as a result of hypophysectomy and seasonal regression (Ghosh et al. 1992). Accumulation of large lipid droplets in the Sertoli cells of treated bonnet monkeys may be the result of ingested residual cytoplasm of degenerating germ cells and alterations in Sertoli cell metabolism within itself. Lipid accumulation, a well-known pathological change; occurs in many cells under a variety of deleterious conditions and other conditions in which spermatogenesis is arrested including cryptorchidism, hypophysectomy, local heating, oestrogen and androgen treatment, testicular feminizing syndrome and various other drug treatments (Chung 1974; Flickinger 1977a, b). Accumulation of cholesterol indicates its reduced metabolization as the Sertoli cells have the capacity to produce cholesterol de novo from acetates (Wilbe and Tilbe 1979) and is the precursor of androgen biosynthesis.

In monkeys, the ectoplasmic junctional complex is found in two locations in Sertoli cells; one at the site of attachment of spermatids and the other at sites of junction between adjacent Sertoli cells forming blood–testis barrier. They consist of a layer of actin filaments together with a cistern of endoplasmic reticulum on one side of the filament layer and regions of the plasma membrane involved with intercellular attachment on the other (Flickinger and Fawcett 1967; Dym and Fawcett 1970; Russell 1977a). The actin filaments are organized into distinct bundles and are crosslinked to each other and to adjacent membrane (Vogl et al., 1985; Groove and Vogl 1989 and Yazama et al., 1991). The natural or pharmacological disruption of actin filaments at apical sites results in a loss of intercellular adhesion (Russell 1977a, b, 1988; Weber et al. 1988). DMPA + TE treatment shows no effect on these blood–testis barriers in our study.

In conclusion, our investigations suggest that DMPA and TE treatment causes a more generalized response on testis, symptomatic of hypophysectomy and blockage of hormones controlling spermatogenesis. It reduces the weight of the testis, affects the Sertoli cell's function, alters the enzymes essential for spermatogenesis, causes a degeneration of specific spermatogenic cells and drops sperm production. The pattern of these effects suggests an antiandrogenic action of DMPA and the dose of testosterone; exogenously administered, is insufficient to maintain the testicular weight and spermatogenesis.

References

Altman K, Gordon GC, Southern AL, Viltek J, Wilker S (1972) Induction of hepatic testosterone, a–ring reductase by medroxyprogesterone acetate. Endocrinology 90:1252

Alvarez-Sanchez F, Faundes A, Brachi V, Leon P (1977) Attainment and maintenance of azoospermia with combined monthly injections of medroxyprogesterone acetate and testosterone enanthate. Contraception 15: 635–648

Amatayakul K, Sivassomboom B, Singkammani R (1979) Effects of medroxyprogesterone acetate on serum lipids, proteins, glucose tolerance and liver function in Thai women. Contraception 21:283

Avari KM, Bhiwgade DA (1992a) Effect of depot medroxyprogesterone acetate and testosterone enanthate on the testis of albino rats: ultrastructural and biochemical studies. Ind J Exptl Biol 30:11181127

Avari KM, Bhiwgade DA (1992b) Effects of medroxyprogesterone acetate and testosterone enanthate on the testis of albino rats: ultrastructural and biochemical studies. Ind J Exp Biol 30:1118

Back DJ, Glover TD, Shenton JC, Boyd GP (1977) Some effects of cyproterone and cyproterone acetate on the reproductive physiology of the male rat. J Reprod Fert 49:237–243

Barbieri RL, Ryan KJ (1980) Direct effects of medroxyprogesterone acetate (MPA) and Megestrol acetate (MGA) on rat testicular steroidogenesis. Acta Endocrinol 94:419

Bardin CW, Musto N, Gansalus G, Kotita N, Cheng SL, Larrea F, Backer R (1981) Extracellular androgen binding proteins. Annu Rev Physiol 43:189

Beach F, Westbrook W (1968) Morphological and behavioural effects of an 'antiandrogen' in male rats. J Endocrinol 42:379–382

Bhiwgade DA, Menon SN, Avari KM (1990) Effect of cyproterone acetate on testis of albino rats: ultrastructural and biochemical approach. Ind J Exp Biol 28: 201–207

Bhiwgade DA et al (1989) Ultrastructural and functional characteristics of anterior pituitary cells in the Indian fruit bat, Rousettus leschenaulti. Acta Anat 175:129–149

Blance A (1980) On the functional significance of LDH-X. John Hopkins Med J 146:231

Bloch GJ, Davidson JM (1971) Behavioural and somatic responses to the antiandrogen cyproterone (1,20 - methylene - 6 - chloroa 6–17) (X - hvdroxyprogesterone). Horm Behav 2:11–25

Brenner PF, Mishell DR, Bernstein GS, Ortiz AO (1970) Study of medroxyprogesterone acetate and testosterone enanthate as a male contraceptive. Contraception 15: 679–691

Chung KW (1974) Fine structure of Sertoli cells and myoid cells in mice with testicular feminization. Fertil Steril 25:325–335

Dorfman RI (1970) Biological activity of antiandrogens. Br J Dermatol 82(Suppl. 6):3

Dym M (1977) The fine structure of the monkey (Macaca) Sertoli cell and its role in maintaining blood testis barrier. Anat Rec 175:639–656

Dym M, Fawcett DW (1970) The blood testis barrier in the rat and the physiological compartmentation of the seminiferous epithelium. Biol Reprod 3:308–326

Fang S, Liao S (1969) Antagonistic action of antiandrogens on the formation of a specific dihydrotestosterone-receptor protein complex in rat ventral prostate. Mol Pharm 5:428

Faundes A, Brache V, Leon P, Schmidt F, Alvarez-Sanchez F (1981) Sperm suppression with monthly injections of medroxyprogesterone acetate combined with testosterone enanthate at a high dose (500 mg) Int. J Androl 4:235

Flickinger CT (1977a) The influence of progestin and androgen on the fine structure of the male reproductive tract of the rat. I general effects and observation on the testis. Anat Rec 187:405

Flickinger CJ (1977b) The influence of progestin and androgen on the fine structure of the male reproductive tract of the rat. I. General effects and observation on the testis. Anat. Rec., 187: 405, II.–Epididymis and sex accessory glands. Anat Record 187:431–462

Flickinger C, Fawcett (1967) The junctional specializations of Sertoli cells in the seminiferous epithelium. Anat Record 158:202–222

Flickinger CJ, Loving CK (1976) Fine structure of testis and epididymis of rats treated with cyproterone acetate. Am J Anat 146:359–384

Fogh M, Corker CS, Hunter WM, McLean H, Philip J, Schou G, Skakkebak NE (1979) The effects of low doses of cyproterone acetate on some functions of the reproductive system in normal men. Acta Endocrinol 91:545–552

Friedl KE, Plymate SR, Paulsen CA (1985) Transient reduction in HDL. Cholesterol following medroxyprogesterone acetate and testosterone cypionate administration to healthy men. Contraception 31:409

Fritz IB (1978) In: Litwack E (ed) Biochemical action of hormones, vol 5. Academic, New York, p 249

Geller J, VanDamne O, Garabieta G, Lou A, Rettura J, Seifter E (1969) Effect of cyproterone acetate on H3-testosterone uptake and enzyme synthesis by the ventral prostate of the rat. Endocrinology 84:1330–1335

Ghosh S, Andrzej B, Grasso P, Reicherter LE, Russel LD (1992) Structural response of the hamster Sertoli cell to Hypophysectomy: a correlative morphometric and endocrine study. Anatomical Reprod 234:513–529

Grootegoed JA, Peters MJ, Mulder E, Rommerts FFG, van der Molen HJ (1977) Absence of a nuclear androgen receptor in isolated and germinal cells of rat testis. Mol Cell Endocr 9:159

Groove BD, Vogl AW (1989) Sertoli cell ectoplasmic specialization. A type of actin associated adhesion junction. J Cell Sci 93:309–323

Hamada H, Neumann F, Junkmann K (1963) Intrauterine antimaskuline beeinflussung von rattenfeten durche in stark gestagen wirksames- steroid. Acta Endocrinol 44: 380–388

Hammerstein J (1974) Quoted from Roy et al. (1976). Contraception, 14(4):403–420

Hansson V, Djoseland O, Reusch E, Attramadal A, Torgersen O (1973) An androgen binding protein in the testis cytosol fraction of adult rats. Comparison with the androgen binding protein in the epididymis. Steroids 21:457–474

Hedman M, Gottlieb C, Svanborg K, Byddeman M, de la Torre B (1988) Endocrine, seminal and peripheral effects of depot medroxyprogesterone acetate and testosterone enanthate in men. Int J Androl 11:265

Heller CG, Clermont Y (1964) Kinetics of the germinal epithelium in man. Recent Prog Horm Res 20:545–575

Heller CG, Moore HC, Su U, Rowley MJ (1970) In: The human testis, ed. E. Rosenberg, C. A. Paulson 249, New York Plenum, 645

Heller CG, Nelson WO, Hall IC, Henderson E, Meddock WC, Jungek EC, Paulson CA, Mortimer GE (1950a) Improvement of spermatogenesis following depression of human testis with testosterone. Fert Steril 1:415

Horn HJ (1974) Administration of antiandrogens in hypersexuality and sexual deviations. In: Eichler O, Farah A, Herken H, Welch AD (eds) Handbook of experimental pharmacology, vol 35/2. Springer, Berlin, p 543562

Horst P, Bader J (1969) Untersuchungen zur bedeutung der jungebermast. 2. Mitteilung: versuche zur unterdruckung des sexualgeruches. Zuchtungskunde 41:248

Jagdeep Kaur P, Ramakrishnan R, Rajalakshmi M (1992) Effect of cyproterone acetate on structure and function of rhesus monkey reproductive organs. Anatomical Reprod 234:62–72

Jeffcoate WJ, Matthews RW, Edwards CRW, Field LH, Besser GM (1980) The effect of cyproterone acetate on serum testosterone, LH, FSH and prolactin in male sexual offenders. Clin Endocrinol 13:189–195

Junkmann K, Neumann F (1964) Mechanism of action of progestagens having an antimasculine effect on fetuses. Acta endocr (Copenhagen) Suppl. 90:139–154

Kalla NR, Bhasin M (1977) Influence of cyproterone acetate on phosphatases and dehydrogenases of rat testis. Ind J Biochem Biophys 14:196

Kerr JB (1991) Ultrastructure of the seminiferous epithelium and intertubular tissue of the human testis. J Electron Microsc Tech 19:215–240

Kerr JB, Millar M, Maddocks S, Sharpe RM (1993) Stage-dependent changes in spermatogenesis and Sertoli cells in relation to the onset of spermatogenic failure following withdrawal of testosterone. The Anat Record 235: 547–559

Koch UJ, Lorenz F, Danehl K, Ericsson R, Hasan SH, Keyserlingk DV, Lubke K, Mehring M, Rommler A, Schwartz U, Hammerstein J (1976) Continuous oral low-dosage cyproterone acetate for fertility regulation in the male? A trend analysis in 15 volunteers. Contraception 14(2):117–135

Koch UJ, Lorenz F, Danehl K, Hammerstein J (1974) Uber the verwendbarkeit von cyproterone acetate zur fertilitatshemmung bein mann. Morphologische veranderungen and einflusse auf the spermien motilitat. 40. Tagung der Deutschen Gesellschaft fur Gynaekologie and Geburtshilfe, Wiesbaden

Lohiya NK, Sharma OP (1983) Effects of cyproterone acetate with combination of testosterone enanthate on seminal characteristics, androgenicity and clinical chemistry in langur monkey. Contraception 28(6): 575–586

Lorenz, F., U. J. Koch, K. Danehl, K. Lubke and J. Hammerstein (1974)

Loving CK, Flickinger CJ (1976) Alterations in the fine structure of the prostate and seminal vesicle of rats treated with cyproterone acetate. Anat Rec 185:13–30

Males JL, Turkington RW (1971a) Hormone control of lysosomal enzymes during spermatogenesis in the rat. Endocrinology 86:579

Males JL, Turkington RW (1971b) Hormonal control of lysosomal enzymes during spermatogenesis in the rat. Endocrinology 88:579

Means R (1975) Biochemical effects of FSH on the Testis. In: Hamilton DW, Greep RO (eds) Handbook of physiology, Vol. 5, Endocrinology. Amer. Physiol. Soc, Bethasda Md, p 203

Mertelsmann R, Keeuzer T, Matthaei H (1968) Hoppeseylers. Z Physiol Chem 349:10

Mills ML, Means AR (1972) Endocrinology 91:147

Moltz L, Rommler A, Schwartz U, Hammerstein J (1978) Effects of cyproterone acetate (CPA) on pituitary gonadotropin release and on androgen secretion before and after LH-RH double stimulation tests in men. Int J Androl Suppl. 2:713

Morse HC, Leach DR, Rowley MJ, Heller CG (1973) Effect of cyproterone acetate on sperm concentration, seminal fluid volume, testicular cytology and levels of plasma and urinary ICSH, FSH and testosterone in normal men. J Reprod Fert 32:365

Mulder E, Peters MJ, van der Molen HJ (1975) Androgen receptors in testis tissue enriched in Sertoli cells. In: French FS, Hansson V, Ritzen EM, Nayfeh SN (eds) Hormonal regulation of spermatogenesis. Plenum Press, New York, p 287

Nag S, Ghosh JJ (1979) Epididymal and testicular enzymes as monitors for assessment of male antifertility drugs. J Steroid Biochem 11:681–686

Neumann F (1972) Use of cyproterone acetate in animal and clinical trials. Gynec Invest 2:150–179

Neumann F (1973) Cyproteronacetat-tierexperimentelle Grundlagen. Medizinische Mitteilungen (Schering AG) 2:2–10

Neumann F (1977) Pharmacology and potential use of cyproterone acetate. Hormone and Metab Res 9:1–13

Neumann F (1994) The antiandrogen cyproterone acetate: discovery, chemistry, basic pharmacology, clinical use and tool in basic research. Exp Clin Endo 102:1–32

Neumann F, Schenck B (1976) New antiandrogens and their mode of action. IPPF congress on agents affecting control of fertility in the male, New Delhi. J. Reprod. Fertil 24:129–145

Neumann F, von Berswordt-Wallrabe R, Elger W, Steinbeck H, Hahn JD, Kramer M (1970) Aspects of androgen-dependent events as studied by antiandrogens. Recent Progr Horm Res 26:337410

Ott F (1968) Hypersexualitat. Antiandrogene and Hodeniunktion Praxis 57:218

Ott F, HofTet H (1968) Beeinflussung von libido, potenz and hodenfunktion durch antiandrogene. Schweiz Med Wochenschr 98:1812–1815

Pangkahila W (1991) Reversible azoospermia induced by an androgen - progestin combination regimen in Indonesian men. Int J Androl 14:248

Prasad MRN, Rajalakshmi M, Gupta G, Markun T (1973) Control of epididymal function. J Reprod Fert 18:215–222

Prasad MRN, Rajalakshmi M, Reddy PRK (1972) Action of cyproterone acetate on male reproductive functions. Gynecol Investig 2:202

Prasad MRN, Singh SP, Rajalakshmi M (1970) Fertility control in male rats by continuous release of microquantities of cyproterone acetate from subcutaneous silastic capsules. Contraception 2:165

Purvis K, Haug E, Aakvaag A, Hansson V (1978) Sites of action of cyproterone and cyproterone acetate in the immature male rat. Int J Androl 1:279–296

Raspe G (1972) Schering Symposium uber sexualdeviationen and ihre medikamentose behandlung, Berlin 1971. Life sciences monographs, 2. Pergamon, Oxford

Rivarols MA, Camacho AM, Migeen CJ (1968) Effects of treatment with medroxyprogesterone acetate on testicular function. J Clin Endocrinol Metab 28:449

Roy S, Chatterjee S, Prasad MRN, Poddar AK, Pandey DC (1976) Effects of cyproterone acetate on reproductive functions in normal human males. Contraception 14(4):403

Russel LD, Clermont (1977) Degeneration of germ cells in normal, hypophysectomized and hormone treated hypophysectomized rats. Anat Rec 198:347–366

Russell L (1977a) Movement of spermatocytes from the basal to the adluminal compartment of the rat testis. Am J Anat 148:313–328

Russell LD (1977b) Observation on rat Sertoli ectoplasmic (junctional) specializations in their association with germ cells of the rat testis. Tissue Cell 9:475–498

Russell LD, Malone JP, Karpas S (1981) Morphological pattern by agents affecting spermatogenesis by disruption of its hormonal stimulation. Tisue and Cell 13: 369–380

Sanborn BM, Elkington JSH, Steinberger A, Steinberger E, Meistrich L (1975) Androgen binding in the testis: in vitro production of androgen binding protein (ABP) by Sertoli cell cultures and measurement of nuclear bound androgen by a nuclear exchange assay. In: French FS, Hansson V, Ritzen EM, Nayfeh SN (eds) Hormonal regulation of spermatogenesis. Plenum Press, New York, p 293

Sanborn BM, Steinberger E (1975) Androgen nuclear exchange activity in rat testis. Endocr Res Commun 2:335

Sanborn BM, Steinberger A, Tcholakian RK, Steinberger E (1977) Direct measurement of androgen receptors in cultured Sertoli cells. Steroids 29:493

Sar M, Stumpf WE, McLean WS, Smith AA, Hansson V, Nayfeh SN, French FS (1975) Localization of androgen target cells in the rat testis: autoradiographic studies. In: French FS, Hansson V, Ritzen EM, Nay'feh SN (eds) Hormonal regulation of spermatogenesis. Plenum Press, New York, p 311

Schenck B, Elger W, Schopflin G, Neumann F (1975) Failure to induce sterility in male rats with microdoses of cyproterone acetate (CPA). Contraception 12(5):517

Schenck B, Neumann F (1978) Some comments on the use of antiandrogens for male contraception. Int J Androl 2:155–161

Schmidtke D, Schmidtke HO (1968) Ein neues antiandrogen beim. Hund Kleintier - Prax 13:426

Steinbeck H, Mehring M, Neumann F (1971) Comparison of the effects of Cyproterone, Cyproterone acetate and Oestradiol on testicular function, accessory sexual glands and fertility in a longterm study on rats. J Reprod Fert 26:65–76

Stern JM, Eisenfeld AJ (1969) Androgen accumulation and binding to macro molecules in seminal vesicles: inhibition by cyproterone. Science 166:233–235

Sugimura Y, Cunha GR, Denjacour AA (1986) Morphological and histological study of castration induced regeneration in the mouse prostate. Biol Reprod 34: 973

Tindall DJ, Miller DA, Means AR (1977) Characterization of androgen rceptor in Sertoli cellenriched testis. Endocrinology 101:13

Tsai YH, Sanborn BM, Steinberger A, Steinberger E (1977) The interaction of testicular androgen-receptor complex with rat germ cell and Sertoli cell chromatin. Biochem Biophys Res Commun 75:366

Tveter KJ, Aakvaag A (1969) Uptake and metabolism in vivo of testosterone - 1, 2 - H by accessory sex organs of male rats; influence of some hormonal compounds. Endocrinology 85:683–689

Tymoczko JL, Liao S (1976) Androgen receptors and the molecular basis for the action of antiandrogens in the ventral prostate. J. Reprod. Fert 24:147–162

Umapathy E, Rai UC (1980) Effects of antiandrogens on the accessory sex glands of castrated rats. Ind J Exp Biol 18:1090–1093

Walsh PC, Korenman SG (1970) Action of antiandrogens: preservation of 5 a.–reductase activity and inhibition of chromatin - dihydrotestosterone complex formation. Clin Res 18:126

Wang C, Yeung KK (1980) Use of low-dosage oral cyproterone acetate as a male contraceptive. Contraception 21(3):245

Weber JE, Turner TT, Tung KSK, Russell LD (1988) Effects of cytochalosin D on the integrity of Sertoli cell (blood testis) barrier. Am J Anat 182:130–147

Whalen RE, Edwards DA (1969) Effects of the antiandrogen cyproterone acetate on mating behavior and seminal vesicle tissue in male rats. Endocrinology 84: 155–156

Whalen RE, Luttge WG (1969) Contraceptive properties of the anti - androgen cyproterone acetate. Nature 223: 633–634

WHO (1992) Special Programme of research, Development, and Research Training in Human Reproduction, Reproductive Health. A key to brighter future Biennial Report, 1990-91

Wilbe JP, Tilbe KS (1979) De novo synthesis of steroids (from acetate) by isolated rat Sertoli cells. Biochem Biophys Res Commun 80:1107

Zucker I (1966) Effects of an antiandrogen in the mating behavior of male guinea pigs and rats. J. Endo. 35:209–210

Ultrastructure of Epididymis in Normal and Experimental Animals; Rat and Bonnet Monkey (*Macaca radiata*)

3

3.1 Introduction

The role of epididymis in the secretion of fluid and ions and the relevance of these to sperm maturation have been supported by the findings of the task force (Wong 1989). Various studies suggest that DMPA and TE treatment induce azoospermia or oligospermia in human beings and animals (Nieschlag et al. 1989; WHO report 1991 & Rao and Roy 1993), and the residual sperms are found to be functionally impaired when tested by hamster oocyte penetration method because of defective metabolism, alteration in morphology, viability and acrosome integrity within the epididymis (Rao and Roy 1993). The present study is aimed to observe the alteration in the ultrastructure and biochemical environment of the different regions of the epididymis following treatment.

Regional differences in the structural features and functions of the principal cells have allowed the epididymis to be subdivided into several distinct regions. However, due to the highly coiled and tortuous nature of the epididymal duct, the exact areas of demarcation between the different regions are difficult to define, and it is not uncommon to find each region further subdivided based on structural and /or functional parameters.

The epididymal tubules of all three regions—the caput, corpus and cauda are surrounded by layers of smooth muscle, their thickness increasing down the duct. Scattered among the smooth muscle cells are fibroblasts, capillaries, small nerves and wandering cells of the stroma.

3.2 Ultrastructure of Epididymis in Normal and Experimental Animals (Rats)

3.2.1 Observations

1. Caput I epididymis of control rat: The caput or the head region is composed of tall columnar epithelium showing four types of cells—the principal cells, the basal cells, the apical mitochondria rich cells and the halo cells (Fig. 3.1). The principal cells dominate the epithelium in number. The basal cells are polygonal in shape and lie opposed to the basement membrane. They can be identified by their pale staining cytoplasm and relatively large nucleus oriented along the circumference of the duct (Fig. 3.1). In the cytoplasm are scattered very few cell organelles like mitochondria, small amount of rough endoplasmic reticulum, dense bodies and polyribosomes. Thin processes of these cells are seen to interdigitate with adjacent thin processes from the principal cells.

The principal cells are tall and columnar (Fig. 3.1) having large nuclei which contain one or two nucleoli, and these cells have euchromatic karyoplasm. At their apical surface, several

D. A. Bhiwgade, S. Menon, *Ultrastructural Investigations on the Pituitary-Gonadal Axis*, https://doi.org/10.1007/978-981-99-3276-4_3

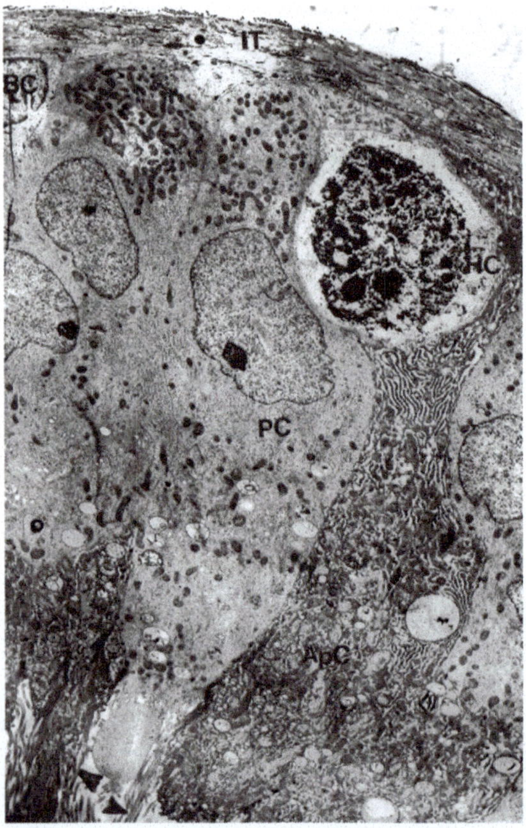

Fig. 3.1 Low-power electron micrograph of caput epididymis of control rat demonstration the various cell types. The basal cells (BC) lying opposed to the basement membrane have a nucleus oriented along the circumference of the duct. The principal cells are tall and columnar cells (PC) having large euchromatic nuclei and stereocilia at their apical ends (arrowheads). The cytoplasm is scattered with mitochondria and rough endoplasmic reticulum. The apical mitochondria-rich cells (ApC) are goblet shaped and have a cytoplasm denser than principal cells. Halo cells (HC) can be identified because of their nuclei surrounded by a pale cytoplasm, which appears empty [X 3600]. Unpublished electron micrograph from Dr. Bhiwgade and Dr. Kadam

stereocilia are present which project into the lumen. At the basal region of the principal cells, the cell membrane is thrown into complex folds and a few junctional specializations are seen. The cytoplasm is filled with scattered mitochondria, rough endoplasmic reticulum, a few lipid droplets and numerous polyribosomes (Fig. 3.1). The nucleus of the principal cells is slightly lobated and is surrounded by profiles of rough endoplasmic reticulum. The latter is very well-developed and occupies a large volume in the cytoplasm (Fig. 3.1). Smooth endoplasmic reticulum is very scanty. The supranuclear region contains Golgi bodies composed of groups of parallel stacked cisternae of smooth surfaced membrane. Each Golgi body is surrounded by a large number of coated and smooth surfaced vesicles of varying sizes (Fig. 3.1). A very prominent feature of the Golgi region is the presence of autophagic vacuoles, containing variable amounts of cell debris. These vacuoles constitute the cytoplasmic autodegradative process and many contain various cell organelles in different stages of degradation. The more advanced ones show their evolution towards lipofuscin pigments (Fig. 3.1).

The apical zone shows the characteristics of absorptive and secretory activities. There is an abundance of pinocytotic vesicles, vacuoles and multivesicular bodies. The luminal surface of the epithelial cells gives out stereocilia and several micropinocytotic infoldings are found at the bases of the stereocilia. Many vacuoles are seen near the apical region lying between the stereocilia (Fig. 3.1).

The apical mitochondria-rich cells are identified by their goblet shape and cytoplasm which is dense than the principal cells (Fig. 3.1). Apical portion of these cells protrudes into the lumen and is packed with several mitochondria and lots of pinocytotic vesicles. Cells also contain pale vacuoles.

Halo cells are found anywhere from the base to the lumen interspersed between the principal cells in the epithelium (Fig. 3.1). They can be identified because of their nuclei surrounded by an empty cytoplasmic area. Ultrastructurally, however, it is seen that these cells do not have a circumnuclear pale area, rather they have an easily discernible cytoplasm. Their pale cytoplasm contains a few mitochondria and scattered profiles of endoplasmic reticulum (Fig. 3.1). The nucleus is heterochromatic and usually indented in nature.

2. Caput I epididymis of CPA + TE treated rat:

The most striking change is the decrease in the concentration of sperms from the lumen of the epididymal tubules (Fig. 3.2a). Though, the cell height appears to have decreased, the principal cells have retained their characteristic columnar shape and the stereocilia. The caput shows the presence of the basal cells and principal cells halo cells; the apical mitochondria-rich cells are not encountered in this plane of the section.

The nuclei of the principal cells show no change due to treatment and are lobated with euchromatic karyoplasm (Fig. 3.2b). The basal region of these cells shows comparatively very few lipid droplets. The RER is as extensive as seen in the control principal cells surrounding the nucleus (Fig. 3.2a, b). The supranuclear region

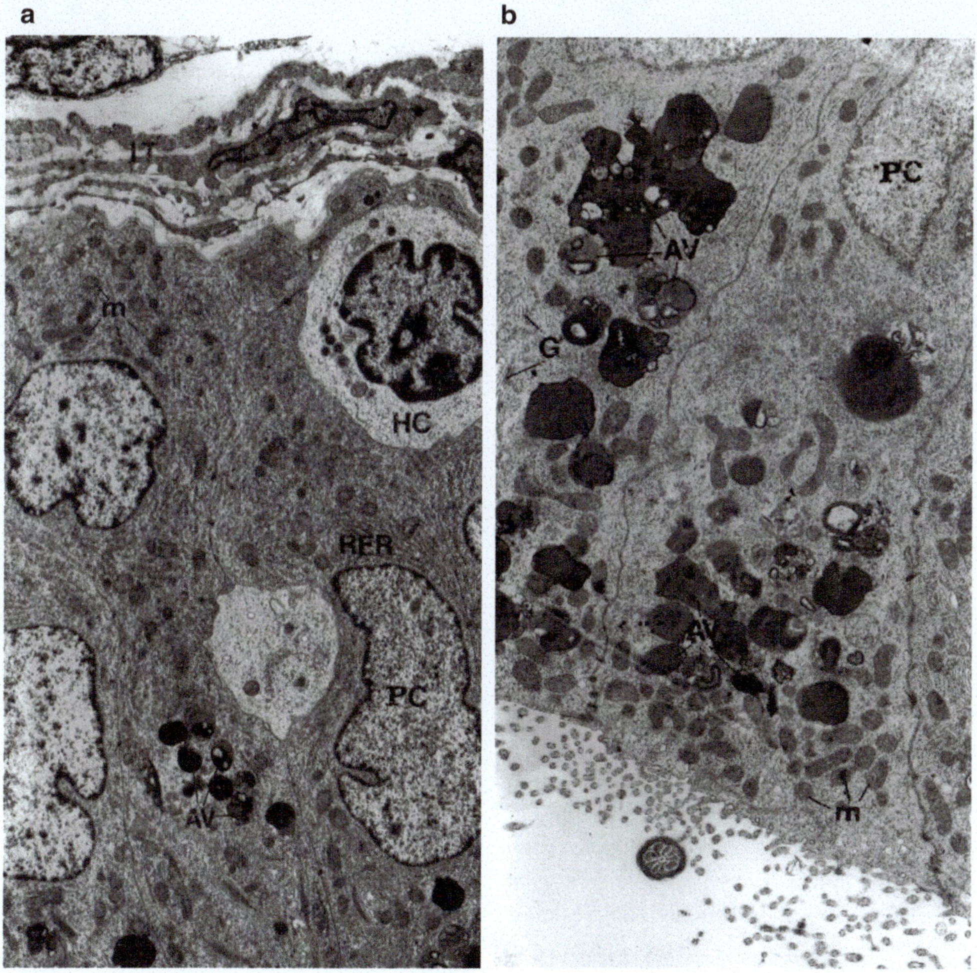

Fig. 3.2 (**a**, **b**) Low-power electron micrograph of the caput epididymis of 60 days CPA + TE treated rat. The intertubular tissue (IT) appears unchanged with the fibroblast cells (FC) and layers of smooth muscle fibres. Note the halo cells (HC) have heterochromatic, indented nuclei and a very pale cytoplasm that contains very less organelles—a few scattered mitochondria and dense bodies. The principal cells (PC) retain their columnar shape and characteristic lobated nuclei. Several mitochondria (m) and profiles of endoplasmic reticulum (RER) are packed into the basal cytoplasm of these cells. Their supranuclear region contains several dense autophagic vacuoles (AV) [X 3600]. Unpublished electron micrograph from Dr. Bhiwgade and Dr. Kadam

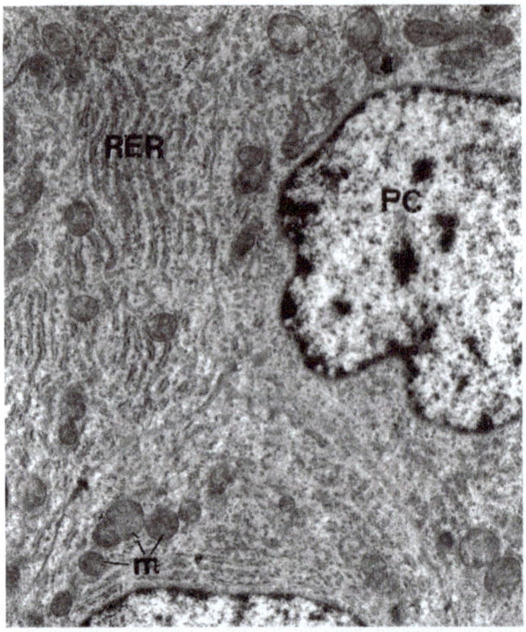

Fig. 3.3 High-power electron micrograph of 60 days CPA + TE treated rat caput epididymis showing the extensive rough endoplasmic reticulum (RER) that surrounds the nucleus of the principal cells (PC). The nuclei is located similar to the control ones. Also seen are mitochondria (m) [X 10, 000]. Unpublished electron micrograph from Dr. Bhiwgade and Dr. Kadam

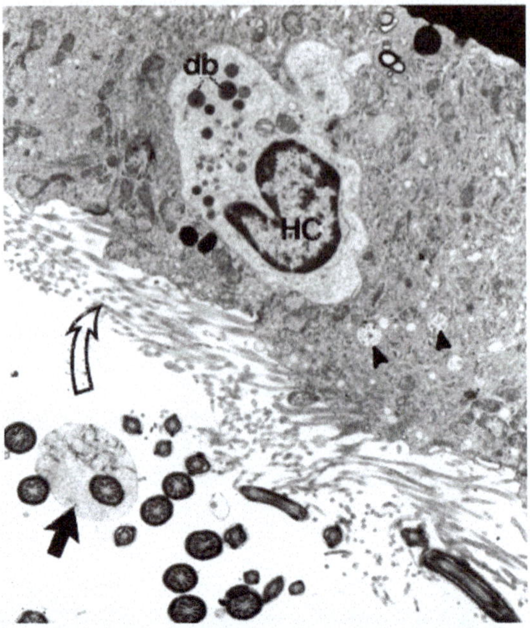

Fig. 3.4 The apical region of the principal cell from 60 days CPA + TE treated rat caput epididymis is illustrated here. Note the scanty amount of micropinocytotic infoldings and vesicles. However, several multivesicular bodies (arrow heads) are found in this region. Also depicted here is a helo cell (HC) with its heterochromatic, indented nucleus and an increased amount of dense bodies (db). The stereocilia (curved arrow) of the principal cells appear unaffected following the treatment. The lumen contains many spermatozoa cut through the tail region and mid-piece region. Some of them have not shed the cytoplasmic droplet (thin arrow) [X 5,000] Unpublished electron micrograph from Dr. Bhiwgade and Dr. Kadam

appears to be reduced in height and the Golgi zone contains many autophagic vacuoles present in close association with the Golgi vesicles (Fig. 3.2b). These numerous autophagic vacuoles occupying the entire supranuclear region are very dense resembling lipofuscin pigments, unlike their control counterparts. A few of these also show the formation of myelin figures within them. In some cells, they so extensive in number that the other organelles appear to have been displaced to the apical zone. This apical zone in the control contains many pinocytotic vesicles, but in the treated caput, there are hardly any vesicles or vacuoles. A few micropinocytotic infoldings are present at the bases of the stereocilia and some coated vesicles line the apical region. Multivesicular bodies, however, have increased and several of them are found in the apical region and in the Golgi zone (Fig. 3.2b).

Halo cells are unchanged in appearance except that they contain many dense bodies in their cytoplasm (Figs. 3.3 and 3.4).

The lumen of the tubules contains many sperm tails, some of them also with cytoplasm droplets. In many of these sperms, the tails have a fragmented axonemal complex and some of the doublets of the 9 + 2 microtubular structure and outer coarse fibres are lost (Fig. 3.5). In some, the mitochondria sheath of the mid-piece surrounding these microtubules has also ruptured thereby causing a loss of the inner contents.

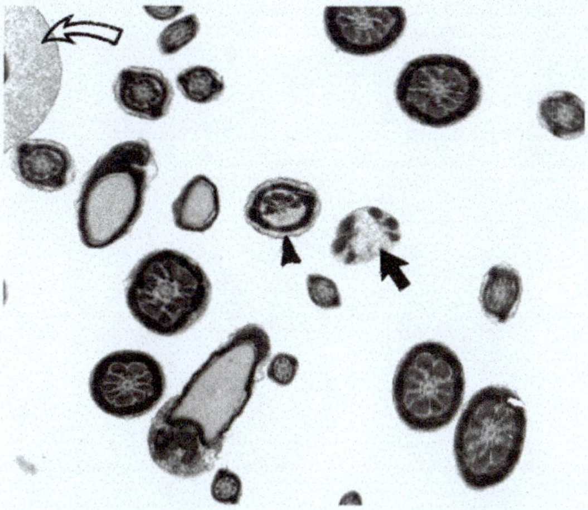

Fig. 3.5 Electron micrograph of 60 days CPA + TE treated captured epididymis taken from the luminal region. Several spermatozoa are present, some with the cytoplasmic droplets (curved arrows). Many sperms contain tails with a fragmented axonemal complex (arrowheads) and a few have lost the outer mitochondrial sheath (thin arrow) [X 6,500] Unpublished electron micrograph from Dr. Bhiwgade and Dr. Kadam

3. Cauda Epididymis of Control Rat

The cell types found in this region are the basal cells, principal cells, halo cells and clear cells. Basal cells are more common in the cauda than in the other regions of the epididymis, and they morphologically resemble their counterparts observed in the head region of caput epididymis. They are polygonal in shape, lying adjacent to the basement membrane and have relatively large nucleus and a pale staining cytoplasm (Fig. 3.6). The cytoplasm contains the usual cell organelles like mitochondria and rough endoplasmic reticulum, but they are very few as compared to the other epididymal cell types.

The principal cells are cuboidal in shape and also more compact (Fig. 3.6). The luminal stereocilia also change and become shorter in size. The cytoplasm is filled with a few mitochondria, rough endoplasmic reticulum and free ribosomes present in the form of rosettes (Fig. 3.6). The nucleus is elongated and surrounded by rough endoplasmic reticulum which is a very developed occupying quite a large area of the cytoplasm. The supranuclear region is filled with many Golgi bodies. Each

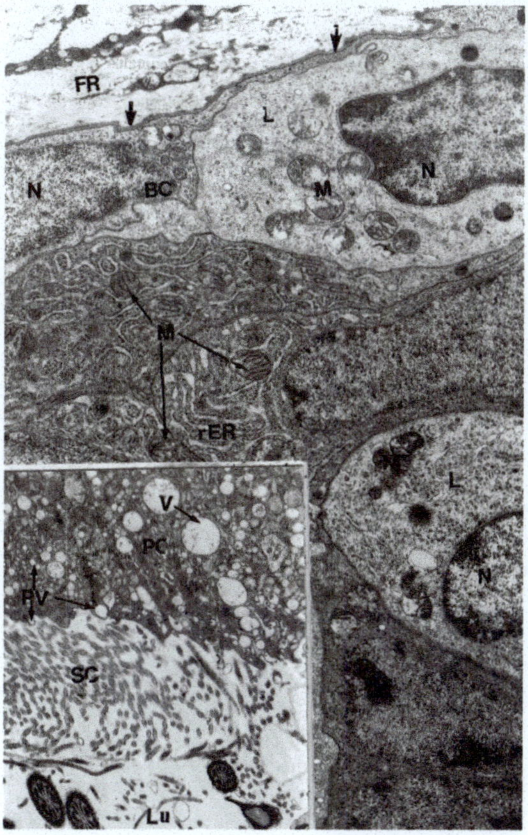

Fig. 3.6 Electron micrograph of the cauda epididymis of a control rat showing the basal region. The basal cell (BC), having a polygonal shape and a relatively large nucleus oriented along the circumference of the duct, lies apposed to the basement membrane. Halo cells (HC) are seen with their heterochromatic nuclei and a pale staining cytoplasm which appears empty due to scanty organelles—mitochondria (m), a small Golgi body (G), scattered rough endoplasmic reticulum (curved arrow) and a few dense granules (arrowheads). The picture also depicts the basal cytoplasm of the principal cells (PC) containing numerous profiles of rough endoplasmic reticulum (RER) and several mitochondria (m) [X 6500] Unpublished electron micrograph from Dr. Bhiwgade and Dr. Kadam

Golgi body is surrounded by a large number of coated and smooth-surfaced vesicles (Fig. 3.7). The apical cytoplasm has an abundance of vesicles of various sizes (Fig. 3.8). Multivesicular bodies and pinocytotic vesicles are also numerous. Micropinocytotic infoldings between the stereocilia indicate absorptive function of the cell.

The clear cells are characterized by the presence of numerous vacuolar elements of different

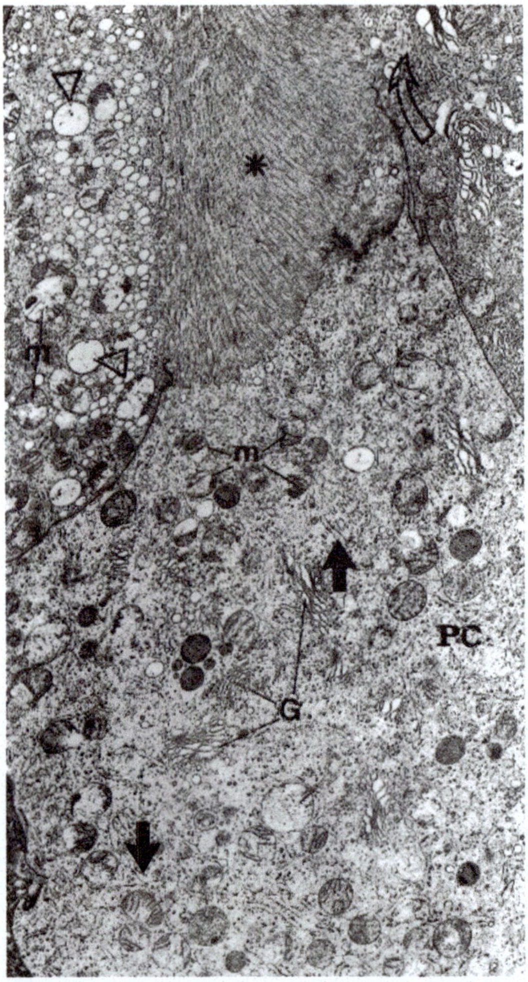

Fig. 3.7 The cauda epididymis of a control rat is seen here. The supranuclear region of the principal cells (PC) is filled with mitochondria (m), several Golgi bodies (G), the rough endoplasmic reticulum is scattered in the cytoplasm (thick arrows). The apical cytoplasm has an abundance of vesicles of various sizes (arrowheads) and the luminal stereocilia are shorter than those found in the caput epididymis (asterisk). Multivesicular bodies (curved arrow) are also seen [X 5,000] Unpublished electron micrograph from Dr. Bhiwgade and Dr. Kadam

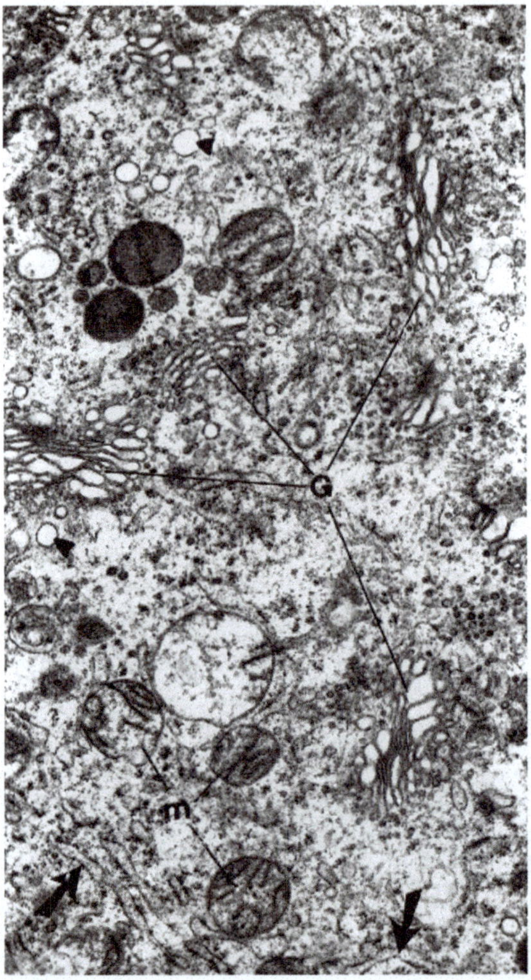

Fig. 3.8 Higher magnification of the Golgi region of the principal cells from a control rat cauda epididymis. Several Golgi bodies (G) are seen with their characteristic two to four parallel stacked cisternae and small smooth-surfaced vesicles surrounding them. Coated vesicles (arrowheads) of Golgi region and a few mitochondria (m) are dispersed in the cytoplasm. The rough endoplasmic reticulum is present in the form of loose scattered cisternae (arrows) [X 13,500] Unpublished electron micrograph from Dr. Bhiwgade and Dr. Kadam

electron opacities (Fig. 3.7). At the light micrographic level, they can be identified by their pale–staining cytoplasm. The basal cytoplasm contains several lipid droplets, which are electron-dense, and lysosomes that resemble dense bodies. The apical regions of these cells contain many large vacuoles of low electron opacity. The electron opacity of the vacuoles increases from the basal region of the cell to the apical region, the vacuoles being more electron dense in the basal region (Figs. 3.7 and 3.8). The Golgi apparatus is small and very few mitochondria are found. There is an abundance of small vacuoles and vesicles near the apical surface, which may be pinocytosed by the cell.

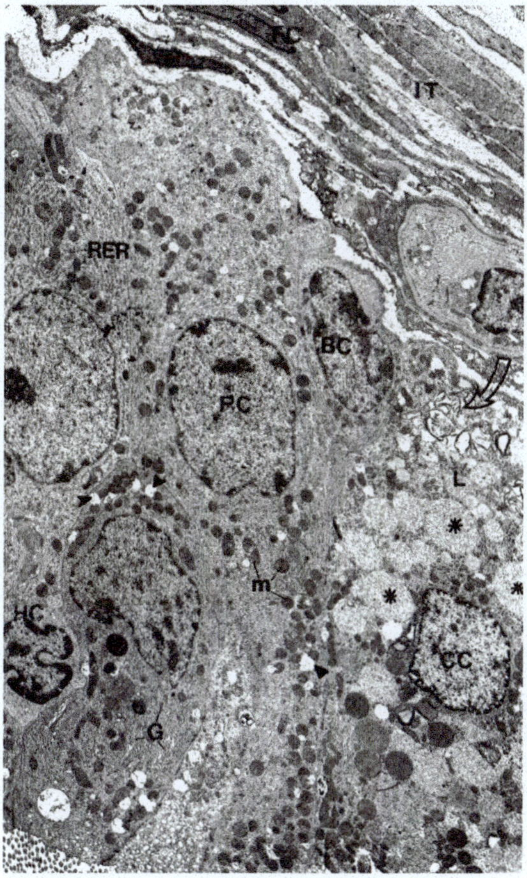

Fig. 3.9 Low- power electron micrograph of the cauda epididymis of rat treated with CPA + TE for 60 days. The intertubular tissue (IT) comprising the fibroblasts (FC) and smooth muscle layers appears uncharged. The basal cell (BC) also does not show any alterations. The principal cells (PC) have the characteristic located nuclei and normal components of rough endoplasmic reticulum (RER), Golgi complex (G) and mitochondria (m). Many of the mitochondria have ruptured or have swelled up and involution has set in (arrowheads). The halo cells (HC) contain the characteristics heterochromatic indented nucleus and a seemingly empty cytoplasm. The clear cells (CC) show prominent changes over controls with myelinization setting in (curved arrows) in the vesicles of the basal region. The vesicles contain particular matter of low electron density (asterisk) throughout the cytoplasm. Lipid droplets (L) are also found in the basal cytoplasm of these cells [X 3600] Unpublished electron micrograph from Dr. Bhiwgade and Dr. Kadam

Short microvilli can be seen on the luminal surface (Fig. 3.7).

Halo cells, which are located anywhere from the base to the lumen between the principal cells, contain nuclei surrounded by a pale cytoplasm (Fig. 3.6). The cytoplasm contains very few organelles, viz., mitochondria, scattered profiles of endoplasmic reticulum, one or two Golgi bodies and a few dense granules. The nucleus, on the other hand, consists of dense heterochromatin and is indented.

4. Cauda epididymis of CPA + TE treated rat: The lumen of the treated cauda epididymis shows a decreased concentration of sperms as compared to the control. The cuboidal structure of the principal cells is unchanged.

Ultrastructurally, the basal cells do not show any post-treatment changes (Fig. 3.9). The principal cells are also not severely affected by the treatment and have their normal components of rough endoplasmic reticulum, mitochondria and Golgi bodies. Some mitochondria, however, have ruptured and involution has set in (Figs. 3.9 and 3.10).

The clear cells show the most prominent changes. The vacuoles and vesicles that are a characteristic feature of these cells contain a granular material of low electron density rather than the electron-dense material found in the control vesicles. Though, these vacuolar elements are numerous even in the treated cauda epididymis, most of them are of low-electron density and resemble giant multivesicular bodies (Fig. 3.12). A few dense bodies are also scattered in the cytoplasm but these vesicles with the granular material outnumber them. Some electron lucent lipid droplets are found in the basal cytoplasm (Fig. 3.12). The degradative process in many of the vesicles has reached an extreme stage and they show the formation of myelin figures within them. The apical region of the clear cells has not undergone any change and several small vesicles and vacuoles are present (Fig. 3.12).

Although many sperms are found in the lumen, quite some have fragmented axial filaments with a loss of the 9 + 2 doublets and also the coarse fibres in the tail piece. Sections through mid-piece region of the spermatozoa exhibit normal axial filaments (Figs. 3.11, Fig. 3.12 and 3.13).

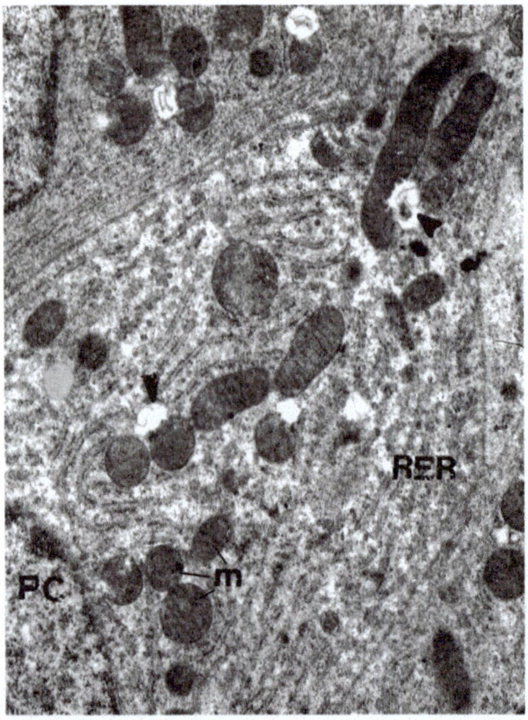

Fig. 3.10 High-power electron micrograph of the principal cell (PC) from 60 day CPA + TE treated rat cauda epididymis. Note the normal components of rough endoplasmic reticulum (RER) and mitochondria (m). A few mitochondria have swollen up and show signs of degeneration (arrowheads) [X 13,500] Unpublished electron micrograph from Dr. Bhiwgade and Dr. Kadam

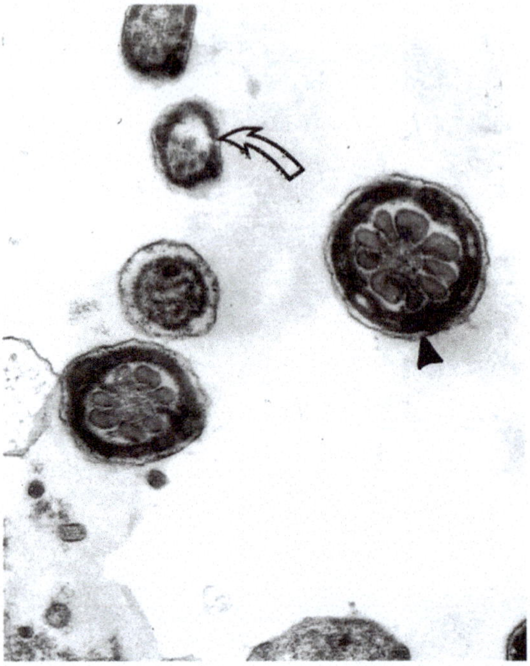

Fig. 3.11 Higher magnification of the spermatozoa from the lumen of 60 days CPA + TE treated rat cauda epididymis. Note the sections through the midpiece region showing the normal axonemal complex and the mitochondrial helix surrounding it (arrowheads). Sections through the tail-piece of the spermatozoa (curved arrow) show a loss of 9 + 2 doublets and the outer coarse fibres [X 25,000] Unpublished electron micrograph from Dr. Bhiwgade and Dr. Kadam

5. Caput and Cauda epididymis of Gossypol treated rat: The ultrastructure of the caput epithelium was not affected. However, the distal caput epididymidis showed a large number of lysosomal phagosomes (digestive vacuoles) and secretory vesicles towards the apical border (Fig. 3.14a). The cauda epididymidis also showed a normal cell architecture as evidenced by well-developed Golgi zone and secretory vesicles (Fig. 3.14a, b). The three types of cells—principal, basal and clear cells were clearly observed (Fig. 3.14c), although the height of the epithelium and also the stereocilia were reduced (Fig. 3.14b, c). The mitochondria of the principal cell appeared hypertrophied (Fig. 3.14b) and autolysis of hypertrophied mitochondria was clearly seen. The major changes were observed in the motor apparatus of the sperm. In the lumen of the caput epididymidis the most common defect in the sperm was the vacuolization (Fig. 3.15a, b, and c) and complete degeneration of the midpiece mitochondria (Fig. 3.15b). Occasional heads of spermatozoa (Fig. 3.15a) clearly showed the above defect to better advantage. In Fig. 3.15c, the plasma membrane was completely missing from large segments of the middle piece, there was a reaggregation of granular substance or satellite fibres in some areas and was absent in other areas, and the mitochondria were missing from these regions (Fig. 3.15c).

Disorganization of the outer dense fibres of the midpiece region and wavy plasma membrane was observed (Fig. 3.16). The cross sections of the

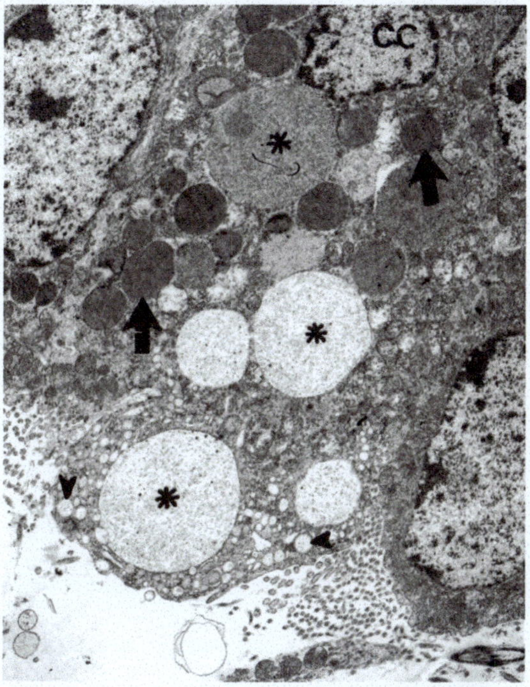

Fig. 3.12 High-power electron micrograph of a clear cell (CC) from the cauda epididymis of rat following 60 days CPA + TE treatment. Note the various vesicles and vacuoles of varying electron densities (asterisk) which appear to be filled with a particulate matter of low electron density compared to their control counter parts. Numerous small vacuoles and pinocytotic vesicles (arrowheads) are found in the apical region of this cell. Smaller vesicle with dense secretory material (thick arrows) are found in lesser numbers [X 13,500] Unpublished electron micrograph from Dr. Bhiwgade and Dr. Kadam

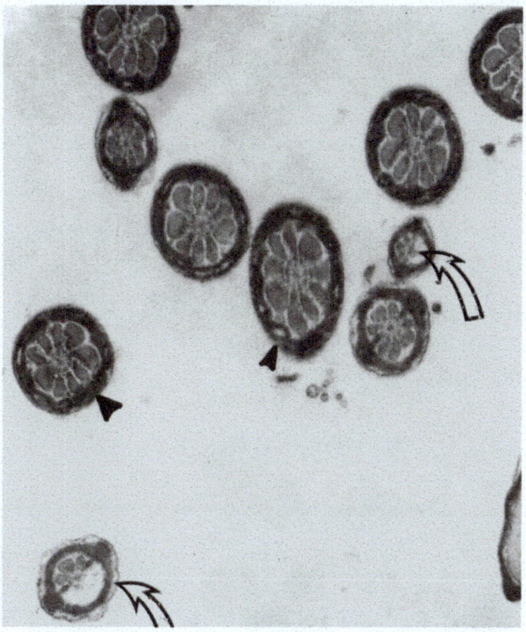

Fig. 3.13 High-power electron micrograph of 60 days CPA + TE treated rat cauda epididymis depicting the spermatozoa in the lumen of the duct. The mid-piece spermatozoa are normal and show no post—treatment changes (arrowheads), but the sections through the tail piece show a fragmentation of the axonemal complex and a loss of the 9 + 2 doublets and the outer coarse fibres (curved arrows) [X 16,300] Unpublished electron micrograph from Dr. Bhiwgade and Dr. Kadam

spermatozoa from the lumen of the cauda epididymidis (Fig. 3.17a, b, and c) also showed missing members of axonemal components. The ultrastructure of the basal epithelial cells was normal (Fig. 3.16), although there was a protein synthesis as seen by the dilated endoplasmic reticulum (Fig. 3.16). The mitochondria appeared hypertrophied (Fig. 3.16). The lumen of the vas deferens was filled with secretion. Different stages of sperm disintegration as seen in the lumen of the epididymis were also well evident in the lumen of the vas deferens (Fig. 3.17c).

6. Seminal Vesicles of control and CPA + TE treated rat: The seminal vesicles are lined with a simple columnar epithelium containing occasional basal cells, which rest on a thin basement membrane (Fig. 3.18a). The nuclei of the epithelial cells containing one or two nucleoli are located basally and are elongated in the direction of the long axis of the cell (Fig. 3.18a, b). At the ultrastructural level, this simple columnar epithelium is seen to rest on a typical basal lamina of connective tissue and a thick coat of smooth muscle. The basal cytoplasm of these columnar cells as also the lateral and apical margins are occupied by abundant rough endoplasmic reticulum composed of parallel cisternae and many mitochondria, which contain a dense matrix. The supranuclear zone of the cytoplasm contains a large Golgi complex (Fig. 3.18a) of flattened membranous sacs, small vesicles and large vacuoles. Most of the vesicles are smooth surfaced spheres

a b c

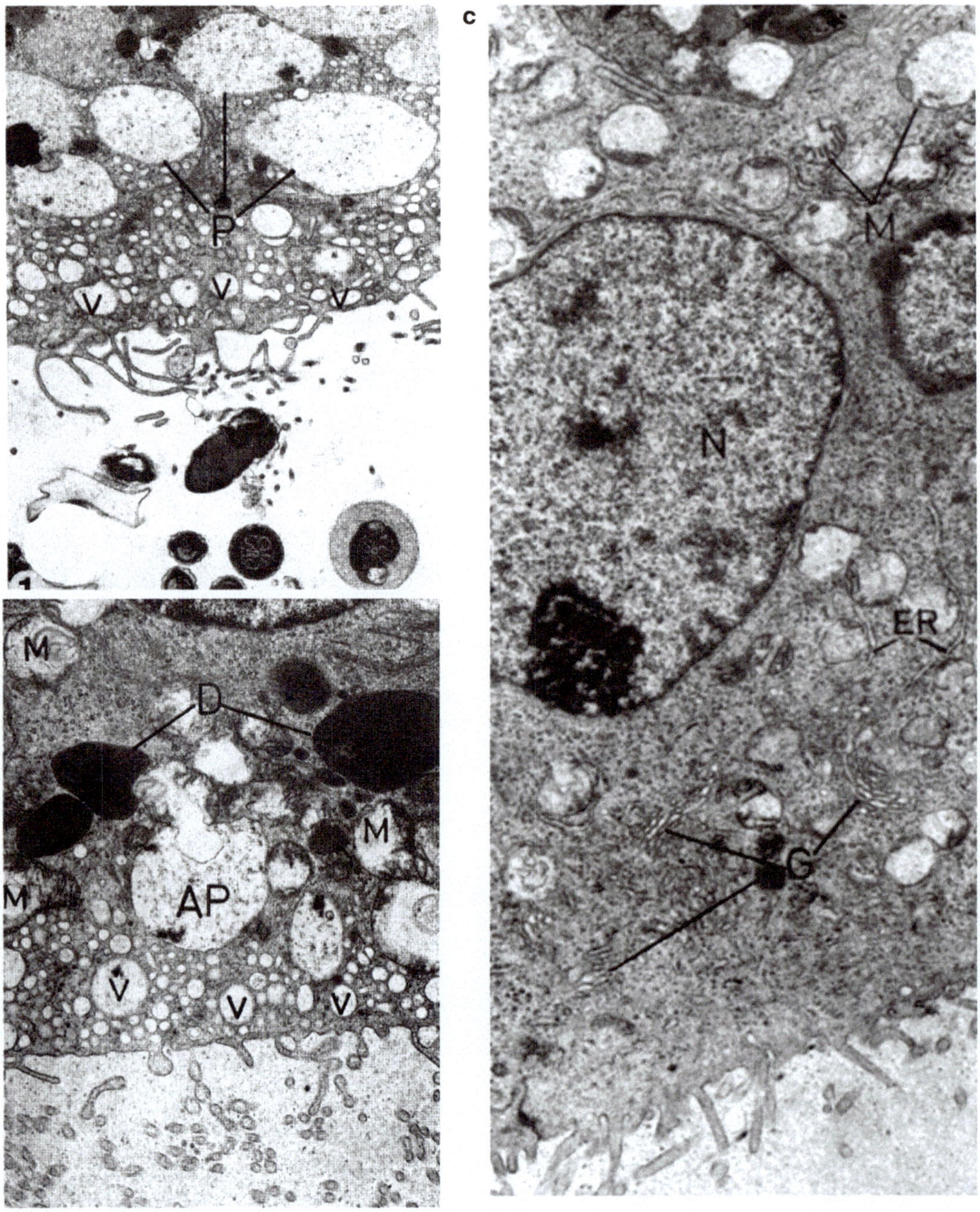

Fig. 3.14 (**a**) Electron Micrograph (EM) of principal cell of distal caput epididymidis of rat treated with gossypol showing phagosome (P) or digestive vacuoles and secretory vesicles (V) [X 6000] (**b**) Electron micrograph of apical region of cauda epididymidis showing dense bodies (Db), secretory vesicles (V), hypertrophied mitochondria (M) and autophagosome (AP) [X 8000] (**c**) Principal cell of cauda epididymidis showing normal complement of Golgi apparatus (G), nucleus (N), ribosome studied endoplasmic reticulum (ER) and hypertrophied mitochondria (M) [X 6000]. Unpublished electron micrograph from Dr. Bhiwgade and Dr. Nair

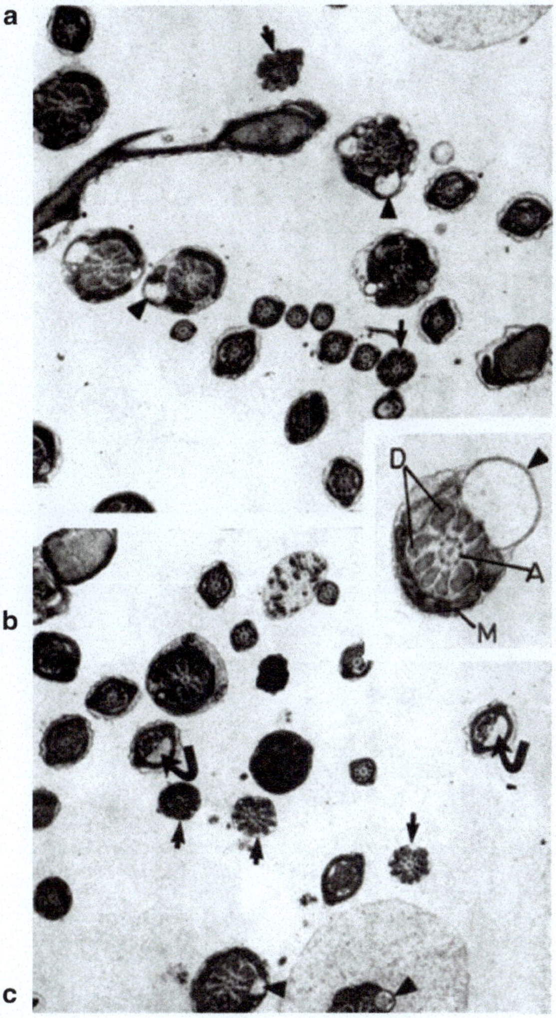

Fig. 3.15 (a, b, c) Cross sections passing through the lumen of caput epididymidis showing vacuolization of mitochondria (arrowheads), missing mitochondrial sheath (arrows) of midpiece spermatozoa, decapitated heads and sections showing missing members of axonemal components and outer dense sheath of the principal region of spermatozoa (curved arrows). Mitochondria sheath (m), outer dense fibres (D) and axonemal components 9 + 2 (A) [X 13,000] [X 20,000] [X 12,000]. Unpublished electron micrograph from Dr. Bhiwgade and Dr. Nair

with a content of moderate electron density. A small number of coated vesicles are also present among them. The flattened membranous sacs are less in number, usually two or three, and align parallel to one another. Other cisternal elements are highly distended and irregular in outline. Many of these flattened and distended cisternal elements have electron-lucent interiors. Some of the Golgi vacuoles and a few of the distended cisternae contain condensed and highly electron-dense secretory granules that occupy only a part of the interior and are surrounded by an electron-lucent zone consisting of fine flocculent material (Fig. 3.18a). Secretory vacuoles, similar in morphology to the Golgi vacuoles, are found in the cytoplasm (Fig. 3.18a). However, they are smoothly contoured and mainly spherical in shape. Numerous short microvilli are present on the apical cell surface (Fig. 3.18b).

The seminal vesicles of treated rats have undergone atrophy (Fig. 3.19a, b, c and d) and histologically, it is observed, that there is an increase in the fibro-muscular region, and the height of the columnar epithelial cells lining the folds of this organ is also decreased. The secretory material, though present, is separated from the surface of the epithelium. Ultrastructurally, the cell height decreased and the normal columnar

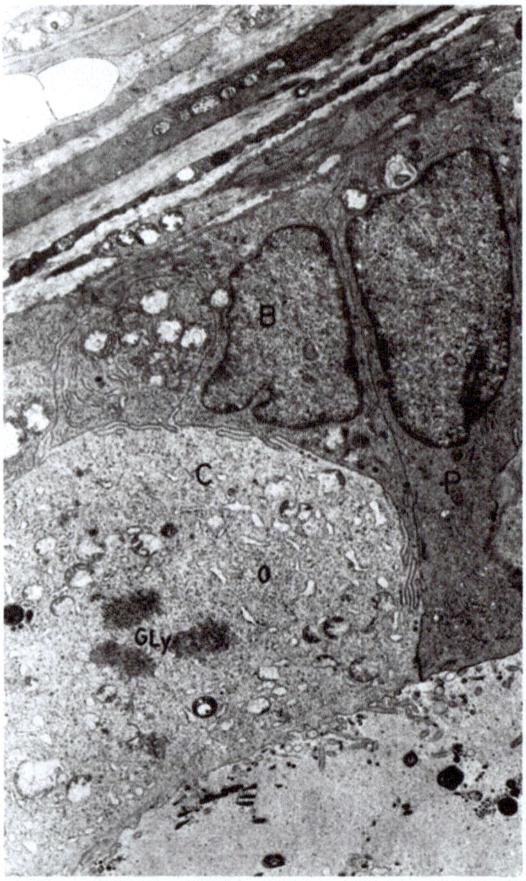

Fig. 3.16 Electron micrograph showing cell (P), basal cell (B) and clear cell (C) of cauda epididymidis. Note glycogen (GLY) and polyribosome (O) [X 6,000] Unpublished electron micrograph from Dr. Bhiwgade and Dr. Nair

shape of the cells becomes cuboidal or of a squamous shape (Fig. 3.19a, b). The cells are dominated by large nuclei, which also become cuboidal in shape. The mitochondria have undergone lysis (Fig. 3.19a, b), and the cisternae of rough endoplasmic reticulum which usually occupy most of the cytoplasm are reduced to only a few narrow and dispersed ones present in the apical region (Fig. 3.19a, b). The supranuclear Golgi apparatus in small with very few distended cisternae (Fig. 3.19a) and vacuoles and the forming of the secretory vacuoles that are normally seen has lessened (Fig. 3.19a and c). Vacuoles, which are apparently not of Golgi origin, are also in the nuclear region (Fig. 3.15b). Very few secretory granules are found in the apical region (Fig. 3.19d).

7. Prostate of control and CPA + TE treated rat: The prostatic acini are lined by a tall columnar epithelium interspersed with a few basal cells (Fig. 3.20). These tall columnar secretory cells reset in a basement membrane, which is opposed to the plasma membrane and merges outwardly, with the ground substance of the interstitial tissue. Fibroblasts and their slender extensions, as well as collagen fibres, are placed concentrically to the glandular units, and the capsule also contains capillary vessels and smooth muscle fibres.

The cytoplasm of the secretory cells has a well-developed, complex rough endoplasmic reticulum, which is arranged principally in the perinuclear zone, but also extends throughout the cells (Fig. 3.20). The parallel cisternae of ER are distended with a flocculent precipitate of newly synthesized secretory protein. A large Golgi complex is localized in the supranuclear region. It is composed of stacks of parallel cisternae, coated and smooth vesicles and vacuoles. Many multivesicular bodies are scattered in the Golgi region and the apical region (Fig. 3.20) and, presumably, are derived from the Golgi. Secretory droplets in the form of coated vesicles are also found in the apical region. The luminal border of the plasma membrane shows minute cytoplasmic projection or microvilli (Fig. 3.20). Secretory material extruded from the cells lies accumulated in the lumen (Fig. 3.20).

The epithelial cells lining the acini, which normally have a tall columnar shape, show a short columnar and cuboidal configuration (Fig. 3.21a). The fibromuscular interstitial tissue has diminished in size. The acini are empty and devoid of secretory material. The reduction in the cell size can be attributed to the decrease in the abundance of rough endoplasmic reticulum (Fig. 3.21b). In contrast to the extensive profiles of this organelle found in the controls, very few scattered cisternae lacking the content of newly synthesized secretory material, are seen in the basal and apical portions of the secretory cells (Fig. 3.21c, d). Mitochondria have undergone

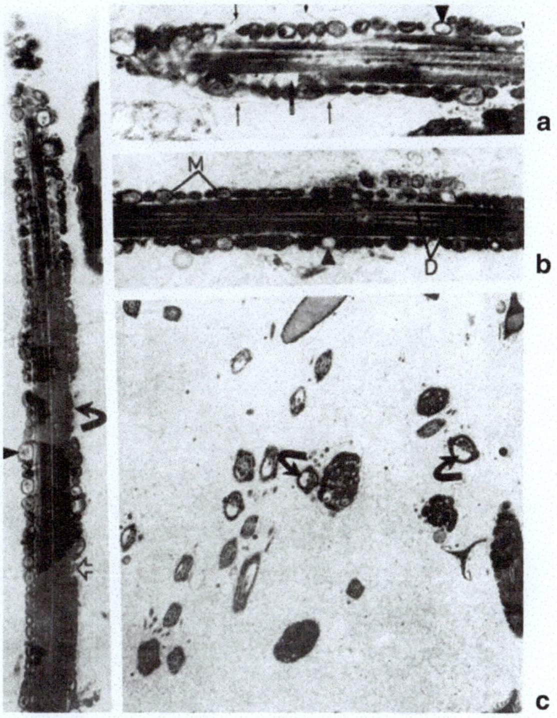

Fig. 3.17 (**a**, **b**, **c**) L. S. of midpiece spermatozoa from lumen of cauda epididymidis showing vacuolization of mitochondria (arrow heads), disintegration of mitochondria (mo), missing cell membrane as seen in Fig. 3.9 and a wavy structure of cell membrane (thin arrow), disorganization of outer dense fib res (arrows), missing granular substance (hollow arrow) and a reaggregation of this granular substance (curved arrow). Mitochondria sheath (M), outer dense fibres (D) and axonemal components 9 + 2 (A) [X 10,000] [X 15,000] [X 12,000]

hypertrophy and appear balloon shaped with loss of cristae (Fig. 3.21d). Golgi complex, though present, has a reduced secretory activity and the coated and smooth vesicles and vacuoles are almost absent (Fig. 3.21c).

The apical region contains very few secretory granules of diminished density in some cells (Fig. 3.21d), whereas they are totally lacking in others (Fig. 3.21d) indicating a hampered secretory activity. The lumen shows the presence of scattered amorphous secretory material of low electron density (Figs. 3.20 and 3.21c).

3.2.2 Discussion

The spermatozoa and the testicular fluids are drained through the rete testis into the epididymis. The sperm remain in the epididymis for varying periods of time before being ejaculated, and during this time they achieve their full fertilizing capacity, undergoing certain physiological, biochemical and morphological changes associated with their maturation.

The ABP is produced by the Sertoli cells and is transported via the rete testis fluid to the epididymis (Hansson et al. 1973). The testosterone in the seminiferous tubular fluid exists in free form and/or bound to the androgen binding protein; both the free testosterone (Voglmayr et al. 1966) as well as the androgen bound to the ABP are also transported to the epididymis via the rete testis fluid (Hansson et al. 1973, 1974). However, after the CPA treatment, since there is an androgen deficit in the testis and also a decline in the ABP level (Schenck and Neumann 1978), the quantity of androgens entering the excurrent duct system is also consequently reduced thereby resulting in an androgen deprivation of the epididymal tissue. The maintenance of a constant internal milieu in the epididymal canal is regulated, in part, by androgens, which influence the capacity of the epididymal epithelium for absorption and secretion. There appears to exist a differential threshold of androgens for regulating absorptive and secretory functions of the epididymis (Jones and Glover 1973; Prasad and Rajalakshmi 1976). The epididymis, especially the initial segment and the

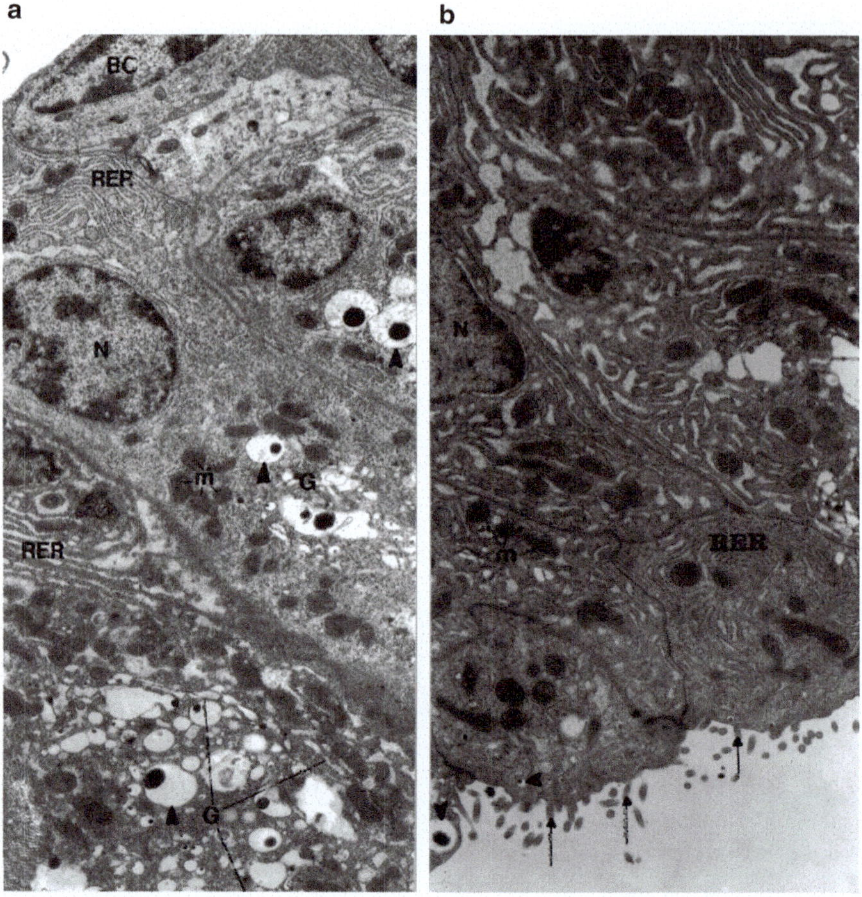

Fig. 3.18 (**a**) Low-power micrograph of the seminal vesicle of a control rat showing the basal cell (BC) resting on the basement membrane with its nucleus elongated in the long axis of the cells. The columnar cells contain abundant rough endoplasmic regions. The supranuclear region contains a large Golgi complex (G) with several small vacuoles and vesicles. Note the numerous secretory granules (arrowheads) in the Golgi zone and the mitochondria (m) [X 5000]. (**b**) Illustrated here is the apical region of the columnar epithelial cells from the seminal vesicle of the control rat. The cells contain abundant rough endoplasmic reticulum (RER), that predominates the basal and nuclear region. The apical region shows several secretory vacuoles (arrowheads). Note the mitochondria (m) and the microvilli (small arrows) at the apical surface [X 6500]. Unpublished electron micrograph from Dr. Bhiwgade and Dr. Nair

caput, transfers fluids from the lumen into the interstitium and absorbs and digests particulate (or macromolecular) material. This is done by the hydrolytic and proteolytic enzymes present in the matrix of the multivesicular bodies (Burgos 1964; Nicander et al. 1965; Nicander et al. 1965; Sedar 1966). In the present study, the micropinocytotic vesicles in the apical region of the principal cells of the caput epididymis appear to have decreased in number, indicating a decrease in the absorptive functions of the cells. However, the autophagic vacuoles in these cells have increased tremendously and almost occupy the entire supranuclear cytoplasm. This increase may be due to the increased digestion of macromolecules and portions of abnormal or dead spermatozoa absorbed from the lumen. Under experimental conditions, the epithelial cells in various regions of the epididymis have been known to be capable of absorbing

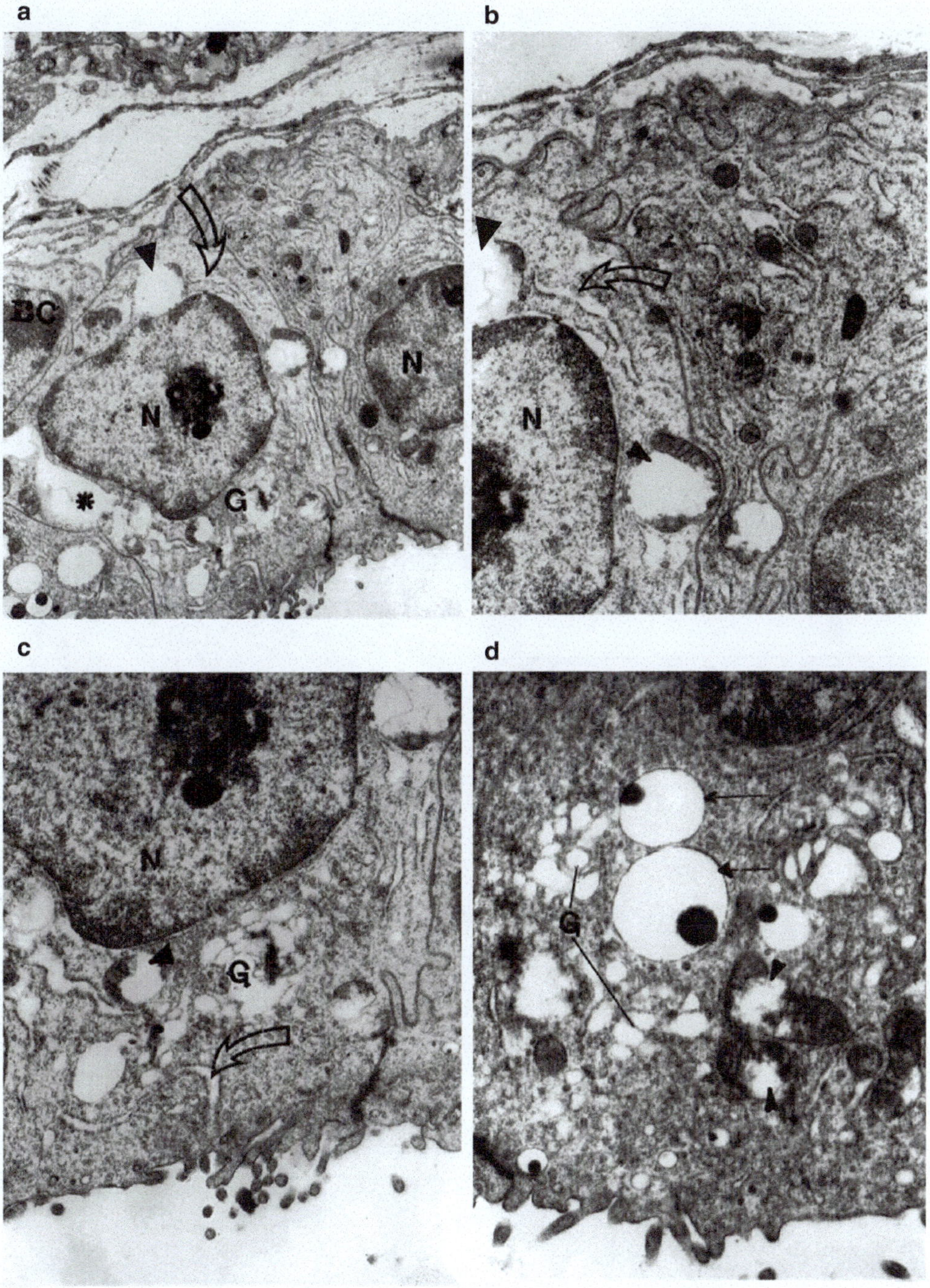

Fig. 3.19 Electron micrograph of the seminal vesicle of 60 days CPA + TE treated rat. (**a**) Note the decrease in the height of the epithelium as compared to the controls and the change of shape from columnar to cuboidal. The nucleus (N), which has also become cuboidal in shape, domains the scene due to scanty cytoplasm. The cytoplasm also vacuolation (asterisk). The basal cells (BC), however, appear uncharged [X 6500]. (**b**) Magnified view of the basal region of the above cells. Note the almost totally depleted rough endoplasmic reticulum (curved arrow) and the mitochondria undergoing ballooning and then lysis (arrow heads) [X 13, 500]. Unpublished electron micrograph from Dr. Bhiwgade and Dr. Kadam. (**c**) Magnified view of the apical region showing the reduced cytoplasmic area. The Golgi complex (G) in the supranuclear region has shrunk in size [X 13,500]. (**d**) Higher magnification of the Golgi zone (G) which appears very much reduced in size with few no vesicles and vacuoles and scanty secretory granules (arrows). Very few secretory vacuoles are found near the apical surface [X 16,300]. Unpublished electron micrograph from Dr. Bhiwgade and Dr. Kadam

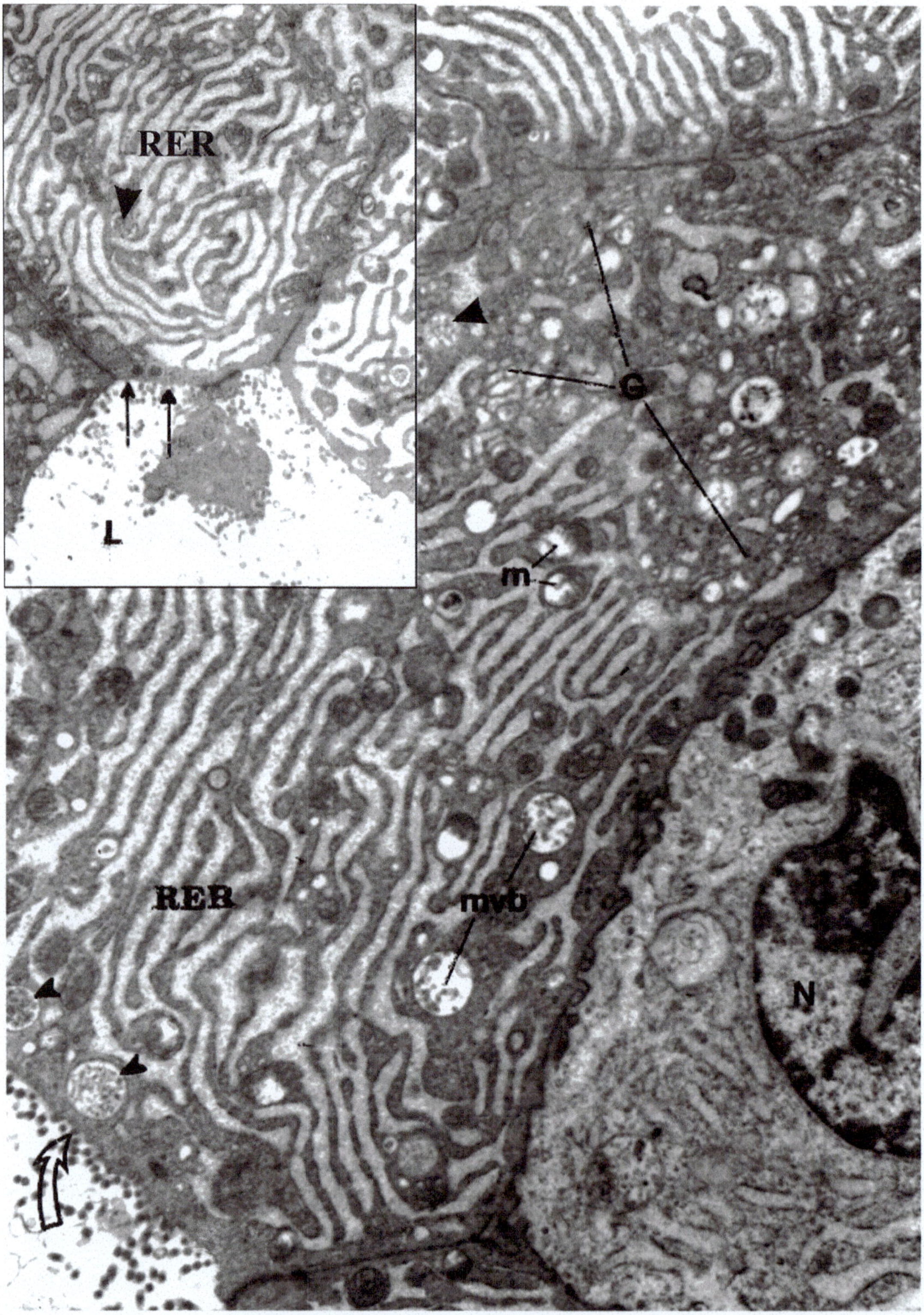

Fig. 3.20 High-power electron micrograph of the secretory cells from the prostate of a control rat. The rough endoplasmic reticulum (RER) predominates the cell and contains a flocculent secretory material. Note the nucleus (N), Golgi complex (G), mitochondria (m) and a few multivesicular bodies (mvb). Several secretory droplets (arrowheads) are seen in the Golgi region and near the apical surface also gives out many microvilli (curved

degenerating spermatozoa and sequestering them in large multivesicular bodies (Hamilton 1975).

Higher androgen levels are required to maintain the secretory functions than the absorptive functions of the epididymis (Jones and Glover 1973). Hence, lack of androgens affects the secretory functions. CPA has an inhibitory effect on the secretory activity of the accessory organs and causes a marked decrease in the levels of bound sialic acid in the epididymis and a concomitant loss in the integrity of the acrosomes of the spermatozoa and their fertilizing ability (Prasad et al. 1972; Ramakrishnan et al. 1990). It also causes a sharp decline in the levels of glyceryl phosphorylcholine (Roy et al. 1976) which is mainly contributed by the epididymis (Mann 1964; Roy and Taneja 1974).

The cauda epididymidis, where sperm maturation is completed, has a higher threshold requirement of androgens for the maintenance of its functions than the caput epididymis (Prasad and Rajalakshmi 1976). A deficit in the androgen levels alters the cytoarchitecture of the clear cells of the cauda as is evident from the present ultrastructural findings. The large vacuoles and vesicles in these cells, which usually contain dense secretory material in the controls, are filled with a particulate matter of low electron density after CPA treatment. This suggests a decrease in the synthetic activity or an altered biosynthesis of the various secretory products. Though biochemical studies have established the various materials secreted by the epididymal epithelium, morphological evidence for any of these secretions is lacking in all the species studied so far. Hence, it is difficult to postulate that which of the various products are secreted by the clear cells of the cauda and are consequently affected by the CPA treatment.

The epididymis being a very important organ for the maturational process of the spermatozoa, any change in its biochemical environment is responsible for aberrant changes occurring in the spermatozoa. Apparently, the altered milieu has caused the disruption of the motility apparatus of the spermatozoa, which appears normal in the sections of the testis, indicative of degeneration rather than deformation. In the lumen of the cauda epididymis, the outer dense fibres of the axonemal complex in the tailpiece of the spermatozoa are fragmented thereby rendering the latter immotile. Our ultrastructural findings are supported by the various investigations on the semen analysis after CPA treatment. CPA reduces the sperm counts to sub-fertile levels, the motility of the spermatozoa is decreased with a distinct drop in their speed and there is an increase in the percentage of non-motile spermatozoa. The antiandrogen also increases the number of abnormal, immature and pathological sperms, and dead spermatozoa and decreases the ability of motile spermatozoa to penetrate a column of cervical mucus (Petry et al. 1972; Morse et al. 1973; Koch et al. 1974, 1976; Lorenz et al. 1974; Hammerstein 1974; Neumann et al. 1976; Neumann and Schenck 1976; Roy et al. 1976; Moltz et al. 1978; Fogh et al. 1979; Wang and Yeung 1980; Lohiya and Sharma 1983). In the present study, androgen supplementation has not succeeded in preventing CPA from exerting its action of rendering sperm immotile.

Since the first report that gossypol is a safe, effective, and reversible contraceptive for the human male, the effects of this compound on the male reproductive tract have been extensively studied. In the present study spermatozoa are examined with the transmission electron microscope, at different points along the reproductive tract and the extent of cellular damage was observed after gossypol treatment. It is found that ultrastructural defects in the spermatozoa are observed in the caput, cauda epididymidis and vas deferens, but the epididymal epithelium nor the vassal epithelium is adversely affected. The appearance of ribosome studded endoplasmic reticulum, Golgi apparatus and secretory vesicles at the cell border of the cauda

Fig. 3.20 (continued) arrows) [X 10,000]. Inset shows the apical region of an epithelial cell with many secretory droplets (arrowheads) and secretory material in the lumen (L). Several microvilli (arrows) are present on the luminal border. Unpublished electron micrograph from Dr. Bhiwgade and Dr. Kadam

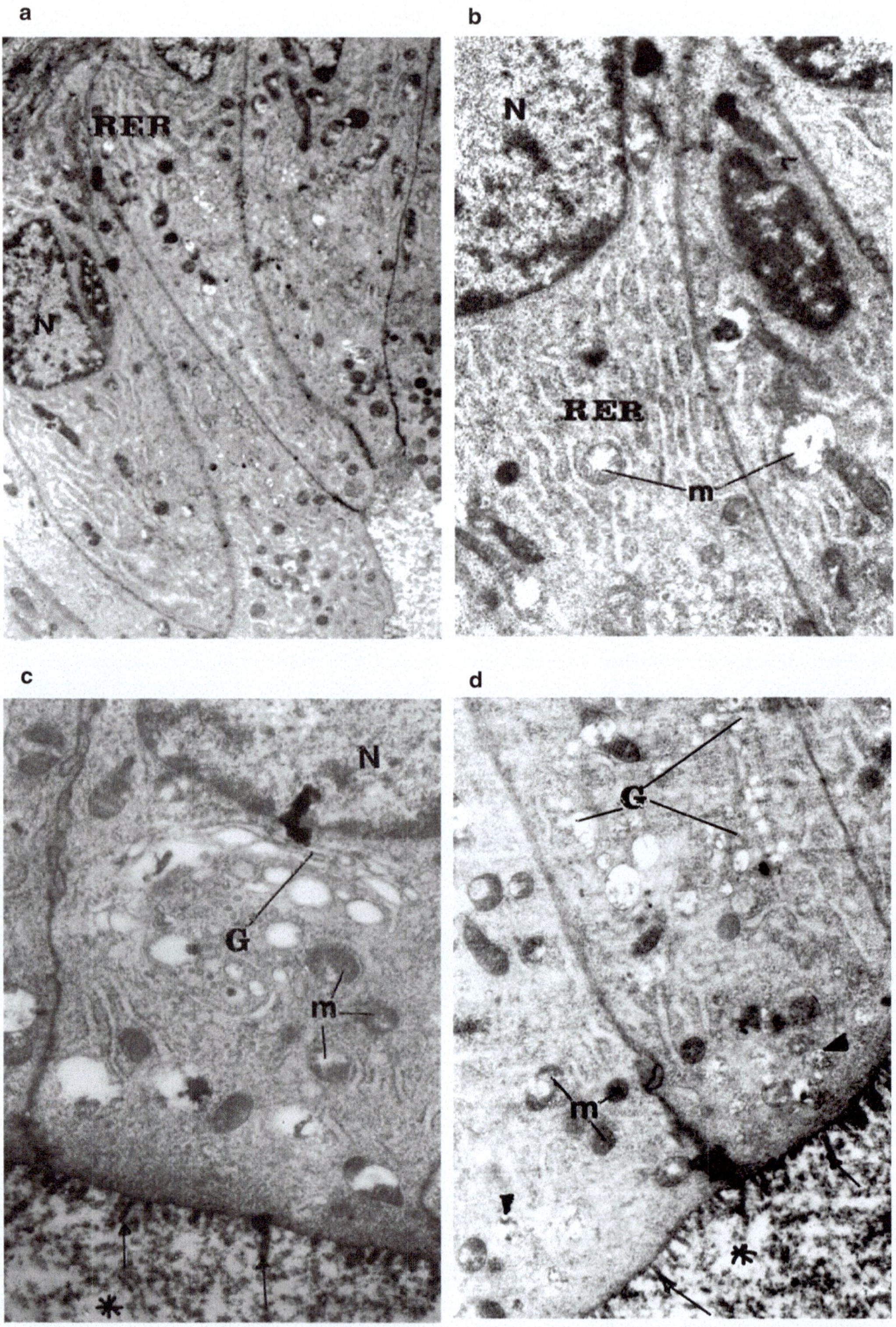

Fig. 3.21 Electron micrograph of the prostate of 60 days CPA + TE treated rat. (**a**) Reduction in cell height, a prominent change following the drug treatment, is apparent in this micrograph [X 5000]. (**b**) Higher magnification of the supranuclear region of the secretory cells of the above, Scanty profile of rough endoplasmic reticulum (RER), with practically empty cisternae, have caused the reduction in the cell height. A few mitochondria (m) have

epididymidis suggests a normal macromolecular production and secretion. The epithelium of the vas deferens shows an enhanced protein synthesis as indicated by the dilated endoplasmic reticulum. However, it is to be noted that the mitochondria from the cauda epididymidis and vas deferens are hypertrophied. Cauda epididymidis is considered to be important in the resorption of non-ejaculated spermatozoa. The autolytic process of the hypertrophied mitochondria as seen in the cauda epididymidis may account for the decrease in the height of the caudal epithelium and the loss of weight of the cauda epididymidis in the present study. The stereocilia of the caudal epithelium is markedly reduced. The microvilli of the vas deferens are also seen to be reduced. However, this cannot be taken as an effect of the drug, as microvilli are sometimes absent or reduced from some areas of the vas deferens. These findings are at variance with a normal ultrastructure of the excurrent duct system. In the present study, the possibility of a prolonged treatment of strain difference and the weight of the animals (200–250 g) may have led to this difference.

Our electron microscopic study of spermatozoa confirms earlier observations that deleterious changes are found in the epididymal and vas deferens spermatozoa of experimental rats. The unequal distribution of granular substance or satellite fibres found between the mitochondrial sheath and outer dense fibres must have led to the dislocation of the fibre bundles. The absence of this cementing layer must have also led to the dislocation of the mitochondrial sheath as seen in cross sections. In addition, it was observed that sperm flagellum or the axonemal component and its outer dense fibres in the principal piece were disintegrated, especially in sections from vas deferens and cauda epididymidis.

3.3 Ultrastructure of Epididymis in Normal and Experimental Bonnet Monkey (*Macaca radiata*)

3.3.1 Observations

In the present study, we have observed the details of ultrastructure of the epididymis of bonnet monkey, *Macaca radiata*, with a special reference to the regional variations of control and DMPA + TE treated group.

3.3.2 Caput I-Epididymis of Control Bonnet Monkey

1. Caput I: Under the electron microscope, the epididymis tubule of the Caput I is surrounded by peritubular tissues containing fibroblasts, collagen, elastic fibres, blood vessels and macrophages followed by a non-cellular basal lamina (Figs. 3.22a and 3.23).

The Caput I in the control Bonnet monkey is composed of four cell types viz. principal, basal, lymphocytes and macrophages. The Apical cells (mitochondria-rich cells) are absent in the Caput I of the control Bonnet monkey, however, few clear cells are observed. Principal cells are the major cell types exhibiting morphological features of a cell involved in secretion and absorption.

The nuclei of the principal cells of control Caput I are almost rounded in shape, smooth in outline and basal in position. A prominent nucleolus is present in the centre of the nucleus. Another feature of the nuclei is the presence of prominent spherical inclusions with low density (Fig. 3.22b).

The rough endoplasmic reticulum studded with ribosomes is well developed in the

Fig. 3.21 (continued) undergone ballooning with a loss of cristae [X 10,000]. (**c**) Magnified view of the apical region of the above, depicting the Golgi complex (G) containing very few cisternae and vacuoles. Secretory droplets are also almost absent. An amorphous secretory material is scattered in the lumen (asterisk). Note the nucleus (N) and the microvilli at the apical surface (small arrows) [X 13,500]. (**d**) Magnified view of the apical region of the secretory cells of the above. The secretory droplets (arrow heads) in this region are very scanty and contain very less secretory material [X 10,000]. Unpublished electron micrograph from Dr. Bhiwgade and Dr. Kadam

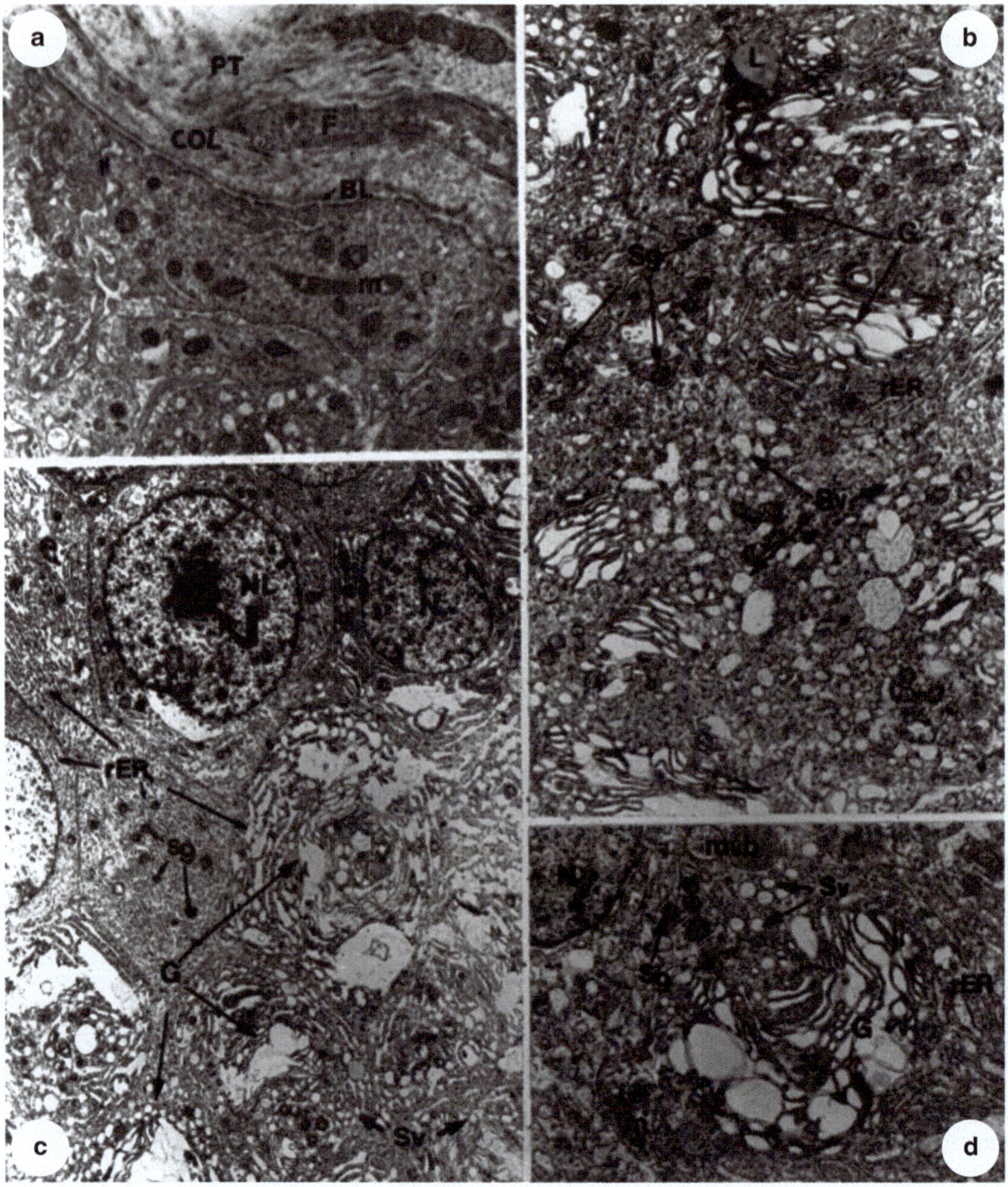

Fig. 3.22 (**a**) Electron micrograph of caput I epididymis from the control Bonnet monkey illustrating mainly the peritubular tissues (PT), containing collagen fibres (Col), fibroblast cells (F) and the Basal Lamina (BL). The infranuclear zone of the principal cell is characterized by the presence of variable mitochondria (m), dense bodies (Db) and Endoplasmic reticulum (ER) [X 16,000]. (**b**) Electron micrograph of the pseudostratified epithelium or the principal cells of the caput I epididymis of the control Bonnet monkey. Note the presence of basal, large and rounded nucleus (N) containing prominent nucleolus (NL). Infranuclear rough endoplasmic reticulum (rER) arranged in parallel rows, dispersed mitochondria (m), supranuclear Golgi complex (G), lysosomes (Ly) and secretory vesicle (SV) [X 13,000]. (**c**) Electron micrograph of apical region from a principal cell of caput I from the control Bonnet monkey showing the extensive supranuclear Golgi apparatus (G), rough and smooth endoplasmic reticulum (rER, sER), mitochondria (m) and lipid droplets (L). Note the different degrees of density of the matrix of the multivesicular bodies (mvb) and secretory vacuoles (SV) [X 13,000] [X 16,000]. Unpublished electron micrograph from Dr. Bhiwgade and Dr. Bansal. (**d**) Electron micrograph of apical region from a principal cell of caput I from the control Bonnet monkey showing the extensive supranuclear Golgi apparatus (G), rough and smooth endoplasmic reticulum (rER, sER), mitochondria (m) and lipid droplets (L). Note the different degrees of density of the matrix of the multivesicular bodies (mvb) and secretory vacuoles (SV) [X 13,000] [X 16,000]. Unpublished electron micrograph from Dr. Bhiwgade and Dr. Bansal

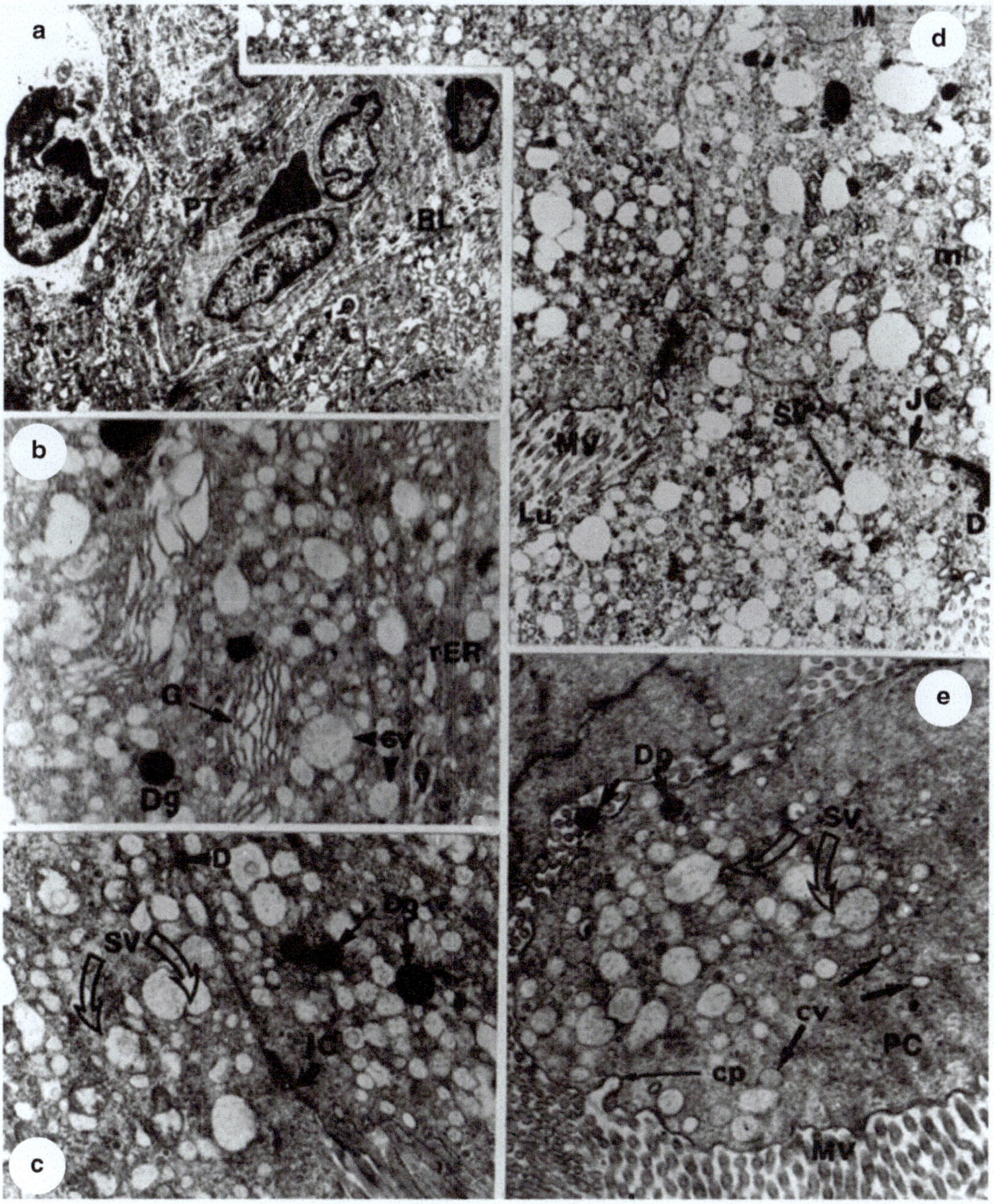

Fig. 3.23 (**a**) Electron micrograph of caput I epididymis from control Bonnet monkey showing peritubular tissues (PT), fibroblast cell (F), basal lamina (BL) and blood capillary (B. Cap) [X 16,000]. (**b**) Electron micrograph of the extensive supranuclear stacks of the Golgi apparatus (G), rough endoplasmic reticulum (rER), secretory vesicles (SV) and the dense granules (Dg) in the caput I of control Bonnet monkey [X 16,000]. (**c**) Electron micrograph of the apical portion of the ciliated principal cells of caput I epididymis of control Bonnet monkey illustrating lumen (Lu), microvilli (mV) secretory vesicles (SV), dense granules (Dg) and smooth endoplasmic vesicles (sER). The junctional complex (JC) provided with desmosomes (D) is also well marked between the adjacent cells [X 16,000] [X 7000]. (**d**) Electron micrograph of the apical portion of the ciliated principal cells of caput I epididymis of control Bonnet monkey illustrating lumen (Lu), microvilli (mV) secretory vesicles (SV), dense granules (Dg) and smooth endoplasmic vesicles (sER). The junctional complex (JC) provided with desmosomes (D) is also well marked between the adjacent cells [X 16,000] [X 7000]. (**e**) Electron micrograph of the apical portion of the ciliated principal cell (PC) of caput I of control Bonnet monkey exhibiting endocytosis, secretory vesicles (SV), dense granules (Dg), microvilli (MV) and the central lumen (Lu). Note the formation of micropinocytotic infoldings, coated pits (CP, long arrows) and coated vesicles (CV, small arrows) [X 16,000]. Unpublished electron micrograph from Dr. Bhiwgade and Dr. Bansal

infranuclear zone, running parallel and closer to the lateral wall of the principal cell. A large number of mitochondria of different shapes and sizes with typical lamellar cristae and dense cytoplasm along with a few dense bodies tend to conglomerate in the basal cytoplasm (Fig. 3.22a and b).

The supranuclear zone, and occasionally the pronuclear cytoplasm contains well-developed Golgi complex with five or more compactly arranged Golgi cisternae. Vesicular and vacuoles are numerous in the vicinity of Golgi complex, some of them are clear while others are of different densities (Fig. 3.22b and c). Multivesicular bodies, lysosomes, membrane bound granules of various densities and free ribosomes are the common features of the apical cytoplasm, especially in the Golgi zone (Fig. 3.22c and d).

The apical cytoplasm of the control Caput I is also characterized by the presence of rough endoplasmic reticulum, smooth endoplasmic reticulum, free and polyribosomes. Lipids droplets and mitochondria are also seen in the supranuclear cytoplasm (Figs. 3.23a, b, c, d and 3.24).

These cells also possess long, branching and irregular stereocilia at the luminal border. The core of each stereocilia contains bundles of closely packed fine filaments, extended deep into cytoplasm (Fig. 3.23c, b). Adjoining Principal cells are separated by occluding junctional complexes (Fig. 3.23c) provided with desmosomes. Below the stereocilia, are many invaginations of the apical plasmalemma, which pinch off as pinocytotic vesicles and coated vesicles into the apical cytoplasm (Fig. 3.23e). A large number of vacuoles containing low-density material are also visible in the apical cytoplasm which seems to be formed from the endoplasmic reticulum and dilated upper most saccule of Golgi complexes (Figs. 3.22c, d and 3.24a).

3.3.3 Caput I-Epididymis Control of DMPA + TE Treated Bonnet Monkey

2. Caput I: Electron microscopic observation of DMPA + TE treated Caput I shows some alterations in the supra nuclear and infra nuclear regions of principal cells. The cell shows intense lysosomal activity following treatment. In treated monkeys, the basal lamina becomes irregular due to the presence of deep invaginations and protrusions as compared to the smooth, thin and straight basal lamina of the control epididymis (Fig. 3.25a, b, c). The interstitium of Caput I grows thick following treatment, due to the accumulation of collagen fibres (Figs. 3.24b and 3.25a).

No detected change is observed in the basal cells, however, the number of macrophage cells is increased following treatment. The macrophage cells are amoeboid in shape and show accumulation of lysosomes, dense bodies, multivesicular bodies and lipid droplets (Fig. 3.25b).

The basal nucleus becomes highly folded and bizarre shaped following treatment as compared to the oval and smooth-walled nucleus of the control animal. The nucleus has one or two nucleolus and peripheral dense heterochromatin (Figs. 3.25c and 3.26a).

The rough endoplasmic reticulum in the infranuclear cytoplasm of the principal cells shows atrophy following treatment (Fig. 3.26b). Another noticeable change is the increase in a number of lysosomes. They are in the form of autophagic and heterophagic vacuoles occupied by membranous material, dense bodies containing amorphous material (Figs. 3.26c, d, and 3.27a, b), multivesicular bodies (Fig. 3.26d) and small, smooth coated vesicles (Fig. 3.27c). Drop in the amount of Golgi cisternae, smooth and rough endoplasmic reticulum are observed following treatment, in the apical cytoplasm (Fig. 3.27b). These organelles are atrophied and disorganized and have lost their special relationship with each other in the cell. Secretory vesicular and vacuoles are lacking in the apical cytoplasm, however, endoplasmic is well marked even after treatment (Figs. 3.27b, c and d). Clear cells do not show any alteration following treatment (Fig. 3.27b).

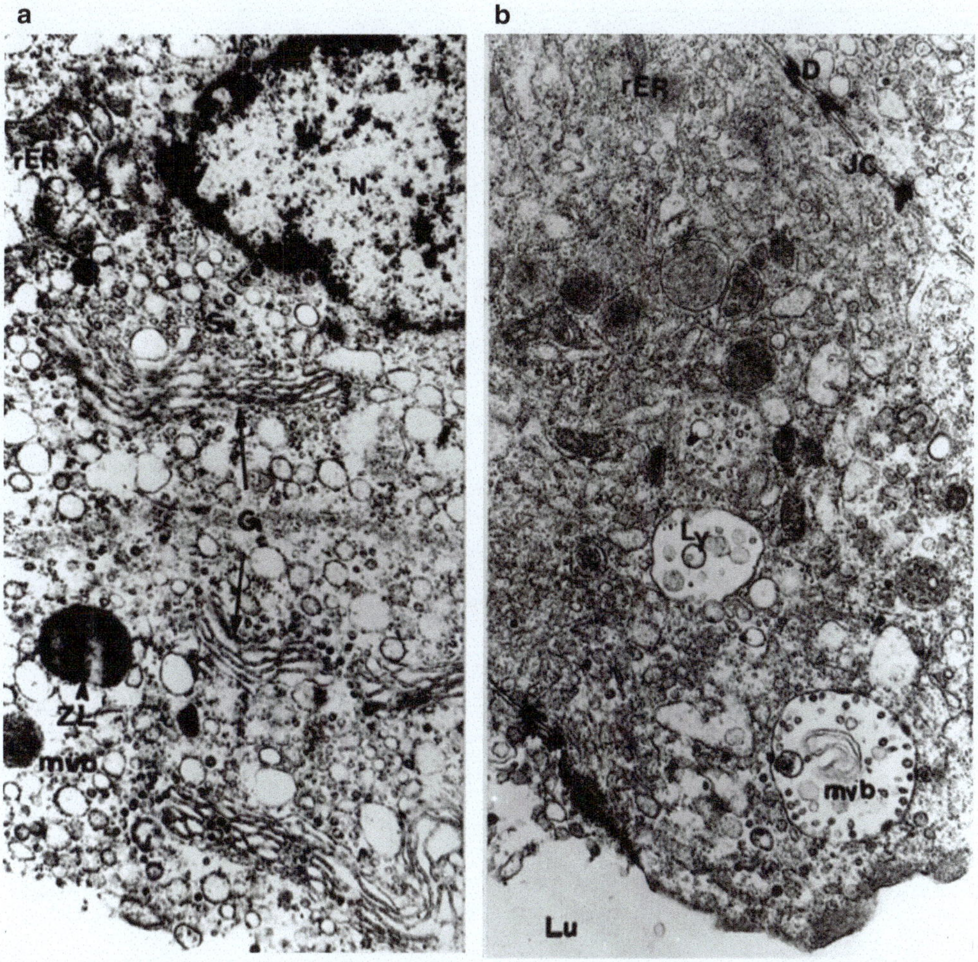

Fig. 3.24 (**a**) Electron micrograph of principal cell from caput I of control Bonnet monkey, note the Golgi although extensive but appears to be composed of many small Golgi bodies (G) rather than a large body. Well-developed smooth endoplasmic reticulum (sER), associated Zebra lipid droplets (ZL), rough endoplasmic reticulum (rER), multivesicular body (mvb), secretory vesicle (SV) and a large nucleus (N) are visible [X 16,000]. (**b**) Electron micrograph of the apical portion of the non-ciliated principal cell from the caput I of control Bonnet monkey illustrating a large number of multivesicular bodies (mvb), some of them containing coated pits. Note the diversity of density of multivesicular bodies. Lysosomes (Ly), mitochondria (m) and a few rough endoplasmic cisternae (rER) are also observed. Note the presence of a Junctional complex (JC) with desmosomes (D) between the neighbouring cells [X 16,000]. Unpublished electron micrograph from Dr. Bhiwgade and Dr. Bansal

3.3.4 Caput II and III: Epididymis of Control Bonnet Monkey

3. Caput II and III: Regions II and III are almost similar in epithelial height, luminal diameter and muscle wall thickness. These regions possess lower epithelium and a greater luminal concentration of sperm in contrast to region I. The epithelium of these regions also contain few apical cells (Fig. 3.28).

Cuboidal basal cells are the small cells lying adjacent to the basal lamina with nuclei, scanty cytoplasm and cell organelles (Fig. 3.28).

Halo cells are scattered throughout the epithelium with an irregular outline and a small amount of electron lucent cytoplasm which is

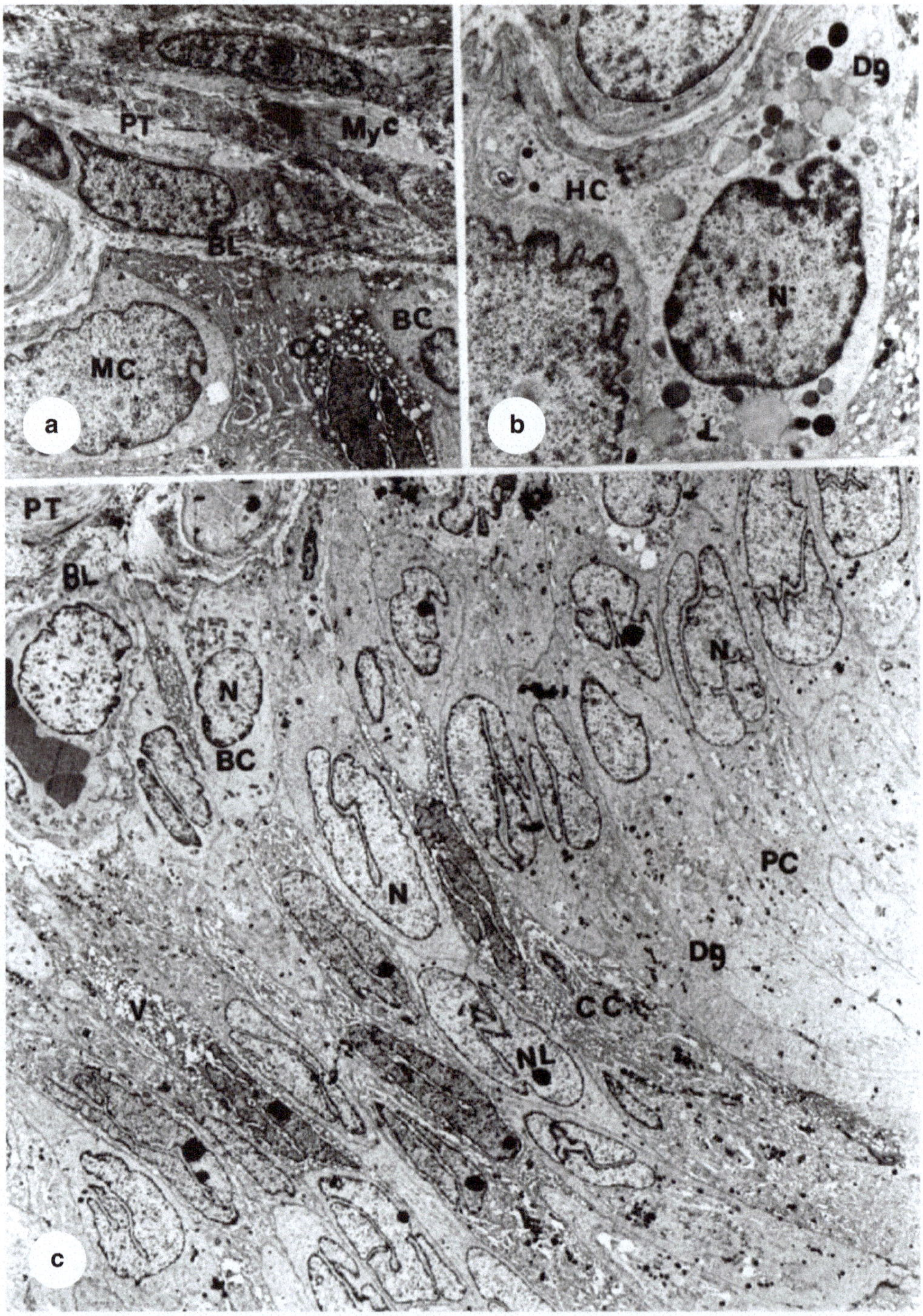

Fig. 3.25 (**a**) Electron micrograph of caput I epididymis of DMPA + TE treated Bonnet monkey showing peritubular tissue (PT) containing increased collagen fibre (COL) fibroblast cells (F), myoid cell (myC) and folded basal lamina (BL). The basal position of one clear cell (CC) and one principal cell (PC) is well marked between macrophage cell (MC) and Basal cell (BC) [X 5000]. (**b**) Electron micrograph of the amoeboid halo cell (HC) or lymphocyte of the DMPA + TE treated caput I epididymis from Bonnet monkey showing a large nucleus (N), accumulation of dense bodies (Db) and lipid droplets (L) [X 30,000]. (**c**) Electron micrograph of caput I epididymis from DMPA + TE treated Bonnet monkey. The section passes through the entire thickness of the epithelium demonstrating peritubular tissues (PT), highly folded basal lamina (BL), Basal cell (BC), clear cell (CC) and the principal cells (PC). Note the presence of bizarrely shaped nucleus (N) with nucleolus (NL), vacuoles cells and the clear cells [X 3500]. Unpublished electron micrograph from Dr. Bhiwgade and Dr. Bansal

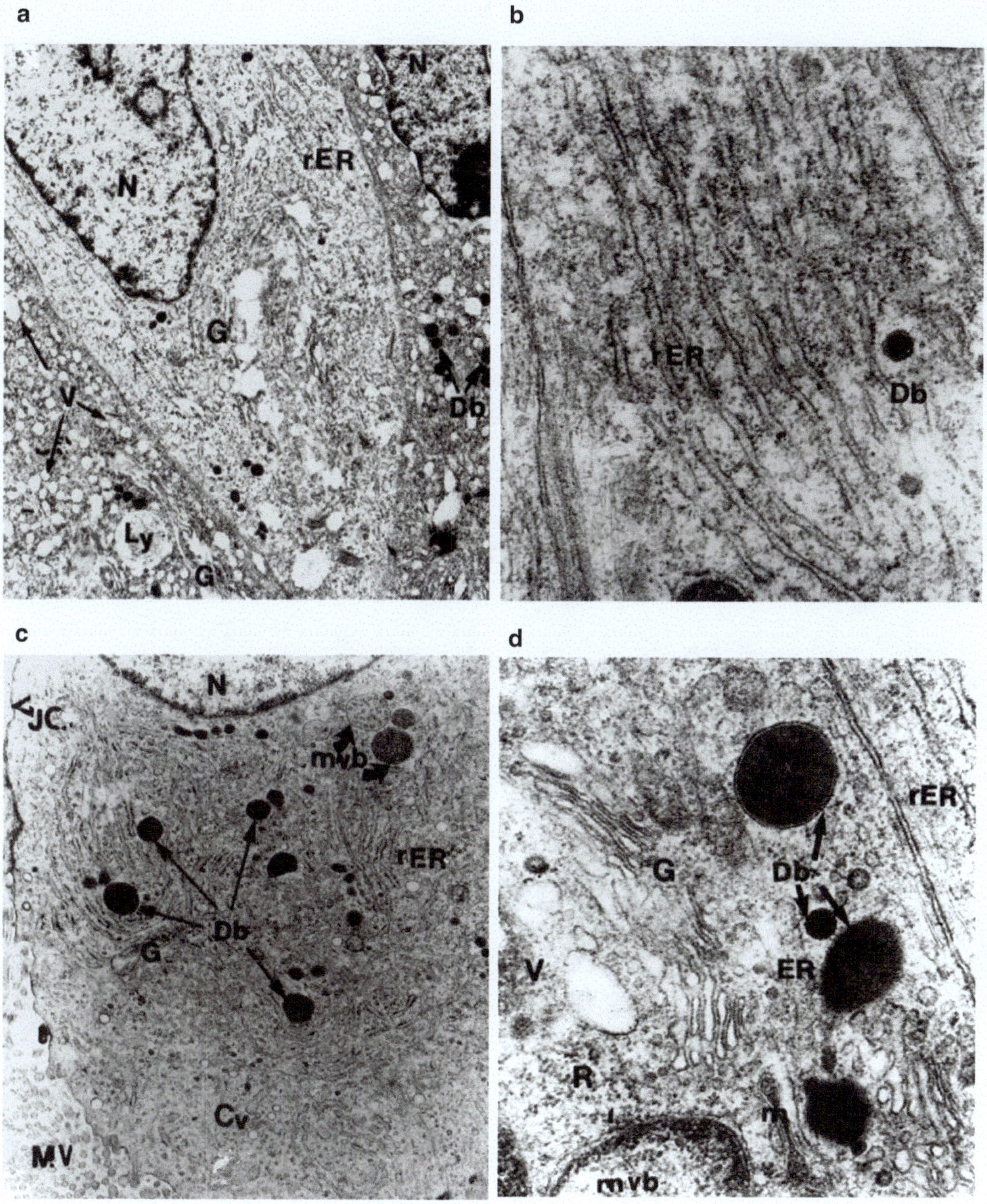

Fig. 3.26 (**a**) Electron micrograph of the basal portion of the principal cells and the clear cell of the caput I epididymis of a Bonnet monkey administrated DMPA + TE for 45 days. A well-developed nucleus (N), atrophied rough endoplasmic reticulum (rER) and Golgi complex are observed in the principal cells. The two clear cells containing their usual organelle with the accumulation of vacuoles (V) dense bodies (Db), lysosomes (Ly) and the Golgi body (G) are seen [X 10,000]. (**b**) Electron micrograph of the infranuclear zone of the principal cells of caput I epididymis from treated Bonnet monkey illustrating vertical rows of retrogressive rough endoplasmic reticulum (rER), dispersed ribosomes and the dense bodies (Db) [X 40,000]. (**c, d**) Electron micrograph of the supranuclear zone of the principal cells of caput I epididymis of treated Bonnet monkey showing atrophy of rough endoplasmic reticulum (rER), dispersed ribosomes (R), hypertrophied Golgi complex (G), accumulation of large dense bodies (Db) containing dark amorphous substance and multivesicular bodies (mvb). Only a few mitochondria (m) and vacuoles (Cv) are observed in the apical cytoplasm. The junctional complex (JC) is well marked [X 13,000] [X 40,000]. Unpublished electron micrograph from Dr. Bhiwgade and Dr. Bansal

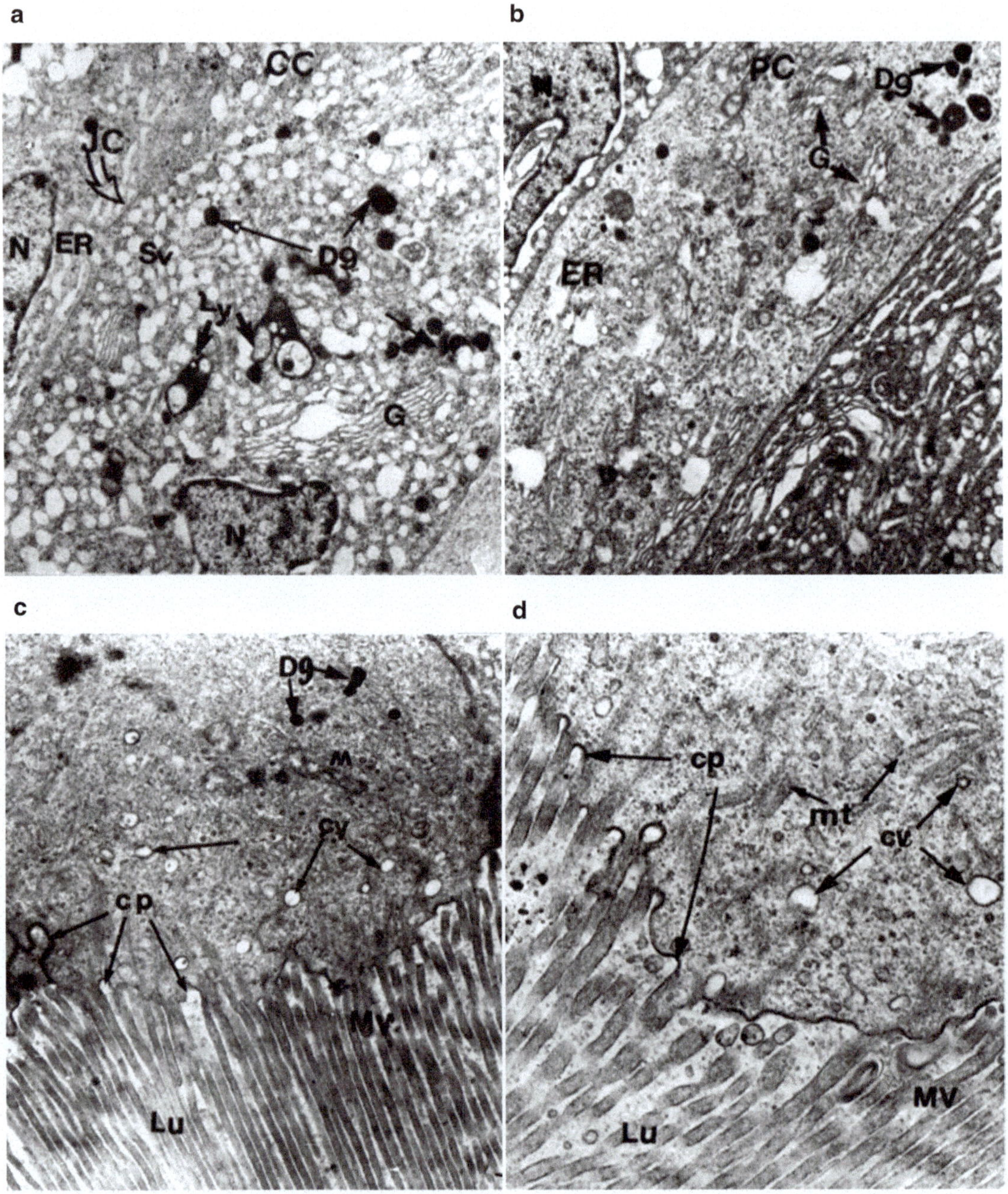

Fig. 3.27 (**a**) Electron micrograph of the clear cell (CC) of the caput I epididymis of a treated Bonnet monkey shows the accumulation of lysosomes (LY) and dense granules (Dg). The other cell organelles, viz. Golgi complex (G), secretory vesicles (SV), endoplasmic reticulum (ER), nucleus (N) and the junctional complex (JC) are normal in appearance [X 13,000]. (**b**) Electron micrograph of two normal-appearing clear cells and one altered principal cell (PC) of caput I epididymis from treated Bonnet monkey. The principal cell (PC) shows atrophied Golgi complex with dilated cisternae (G), atrophied endoplasmic reticulum (ER), accumulation of lysosomes and dense granules (Dg) [X 10,000]. (**c**, **d**) Electron micrograph of apical portion of the principal cells of caput I epididymis from treated Bonnet monkey illustrating different stages of formation of coated pits (CP) and coated vesicles (Cv). Only a few vacuolated mitochondria (m), many dense bodies (Db) and microtubules (mt) are observed in the apical cytoplasm. The central lumen (Lu) contains well-developed microvilli (Mv) following treatment [X 13,000] [X 25,000]. Unpublished electron micrograph from Dr. Bhiwgade and Dr. Bansal

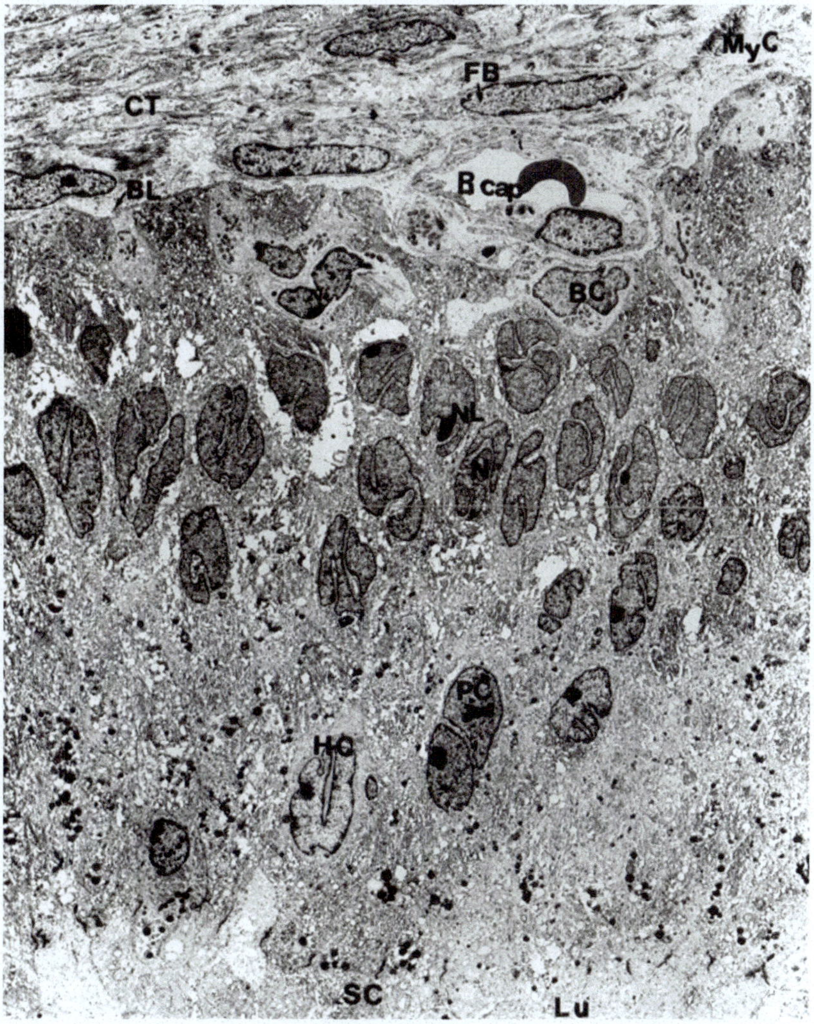

Fig. 3.28 Electron micrograph of caput II epididymis of the control Bonnet monkey passing through the entire thickness and showing the columnar epithelial cells and small triangular basal cells resting over the basal lamina. The surrounding peritubular tissues (PT), contain the connective tissue (CT), fibroblast cells (FB), Myoid cells (MYC) and the blood capillary (B. Cap). The epithelial cells are characterized by the presence of basal folded nucleus (N) and infranuclear and supranuclear dense granules. Few basal cells (BC), infraepithelial lymphocytes and the halo cells (HC) are also seen. Central lumen (Lu) contains the sterocilia (Sc) at the apical margin of the principal cells and the testicular fluid [X 3500]. Unpublished electron micrograph from Dr. Bhiwgade and Dr. Bansal

characterized by the presence of membrane bound granules (Fig. 3.28).

The interstitium consists of much folded basal lamina containing micropinocytotic infoldings (Figs. 3.28 and 3.29c), and thick lamina propria. The lamina propria is many layered, thick and composed of fibroblast cells with an elongated nucleus, cell organelles, collagen fibres and sub-epithelial blood capillaries close to the basal lamina (Figs. 3.28, 3.29a and 3.33a). The infra nuclear cytoplasm in Caput II and III is filled with rough endoplasmic reticulum, dense bodies, variable mitochondria and only a few vacuoles (Figs. 3.29a, b, and c).

The nucleus is irregular and folded as compared to the oval nucleus of the Caput I and is basal in position with a well-developed nucleolus and dispersed heterochromatin (Figs. 3.28 and 3.34c).

The supranuclear cytoplasm of the principal cells in Caput II and III of control Bonnet monkey is packed with a variety of common cell organelles. Well-developed Golgi apparatus with 4 or 5 closely packed dilated Golgi cisternae running vertically and parallel to the longitudinal axis and associated secretory vacuoles and vesicles are well marked (Figs. 3.30a, d, and 3.34a, b). Numerous large-sized multivesicular bodies filled with coated vesicles are a common

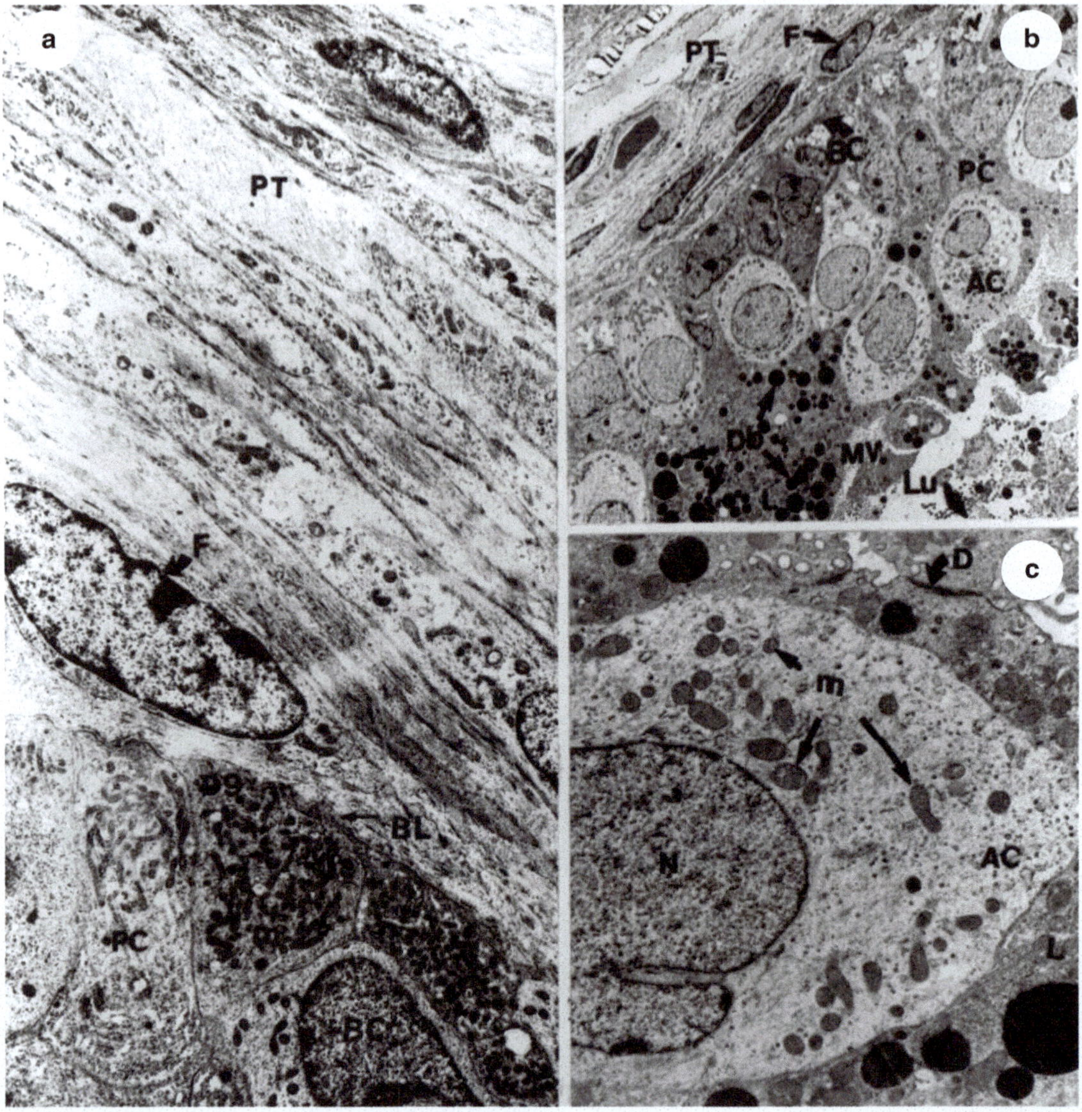

Fig. 3.29 (**a**) The infranuclear region of principal cells with basal lamina and peritubular tissue, is seen in this electron micrograph of caput II from the control Bonnet monkey. Small dense granules (Dg) and mitochondria (m) are characteristic features of the basal cytoplasm of the principal cells. Note the well-developed smooth and rough (sER and rER) endoplasmic reticulum. The basal cell (BC) with lobed nucleus and cell organelle is also visible. Basal lamina (BL) is slightly folded. The peritubular tissue is composed of fibroblast cells (F) and collagen fibres [X 5000]. (**b**, **c**) Electron micrograph of caput II epididymis from treated Bonnet monkey illustrating outer peritubular tissues (PT) with fibroblast cells (F), the basal lamina (BL) and the epithelial cells regressed in height. Epithelium is composed of basal cells (BC), principal cells (PC) and Apical cells (AC). Principal cells are characterized by the presence of basal nucleus and Apical cells with medially placed nucleus. The micrograph shows the accumulation of dense bodies (Db) in the apical cytoplasm and formation of vacuoles (v) in the infranuclear zone. Central lumen (Lu) shows the presence of microvilli (MV) and testicular debris [X 3500]. Unpublished electron micrograph from Dr. Bhiwgade and Dr. Bansal

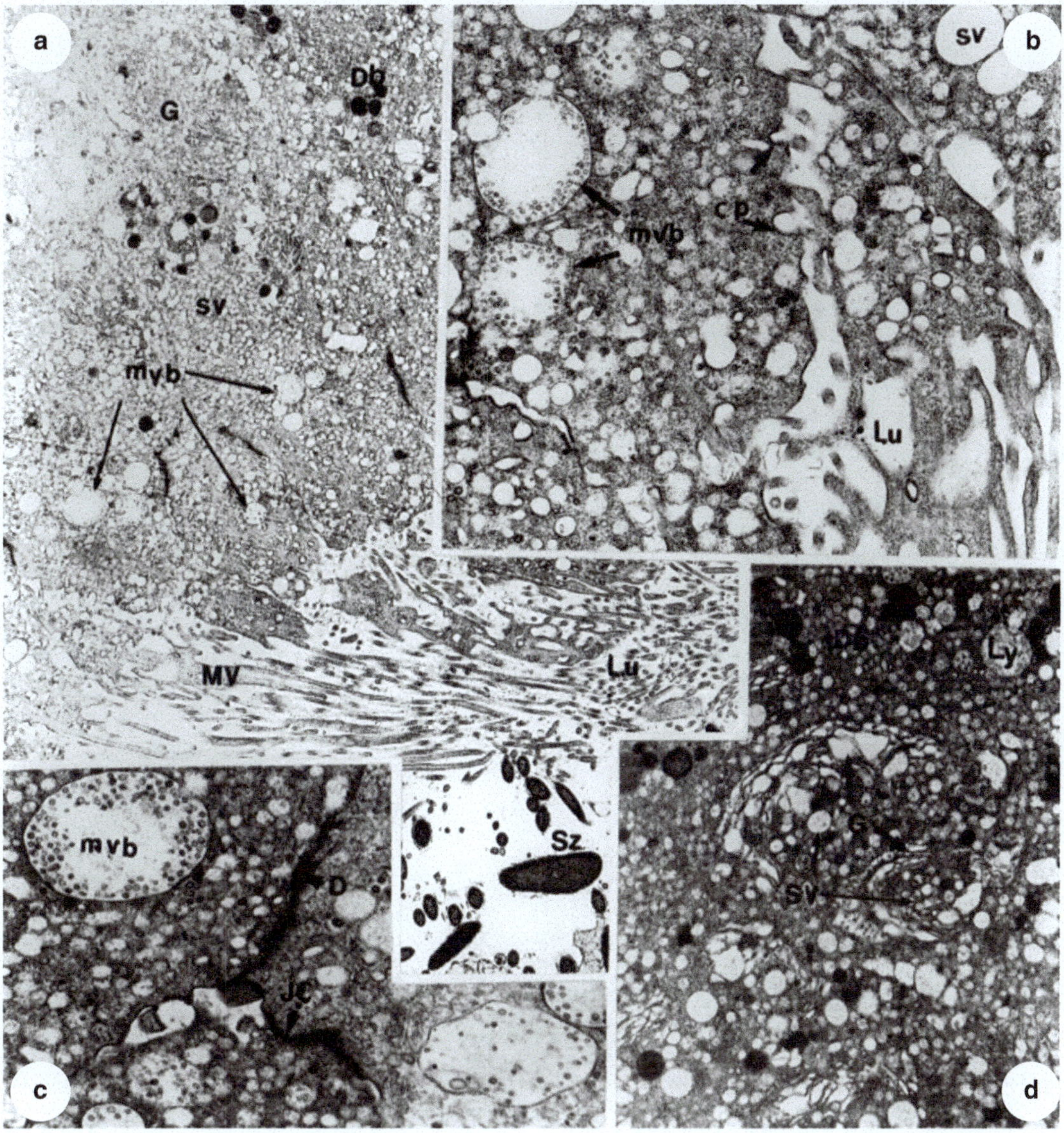

Fig. 3.30 (**a**, **b**, **c**) Electron micrograph of the apical cytoplasm of the principal cells of the caput II epididymis from the control Bonnet monkey. The apical surface of the cell is covered with long stereocilia (mv). Coated invaginations of the plasma membrane (CP), coated vesicles (CV), multivesicular bodies (mvb), dense bodies (db), secretory vesicles (Sv) and numerous profiles of Golgi apparatus (G) are seen in the underlying cytoplasm. Note the Junctional complex (JC) with brush-like desmosomes (D) between the two neighbouring cells and spermatozoa in cross section and longitudinal section (SZ) in the central lumen (Lu) [X 3500] [X 10,000] [X 16,000]. (**d**) Electron micrograph of a portion of apical cytoplasm of the caput II epididymis from control Bonnet monkey illustrating Golgi bodies (G) with fenestrated cisternae, associated secretory vesicle (Sv) and mitochondria (m). Note the presence of lysosomes (Ly) multivesicular bodies (mvb) and dense bodies (Db) [X 10,000]. Unpublished electron micrograph from Dr. Bhiwgade and Dr. Bansal

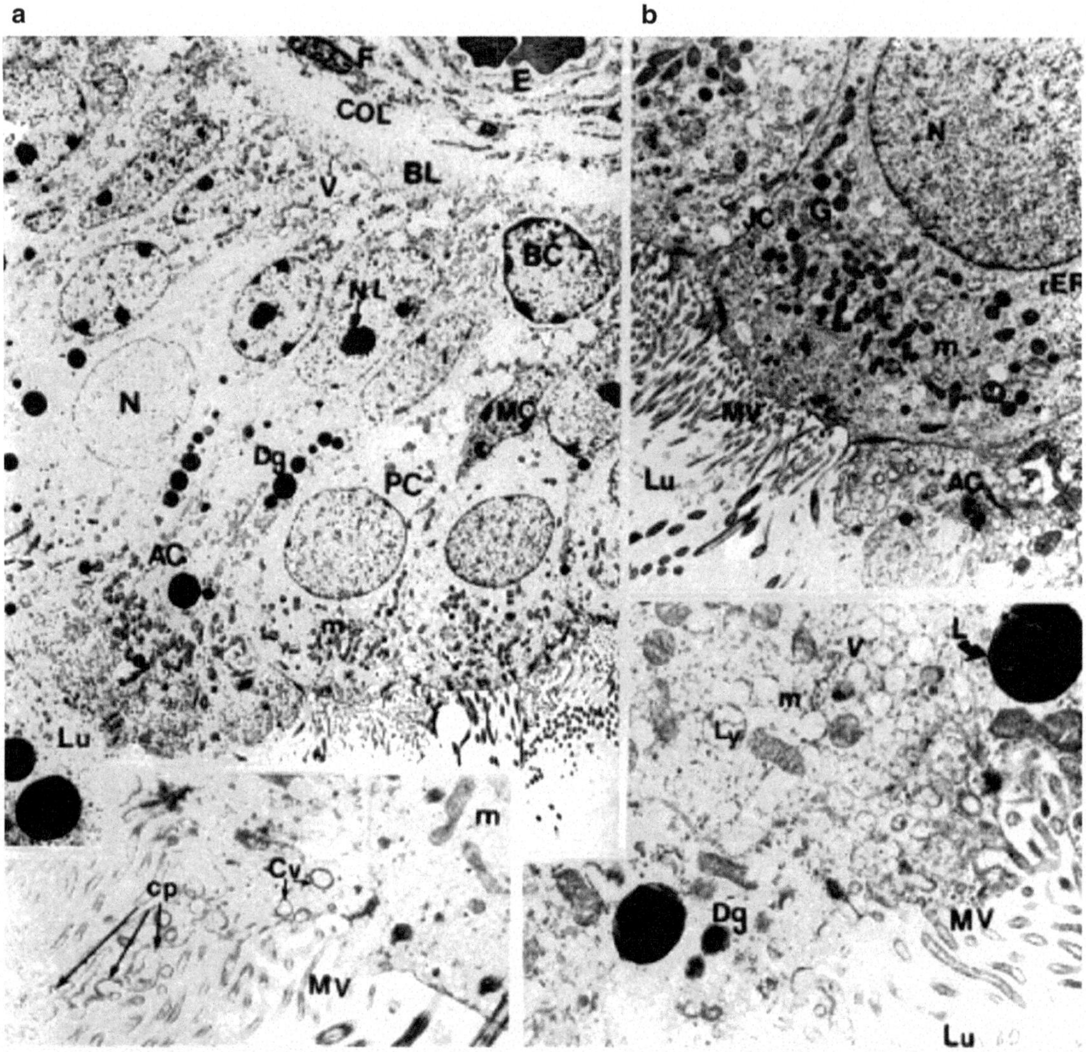

Fig. 3.31 (**a**) Electron micrograph of the caput II from the 45 days treated Bonnet monkey passing through the entire width of the epididymis, illustrating peritubular tissues, regressed epithelial layer and well-marked highly folded basal lamina (BL). Peritubular layer shows an accumulation of collagen fibres (COL), reduced fibroblast cells (F) and normal appearing endothelium (E) of the blood capillary. Note the presence of normal-appearing Basal cell (BC), macrophage cells (MC) and apical cells (AC) and altered principal cells (PC) which shows an accumulation of dense bodies (Db), basal and oval nucleus (N) and profuse vacuolation (V) in the infranuclear zone. Central lumen (Lu) contains testicular debris [X 3500]. (**b**) Electron micrograph of the Apical cell (AC) of caput II epididymis from the treated Bonnet monkey. Apical cell seems to be unaltered following treatment and characterized by the presence of apical, large, oval nucleus (N), apical variable mitochondria (m) and few cisternae of rER. Apical free margin of the cell is provided with microvilli (mv) while the central lumen contains a few membrane bound spheres and spermatozoa in cross sections [X 6000]. Unpublished electron micrograph from Dr. Bhiwgade and Dr. Bansal

feature of apical cytoplasm of Caput II (Fig. 3.30a, b and c). Many lysosomes, membrane bound granules of different sizes and densities, free and polyribosomes and mitochondria of different shapes are also observed (Figs. 3.30d, and 3.34a, b). A distinct feature of the apical

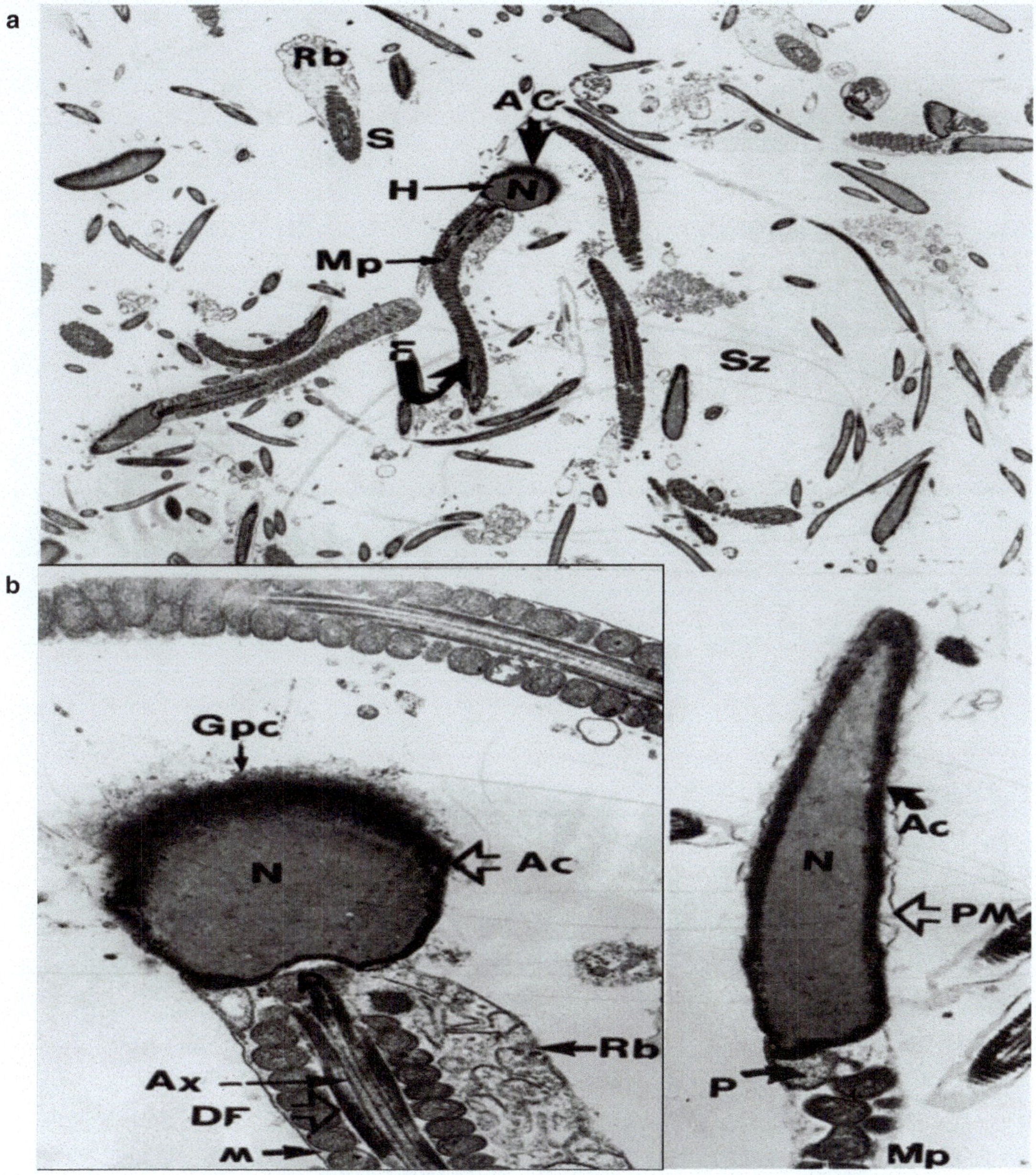

Fig. 3.32 (**a**, **b**) High-power electron micrograph of spermatozoa (SZ) in longitudinal section and cross section in the caput II epididymis of the control Bonnet monkey. The spermatozoa in the longitudinal section are composed of anterior most head (H) with a central nucleus (N)—surrounded by a plasma membrane (PM) and acrosome (AC) with the deposition of glyceryl phosphoryl choline (GPC), followed by a middle piece which is composed of a central row of axoneme (AX) surrounded by a mitochondrial sheath (M), followed by a residual body (Rb) containing cytoplasmic organelle [X 3500] [X 16,000]. Unpublished electron micrograph from Dr. Bhiwgade and Dr. Bansal

a

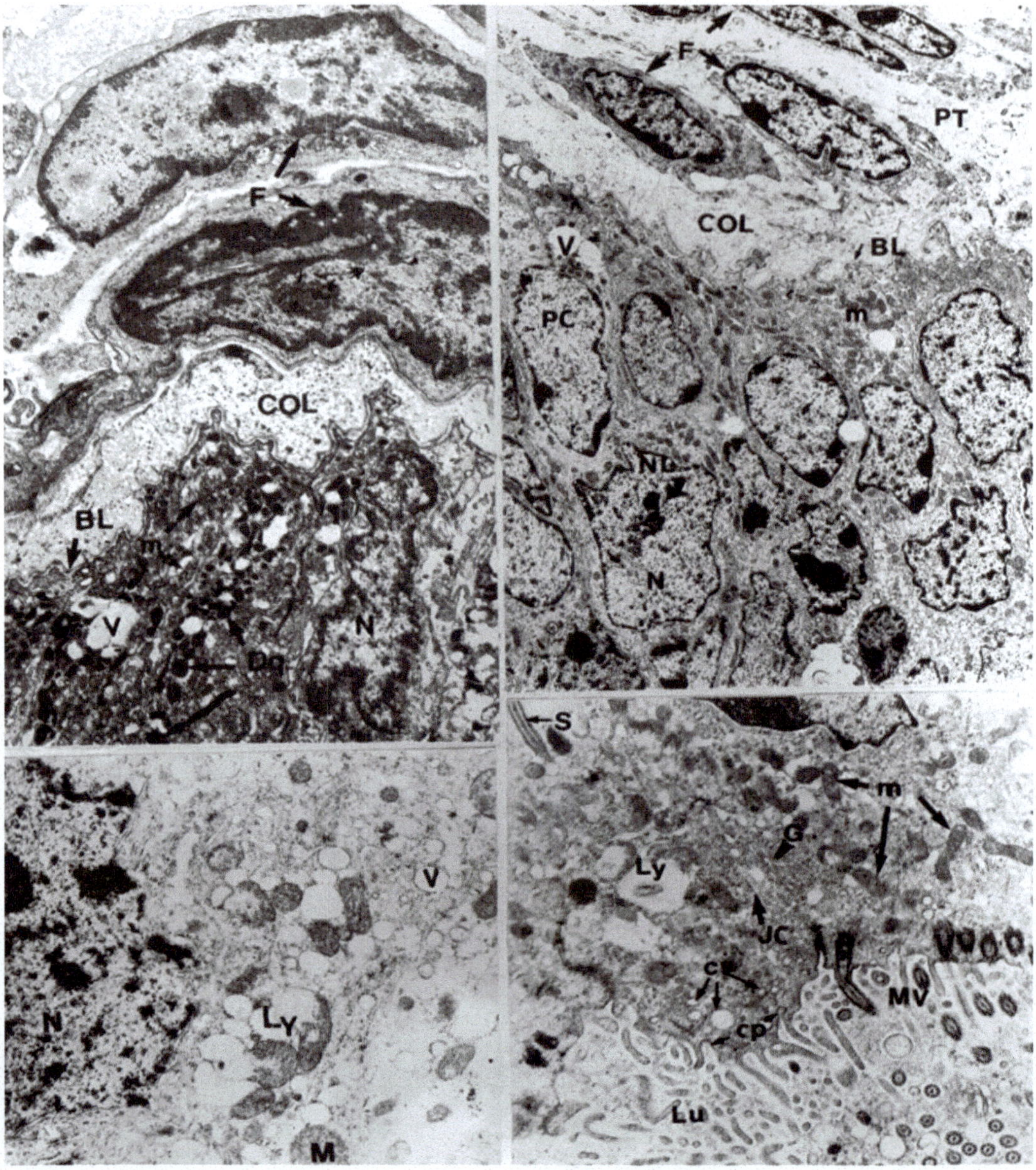

b

Fig. 3.33 (**a**) The electron micrograph depicts the basal region of several principal cells of caput III epididymis from a control Bonnet monkey resting over the highly folded basal lamina (BL). This is followed by a well-developed peritubular tissue, composed of fibroblast cells (F) containing elongated nucleus and other cell organelles and collagen fibres (COL). Note the abundant small membrane bound dense granules (Dg) and clusters of mitochondria (m) along with few vacuoles (V). The dense granules resemble the secretory granules of endocrine glands. Note the presence of folded basal nuclei in the principal cell [X 16,000]. (**b**) The electron micrograph of the basal region of several principal cells of caput III epididymis from treated Bonnet monkey depicting few alterations as compared to the control. Peritubular tissue (PT) exhibits accumulation of collagen fibres (COL) and reduction of fibroblast cells (F). The basal lamina (BL) exhibits micropinocytotic folding. The basal nucleus of the principal cells is almost smooth surfaced. The infranuclear zone contains the mitochondria (m) and a few endoplasmic reticulum (ER). The dense granules, however, are absent [X 3500]. Unpublished electron micrograph from Dr. Bhiwgade and Dr. Bansal

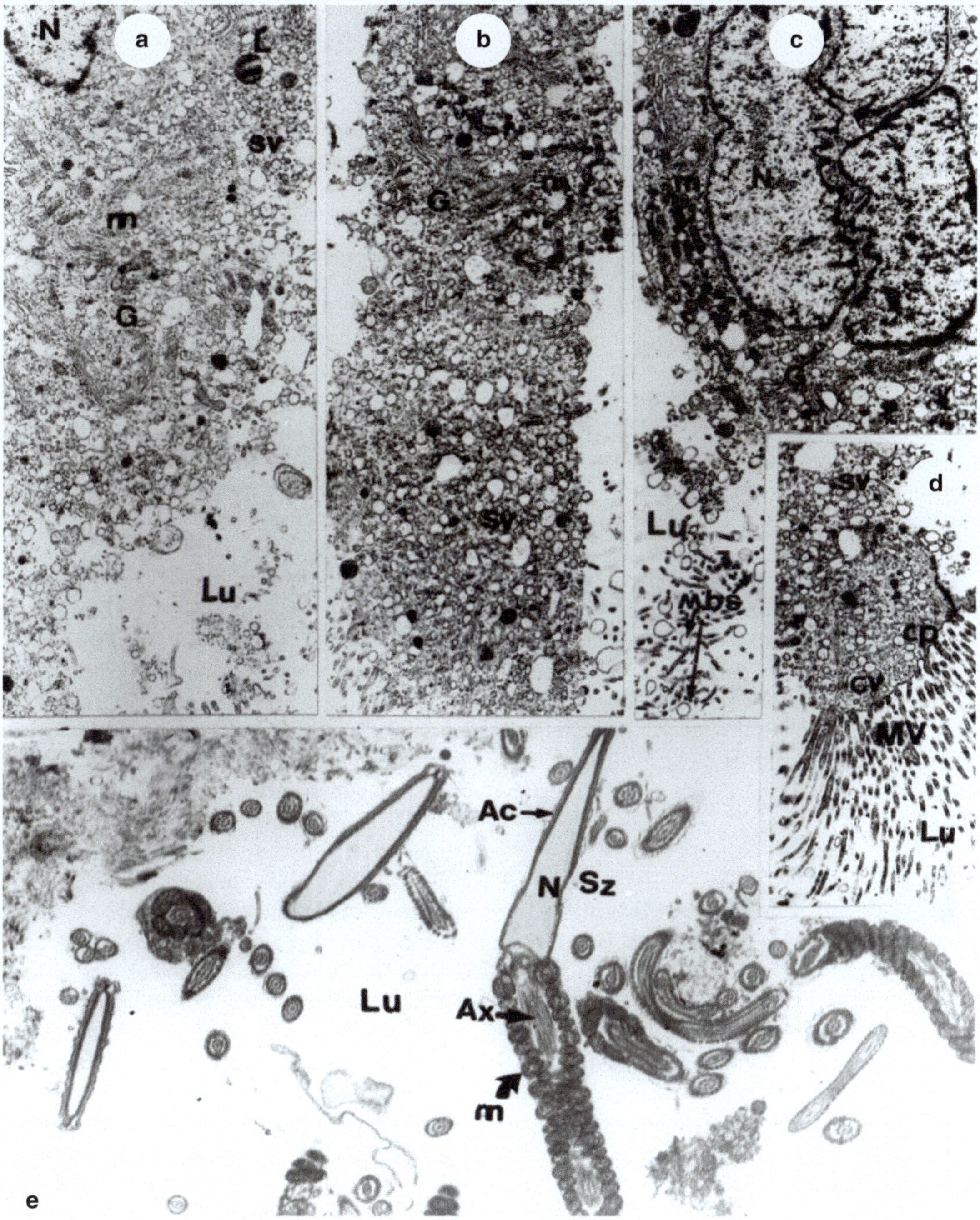

Fig. 3.34 (**a**, **b**, **c**) The extensive Golgi with compactly arranged Golgi cisternae (G) of the principal cells is shown in these electron micrographs of the caput III from the control Bonnet monkey. Note the dense granule, secretory vesicle and vacuoles (Sy) in the Golgi region. The variable mitochondria (m) and the lipid droplets (L) are also seen [X 6000] [X 6000]. (**d**) Electron micrograph of the basal portion of principal cell of caput III from control Bonnet monkey illustrating highly folded nucleus (N) surrounded by variable mitochondria (m), endoplasmic reticulum (ER) and Golgi (G). Membrane bound spheres (mbs) are also noted [X 6000]. (**e**) Electron micrograph of the apical surface of principal cell of control Bonnet monkey. Apical surface is covered with microvilli (MV), coated invaginations of plasma membrane (CP) and coated vesicles (CV). Dense bodies, secretory vesicles and profiles of endoplasmic reticulum are visible in the underlying cytoplasm. Note the presence of membrane bound spheres between the stereocilia within the central lumen [X 10,000]. Unpublished electron micrograph from Dr. Bhiwgade and Dr. Bansal

cytoplasm of control Caput II and III is the presence of numerous rounded, membrane bounded vacuoles containing light flocculent material. These are variable in size and the smaller ones are located between the Stereocilia, whereas the large ones are seen in the tubular lumen (Figs. 3.30b, and 3.34b, d). The apical membrane of the Principal cells of Caput II and III is thrown into many invaginations, coated with a fuzzy material, indicating a stage of formation or breakdown of the numerous micropinocytotic coated vesicles present in apical cytoplasm (Figs. 3.30b and 3.34a). The apical surface of the principal cells is covered with long stereocilia (Figs. 3.30, 3.31a, b, 3.32a, b, 3.33a, b, 3.34a, b, c and d).

Adjoining principal cells are connected, at their luminal border, by occluding junctional complexes. Both tight and gap junctional complexes are observed in the apical zone. Desmosomes and associated filaments are seen along the lateral cell membranes, more frequently in the upper two-thirds of cells (Figs. 3.30a and 3.34e).

A large number of sperms in cross section and longitudinal section are also present in the central lumen of Caput II and III along with testicular debris and residual bodies (Figs. 3.30a and 3.34e).

3.3.5 Caput II and III: Epididymis of DMPA + TE Treated Bonnet Monkey

4. Caput II and III: Following DMPA + TE treatment, the interstitium exhibits deterioration in the form of an increase in the foldings of basal lamina, accumulation of collagen fibres and narrowing of fibroblast cells (Figs. 3.31a and 3.33b).

Principal cell in Caput II and III undergoes changes almost similar to those in the Caput I—following DMPA + TE treatment. Both, length and width of the principal cell are reduced following treatment, and the columnar cell assumes a cylindrical shape (Fig. 3.31a).

The nucleus of Caput II and III shows pycnosis, nuclear membrane loses its foldings and acquires an oval shape (Figs. 3.31a and 3.33b).

The significant change observed in Caput II following treatment is the formation of large vacuoles in the infranuclear zone, where the mitochondria, rough endoplasmic reticulum and dense granules are being accumulated as seen in the control monkey (Fig. 3.31a). This change is not well marked in Caput III (Fig. 3.33b).

The mitochondria occupy a greater proportion of the cytoplasm as there is a rise in the mitochondrial bulk following treatment. These are more dense as compared to those in the control animals (Figs. 3.31a and 3.35a). The small and large vesicles are found in less number in the apical cytoplasm after treatment. Qualitative changes are characterized in the apical region, caused by disappearance of the small vesicles from the apical border of the cell after DMPA + TE treatment (Figs. 3.31a and 3.35b). The Golgi complex shows marked atrophy and is found in the form of 2 or 3 compactly packed thin Golgi saccules (Figs. 3.31a and 3.35a). No secretory vesicles and vacuoles are observed in Golgi vicinity. Apical cytoplasm shows intense lysosomal activity and at one instance, a sperm head is also observed in the apical cytoplasm of the principal cells (Fig. 3.34a, b). The rise in the lysosomal activity is evident by the increase in large autophagic vacuoles and dense bodies (Figs. 3.30c and 3.35b). The multivesicular bodies and coated vesicles have been reduced slightly in Caput II and III. Membrane bound spheres are observed in the central lumen in Caput II and III following treatment (Figs. 3.31b and 3.34b).

Few apical cells (mitochondria-rich cells) are also encountered in tubular cross sections (Fig. 3.31a). These cells are characterized by a wide apical part, a narrow body, an apical location of nucleus and a dome-shaped luminal border with a few short microvilli. A distinctive feature of these cells is the pale cytoplasm, containing numerous mitochondria in the wide portion. Apical cells, basal cells and the intraepithelial lymphocytes and macrophages are almost same as seen in the control epididymis after treatment.

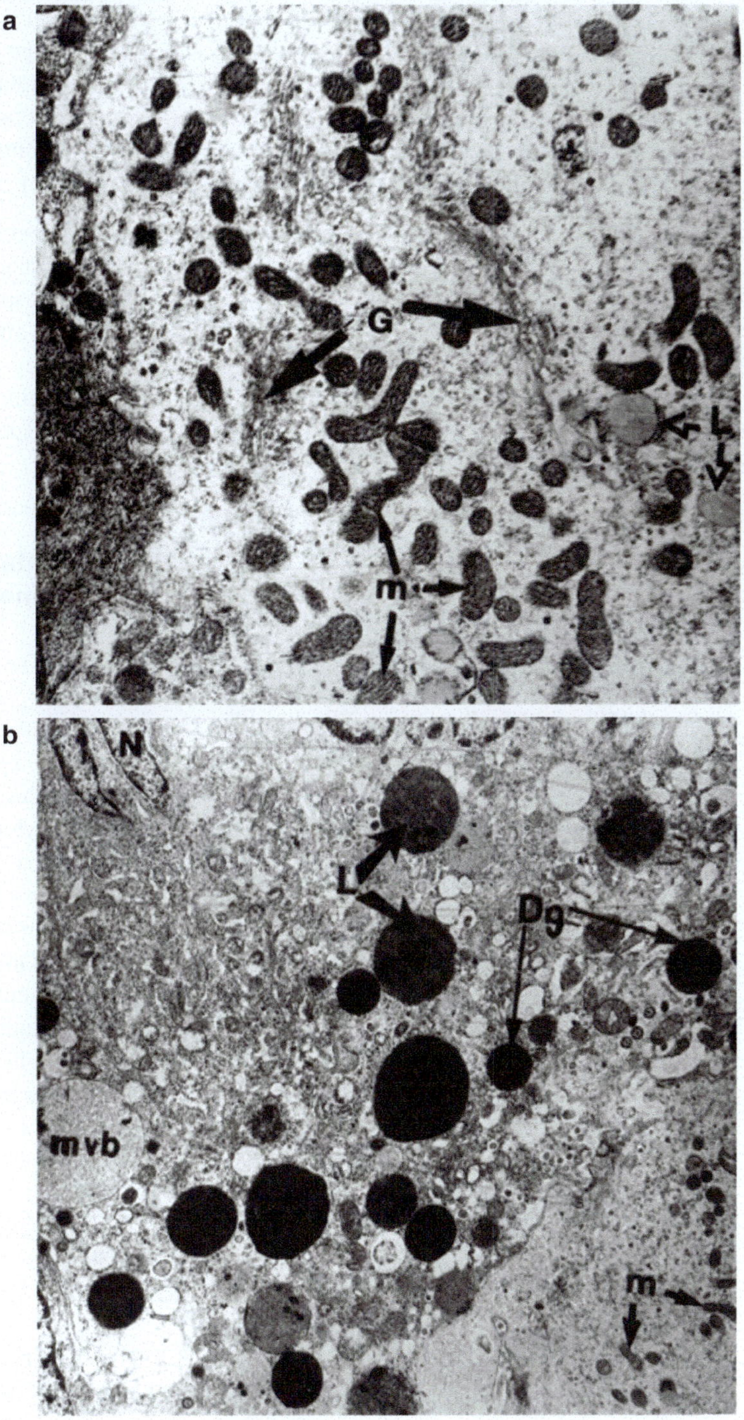

Fig. 3.35 (**a**) Electron micrograph of a portion of apical cytoplasm of caput III epididymis from treated Bonnet monkey demonstrating atrophied Golgi (G), accumulation of lipid droplets (L), dispersed ribosomes and normal appearing mitochondria (m) [X 16,000]. (**b**) A portion of apical cytoplasm of the principal cells of caput III from a treated monkey illustrates accumulation of lipid droplets (L), dense granules (Dg), multivesicular bodies (mvb), endoplasmic reticulum and mitochondria (m) [X 10,000]. Unpublished electron micrograph from Dr. Bhiwgade and Dr. Bansal

No spermatozoa are observed in the central lumen of Caput II and III following treatment as compared to the control. However, it is filled with a large amount of cellular debris containing spheres of cytoplasm and parts of sperms (Figs. 3.31b and 3.34b).

3.3.6 Cauda I–Epididymis of Control Bonnet Monkey

5. Cauda I: The epithelial cells of control monkey's Cauda I epididymis are tall and separated from the adjoining cells by the well-developed junctional complexes provided with desmosomes (Fig. 3.36a). The principal cells of control Cauda I exhibit the characteristics of cells involved more in absorptive actively rather than secretory. The apical cytoplasm consists of mainly the components of endocytotic apparatus including pinocytotic vesicles, coated vesicles, subapical vacuole and multivesicular bodies (Fig. 3.36a). Many autophagic vacuoles are also observed in the apical cytoplasm. Fragments of tubular rough endoplasmic reticulum, vesiculated ER, associated with mitochondria (Fig. 3.36c), free and polyribosomes are also seen (Fig. 3.36a and c). Golgi apparatus lies deep in the perinuclear zone of principal cells and consists of 3–4 Golgi saccules with secretory vesicles (Fig. 3.36b and c). The other differences from Caput I, II and III include the deeper invaginations of the nuclear membrane in the Cauda I epididymis (Fig. 3.36b).

3.3.7 Cauda I–Epididymis Control of DMPA + TE Treated Bonnet Monkey

6. Cauda I: Following DMPA + TE treatment, the major detectable change observed in Cauda I, is the hypertrophy of Golgi complex and endoplasmic reticulum. The entire cell seems to be filled with distended cisternae of rough endoplasmic reticulum, containing low-density material and pigmented granules. The infranuclear cytoplasm contains a large number of mitochondria besides the enlarged rER and dense granules (Fig. 3.37a). The hypertrophied Golgi complex appears in the form of extremely dilated membranous saccules, intermingled with endoplasmic reticulum and cytosol rather than a large conspicuous organelle. Lipid droplets, light-dense bodies and pigmented vesicles are also observed in the vicinity of Golgi complex in treated animals (Fig. 3.38b).

The nucleus exhibits pycnosis and instead of being bizarre shaped, is large oval and basal in position with dispersed heterochromatin and one or two nucleoli after treatment (Fig. 3.37a and c). The morphology of the basal lymphocytes and macrophage cells is the same as described for the previous regions (Figs. 3.37b, c, and 3.38a, b, c, d).

3.3.8 Cauda II–Epididymis of Control Bonnet Monkey

7. Cauda II: The terminal segment of the normal epididymis, i.e. Cauda II is characterized by thinner epithelium as compared to the other regions of epididymis. The principal cells contain dense granules and vacuoles, beside the endocytotic apparatus. Golgi complex and cisternae of rough endoplasmic reticulum are well developed with free ribosomes and ribosomal rossetters (Fig. 3.39a, b). Multiple stacks of Golgi and associated vesicles are the prominent feature of supranuclear cytoplasm. Cauda II of normal monkey also contains light or clear cells, so called because of their less dense cytoplasm as compared to the adjacent principal cells. These cells have a large number of electron lucent vesicles and vacuoles in the apical as well as basal cytoplasm, which resemble the lysosomes (Figs. 3.39a, b, c and 3.40a).

3.3.9 Cauda II–Epididymis of DMPA + TE Treated Bonnet Monkey

8. Cauda II: Following DMPA and TE treatment, there is a further decline in the volume of the principal cell and the cells appear narrow. One of the important changes observed

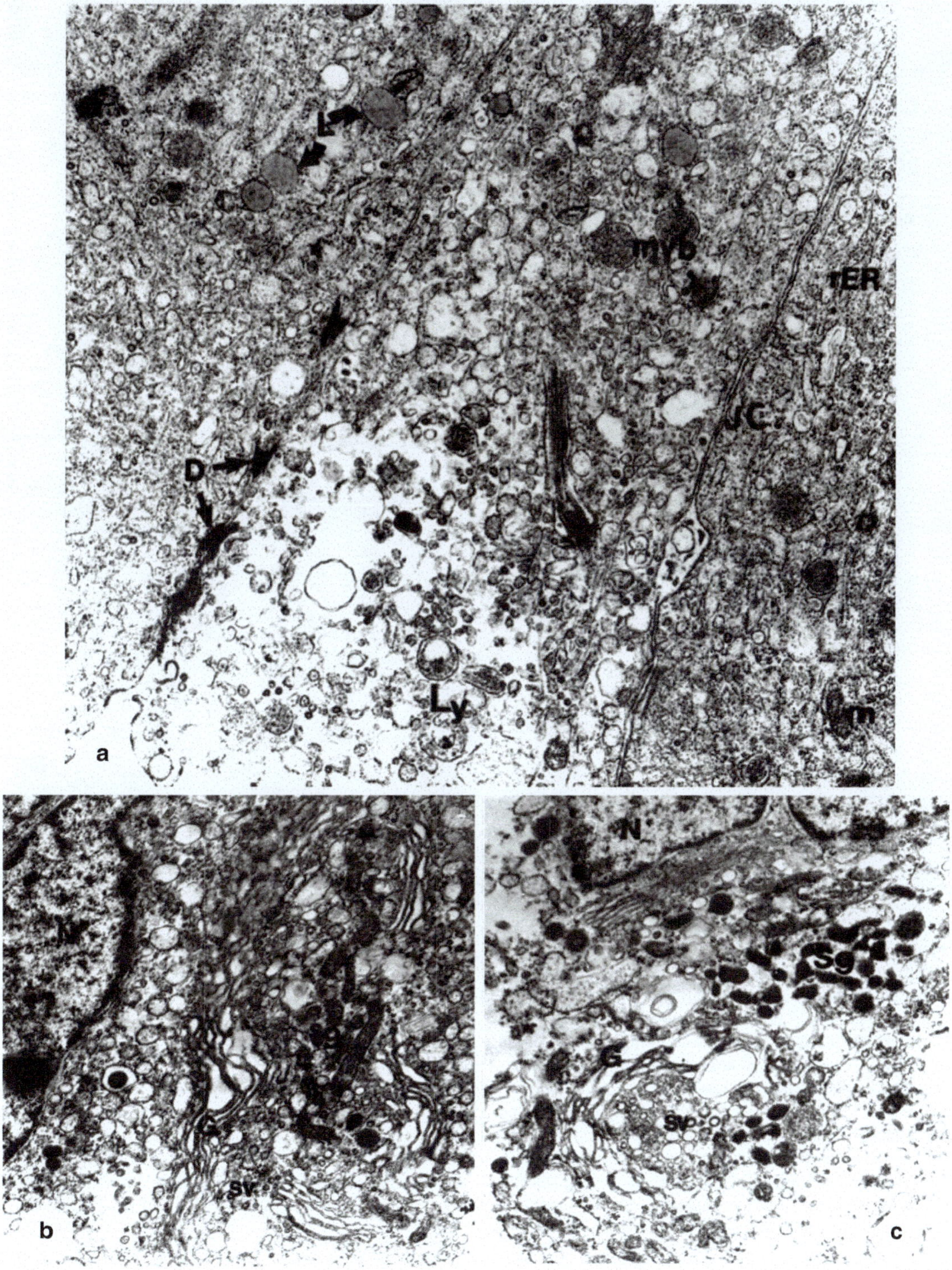

Fig. 3.36 (**a**, **b**, **c**) Electron micrograph of the principal cells from the cauda I epididymis of a control Bonnet monkey, illustrating Junctional complexes (JC) provided with desmosomes (D) between the neighbouring cells, flattened and vesiculated smooth endoplasmic reticulum (sER), lipid droplets (L), lysosomes (Ly), multivesicular bodies (mvb) and mitochondria (m) in the apical cytoplasm. Well-developed Golgi body (G) with fenestrated cisternae, secretory granules (Sg) and a portion of the nucleus (N) are also seen [X 10, 000] [X 16,000] [X 20,000]. Unpublished electron micrograph from Dr. Bhiwgade and Dr. Bansal

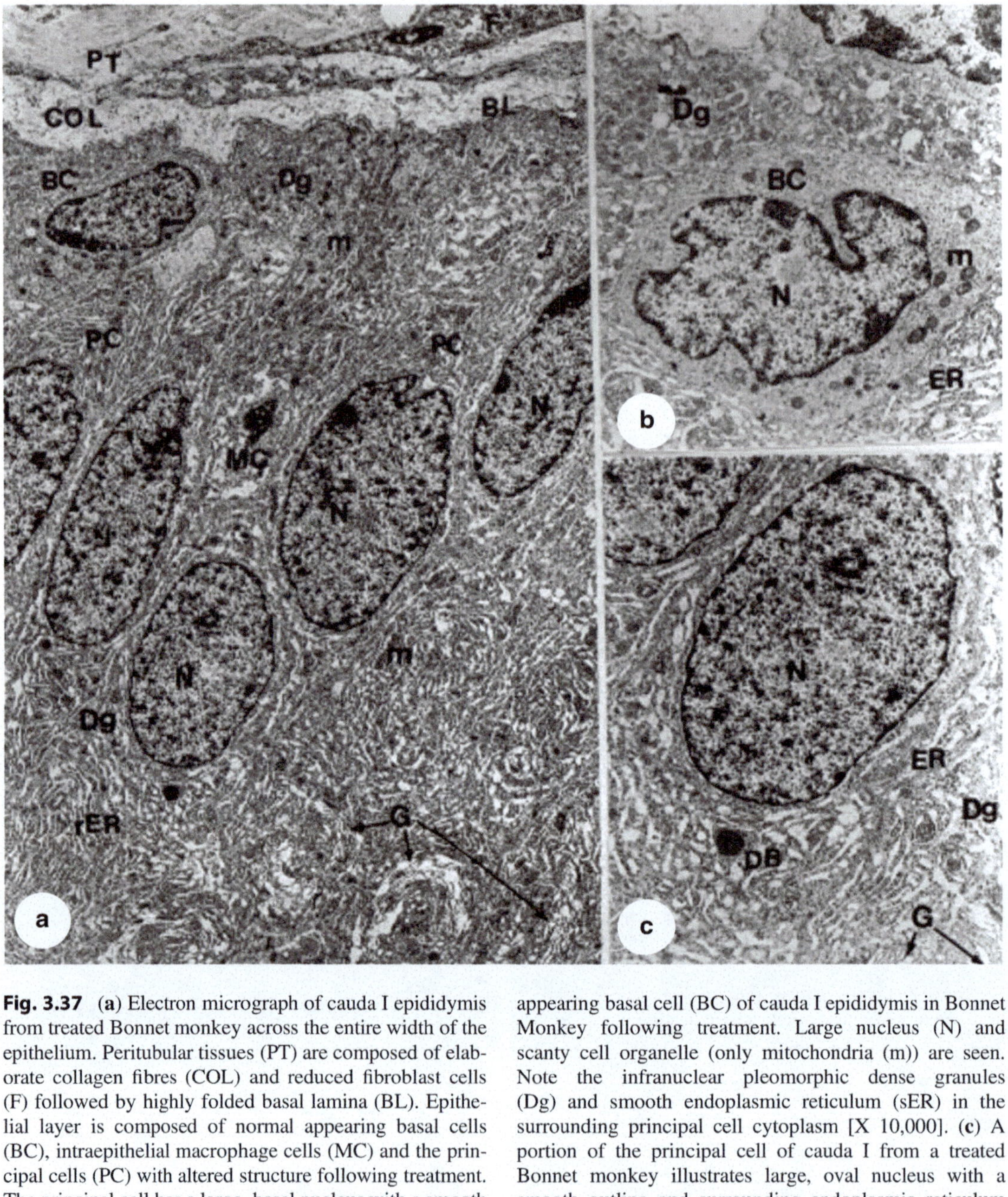

Fig. 3.37 (**a**) Electron micrograph of cauda I epididymis from treated Bonnet monkey across the entire width of the epithelium. Peritubular tissues (PT) are composed of elaborate collagen fibres (COL) and reduced fibroblast cells (F) followed by highly folded basal lamina (BL). Epithelial layer is composed of normal appearing basal cells (BC), intraepithelial macrophage cells (MC) and the principal cells (PC) with altered structure following treatment. The principal cell has a large, basal nucleus with a smooth outline, hypertrophied endoplasmic reticulum (ER), hypertrophied Golgi (G) with fenestrated cisternae, mitochondria (m) and small dense granules (Dg) [X 3500]. (**b**) Electron micrograph of normal appearing basal cell (BC) of cauda I epididymis in Bonnet Monkey following treatment. Large nucleus (N) and scanty cell organelle (only mitochondria (m)) are seen. Note the infranuclear pleomorphic dense granules (Dg) and smooth endoplasmic reticulum (sER) in the surrounding principal cell cytoplasm [X 10,000]. (**c**) A portion of the principal cell of cauda I from a treated Bonnet monkey illustrates large, oval nucleus with a smooth outline and surrounding endoplasmic reticulum (ER). Dense body (Db), mitochondria (m) and the Golgi apparatus (G) are also seen [X10,000]. Unpublished electron micrograph from Dr. Bhiwgade and Dr. Bansal

in Cauda II after treatment is the formation of protrusions within basal lamina in the proximity of the blood capillary (Fig. 3.38a). The principal cells of Cauda II appear to be altered by the treatment since they exhibit significant accumulation or large dense

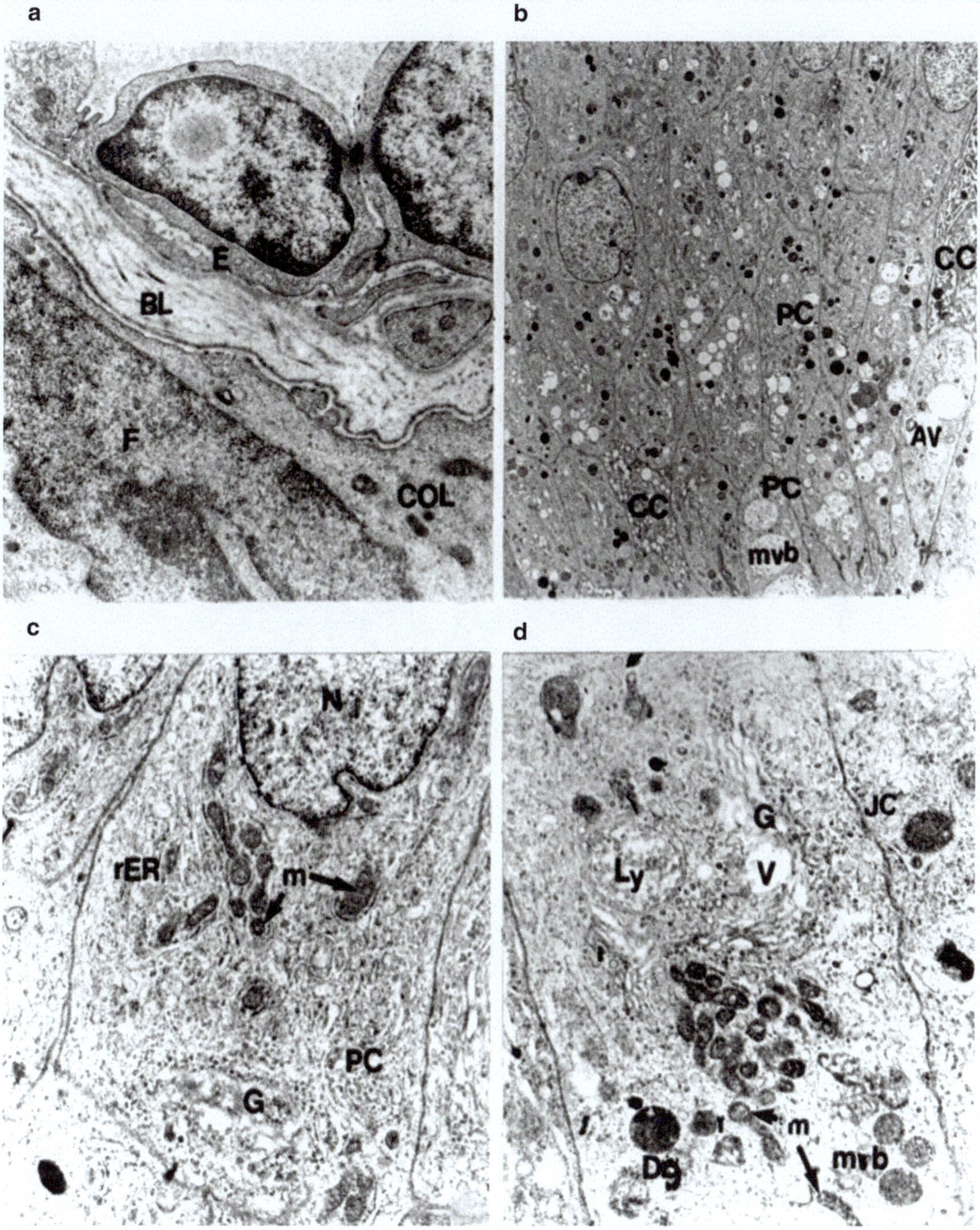

Fig. 3.38 (**a**) High-power electron micrograph of cauda II epididymis from treated Bonnet monkey showing that the small exopinocytotic body formed from basal lamina (BL) towards the endothelium of the blood capillary (E). Note the presence of a small portion of fibroblast cells (F) and collagen fibres (COL) of the peritubular tissue [X 16,000]. (**b**) Electron micrograph of the pseudostratified epithelium of the cauda II Epididymis from treated Bonnet monkey. Observe the morphological features of the clear cells (CC) and the principal cells (PC). Note the round to ovoid euchromatic nucleus with prominent nucleoli, accumulation of large-sized multivesicular bodies (mvb) autophagic vacuoles (AV) and the dense granules (Dg) within the principal cells [X 3500]. (**c**) Infranuclear zone of the principal cells (PC) of cauda II epididymis from the treated Bonnet monkey illustrating a portion of the nucleus (N); deteriorating Golgi (G), rough endoplasmic reticulum (rER), dispersed ribosomes and the normal appearing mitochondria (m) [X 16,000]. (**d**) A portion of principal cell cytoplasm of cauda II epididymis of the treated monkey depicting deteriorating Golgi (G) with small and large faded vacuoles (V), accumulated multivesicular bodies (mvb), dense granules (Dg), lysosome (Ly) and mitochondria (m) [X 16,000]. Unpublished electron micrograph from Dr. Bhiwgade and Dr. Bansal

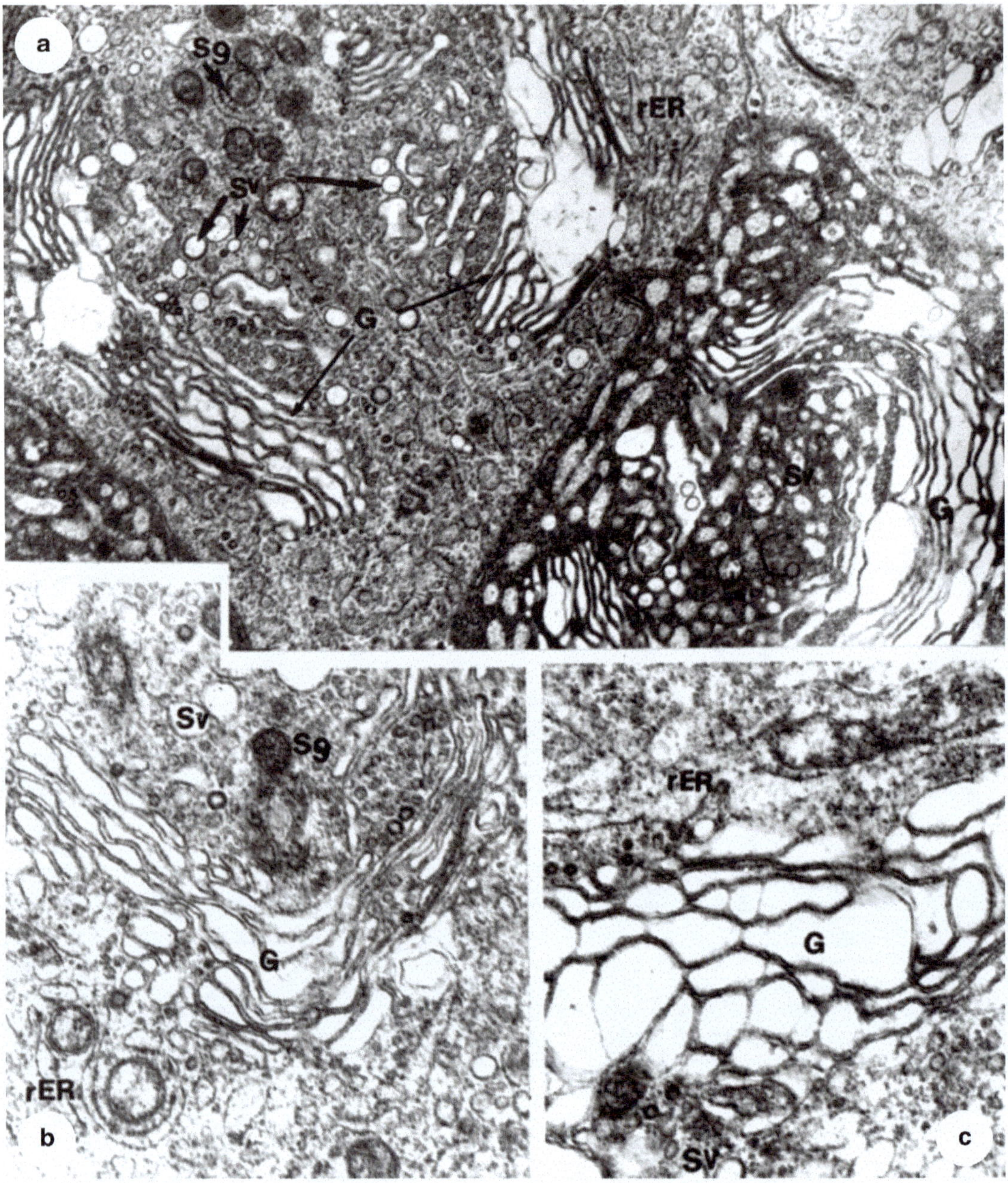

Fig. 3.39 (**a**, **b**, **c**) High-power electron micrograph of the cauda II epididymis from the control bonnet monkey depicting active phase of Golgi both in the principal cells (PC) and the clear cell (CC). Golgi complex (G) appears as patches of membrane and fenestrated cisternae intermingled with endoplasmic reticulum (rER) and cytosol. Note the small vesicles present within the cylinder of cisternae whereas larger vesicles (SV) containing spare flocculent material are located at the apical and basal side of the Golgi. Note the presence of mitochondria (m) sometimes encircled by endoplasmic reticulum [X 13,000] [X 16,000] [X 25,000]. Unpublished electron micrograph from Dr. Bhiwgade and Dr. Bansal

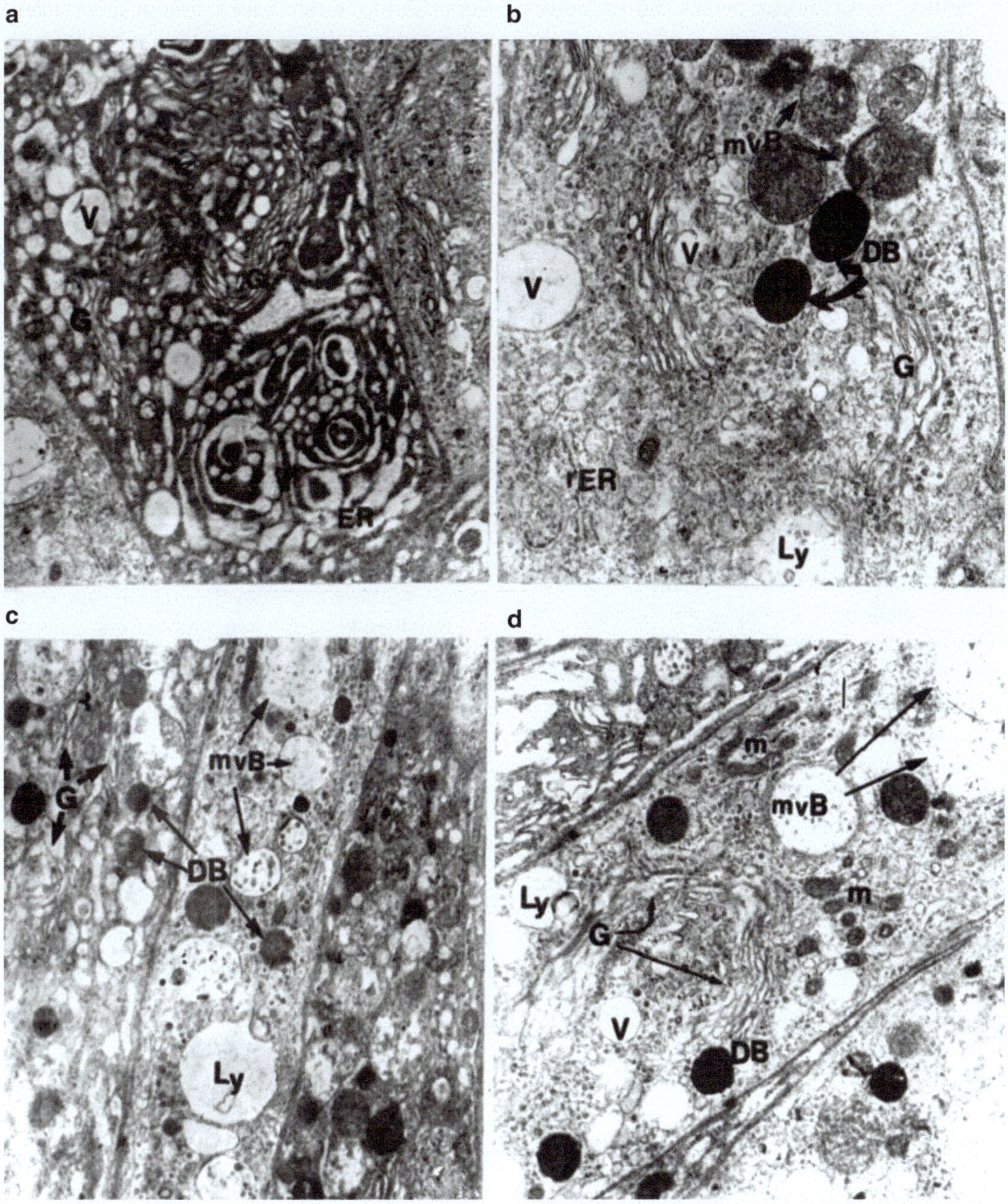

Fig. 3.40 (**a**) Electron micrograph of the cauda II epididymis from Bonnet monkey administered DMPA and TE for 45 days. The clear cells contain apical vacuoles (V) normal appearing Golgi (G) with surrounding endoplasmic reticulum (ER) and dense bodies (Db) [X 13,000]. (**b**) A portion of principal cell of cauda II epididymis from a treated Bonnet monkey illustrating atrophied Golgi (G), reduced number of secretory vesicles and vacuoles (SV), accumulation of multivesicular bodies of variable density (mvb), dense bodies (Db) lysosomes (Ly) and unaltered variable mitochondria (m) [X 25,000] [X 13,000] [X 16,000]. Unpublished electron micrograph from Dr. Bhiwgade and Dr. Bansal. (**c**)A portion of principal cell of cauda II epididymis from a treated Bonnet monkey illustrating atrophied Golgi (G), reduced number of secretory vesicles and vacuoles (SV), accumulation of multivesicular bodies of variable density (mvb), dense bodies (Db) lysosomes (Ly) and unaltered variable mitochondria (m) [X 25,000] [X 13,000] [X 16,000]. Unpublished electron micrograph from Dr. Bhiwgade and Dr. Bansal. (**d**) A portion of principal cell of cauda II epididymis from a treated Bonnet monkey illustrating atrophied Golgi (G), reduced number of secretory vesicles and vacuoles (SV), accumulation of multivesicular bodies of variable density (mvb), dense bodies (Db) lysosomes (Ly) and unaltered variable mitochondria (m) [X 25,000] [X 13,000] [X 16,000]. Unpublished electron micrograph from Dr. Bhiwgade and Dr. Bansal

bodies, multivesicular bodies and lysosomes (Figs. 3.38d, and 3.40b, c, d). The Golgi complex and endoplasmic reticulum show atrophy as compared to the control animal (Fig. 3.38c and d). Mitochondria exhibit mild atrophy and vacuolation in the centre.

The clear cells of Cauda II epididymis do not show any changes following DMPA + TE treatment. Clear cells show intense lysosomal activity following treatment like the normal one (Fig. 3.40a).

3.3.10 Discussion

Electron microscopic study of epididymis of a male Bonnet monkey, *Macaca radiata*, reveals that epididymal epithelium is composed of four main cell types; Principal, Basal, Apical and Macrophage cells which are region and function specific. The epididymal epithelium carries out an intensive process of absorption and secretion in order to maintain the adequate environment necessary for sperm maturation. The centre of these processes is the Golgi lysosomal system (Robaire and Hermo 1988) which exhibits regional differences. The present study investigates the ultrastructural changes in epididymis of Bonnet monkeys with particular reference to its secretory and absorptive functions following DMPA + TE treatment. The important changes are observed in principal cells of different epididymal segment, however, basal and apical cells do not alter much as compared to the control animal following treatment since their function is not androgen dependent (Palacios et al. 1991).

In the control Bonnet monkey, a well-developed rough endoplasmic reticulum, Golgi complex and associated secretory vesicle and vacuoles in the three segments of caput, indicate that these cells are actively involved in the synthesis and secretion of various substances such as glyceryl phosphoryl choline (Dawson and Rowland 1959 & Scott et al. 1962), Sialic acid (Rajalakshmi and Prasad 1968), specific portions (Lea et al. 1978 & Brooke and Higgins 1980) and acidic glycoproteins (Lea et al. 1978; Robaire and Hermo 1988), which have a role in sperm maturation and attaining sperm motility (Orgebin et al. 1975; Hamilton 1975; and Robaire and Hermo 1988). Endoplasmic reticulum synthesizes proteins and transfer them to Golgi complex where glycogen is added. Chemicals bound in the secretory granules then move towards the apical surface of principal cells and are finally released into the central lumen (Hermo et al. 1991a, b). The glycoproteins coat spermatozoa as they pass through the epididymal duct and provide functional ability to the spermatozoa (Moore 1980). Goyal (1991) has suggested that various proteins including forward motility proteins, acrosomal stabilizing proteins and immobilien (Usselman and Cane 1983) bind to the sperms and bring about functional changes within bovine epididymis. In the control Bonnet monkey, the principal cells of the proximal cauda epididymis have a less developed Golgi apparatus and rough endoplasmic reticulum as compared to the three segments of the caput and cauda II. Presumably, in this region, the principal cells are predominantly absorptive rather than secretory in function.

The apical cytoplasm of the principal cells also contains a large number of secretory vesicles in all the segments of the epididymis associated with the exocytosis of the substances secreted by Golgi and rER. The electron microscopic study also suggests that the apical margins of the principal cells in the control Bonnet monkey are irregular due to the presence of a large number of protrusions which ultimately form membrane bound spheres between stereocilia or immediately over them within the lumen. They are more common in close proximity to spermatozoa in caput II and III and cauda I. In Rhesus monkeys, these membrane bound spheres are possibly involved in exocytosis or endocytosis or both (Ramos and Dym 1977) and are supposed to transport the cellular products into the lumen. Fornes et al. (1991) suggested that the membrane bound spheres are the possible forms of secretion of epididymal epithelium and contribute to the intraluminal environment. Davis (Davis and Mungund 1976; Davis 1978) reported that the membrane bound vesicles caused sperm decapacitation and inhibited the fertilizing capacity of the rabbit spermatozoa in caput epididymis. Yanagimachi et al. (1985) have suggested that the membrane bound spheres in the hamster contained cholesterol which mingled with sperm plasma membrane and stabilized the sperms. They also carry enzymes and glycoproteins that altered physio-chemical properties of the membrane of the maturing spermatozoa in epididymis. Aggarwal and Partulla (1987) have observed that they caused hyperactivation and acrosome reactions in bovine sperms. Hinton et al. (1991) and Carlos et al. (1992) have suggested the release of some lysosomal proteins through membrane bound vesicles in rat caput epididymis lumen. Following DMPA + TE treatment, a lesser number of membrane bound spheres are observed in caput II, III epididymis as compared to the control Bonnet monkey which is an indication of reduced secretory activity of principal cells in caput epididymis after treatment. This may be due to the absence or the presence of the less number of sperms and testicular fluid in caput epididymis following treatment as it has been well documented that the epididymal segments functionally depend on testicular fluid contained in the lumen (Abe et al. 1982 and Fawcett and Hoffer 1982).

Following treatment, the other changes observed are the marked atrophy of the Golgi complex, rough endoplasmic reticulum, especially in the three segments of caput and cauda II. There are clear evidences of reduced synthetic activity within these segments or the epididymis. In contrast to the other segments, endoplasmic reticulum shows hypertrophy in cauda I epididymis, which is an indication of deteriorating effect of DMPA on rER and Golgi complex in the proximal segment of cauda epididymis. These changes are similar to those described after castration (Hamilton et al. 1969; Orgebin Crist and Davis 1974). Similar changes have been observed in humans following CPA treatment by Lopez and Cerda (1993). Atrophied rER in the three segments of the caput II epididymis and hypertrophied rER in cauda I epididymis also confirm the findings of Orgebin Crist et al. (1987) that some specific proteins of caput and cauda II epididymis appeared to be more androgen dependent than the cauda I epididymal proteins. The presence of reduced secretory vesicles in the apical cytoplasm of the principal cells also reflects the reduced secretory activity similar to that observed in the castrated rats (Moor and Bedford 1979 a, b). Biochemical estimations of epididymis in our study also confirm the ultrastructural findings. Proteins and sialic acids, the main components secreted by rER and Golgi apparatus and conveyed by membrane bound spheres to the central lumen have been found to be reduced significantly following treatment. Changes in proteins and sialoproteins contents, cause maturational defects in sperms since the epididymal milieu is affected (Robaire and Hermo 1988) and thereby hampering the sperm fertilizing capacity (Rao and Roy 1993) in DMPA treated rats. Reduced synthesis of proteins by rER within epididymis is not only the reason for significant reduction of proteins in the epididymis following DMPA + TE treatment, but the other reason may be due to reduced testicular fluid coming to the epididymal lumen as this fluid contributes to the major bulk of epididymal proteins. Secretory activity of the epididymis is regulated by androgen (Rajalakshmi 1985) and reduced secretory activity of the epididymis following treatment could be due to insufficient supply of Androgen Binding Proteins from the testis and also because of high affinity of DMPA for androgen receptors than the androgen itself. (Brook 1979). Castration (Rajalakshmi and Prasad 1968), treatment with luteinizing hormone antiserum, treatment with androgen antagonist, or with cyproterone acetate; all induced a reduction in sialoproteins in the epididymis (Rajalakshmi and Prasad 1977; Gupta et al. 1974a, b; Bose et al. 1975 and Sumitra Sumitra and Ghosh 1979). These observations are all in accordance with our findings.

Abundance of smooth endoplasmic reticulum in association with mitochondrial clusters, Golgi cisternae and lipid droplets in the apical cytoplasm of the principal cells of caput I and cauda I epididymis of control Bonnet monkey suggest an additional secretory function of testosterone synthesis (Hamilton 1971; Goyal 1985; Paulson and Dym 1985). Morphologically similar structures with an endocrine function are also observed in the infranuclear zone of cytoplasm in the form of abundant dense granules which do contain glycoproteins as indicated by positive periodic acid Schiff reaction (Ramos and Dym 1977) and clusters of mitochondria in close proximity of sub-epithelial and peritubular connective tissue capillaries (Ramos Jr. 1980). Ultrastructural studies reveal the presence of exocytosis or blebbing of these granules into the basal membrane which confirms its endocrine nature. Since the epididymal epithelium is regressed following treatment in response to testicular androgen deprivation, the epididymal steroidogenesis is hampered. Smooth endoplasmic reticulum and lipid droplets are reduced drastically in the apical cytoplasm and several vacuolations is observed in the infranuclear cytoplasm of three segments of caput and cauda I epididymis. Experiments on Guinea pigs have shown similar results following Orchidectomy (Oritz 1953; Danza and Orgebin Crist 1974), & after efferentectomy (Fawcett and Heffer 1979; Goyal 1983).

Biochemical observation also suggests an accumulation of cholesterol and reduction of lipids in epididymis following treatment due to arrest of testosterone synthesis as indicated by reduced Smooth Endoplasmic Reticulum in caput I and cauda I epididymis in Electron microscopic study. Our study also demonstrates the presence of ultrastructural features which are characteristics of absorptive activity like the apical pinocytotic infoldings, extensive population of coated and uncoated vesicles and subapical vesicles in the apical cytoplasm of principal cells of all the three regions of caput epididymis. Many lysosomes are observed in the form of vacuoles, multivesicular bodies and dense granules in the three segments of caput and cauda I with predominance in cauda II suggesting the transportation and digestion of particulate material (Proteins and Carbohydrates) from the lumen (Hamilton 1975; Ramos and Dym 1977). A significant increase in the number of lysosomes, autophagic vacuoles, multivesicular bodies and dense granules have been observed in the present study in the supranuclear regions of the Principal cells of these segments of caput and cauda II epididymis following treatment suggesting an increase in activity. Fragments of degenerating spermatozoa can also be seen at a few places in cauda I epididymis of the treated Bonnet monkey. Similar results are observed by Flickinger et al. (1993) in vasectomized rats. Biochemical parameters also support the ultrastructural data in our study. The significant increase in acid and alkaline phosphatase activity of epididymis after treatment is in accordance with the earlier findings and suggests increased lysosomal activity in epididymis and increased removal of the defective or dead spermatozoa within the epididymis (Rao and Roy 1993). In normal Bonnet monkey, the epididymis shows a marked acid and alkaline phosphatase activity suggestive of sperm maturation and resorption. The bulk of defective sperms are removed in cauda epididymis which is the site of high alkaline phosphatase activity (Roussel and Stallcup 1966), however, the highest acid phosphatase activity is observed in caput epididymis (Riar et al. 1973) and is indicative of active resorption of testicular material occurring in this region. Ultrastructural studies also show the presence of pinocytotic infoldings, multivesicular bodies, dense granules correlating with high acid phosphatase activity. The multivesicular bodies are lysosomal and digest proteins and carbohydrates of testicular origin after absorption by epididymal epithelium (Waits and Setchell 1969 & Glover and Nicander 1971). According to Linnetz and Amann (1968), the occurrence of acid phosphatase containing organelle in epididymal cells of vasectomized rabbits is higher than that of the intact animal as observed by us. The phagocyte cells with more vacuoles and lysosomes, containing acid phosphatase enzyme are more numerous in rats' epididymis after vasectomy (Flickinger et al. 1993).

The activity of the mitochondrial enzymes viz. LDH and SDH in the epididymis are also androgen dependent (Prasad et al. 1972; Brook 1979), which are responsible for attaining the motility of spermatozoa within the epididymis. The reduced LDH activity observed in our study following DMPA + TE treatment indicates the reduction in sperm metabolic activity because of its depleted number in the epididymis. The elaborate infoldings of the nuclear membranes with large nucleoli in two segments of cauda epididymis of control Bonnet monkey and three segments of the caput epididymis of the DMPA + TE treated monkey, may be interpreted as a sign of intense metabolic activity (Ramos and Dym 1977). The relative absence of nuclear infoldings in principal cells of cauda I and II after treatment may indicate a decrease metabolic activity due to androgen deprivation. Kaur et al. (1992) have observed similar changes in CPA treated Rhesus monkey. Thickening of basement membrane is the noteworthy finding after DMPA+TE treatment. The basement membrane is irregular due to the presence of deep invaginations and protrusions containing coated pits and vesicles presumably as a result of tubular shrinkage. Paucity of coated pits and vesicles along the basolateral surface of the principal cells of distal caput and cauda epididymis, may be suspected as an increase in the endocytotic activity at the base of the cell. This change may be in compensation for the fact that the proteins and other substances, no longer received in sufficient quantity at the luminal surface are received through the basal surface. Similar results are observed in efferent duct ligated rats—Paulson and Dym (1985) and vasectomized Rhesus monkey, Ratna Kumar et al. (1990).

The monkey's epididymis contains blood capillaries and the tubular myoid cells in the peritubular tissues as well as elaborate tight junctions, gap junctions and desmosomes in the epithelium. The intercellular contacts between capillary, epithelial cells and peritubular myoid cells as well as those between epididymal epithelial cells function as Blood Epididymis Barrier (BEB). No significant difference is observed in BEB in all five regions of epididymis following DMPA + TE treatment. Increasing thickness of the myoid layer can hinder the passage of different substances through BEB. Similar results are observed by Hoffer and Hinton (1984) in case of Gossypol treated rats. A decrease in the number of sperms in the lumen of different segments of epididymis of the treated Bonnet monkey may be due to diminished production by testis (Schulze 1977). Previous results of DMPA + TE in the testis of rat have shown that spermatogenesis is arrested in the testis at the early spermatid stage (Flickinger 1979). The spermatozoa in the epididymal lumen are phagocytosed by macrophage cells. The intraepithelial lymphocytes or pleomorphic macrophage cells are found at the base of the epididymal epithelium. These are strongly Periodic Acid Schiff (PAS) positive and acid phosphatase positive. Interestingly their number increased markedly in the treated Bonnet monkey. Similar results were observed in orchidectomised bull (Goyal 1983). The intraepithelial cells are involved in sperm phagocytosis.

It is obvious from the above discussion that all these effects finally impair the fertility of treated animals within the epididymis. Thus, DMPA + TE combination treatment proves to be a better contraceptive, not only acting on spermatogenic process at the testicular site but also it modifies the epididymal physiology, affecting sperm maturation process in it and making them non-viable.

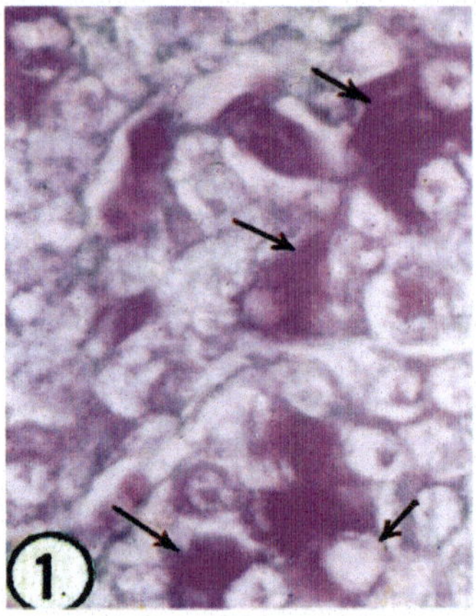

Lactotrophic (LTH) cells (arrows) with Crossmon's stain(600X)

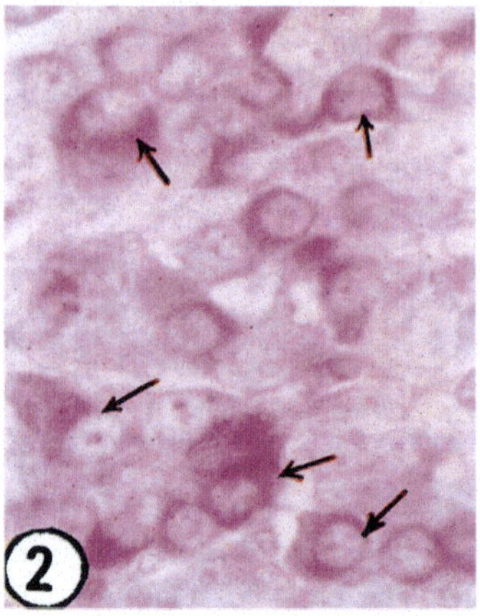

Lactotrophic (LTH) cells (arrows) with Martius Scarlet Blue stain (600X)

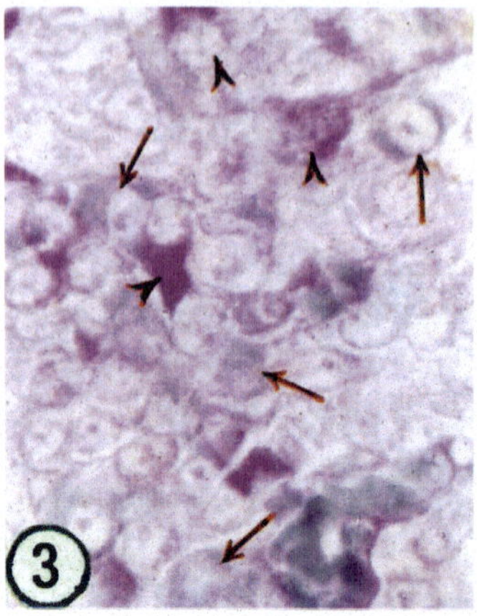

Somatotrophic (STH) cells (arrows) and Gonadotroph cells (arrow heads) with Luxol Fast Blue / Periodic acid Schiff /Orange G stain (600X)

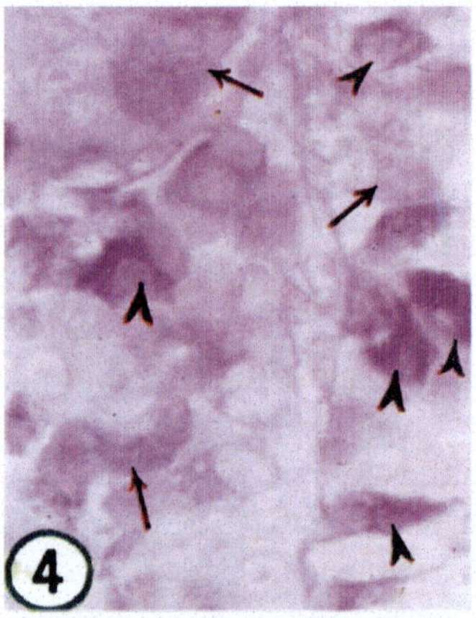

FSH cells (arrows) and ICSH cells (arrow heads) with Periodic Acid Schiff / Orange G stain (600X)

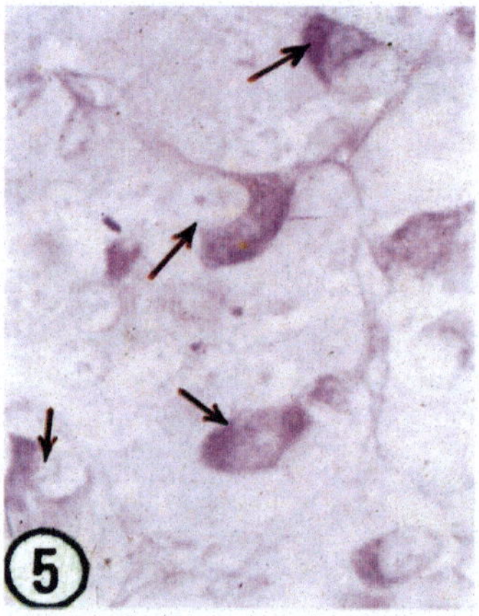

TSH cells (arrows) with Aldehyde Thionin /Periodic Acid Schiff/Orange G stain (800X)

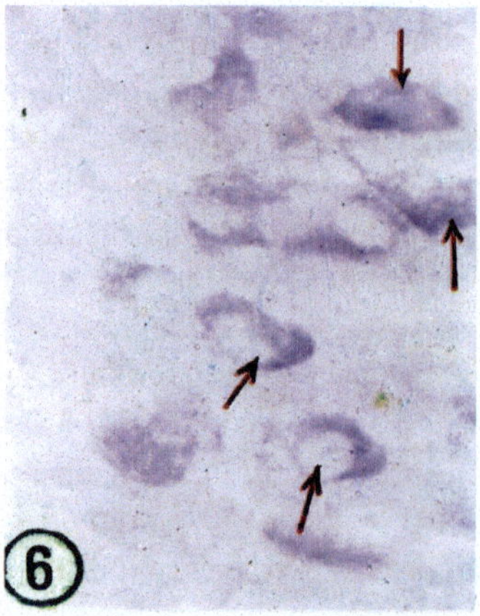

TSH cells (arrows) with Aldehyde Fuschin /Light Green/ Orange G stain (800X)

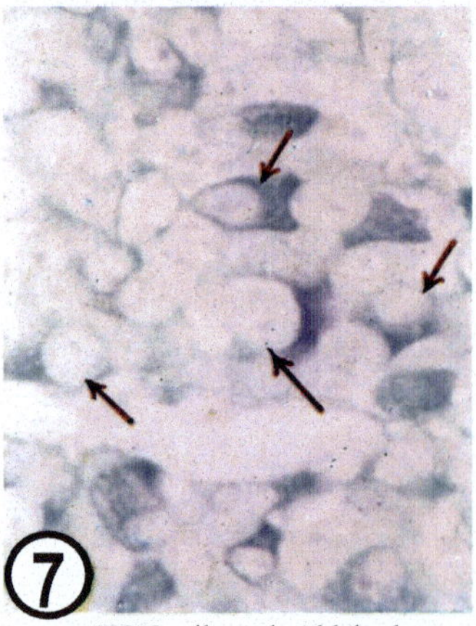

8- TSH cells with Aldehyde Fuschin/Light Green/Orange G stain (800X)

Cells of anterior pituitary in Five Striped Palm Squirrel (*Funambulus pennantii*)

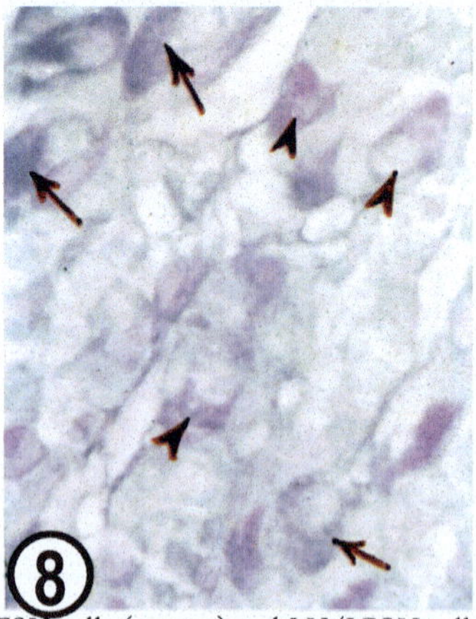

FSH cells (arrows) and LH/ICSH cells (arrow heads) with Aldehyde Thionin /Periodic Acid Schiff/Orange G stain (600X)

References

Abe K, Takano H, Ito T (1982) Response of epididymal duct in corpus epididymis to efferent or epididymal duct ligation in the mouse. J Reprod Fertil 64:69–72

Aggarwal Y, Partulla TV (1987) Effect of secretory bovine seminal vesicle secretion, sperm motility and acrosome reaction. J Reprod Fertil 79:409–419

Bose TK, Prasad MRN, Moudgal NR (1975) Ind J Exp Biol 13:8

Brook DR (1979) Biochemical environment of sperm maturation. In: Fawcett DW, Bedford JM (eds) The spermatozoa. Urban & Sehwarzenberg, Baltimore, pp 23–34

Brooke DE, Higgins SJ (1980) Characterization and androgen dependence of proteins associated with luminal fluid and spermatozoa in the rat epididymis. J Reprod Fertil 59:363–375

Burgos MH (1964) Uptake of colloidal particles by cells of the caput epididymidis. Anat Rec 148:517525

Carlos A, Quian S, Jelesoff N, Byers SW (1992) Lysosomal integral membrane proteins exhibit region and cell type specific distribution in the epididymis of the adult rat. Anat Record 232:85–96

Danza BJ, Orgebin Crist MC (1974) Androgen regulation of epididymal function. In: Sperm transport, survival and fertility ability in vertebrates. E. S. E. and C. Thibault Inserm, Paris, pp 61–87

Davis BK (1978) Inhibition of fertilizing capacity in mammalian spermatozoa by natural and synthetic vesicles. In: Symposium on the pharmacological effects of lipids: monograph 5. AOCS, Champaign II, pp 145–158

Davis BK, Mungund BJ (1976) Effect of modified vesicles from seminal plasma on the fertilizing capacity of rabbit spermatozoa. Biochem Biophy Res Commun 64:1004

Dawson RMCT, Rowland IW (1959) Glyceryl phosphorylcholine in the male reproductive organs of rats and Guinea pig. O J Exp Physiol 44:26–34

Fawcett DW, Heffer AP (1979) Failure of exogenous androgen to prevent regression of the initial segment of the rat epididymis after efferent duct ligation or orchidectomy. Biol Reprod 20:162–181

Flickinger CJ (1979) Synthesis transport and secretion of protein in the initial segment of the mouse epididymis as studied by electron microscope radio- autography. Biol Reprod 20:1015–1030

Flickinger CJ, Herr JC, Sisak JR, Howards SS (1993) Ultrastructure of Epididymal interstitial reaction following vasectomy and vasovasostomy. Anat Rec 235: 61–73

Fogh M, Corker CS, Hunter WM, McLean H, Philip J, Schou G, Skakkebak NE (1979) The effects of low doses of cyproterone acetate on some functions of the reproductive system in normal men. Acta Endocrinol 91:545–552

Fornes MW, Rosas D, Juan C (1991) Interaction between rat epididymal epithelium and spermatozoa. Anat Record 231:193–200

Glover TD, Nicander L (1971) Some aspects of structure and function in the mammalian epididymis. J Reprod Fertil 13:39

Goyal HO (1983) Histoquantitative effects of orchidectomy with and without testosterone enanthate treatment on the bovine epididymis. Am J Vet Res 44: 1085–1090

Goyal HO (1985) Morphology of bovine epididymis. Am J Anat 172:155–172

Goyal H.O. (1991)

Gupta G, Rajalakshmi M, Prasad MRN (1974b) Regional difference in androgen threshold of epididymis of castrated rats. Steroid 24:55

Gupta G, Rajalakshmi M, Prasad MRN, Moudgal NR (1974a) Effects of antiserum to luteinizing hormone (LHAS) on the physiology of epididymis and accessory glands of albino rats. Contraception 10:491

Hamilton DW (1971) Steroid function in the mammalian epididymis. J Reprod Fertil 13:89–97

Hamilton DW (1975) Structure and function of the epithelium lining the ductus efferentes, ductus epididymis, ductus deference in the rat. In: Hamilton DW, Greep RO (eds) Handbook of physiology, section 7, Vol. 5. American Physiological Society, Washington, pp 259–301

Hamilton DW, Jones AL, Fawcett DW (1969) Cholesterol biosynthesis in the mouse epididymis and ductus deference. A biochemical and morphological study. Biol Reprod 1:167–184

Hammerstein J (1974) Quoted from Roy et al. (1976). Contraception 14(4):403–420

Hansson V, Djoseland O, Reusch E, Attramadal A, Torgersen O (1973) An androgen binding protein in the testis cytosol fraction of adult rats. Comparison with the androgen binding protein in the epididymis. Steroids 21:457–474

Hansson V, Trygstad O, French FS, McLean WS, Smith AA, Tindall DJ, Weddington SC, Petrusz P, Nayfeh SN, Ritzen KM (1974) Androgen transport and receptor mechanisms in testis and epididymis. Nature (London) 250:387–391

Hermo L, Wright J, Oko R, Morales CR (1991a) Role of epithelial cells of the male excurrent duct system of the rat in the endocytosis or secretion of sulfated glycoprotein. Biol Reprod 44:1113–1131

Hermo L, Green H, Clermont I (1991b) Golgi apparatus of epithelial principal cells of the Epididymal initial segment of the rat: structure relationship with endoplasmic reticulum and role in the formation of secretory vesicle. Anat Record 229:159–176

Hinton BT, Palladino MA, Mattnueller DR, Bard D, Gond K (1991) Expression and activity of gamma-glutamyl transpeptidase in the rat epididymis. Mol Reprod Dev 1:40–46

Hoffer AP (1982) Ultrastructural studies of the seminiferous and epididymal epithelium and epididymal sperm in rats treated with gossypol. Arch Androl 9:35

Hoffer AP, Hinton BT (1984) Morphological evidence for a blood-epididymis barrier and the effects of gossypol on its integrity. Biol Reprod 30:991–1004

Jones R, Glover TD (1973) The effects of castration on the composition of rabbit epididymal plasma. J Reprod Fertil 34:405–414

Kaur J, Ramakrishnan PR, Rajalakshmi M (1992) Effect of cyproterone acetate on structure and function of rhesus monkey reproductive organs. Anat Recd 236: 62–72

Koch UJ, Lorenz F, Danehl K, Ericsson R, Hasan SH, Keyserlingk DV, Lubke K, Mehring M, Rommler A, Schwartz U, Hammerstein J (1976) Continuous oral low-dosage cyproterone acetate for fertility regulation in the male? A trend analysis in 15 volunteers. Contraception 14(2):117–135

Koch UJ, Lorenz F, Danehl K, Hammerstein J (1974) Uber the verwendbarkeit von cyproterone acetate zur fertilitatshemmung bein mann. Morphologische veranderungen and einflusse auf the spermien motilitat. 40. Tagung der Deutschen Gesellschaft fur Gynaekologie and Geburtshilfe, Wiesbaden

Lea OA, Petrusz P, French FS (1978) Purification and localization of acidic epididymal glycoproteins (AEG): a sperm coating protein secreted by the rat epididymis. Int J Androl 21:592–607

Linnetz LJ, Amann RP (1968) Histochemistry of epididymis and ampulla as influenced by sperm output. J Reprod Fertil 16:343–350

Lohiya NK, Sharma OP (1983) Effects of cyproterone acetate with combination of testosterone enanthate on seminal characteristics, androgenicity and clinical chemistry in langur monkey. Contraception 28(6): 575–586

Lopez AML, Cerda MC (1993) Effects of cyproterone acetate on the ultrastructure of the human epididymis. Andrologia 25:35–45

Lorenz F, Koch UJ, Danehl K, Lubke K, Hammerstein J (1974) Uber the verwendbarkeit von cyproterone acetate zur fertilitatschemmung beim mann, Physiologische and biochemische veranderungen. 40. Tagung der Deutschen Gesellschaft fur Gynaekologie and Geburtshilfe, Wiesbaden

Mann T (1964) The biochemistry of semen and of the male reproductive tract. Wiley, New York

Moltz L, Rommler A, Schwartz U, Hammerstein J (1978) Effects of cyproterone acetate (CPA) on pituitary gonadotropin release and on androgen secretion before and after LH-RH double stimulation tests in men. Int J Androl Suppl. 2:713

Moore HDM (1980) Localization of specific glycoproteins secreted by the rabbit and hamster epididymis. Biol Reprod 22:705–718

Moore HDM, Bedford JM (1979a) The differential absorptive activity of epithelial cells of rat before and after castration. Anat Record 193:313–328

Moore HDM, Bedford JM (1979b) Short term effect of androgen withdrawal on the structure of different epididymal cells in the rat epididymis. Anat Record 193: 293–312

Morse HC, Leach DR, Rowley MJ, Heller CG (1973) Effect of cyproterone acetate on sperm concentration, seminal fluid volume, testicular cytology and levels of plasma and urinary ICSH, FSH and testosterone in normal men. J Reprod Fertil 32:365

Neumann F, Diallo FA, Hasan SH, Schenck B, Traore I (1976) The influence of pharmaceutical compounds on male fertility. Andrologia 8(3):203–235

Neumann F, Schenck B (1976) New antiandrogens and their mode of action. IPPF congress on agents affecting control of fertility in the male, New Delhi. J Reprod Fertil 24:129–145

Nicander L, Paulsson S, Selander U (1965) An electron microscopical study of iron absorption in the epididymal tail of rabbits. Scand Congr Cell Res 4th:51–52

Nieschlag E, Behre HM, Weinbaner GF (1989) In perspectives in andrology. Raven Press, New York. 517

Orgebin Crist MC, Davis J (1974) Functional and morphological effects of hypophysectomy and androgen replacement in the rabbit epididymis. Cell Tissue Res 148:188–201

Orgebin Crist MC, Hoffman LH, Olson G, Shuklarek MC (1987) Secretion of proteins and glycoproteins by perfused rabbit corpus epididymal tubules: effects of castration. Am Anat 180:49–68

Orgebin C, Danze MC, B. J. and Davier J. (1975) Endocrine control of the development and maintenance of sperm fertilizing ability in the epididymis. In: Hamilton DW, Greep RO (eds) Handbook of physiology sec. 7, Vol. 5. American Physiological Society, Bethesda, M. D, pp 319–339

Oritz E (1953) Effect of castration on the reproductive system of the golden hamster. Anat Record 117:91

Palacios J, Regadera J, Manuel N, Paniagua R (1991) Apical mitochondrial rich cells in the human epididymis: an ultrastructural enzyme histochemical and immunohistochemical study. Anat Record 231:82–88

Paulson HL, Dym M (1985) Morphometric analysis of coated pits and vesicles in the proximal and distal caput epididymis. Biol Reprod 32:191–202

Petry R, Mauss J, Rausch-Stroomann JG, Vermeulen A (1972) Reversible inhibition of spermatogenesis in men. Horm Metab Res 4:386–388

Prasad MRN, Rajalakshmi M (1976) Comparative physiology of the mammalian epididymis. Gen Comp Endocrinol 28:530–537

Prasad MRN, Rajalakshmi M, Reddy PRK (1972) Hormones and antagonists. In: Hubinont PO, Mendeles SM, Preumont P (eds) , vol 2. Karger, Basel, pp 202–212

Rajalakshmi M (1985) Physiology of epididymis and spermatozoa. J Biosci 7(2):191–195

Rajalakshmi M, Prasad MRN (1968) J Endocrinol 41:471

Rajalakshmi M, Prasad MRN (1977) In Reproductive physiology (ed. R.O. Greep). University Park Press, Baltimore, p 153

Ramakrishnan PR, Kaur J, Rajalakshmi M (1990) Effect of non-aromatizable androgens on testicular and accessory gland functions in rhesus monkeys. (Macaca-mulatta). J Fertil 89:69–76

Ramos AS, Dym M (1977) Fine structure of the monkey epididymis. Am J Anat 149:501–532

Ramos AS Jr (1980) Ultrastructure and histological observations on the principal cells of monkey epididymis. Arch Androl 5:159–168

Rao MV, Roy GK (1993) Biochemical and morphological changes of spermatozoa in progestic androgen injected rats. Ind J Exp Biol. 31:12–15

Ratna Kumar BV, Shipstone AC, Setty BS (1990) Effects of vasectomy on the ultrastructure of epididymal epithelium in rhesus monkey. Int J Fertil 35(3):180–191

Riar SS, Setty BS, Kar AB (1973) Studies on the physiology and biochemistry of mammalian epididymis. Biochemical composition of epididymis. A comparative study. Fert Steril 24:355–363

Robaire B, Hermo L (1988) Efferent ducts, epididymis and vas deference: structure, function and their regulation. In: Knobil E, Neill JD (eds) The physiology of reproduction, vol 1. Raven Press, New York, pp 999–1080

Roussel JD, Stallcup OT (1966) Relationship between phosphatase activity and other characteristics in bull semen. J Reprod Fertil 12:423–429

Roy S, Chatterjee S, Prasad MRN, Poddar AK, Pandey DC (1976) Effects of cyproterone acetate on reproductive functions in normal human males. Contraception 14(4):403

Roy S, Taneja SL (1974) Vasectomy, vasocclusion and vasanastomosis: a critical appraisal. Monograph no. 23. National Institute of Family Planning, New Delhi

Schenck B, Neumann F (1978) Some comments on the use of antiandrogens for male contraception. Int J Androl 2:155–161

Schulze W (1977) Licht–und elektronmikroskopische studien an den A-Spermatogenien von Mannernmt intakter- spermatogenese and bei patienten. Andrologia 10:307–320

Scott TW, Wales RG, Wallace JC, White (1962) Composition of ram epididymal and testicular fluid and the biosynthesis of glyceryl phosphorylcholine by the rabbit epididymis. J Reprod Fertil 6:49–59

Sedar AW (1966) Transport of exogenous peroxidase across the epithelium of the ductuli efferentes (abstract). J Cell Biol 31:102 A

Sumitra N, Ghosh J (1979) Epididymal and testicular enzymes as monitors for assessment of male antifertility drug. J Steroid Biochem 11:681

Usselman MC, Cane RA (1983) Rat sperm are mechanically immobilized in the canal epididymis by immobilin a high molecular weight glycoprotein. Biol Reprod 29:1241–1253

Voglmayr JK, Waites GMH, Setchell BP (1966) Studies on spermatozoa and fluid collected directly from the testis of the conscious ram. Nature (London) 210:861–863

Waits GMH, Setchell BP (1969) Physiology of the testis, epididymis and serotum. In: McLaren A (ed) Advances in reproductive physiology, vol 4. Logos Press, London, pp 1–63

Wang C, Yeung KK (1980) Use of low dosage oral cyproterone acetate as a male contraceptive. Contraception 21(3):245

WHO (1991) Annual technical report, Geneva, Switzerland

Wong PYD (1989) Electrolyte and fluid transport by the epididymis. In: Young JA, Wong PYD (eds) Epithelial secretion of electrolyte and wate. Springer Verlag, Heidelberg

Yanagimachi R, Kamiguchi Y, Mikamo K, Suzuki F, Yanagimachi H (1985) Maturation of spermatozoa in the epididymis of the Chinese hamster. Am J Anatomy 172:317–330

4 A Summary of Placental Embryology

4.1 Mechanism of Ovulation

4.1.1 Introduction

The earliest record of the direct observation of the process of ovulation was provided, in *Rana temporaria*. Rugh believes that 'ovulation is a dual process, involving first the rupture of the follicle and subsequently the active contraction of follicular muscle cells forcing the egg out through the area of rupture'.

Ovulation occurs at estrus in mammals. Subsequently, the final growth of the follicle is very rapid. Thus, ovulation is the climax of the final period of accelerated growth of follicles.

In a great majority of mammals, ovulation occurs spontaneously at oestrus, but in a few cases, ovulation occurs at a definite interval after copulation. The rabbit and ferret are examples in which ovulation follows 10 and 30 h after copulation, respectively. The actual mechanism of ovulation can be summarized conveniently under three main headings.

(a) *Enzymes Action*: demonstrated that the pressure of proteolytic enzymes in the Liquor folliculi. He suggested that these enzymes may play a part in bringing about perforation of the follicle by autolysis of its wall. It should be remembered, however, that there is no evidence that the enzymes shown by Schochet are proteolytic in vitro or proteolytic in vivo.

(b) *Internal Pressure*: According to second explanation, the pressure within the Graffian follicles rises until its wall bursts at the weakest point. Such a rise in the internal pressure must be brought about either by increasing the content of the follicle or by the pressure exerted on it by the follicular wall.

   It is quite possible that the rise in internal pressure is brought about by the active secretion of liquor folliculi into the autrum, either by membrana granulosa or by theca interna or by both.

(c) *Growth Pressure*: It was suggested, that the increased intrafollicular pressure is brought about in the mature follicle by hypertrophy of the cells of theca interna.

It is suggested that the rupture is caused by the richly vascularized in-growth of theca interna cells which pushes the follicular content toward the thinnest part.

The rupture of a mature follicle was observed by Clark in 1898, in an ovary injected with a carmine-gelatine mass. He attributed ovulation as a consequence of increased blood-pressure and the specific arrangement of blood vessels of the follicular wall.

In order to form a reasoned opinion of the cause of ovulation it is necessary to obtain a picture of early events within the follicle just prior to its rupture.

D. A. Bhiwgade, S. Menon, *Ultrastructural Investigations on the Pituitary-Gonadal Axis*,
https://doi.org/10.1007/978-981-99-3276-4_4

It is a point of controversy, whether the follicle rupture is a result of pressure from within or whether the rupture is due to erosive factors. There have been various theories put forward to explain this phenomenon.

(a) *Mechanical Theory*: Most of the earlier suggestions have stressed the role of increasing intra-follicular pressure as the cause of rapture.
Heape in 1905 suggested that vasodilation and bursting blood vessels added to the pressure within the follicle at the time of ovulation.
Thomson in 1919 pointed out that increased secretion of liquor folliculi might be the immediate cause of ovulation.
Smith in 1938 suggested that the increase in the liquor foliculi was due to the liberation of an osmotically active substance possibly a carbohydrate, into the maturing follicle.
Objection to mechanical theory:—Lots of evidence has been cited to generate a trend of opinion away from the pressure theory.
One reason for this is the form of rupture point.
Erams and Cole in 1931, observed that in dog, the follicle wall becomes folded in a complex manner before ovulation. This does not suggest increased tension, since this might be expected to expand the wall.
In Bat, Wimsatt (1944) found that there is a decrease in the size of follicle and folding of wall immediately before ovulation.
All these observations suggest that rapid secretion of liquor folliculi and perhaps, hypertrophy of the follicle cells may cause the wall to weaken.
(b) *Hormonal Theory*: Fairly critical evaluation of various evidence has been made by Kraus in 1947. Kraus made direct observation on rabbit by using suitable gonadotropin hormones to induce ovulation.
Attempts were made by Kraus to produce ovulation by adding stimulants of smooth muscle to the fluid in which excised ovary was placed.
As a result of some extensive work, Kraus, concluded that neither the pressure nor the enzymes theory filled all the gaps and the immediate cause of ovulation remained unknown.
Kraus suggested that the morphological changes in the follicle induced by the action of gonadotropin or another hormone are responsible for this phenomenon.
Maxwell in 1960 grafted rabbit ovarian follicles into the anterior chamber of the rabbit eye. Has observed the behaviour of follicle grafted into Rabbit. When LH is injected, a wave of contraction was seen to pass over the surface of the follicle. The author has attributed such waves of contraction within the muscle cells of externa to be an important factor in ovulation.

Control of ovulation: Ovulation is identified as a dynamic process that is continuously regulated by a complex integration between the neuronal signals, secretions from the pituitary, secretions and structural features of the ovary and the external environment.

A complicated reciprocal relationship exists between the two main ovarian hormones, estrogen and progesterone, and the hypophysis, which produces three gonadotrophins: FSH, LH and LTH or luteotropin.

The exact role of these hormones in the control of ovulation is difficult to assess. Induction of ovulation starts with a sudden surge in LH level. This surge in LH may be triggered either by the stimulus of coitus or it could be due to a neural clock in animals with spontaneous ovulation. The higher neural centres are responsible for the various physiological events related to ovulation. It is important to consider the role of various hormones that are involved in ovulation, as depicted in the diagram, like the Follicle Stimulating Hormone, the Luteinizing Hormone, the Luteotropic Hormone and the Estrogen.

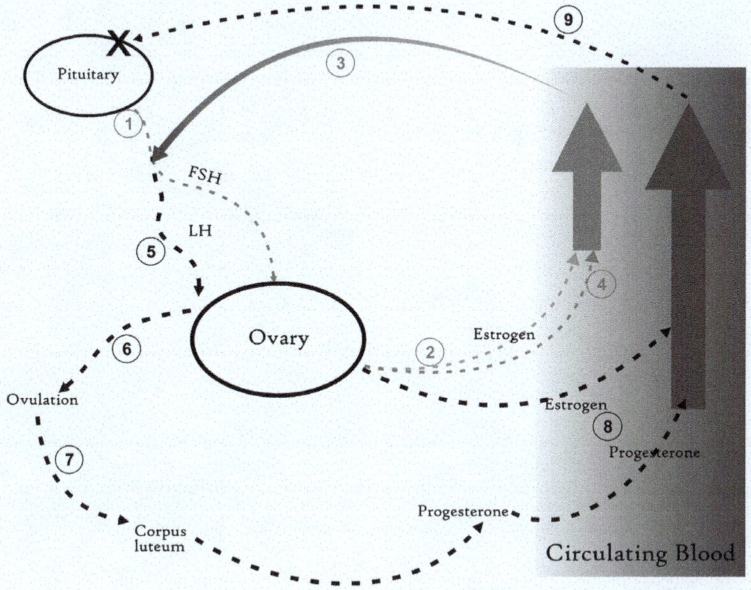

(1) FSH from pititary stimulates ovary.
(2) Ovary Secretes estrogen and its level increases in circulating blood
(3) Increased estrogen stimulates release of FSH - LH from pituitary
(4) Ovary secrete more Estrogen
(5) Rising estrogen levels stimulates release of more FSH - LH
(6) Results in Ovulation
(7) Corpus Luteum develops
(8) Surge in Progesterone and estrogen levels in circulating blood
(9) Increased levels of Progesterone and estrogen exerts a negative feedback on pituitary secretion of FSH & LH

Diagram: Endocrine regulation of Ovarian Function (Fig. 4.1)

4.2 Development of the Embryo

The structure of ovum in bat family resembles other mammals. It is round or oblong in shape. The microlecithal is about 65–75 µm in diameter (Gopalakrishna et al. 1974) and has an eccentric nucleus. It is surrounded by zona pellucida which is reported to be eosinophilic and PAS-positive. The ovum shrinks after ovulation, but it remains in its shrunken size till the time of hatching. Information on the maturation of ovum and the process of fertilization in bats is scanty and is available only for a few species of bat (Rasweiler IV 1979). Observation on the first meiotic division is available for the mature preovulatory follicles of *Carollia perspicillata* and *Desmodus rotundus* (Bonilla and Rasweiler IV 1974), and *Noctilioal biventris* (Rasweiler IV 1977).

Further development of the ovum is stopped at the metaphase stage of the second division, till ovulation, as observed in other mammals. In *Vesperugo noctule*, the elimination of the first polar body and the formation of second meiotic spindle has been reported (van der Stricht 1909). Ovulation is reported to be spontaneous in Pteropus species. (Martin et al. 1987), *Glossophaga soricina* and *Molossus ater*. It is reported to be induced in Carollia and a few other bat species. In *Myotis lucifugus* (Wimsatt 1944a), the ovum at ovulation has been reported to be surrounded by cumulus cells. The ontogeny of Chiropteran embryos shows variations that are very interesting.

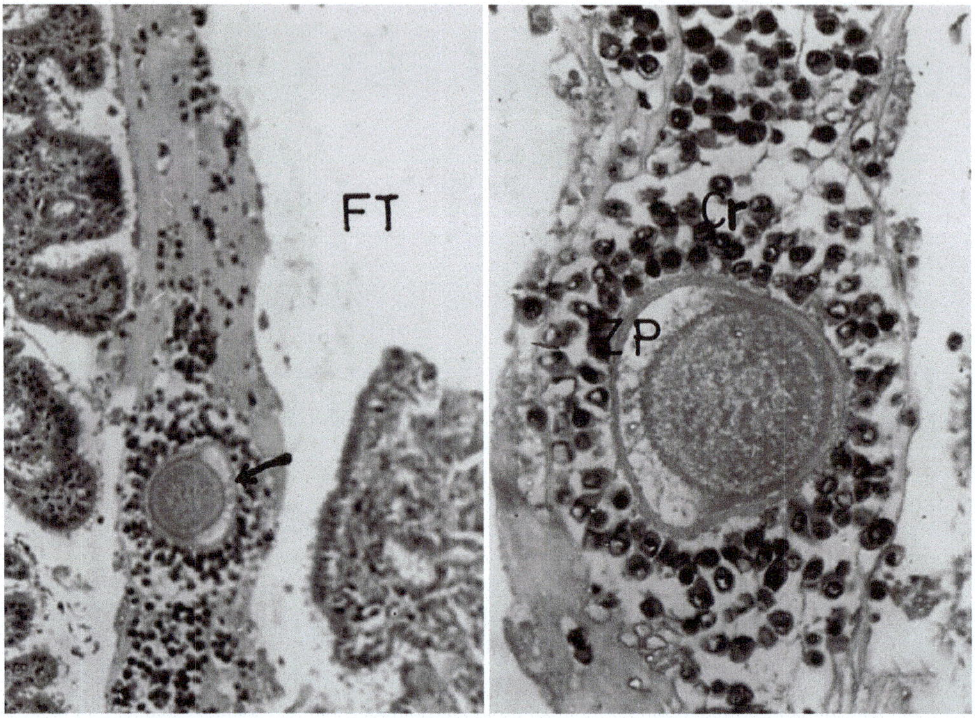

Fig. 4.1 Section of Fallopian tube (F.T.) showing the released ovum (arrow). Note the zona pellucida (zp) and corona radiata cells (cr). [X 100], [X 400]. Unpublished light electron micrographs from Bhiwgade and Banerjee

4.2.1 Fertilization

Krutzsch and Crichton (1991) have provided an extensive review of fertilization in various groups of bats. It is only when the ovum is fertilized that the second phase of maturation or the meiotic division is completed. This marks the release of second polar body. In *Myotis lucifugus* (Wimsatt 1944b). *Pipistrellus mimus* (Karim 1975) and in some other vespertilionid bats, fertilized eggs with two pronuclei have been reported. Different sites of fertilization have been reported in bats. Generally, it is in the infundibulum of the oviduct but oviductal ampulla has also been reported in some bats. Uchida 1953 and Karim 1975 reported that fertilization occurs in the periovarian space in Pipistrellus species. Fertilization by multiple sperms is blocked in bats too. Probably two strategies enable this, as seen in, *Pipistrellus mimus* by blocking of zona pellucida (Karim 1975) or by the vitelline block as observed in *Miniopterus schreibersii* (Mori and Uchida 1981).

4.2.2 Cleavage

The egg starts to divide after fertilization as shown in Fig. 4.2a–e. The cleavage has been reported to be holoblastic and equal. This has been observed in *Pipistrellus* where the development of fertilized eggs up to two cell stages has been recorded. The fertilized egg, as it undergoes division is transported down the oviduct. In bats held in captivity, it has been observed that the fertilized egg remains in the oviduct only for a short time. The travel time within the oviduct differs in various bat species; 4–5 days in the little brown bat *Myotis lucifugus* (Rasweiler IV 1979), 12–17 days in *Glossophaga* and *Desmodus* (Qivntero and Rasweiler 1974). Blastocysts have been observed in the oviduct on 16th day after coitus in *Carollia* and on 22nd day after coitus in *Desmodus* (Rasweiler IV 1979). It has been reported that in some bat species like *Glossophaga soricina*, *Noctilioal biventris* and *Peropteryx kappleri* the degenerating ova are retained within the oviduct and do not reach the uterus.

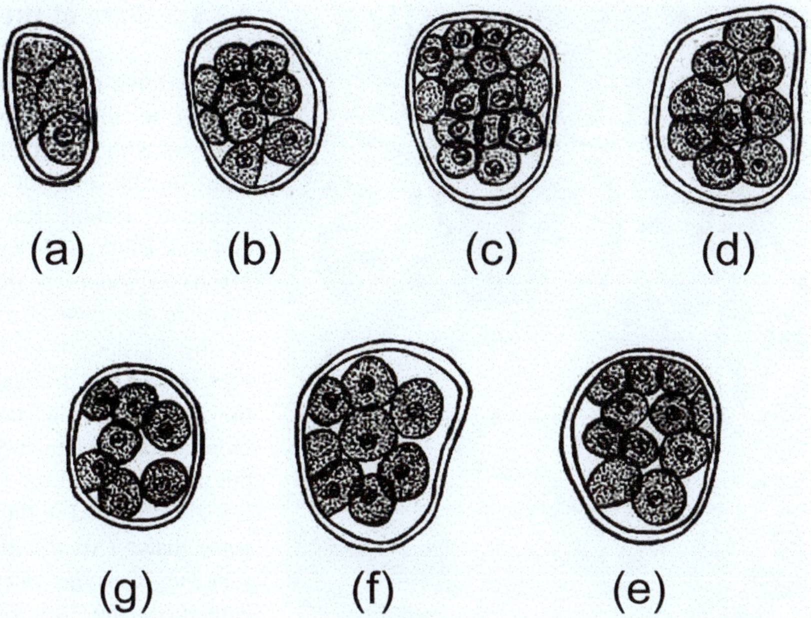

Fig. 4.2 The figure diagrammatically depicts the fertilized egg as seen as a blastocyst in the oviduct of *Rhinolophus rouxi* (**a**–**e**). The blastocyst is seen in an enlarged pocket of uterus around the middle of uterine horn. The free uterine morula shows a group of numbering up to 48 cells with a distinct envelope of zona pellucida. The cells of the morula show granular cytoplasm with vesicular nuclei. Most advanced of these blastocysts showed a flattened inner mass of cells numbering about 80 cells. Bhiwgade 1976

4.2.3 Preimplantation Development of the Embryo

The development of embryo before implantation differs in different bat species. Studies have been done on 50 bats species belonging to nine families (Karim 1976; Rasweiler IV 1979; Gopalakrishna and Karim 1980). The unilaminar blastocyst in bats resembles the mammalian blastocyst except in the bat, *Noctilio albiventris* (Rasweiler IV and Badwaik 1996a, b) where the trophoblastic cells are arranged either fully or partially as a monolayer and the embryonic cell mass is discernible only after the attachment of blastocyst to the uterine wall. In the bat, *Desmodus rotundus,* the endoderm is also differentiated before the embryo reaches the uterus (Wimsatt 1954, 1975). In bats of the families, Rhinolophidae, Vespertilionidae and Molossidae, the embryos at the early cleavage or at the morula stage itself reach the uterus (Rasweiler IV 1993) whereas in the bats *Rousettus leschenaulti* and *Desmodus rotundus* the bilaminar embryo has been observed to lie free in the uterine lumen.

The duration that the embryo spends in the uterine lumen is very small but in some bats like, *Miniopterus* the embryo spends extended periods lying free in the uterine lumen as part of delayed implantation, the extension of the delay, however, has not been estimated (Figs. 4.3 and 4.4).

4.2.4 Formation of Endoderm in the Bilaminar Blastocyst

The blastocyst which is with a single layer starts becoming bilaminar. This process is similar to the one observed in other mammals. The endoderm forms by the migration of cells on the ventral side from the inner cell mass. These migrating cells line the interior of the blastocyst cavity (Fig. 4.3) giving rise to the yolk sac. In some bat species like *Glassophaga soricina*, *Haplonycteris fischeri* (Heideman 1989), *Carollia perpicillata* (Badwaik et al. 1997) and *Desmodus rotundus* (Wimsatt 1954) the cells from the inner cell mass also differentiate along the outside to form an extraembryonic mesoderm. In *Glassophaga soricina* and *Carollia perspicillata* (Badwaik et al. 1997), interestingly, the primitive endoderm appears like a meshwork, a feature seen in primates like bushbabies (Galagos), Chimpanzees and human.

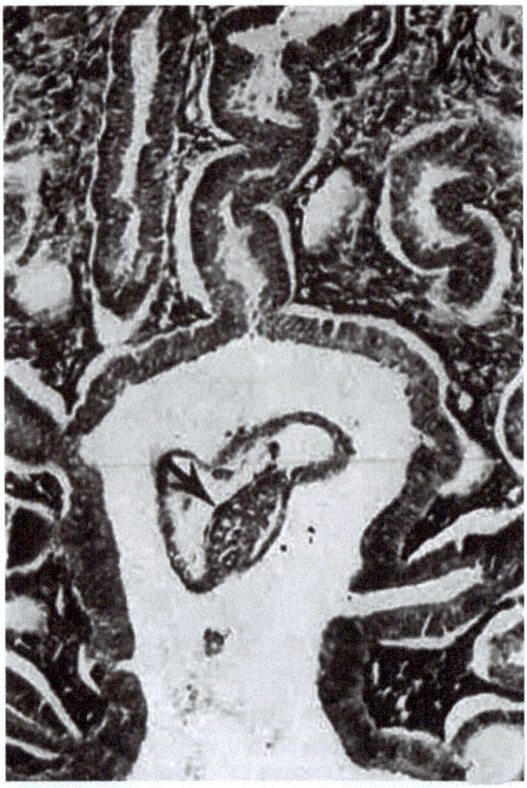

Fig. 4.3 Image under the light microscope of the blastocyst lying free in the uterus lumen of the bat, *Hipposideros lankhadiva*. The arrow shows the inner mass of endodermal cells below the embryonic disc [X 140]. Published light micrograph from Bhiwgade 1979

4.2.5 Fate of the Zona Pellucida

Various investigators have looked at the stages leading to the loss of zona pellucida. Early embryos with zona pellucida have been reported both in the oviduct and the uterine lumen (Wimsatt 1944a; Rasweiler IV 1979, 1993; Gopalakrishna and Karim 1980). In the bat, *Myotis lucifugus lucifugus*, the zona pellucida expands and then thins out to finally disappear as the blastocyst grows (Wimsatt 1944a). In *Pipistrellus pipistrellus* (Potts and Racey 1971), the zona pellucida starts to break in the early morula stage when the embryo is in the uterus but the remains of the fragmented zona pellucida are still observed in the primitive streak stage. In some bats, (*Noctilioal biventris*, *Glassophaga soricina*, *Carollia perspicillata* and *Desmodus rotundus*), the loss of zona pellucida occurs before the blastocyst enters the uterus and gets eliminated at the morula or early blastocyst stage (Rasweiler IV 1979). In *Miniopterus australis*, the discarded zona pellucida has been reported in the uterus that contained a unilaminar blastocyst (Richardson 1977).

Figure 4.5 shows a section through the fallopian tube of *Hipposideros lankhadiva*. The zona pellucida is visible in several processes that can be seen arising from the zona pellucida and extending towards the ne epithelium. Figure 4.6

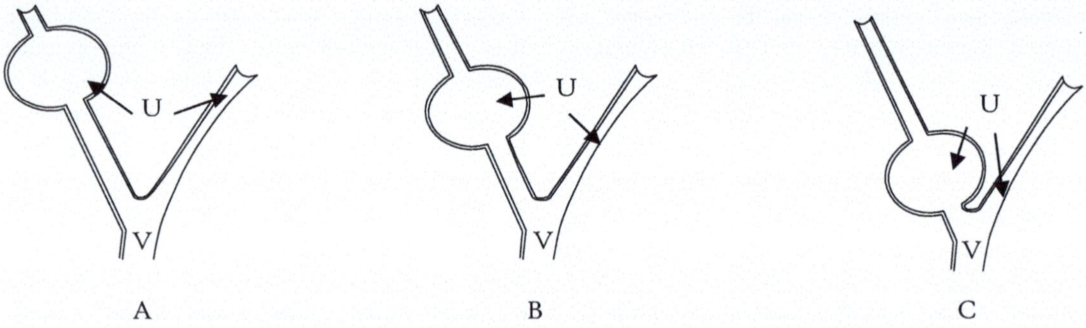

Fig. 4.4 Diagrammatic representation of various implantation sites observed in the bicornuate uterus (U) where (V) is the vaginal end. (**a**) Shows the Implantation occurring at the cranial end of the uterus. In (**b**), the Implantation is in the middle of the uterus while in (**c**) the implantation is towards the vaginal end

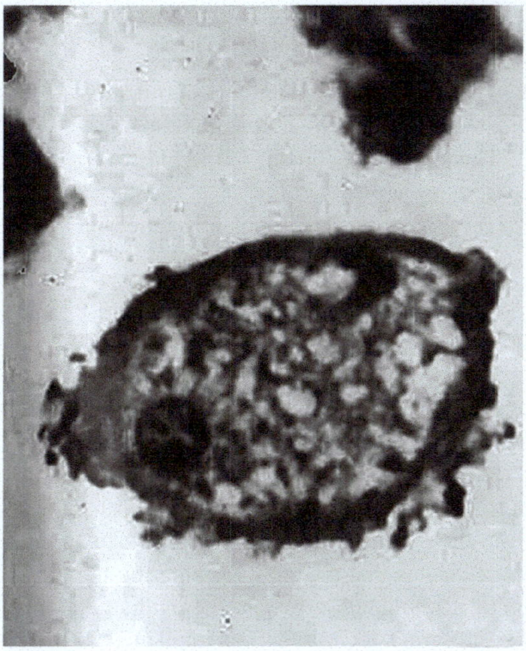

Fig. 4.5 Section of the fallopian tube in *Hipposideros lankhadiva* showing embryo in morula stage. Note the presence of processes arising from zona pellucida. The remnants of cumulus cells can be seen adhering to the outer surface of zona pellucida [X400]. Bhiwgade 1979

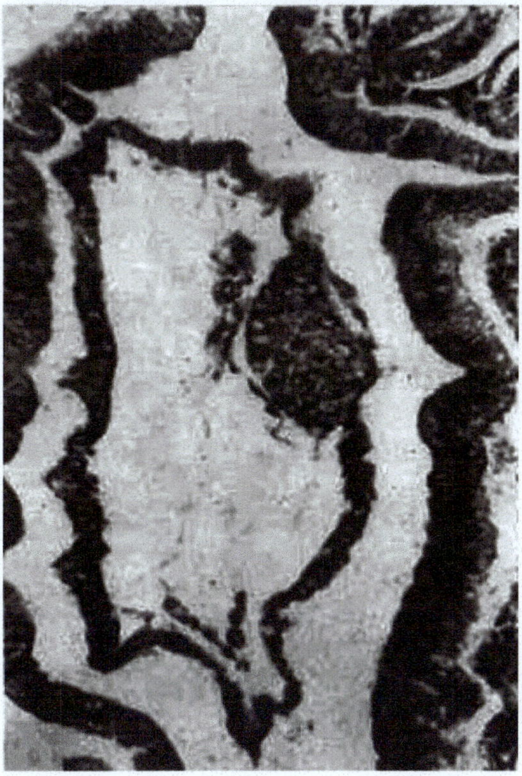

Fig. 4.6 Section of uterus showing the late free bilaminar blastocyst in *Rhinolophus rouxi.* Pseudopodia-like processes from the blastocyst are seen extending towards the uterine wall on antimesometrial side [X 380]. Bhiwgade 1976

depicts a bilaminar blastocyst seen within the uterine lumen of *Rhinolophus rouxi.* A disintegrating zona pellucida is seen with processes extending towards the uterine wall.

Figure 4.7 is a section of the uterus showing an early implanting blastocyst in *Rhinolophus rouxi.* The cavity of blastocyst is enlarged pushing the inner cell mass to one side. The processes from the blastocyst are seen extending to the uterine epithelium at the anti-mesometrial side. It is interesting to note that the uterine epithelium is discontinuous in several places.

4.3 Implantation

The blastocyst invades the uterus by disrupting the uterine tissue to make a lifeline connection between the blastocyst and the uterine tissue. This process is implantation by which the blastocyst gets the much-needed nourishment and continues its further development as a parasite in the uterus.

The blastocyst continues its life as an obligate parasite to develop complex structures without affecting any other organs of the maternal environment. The process of implantation involves complex adaptations in mammals, at morphological, physiological and psychological levels in the mother for providing a conducive environment for the developing embryo.

The process of implantation involves several interrelated factors that influence its success:

1. Positioning of blastocysts in the uterus to ensure adequate space between them.
2. The formation of an efficient attachment between the blastocyst and the uterine epithelium.
3. The invasion of the blastocyst into the endometrium by cytolysis of the uterine epithelium

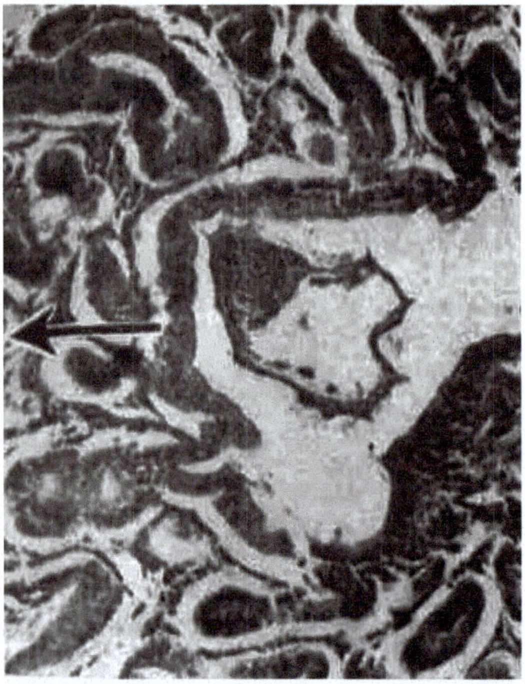

Fig. 4.7 Section of the uterus with an early implanting blastocyst in *Rhinolophus rouxi.* The embryonic disc is multilayered and the endodermal layer is seen as a distinct lining on the ventral surface of the embryonic disc. The arrow points to the mesometrial side [X 80]. Bhiwgade 1976

and the subsequent changes in the endometrium for the implantation of blastocyst.

The relationship between the blastocyst and uterus varies in different mammals and can be evaluated at three distinct levels:

1. The morphological and topographical factors that influence the interaction between the blastocyst and the uterus.
2. Physiological factors that are involved in the interaction between blastocyst and uterus.
3. Endocrine factors that prepare the uterus for implantation and support the maintenance of the attached blastocyst.

The morphological factors that influence implantation have been studied and reported (Mossman, 1937). There exist a lot of variations in the process of implantation or nidation in different mammals. These differences are generally related to the following aspects:

1. The time of implantation as compared to the time of ovulation and copulation.
2. The site of attachment of the blastocyst in the uterus.
3. The extent of development of the embryo at the time of arrival in the uterus.
4. The actual site of implantation.
5. The orientation of the embryonic disc.

It has been observed that in a bicornuate uterus (e.g. pig), during the first two weeks of gestation, the embryos that are implanted nearer to the oviduct show more advanced development as compared to those embryos implanted further away from the oviduct towards the cervix. This probably indicates that the embryos entering the uterus first get implanted nearer to the oviduct and then the subsequent embryos get implanted in series away from the oviduct. If the embryos enter together, the ones that implant towards the cervix show a slower rate of development. This difference in the development could be more related to the site of implantation rather than the time of entry of embryos.

The uterus is equally favourable for implantation throughout. The specific reason for the embryos to prefer the sites near the oviduct has not been fully understood. If the site near the oviduct is more preferred, then all embryos may try to bunch there. This does not happen. Mosmann suggested that the first embryo that enters the uterus implants itself into the nearest uterine space adjacent to the oviduct. The implementation of the embryo makes the nearby uterine mucosa to be physiologically unconducive for the other embryos to implant, forcing them to move further away to a site where the physiological influence of the implanted embryo is absent. Thus, the embryos get spaced out till the end near the cervix of the uterine horn. After this, further incoming embryos enter and are forced to migrate to the next uterine horn if it is free.

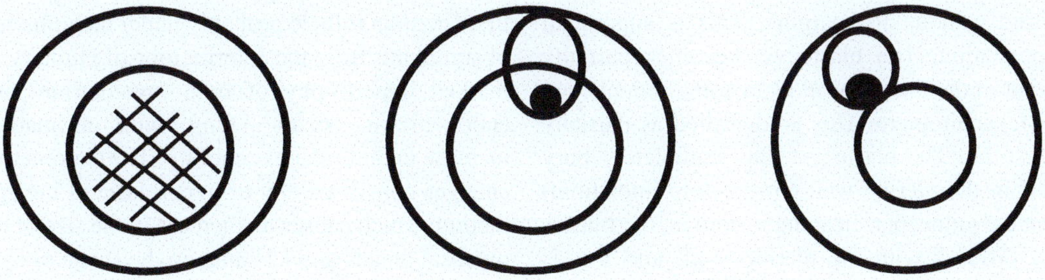

Central Partial Interstitial Completely Interstitial

Mclaran and Witchi in 1959, evaluated Mossman's hypothesis. According to them the peristaltic movement of the uterus stirs up embryos to space them out. They found no refractory zones in the uterus as suggested by Mossman.

In ruminants, the site of implantation is symmetrically in the middle of the uterine horn. In Armadillo, sloth, monkey and human, there are definite sites in the uterus where implantation occurs. It appears that, probably a combination of mechanical and physiological factors finally decides the implantation site in the uterus.

Types of Implantations

Three main types of implantations have been described by Bomet in 1882:

1. *Central*: The blastocyst remains in the uterine cavity and then grows separately to fill the uterine lumen (Catarrhine and Platyrrhine monkeys, some Insectivores, Edentates and Vespertilionid bats, Ungulates, Carnivores etc.)
2. *Concentric*: In many rodents, the blastocyst implants into a cleft within the uterine lumen.
3. *Interstitial* (which depends on the relationship of the blastocyst to the uterine cavity): The blastocyst enters the uterine epithelium and is completely cut off from the uterine lumen. This is seen in Cavia, certain Chiroptera, Rhinocerous, Chimpanzees and human.

If the morphological association of the blastocyst and uterus is considered, then blastocyst can be mesometrial as in fruit bats or antimesometrial as in the case of Rodents and many bats. The blastocyst can be lateral as in the case of Emballomurid bats.

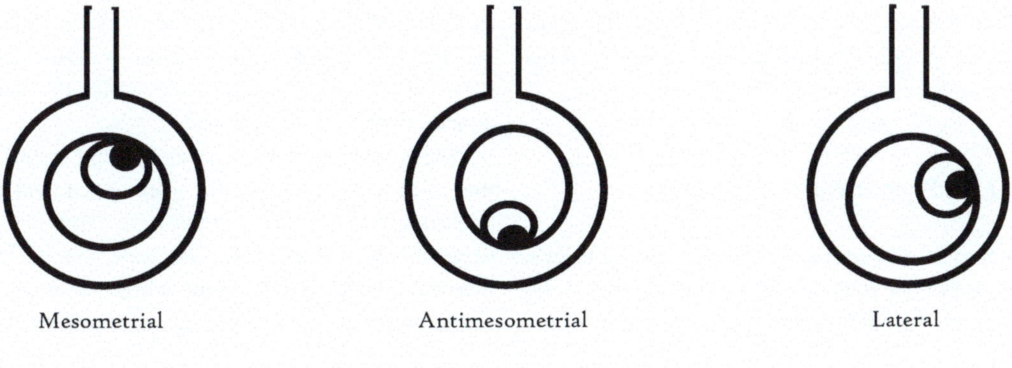

Mesometrial Antimesometrial Lateral

In squirrels, carnivores and majority of insectivore bats, the blastocyst is in contact with the uterine wall on its margins. This is superficial implantation. The blastocyst becomes partially buried in the uterine wall as in some rodents and in Megachiroptera. This is identified as partially interstitial. The blastocyst may completely bury itself in the uterine wall (completely interstitial) as in the guineapig. In some instances, the blastocyst diffuses with the uterine wall, and this is identified as diffused implantation or circumferential implantation.

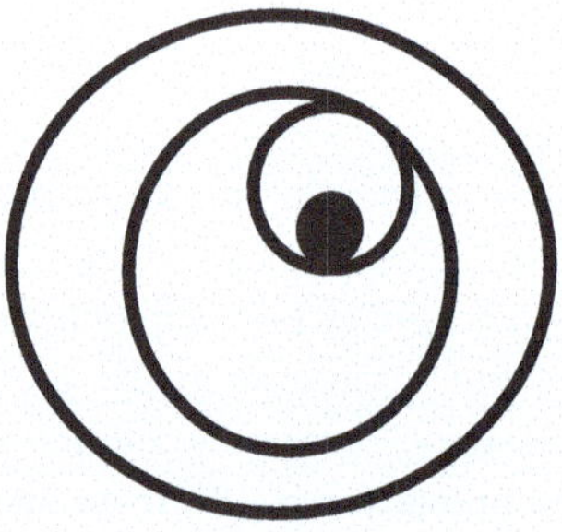

Superficial

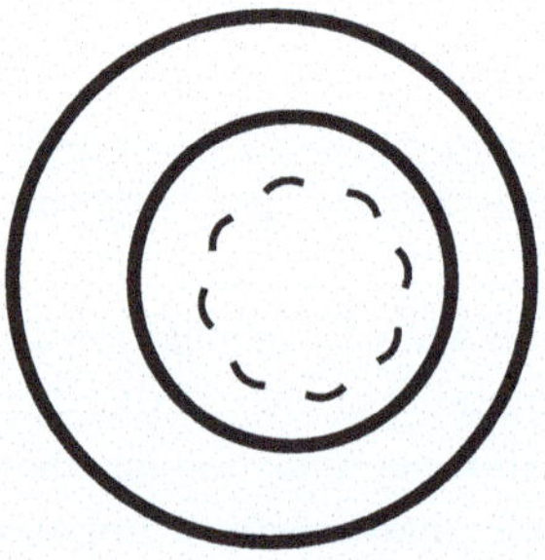

Diffused

4.3.1 Physiology of Implantation

The blastocyst undergoes several changes in orientation and in development pattern, at the time of actual attachment to uterine wall. Thus, the embryo becomes so arranged at the locus of the attachment that its specific characters are defined. But to understand the implantation process we must, not only consider the preparation of the embryo but also that of the uterus itself. Implantation is carried out at a site where interactions between the embryo and the uterus take place.

4.3.2 Preparation of Uterus

The uteri of all mammals before puberty are relatively small due to the embryonic condition of the two fundamental layers. The musculature or myometrium layers and the mucosa or endometrium undergoes changes with the ripening of the first Graafian follicle probably under the influence of stimulants from the anterior lobe of pituitary. A marked hypertrophy of both myometrium and endometrium occurs along with a marked increase in muscular excitability. The endometrial changes consist chiefly of hyperplasia of the epithelium which causes an increase in the size of the mucosal glands and a change in the shape of cells from cuboidal to columnar. Similar changes also take place in the living epithelium as well as in uterine glands. Consequently, there is significant hyperplasia of the connective tissue and an increase in the blood supply and new growth of blood vessels.

The mammalian embryos themselves are much alike up to the implantation period and then begin to diverge. This double parallelism and divergence are itself a definite indicator that the embryo and the uterus are interdependent systems. While both the blastocyst and the uterus are ready, the question is, who is responsible between the uterus and the embryo for the actual attachment of the blastocyst?

There are two contrasting views held regarding the respective roles of the trophoblast and the endometrium each supported by valid experiments. According to one view, the blastocyst is an active agent while the maternal tissue remains passive. Many researchers have supported this view. They regard blastocyst as a parasite invader in the uterine endometrium and the blastocyst literally burrows into the uterine mucosa. In the guinea pig blastocyst, Van Spee in 1901, has described pseudopodia-like projections of the implantation pole penetrating the epithelium and Ashton in 1894 has explained implantation as the effect of expansion of blastocyst in the uterine epithelium. It has been accepted for many years that the implanting blastocyst destroys the uterine epithelium in order to gain access to the underlying tissue.

While studying the implantation in the bat, Wimsatt (1944) concluded that the changes in the epithelium to form a pocket into which the blastocyst implants, are a result of local physiological responses of the uterus to some unknown chemical signals released by the blastocyst. These signals could act locally in the epithelium or may cause local relaxation of uterine musculature. People have even gone so far as to say that the trophoblast secretes enzymes that digest the uterine epithelium. These statements have been widely accepted. Some thought that the attachment was due to some enzyme secreted by the trophoblast. The case which drew Mossman's attention in 1937 was the observation in the Chipmunk (*Tamias striatus*). In this animal, there are several abembryonic trophoblastic 'attachments' which penetrate the uterine epithelium on the antimesemetrial side. There is no suitable explanation given for this.

One of the most convincing pieces of evidence that the embryo is the main factor for implantation of blastocyst was given by Rummer in 1947. He took blastocyst of rats and transplanted them into the anterior chamber of the eye. The blastocysts developed within the eye and underwent all the changes as though they were in a normal uterus. This proves that the embryo is responsible for the attachments of blastocyst.

In Figs. 4.6 and 4.7, it is visible that the blastocyst wall at the abembryonic pole where attachment takes place, producing pseudopods-like protoplasm that begin to touch the uterine wall. They contain an enzyme that acts on the uterine wall and destroys it. The blastocyst probably acts on uterine wall in two ways:

(i) Production of proteolytic enzymes by pseudopodia
(ii) Phagocytosis

These actions are, however, contradictory. The embryo is capable of implanting in any place thereby proving that the embryo is the active agent during implantation.

The second view is that the process of Implantation is controlled by the uterus. Mossman (1937), Wimsatt (1944) and others supported this view. They said that the embryo has nothing to do with the attachment but the change in the process of implantation are innate qualities of uterus activated by endocrine factors. The embryo, therefore, plays no part whatsoever apart from the possible mechanism of irritating the uterine wall at the site of attachment. Blandau in 1947 has shown experimentally in the rat that the uterine mucosa develops all the progestational changes in the presence of artificial ova. He inserted artificial ova made up of glass and paraffin beads into the uterine of rat and mouse. He found that these artificial ova got implanted as though they were normal embryos. In other words, the uterus underwent all the changes necessary for implantation as though they were for normal embryos. The changes in uterus are independent of the presence of ova and are controlled by endocrine secretions.

Both these contrasting views garnered equal support. Bryce in 1935 maintained that there is complimentary activity of both uterus and the blastocyst at the time of attachment.

The exact roles of the blastocyst and the uterine wall during implantation are unknown. It is probable that the egg exerts some influence, both chemical and physical, on the endometrium. It seems reasonable to presume that a cytolytic substance is responsible for the actual erosion at the site of attachment. There is very little proof, however, that the trophoblast does actually secrete cytolytic enzymes.

Lutwakman in 1958 observed that during the initiation of implantation, the blastocyst respiratory activity is highest and consequently there is a sudden outburst of CO2. This CO2 is converted into carbonic acid by the interaction of the fluid of the uterine glands and the carbonic acid acts as a converting agent. It is to be noted that, the endometrium is able to implant an inert object. The ovum is also capable of imbedding itself in extrauterine sites.

4.3.3 Endocrine and Biochemical Aspect

There is a general agreement that the presence of luteal hormone is eventual for the successful implantation of the blastocyst but the mechanism which makes implantation possible is not known. A prominent exception to the generalized fact is that in armadillo, implantation occurs in the absence of hormones.

Corner in 1928 found that progesterone is necessary not only for implantation but also for the nutrition of the blastocyst during the three or four days when they are free in the uterus. In the mouse, Bloch reported in 1939, an osmophilic substance being secreted by the free blastocyst. Markee in 1840 reported an increased blood supply below the uterine epithelium just before the implantation of blastocyst. This ensures an adequate supply of nutritive materials to the endometrium. It is possible that some chemical or hormonal substance is produced by the fertilized ovum which induces the continued activity of the corpus luteum. During the pregestational period, there is a decrease in the muscular activity of the uterus, and this facilitates the retention of the ova in the uterus. It is also found that prolactin and serotonin could also bring about implantation.

Explaining the biochemistry, Lutwakmann described the rabbit blastocyst on sixth or seventh day as a fluid-filled vesicle. When dried on a glass slide, it showed many crystals in the abembryonic region, which was potentially alkaline. Similar crystals were obtained in greater quantity by drying the fluid from the blastocyst. Analysis of the dried blastocyst fluid by X-ray crystallography showed sodium potassium, chlorides in abundance and a small amount of calcium.

Lutwakmann demonstrated that in rat blastocysts there is a high concentration of calcium in bicarbonate form before implantation. He also demonstrated that the uterine endometrium, on the other hand, becomes rich in carbonic anhydrase at the time of implantation. This occurs mostly, intracellularly, in secretory cells and is absent from cell-free body fluids. This enzyme is present mainly in uterine endometrium, placental tissue, fallopian tubes and also in low concentration in the non-pregnant uterus. In hamsters, guinea pigs and rats, the enzyme has been reported from the maternal side of the placenta. In non-pregnant sheep, carbonic anhydrase was found in the uterine mucosa mainly in the intercotyledanary areas and in the uterine region next to the fallopian tube. This persisted after ovariectomy in the uterus.

It is suggested that in the rabbit, carbonic anhydrase contributes to the maintenance of bicarbonate in the blastocyst fluid. This may be an important factor in blastocyst metabolism as indicated by the rapid changes in maternal blood sugar levels. Adams and Lutwakmann in 1956 stated that the carbonic anhydrase levels at different regions of the rabbit endometrium agreed closely with the proliferation site identified by bioassay for progesterone. The enzyme ranges in amount from 10 to 15 IU μ/g in uterine mucosa of non-pregnant rabbit to 78–140 IU μ/g in pregnancy and pseudopregnancy or in response to gonadotrophin or progesterone. This enzyme is responsible for decarbonification in the blastocyst rather it catalyzes both the phases of reversible reaction:

$$H2CO3 \leftrightarrows CO2 + H2O.$$

Using this evidence Boving in 1959 suggested a possible hypothesis to explain the changes preceding implantation:

Progesterone→Carbonic anhydrase→Bicarbonate conversion→Alkalinity→attachment

According to him, progesterone promotes rabbit blastocysts for antimesenterial implantation by augmenting the carbonic anhydrase concentration of the uterine epithelium thereby avoiding the transfer of carbonate from the blastocyst to the maternal circulation by accelerating carbon dioxide ‘blow off’ from carbonic acid. As a consequence of its effects on the equilibrium favouring conversion for the more alkaline Sodium Carbonate so that there is an increase in pH at the site of transfer. The alkalinity thereby renders the blastocyst sticky and favours dissociation of the cellular uterine epithelium thus promoting first adhesion and then penetration.

Interesting observation on the biochemistry of the blastocyst fluid has been demonstrated by Lutwakmann in the unattached blastocyst of the sixth day. There is little N2 and phosphorus containing material, and these do not reach the same level of maternal plasma on the eighth day. The sodium chloride and potassium ions occur in concentrations similar to those of maternal circulation. There is a steady increase of glucose from maternal plasma to the blastocyst by the eighth day. The change in the glucose level in the mother does not relate to the glucose concentration of the un-implanted rabbit blastocyst (5–sixth day) and implanted ones (7–eighth day) and also of the uterine endometrium. In the early pregestational phase the uterine secretion contained 1.3 mg /g of vitamin B12.

The injection of gonadotrophin to immature virgin rabbit causing multiple ovulation raised the vitamin B12 content of the uterus after 2 days to 3 mg/g and after 3 days to as high as 8 mg/g. The amount declined after the fifth day to a level characteristic of the normal luteal phase. All the above chemicals which are of great metabolic value, help in the attachment of the blastocyst by progressively becoming of the same concentration as that of the maternal level during implantation.

Due to wide variations amongst different species, it is not easy to generalize the process involved in the process of implantation. There has been great advancement in the research on implantation among many species especially primates, farm animals and laboratory animal species (Koji, 2018). Hertiget and his co-workers in 1956, studied the human ova in the first 2 weeks and reported the implantation of embryo at the end of the first week to be superficial and maternal endometrium covers it partially. They found that the relative position of the trophoblast to the maternal tissue, influenced activities in both the embryo and the maternal tissue. The direct contact with the maternal tissue, presumably, stimulated the trophoblast cells. They suggested that an induction factor from the endometrium triggered the proliferation and differentiation of blastocyst wall that was in direct contact with the uterus (Hertig et al. 1956). Immunohistochemical studies showed that, from the first trimester, integrin subunits *alpha* 6 and *alpha* 4 are secreted by the villus trophoblast and this probably enables the anchoring of trophoblast to basement membrane (Aplin 1993). Many galectins have been reported to play an important role in the oligosaccharide-based ligand system active at the feto-maternal interface and endometrium, in humans. They have been hypothesized to significantly contribute to receptivity of endometrium and pregnancy physiology (Jeschke et al. 2013).

Dickmann and Noyes in 1960 carried out several embryo transfer studies in rats. They observed that endometrial maturation is a prerequisite before the blastocyst can get attached whereas, the blastocyst may wait till the endometrium matures (Dickmann and Noyes, 1960). Morphological changes, indicating the readiness of uterine wall for implantation include bulging out of the uterine luminal epithelial surface and the appearance of a channel system within the nucleolus of human uterine epithelial cells, which is influenced by progesterone (Nikas 1999). In vitro studies on Human uterine cell lines, indicated that trophoblasts secrete human chorionic gonadotropin (hCG) causing a local surge in its levels as the trophoblast approaches the receptive regions of the endometrium. This results in inducing trophinin on the cell surface of luminal epithelium (Fukuda et al. 1995).

Turco and his co-workers used Human endometrial organoids and experimentally showed that when exposed to human chorionic gonadotropin and human placental lactogen and prolactin, the organoids can be made to acquire characteristics of gestational endometrium (Turco et al. 2017). Kao and his co-workers, carried out gene profiling of human endometrial samples at the time of implantation and showed more than 150 upregulated genes. These genes were related to synthesis of neurotransmitters, prostaglandin, proteoglycan and secretory proteins involved in signal transduction and cholesterol transport. Genes related to synthesis of extracellular matrix components, receptors, numerous immune modulators were also upregulated (Kao et al. 2002). Genes involved in detoxification, and those involved in water and ion transport were also expressed. Trophoblast-originated interferon tau has been identified as a unique interferon that is expressed only by trophectoderm of the blastocyst. It is a significant molecule involved in signalling between embryo and mother to maintain pregnancy (Roberts 1996). This discovery had a significant influence on design and planning of animal breeding systems, as well as in ensuring human health and well-being. With the increasing progesterone levels and additional Interferon-tau from the embryo, at early stages of gestation, the endometrial glands are stimulated to release secretions that are essential for development of the embryo (Spencer 2014).

Molecular signalling has a significant influence on the successful implantation of the blastocyst in the endometrium. Blastocyst adhesion to the uterus in human involves a ligand, trophoblast L-selectin, interacting with its ligand receptor (Genbacev et al. 2003). Several such adhesion molecules have been identified at the site of implantation like glycodelin, tumour necrosis factor-alpha (TNF-alpha), Interluekin-1, Leptin and Insulin-like growth factor (IGF) (Van Mourik, et al. 2009). The specific molecular pathway, which involves these signalling molecules, leading to the blastocyst adhesion remains to be unravelled. Xie and his co-workers showed in experimental mice, the significance of Heparin Binding Epidermal Growth Factor (HB-EGF) in ovarian estrogen secretion and also in the uterine–blastocyst interactions (Xie et al. 2007). Chakraborty and his co-workers, demonstrated the changes in COX1 and COX2 gene expression during implantation (Chakraborty et al. 1996) and later Lim and co-workers showed that the disruption of COX2 resulted in failure of female reproduction including implantation (Lim et al. 1997). The presence of high mobility group protein 17 (hmg17) which can affect several DNA-dependent activities was demonstrated by Li and co-workers (Li et al. 2017). The presence of this protein in the luminal epithelium can significantly influence the dynamic changes that occur in the endometrium. Monsivaias and co-workers showed the role of molecular signalling in the transition of non-receptive uterus to receptive uterus and elucidated the role of ALK3 (Activin-like kinase 3) in uterine receptivity (Monsivaias et al. 2015). The progesterone action in uterus is mediated through progesterone receptors and this molecular signalling significantly influences the uterine response to progesterone. Progesterone also has effect on the maternal immune cells and can change their character during pregnancy. The molecular aspects that lead to the change in character of immune cells and their significance in early gestation for uterine transformation and implantation, remain an enigma (Koji 2018).

From the foregoing discussion, it can be said that implantation is as yet an imperfectly understood process. From the available experimental evidence, it can be said that blastocyst and the endometrium both have to react to stimulate and initiate the process of implantation independently. However, at present it can only be said that for perfection of the process both are equally important and are complementary to each other.

4.4 Placentation

Bats provide a very fascinating aspect of embryonic development with different bat species showing different types of placentas enabling an interesting comparison of placental biology between species. During development, in most bats, three types of placentas occur in chronological sequence and the same uterus may show more than one type of placenta in different regions of the uterus.

A) Trophoblastic Placenta

The initiation of establishing the bridge between maternal and foetal environment is marked by the proliferation of trophoblast cells surrounding the blastocyst to produce a syncytium of trophoblast cells recognizable as the syncytiotrophoblast. Soon after the implantation of the blastocyst, the syncytiotrophoblast extends and invades the endometrium to form an attachment with the uterine wall. At tissue level, trophoblastic placenta shows a mass of syncytiotrophoblast that surrounds the maternal blood capillaries after having destroyed the connective tissue of the endometrium. The sides of the blastocyst that is involved in the invasion by the syncytiotrophoblast decide the nature of attachment of the blastocyst, based on which the trophoblastic placenta develops. In bats like *Rousettus, Cynopterus*, *Taphozous*, *Megaderma*, *Hipposideros* and *Tadarida*, the trophoblastic placenta develops on all sides of the implantation site and appears like a spherical shell. In other bats like *Pteropus*, *Noctilio labialis minor* (Anderso and Wimsatt 1963) and all Vespertilionids, the trophoblastic placenta develops only in the embryonic region of the implantation chamber. The implantation is interstitial in D*esmodus rotundus murinus* (Wimsatt 1954), all Phyllostomids (Hamlett 1935; Rasweiler IV 1974) and it is superficial in the bat, *Miniopterus schreibersii.* In these bat species though the entire blastocyst wall is in contact with the endometrium, only the trophoblast at the embryonic region invades the endometrium. With the involvement of mesoderm in placentation, the trophoblastic placenta is converted into the chorionic placenta. This structure is limited to the dorsal side of the embryo due to the hindrance by the yolk sac wall on the other side.

B) Yolk Sac Placenta

As the yolk sac wall meets the uterine wall, the yolk sac placenta is established. This placenta is initially non-vascular but as the vascularisation extends into the yolk sac wall, the placenta gets converted into the chorio-vitelline placenta. In the bats, as the exocoelom extends into the yolk sac wall, the chorion and the vascular splanchnopleure get separated. This causes the disappearance of chorio-vitelline placenta from the embryonic region. In fact, in *Desmodus rotundus murinus* and Phyllostomids, the chorio-vitelline placenta is not formed because the extraembryonic mesoderm does not extend beyond the edges of the placental disc (Wimsatt 1954).

C) Chorio-allantoic placenta

In bats, as the allantois grows, it carries with it the foetal blood vessels and finally meets the existing chorionic placenta on the dorsal side of the embryo. The final placenta in bats is formed on the side opposite to where the yolk sac placenta is formed. In *Tadarida brasiliensis cynocephala* and *Tadarida aegypiaca*, however, a thick pre-placental pad is formed on the mesometrial side where the blastocyst had made its first contact with the uterine wall (Stephens 1962). As the allantois grows, it causes the separation of splanchnopleure of the yolk sac from the chorion, and the allantois extends beyond the exocoelom to attach with the placental pad. In Megachiropteran bats, and in *Rhinopoma*, *Taphozous*, *Megaderma*, *Rhinolophus*, Hipposiderids and *Tadarida*, during early stages of development, the allantois spreads across the embryo but as the placenta develops into a discoidal structure the allantoic vesicle shrinks or disappear completely.

Figure 4.8 shows the cross section of uterus of a pregnant bat showing the chorioallantoic placenta with the arrangements of foetal membranes. Endometrium has two regions the decidua basalis, and the decidua parietalis. The decidua basalis consists of para-placental layer which is more superficial and the deeper zone which is wider and makes up the rest of the basalis layer. The deeper region gets reduced as the gestation advances due to significant cell deaths occurring between the two zones of basalis. The placental tubule encloses a maternal capillary and is lined by two layers of trophoblast. The allantois arises from the posterior region of the embryo and extends as a diverticulum. The yolk sac is seen on the mesometrial side while the uterine glands are seen anti-mesometrially within the decidua parietalis.

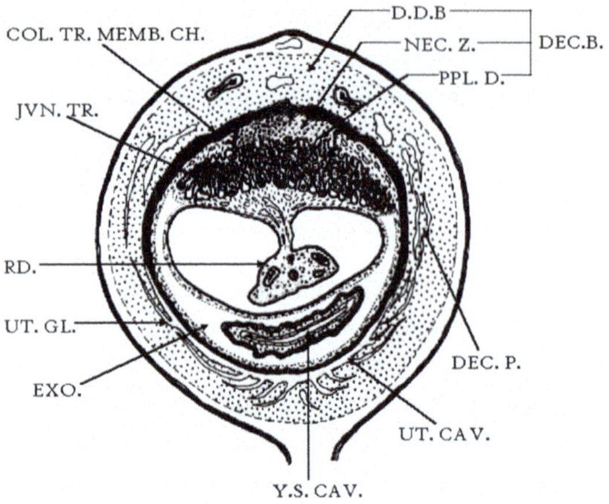

Fig. 4.8 Diagrammatic cross section of the pregnant uterus of bat showing the chorioallantoic placenta and the definitive arrangement of the foetal membranes. *D.D. B.*, deeper zone of decidua basalis; *NEC.Z* necrotic zone; *DEC.P* decidua parietalis; *UT. CAV* uterine cavity; *COL. TR.MEMB.CH* columnar trophoblast, of membranous chorion; *SYN.TR* syncytial trophoblast; *A.D* allantoic diverticulum; *UT.GL* uterine gland; *EXO* exocoelom; *Y.S.CAV* yolk sac cavity

D) Histogenesis of the placenta

As the placenta develops, in bats, four distinct stages can be identified though the stages may overlap considerably. Several workers have reported the histological changes that take place in the placenta of various species of bats. Development of haemochorial placenta has been reported in *Desmodus rotundus murinus* (Wimsatt 1954; Bjorkman and Wimsatt 1968), *Myotis lucifugus lucifugus* (Wimsatt 1945a, b; Enders and Wimsatt 1968), *Scotophilu stemmincki* (Gopalakrishna 1950a) and *Tadarida brasiliensis cynocephala* (Stephens 1962, 1969). In *Hipposideros bicolor pallidus* (Gopalakrishna and Moghe 1960) and *Megaderma lyra lyra* (Gopalakrishna and Khaparde 1978a, b), development of endotheliochorial placenta has been reported. The distinctive feature that separates the development of these two placental types is the fate of endothelium of the maternal blood vessels; whether it is lost or persisting.

The placental histogenesis commences with the blastocyst implantation and the first stage is completed with the establishment of trophoblast placenta. Figure 4.9a, b and c, gives a schematic representation of the development of placental tubules in three stages as it progresses. In the first stage, the trophoblast layer of the embryo proliferates starts invading the uterine epithelium. The invading trophoblast cells are in syncytial layer and cause histolysis of the decidual layer. The syncytial trophoblast layer proliferates and surrounds the maternal blood capillaries and within a short span of time after implantation, most of the region of decidua basalis is occupied by the trophoblast. The invading trophoblast layer soon forms a shell around the embryo. It has been observed that the activity of trophoblast layer is not uniform around the embryo. In Vespertilionids (Wimsatt 1945a, b; Gopalakrishna 1949) and *Desmodus rotundus murinus* (Wimsatt 1954), the trophoblast invasion is noticeably more along the embryonic pole of the blastocyst. In bats with circumferential implantation, the activity of trophoblast is uniform around the blastocyst with comparatively higher invasion seen along the embryonic pole region. In *Tadarida brasiliensis cynocephala* (Stephens 1962) and *Tadarida aegyptiaca*, however, the trophoblasts are more active in the abembryonic side of the blastocyst (Fig. 4.9a).

The second stage of placental histogenesis is marked by the proliferation of the basal cytotrophoblast cells that border the syncytiotrophoblast layer. These cytotrophoblast cells proliferate and start growing into finger-like solid columns penetrating the mass of syncytiotrophoblast layer in front of them. This stage of histogenesis ends with extensive ingrowth of cords of cytotrophoblast cells deep into the syncytiotrophoblast layer (Fig. 4.9b).

In the third stage, the cytotrophoblastic cords hollow out from the inside at the foetal end. The foetal mesenchyme starts entering these hollows from the foetal side. Due to the entry of the foetal mesenchyme cells, hollows within the cytotrophoblast cords become wider. In cross sections, the maternal vascular channels appear as tubules hanging from the uterine wall. The maternal vascular channels appear with their surrounding endothelium around which the syncytiotrophoblast forms a layer. The syncytiotrophoblast is encircled by the

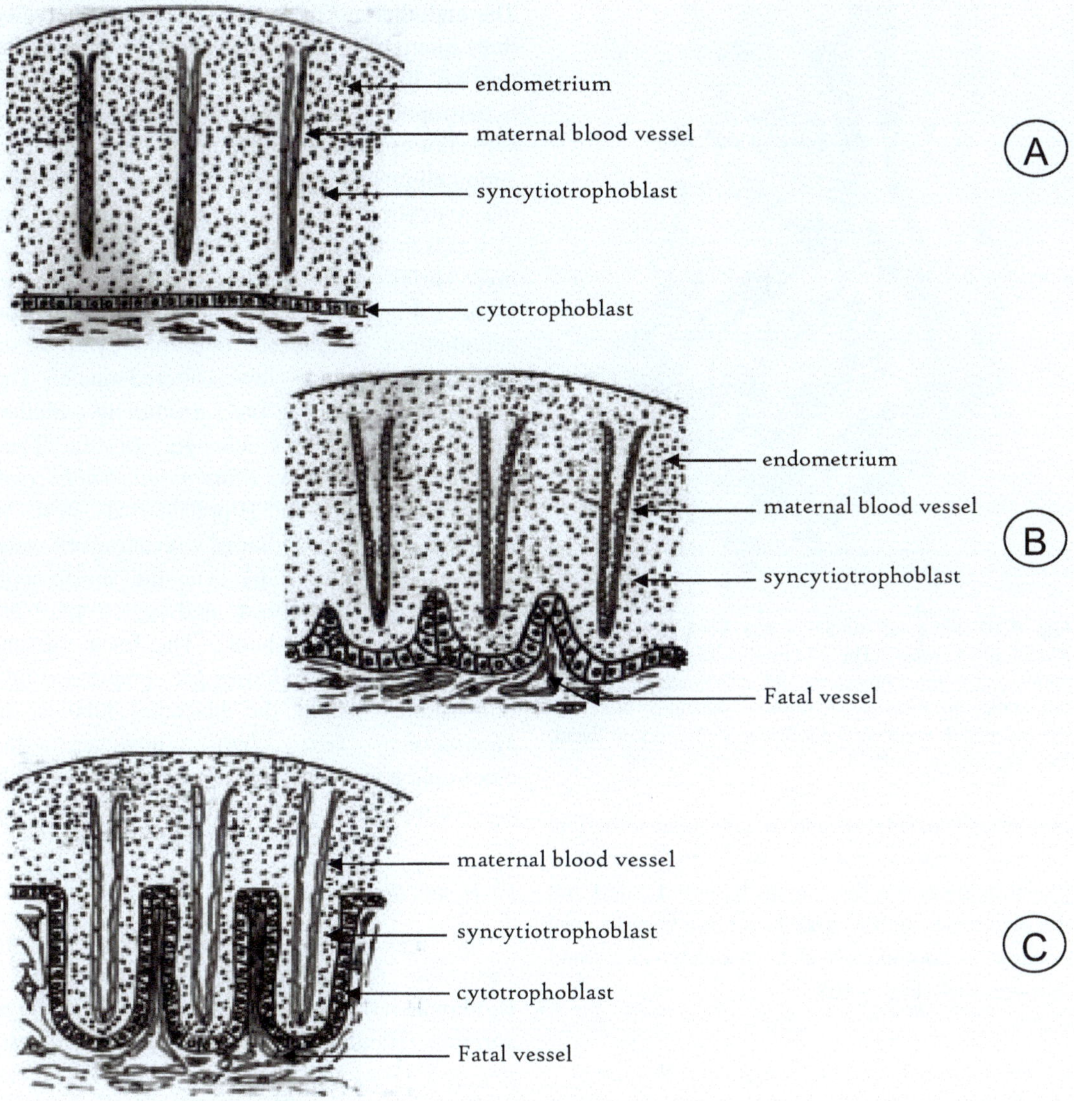

Fig. 4.9 Schematic representation of the development of placental tubules (**a**—Stage 1, **b**—Stage 2, and **c**—Stage 3 as per text)

cytotrophoblast and then by the mesenchyme cells. Towards the end of the third stage, the foetal capillaries start entering the mesenchymal cores within the cytotrophoblast cords. This marks the initiation of foetal vascularization of the placenta (Fig. 4.9c). The placental tubules can now be recognized as a tier of cellular layers starting from the maternal blood followed in sequence with maternal endothelium, syncytiotrophoblast, cytotrophoblast, foetal mesenchyme, foetal endothelium and foetal blood. The placental tubules extended and started branching and join with neighbouring tubules to form a network to eventually develop into the placental labyrinth.

The histological changes that take place within the placental labyrinth concern the establishment of the definitive structure of the placenta. The definitive structure of the placenta is based on the fate of the maternal endothelium within the placental labyrinth. Based on these changes, the

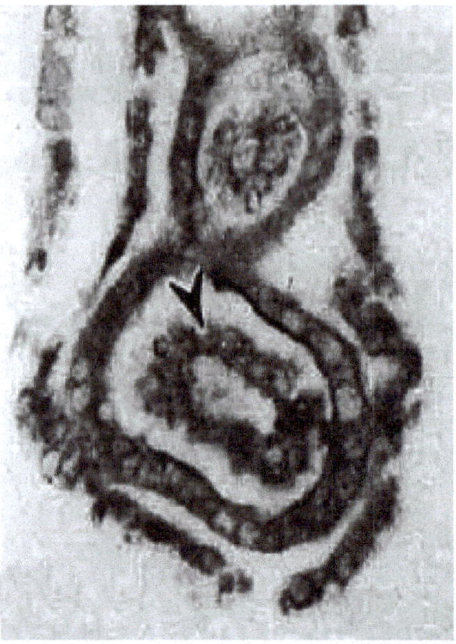

Fig. 4.10 Placental tubule in transverse section from *Rhinolophus rouxi*. The chorioallantoic placenta has undergone extreme branching and anastomosis of placental tubules to form a labyrinthine structure. Note the hypertrophied maternal endothelial cells (arrow head) [X 600]. Bhiwgade 1977

definitive placental structure gets determined in various bat species. Studies on the final nature of placenta in various bat species have indicated that the placentae can be divided into two broad types, namely, endotheliochorial (vasochorial), and hemochorial (Fig. 4.10).

E) Endotheliochorial or Vasochorial Placenta

In bats *Pteropus giganteus*, *Cynopterus sphinx*, *Rhinopoma kinneari* (Srivastava 1952; Gopalakrishna 1958), *Taphozous longimanus* (Gopalakrishna 1958), *Taphozous melanopogon*, *Noctilio labialis minor* (Anderso and Wimsatt 1963), *Megaderma lyra. Lyra* (Wimsatt 1958; Gopalakrishna 1958; Gopalakrishna and Khaparde 1978a, b), *Rhinolophus rouxi* (Gopalakrishna and Bhiwgade 1974; Bhiwgade 1977) and all Hipposiderid bats (Gopalakrishna 1958; Gopalakrishna and Moghe 1960; Karim 1972a, b), the maternal endothelial cells are reported to be persistent till the full term. These bats, therefore, show endotheliochorial placenta. The endothelium in these bats, during the early three-quarters of the gestation, is seen as a distinctive lining made up of closely arranged hypertrophied cells. At this time, the inner syncytiotrophoblast surrounding the maternal endothelium and the cytotrophoblast layer around the syncytiotrophoblast is clearly discernible. Towards final quarter of gestation, the endothelial layer shrinks with its cells becoming lesser in number (Fig. 4.10). Little ahead of the birth, the endothelium is reduced to a thin cytoplasmic lamina which shows few scattered nuclei. The trophoblastic covering the surrounding endothelium also undergoes change. In the bats, *Pteropus, Cynopterus*, *Rhinopoma*, *Taphozous*, *Rhinolophus rouxi* and Hipposiderieds, near the term, the maternal capillaries with shrunken cells, appear to be covered by syncytial sheath with both syncytiotrophoblast and cytotrophoblast becoming indistinguishable. The foetal mesenchyme and foetal capillaries get compressed into thin strands within the placental tubules. In *Megaderma lyra lyra,* however, the cytotrophoblast cells transform into giant cells and are incorporated into the syncytial layer.

F) *Hemochorial placenta*

In the bats, *Rousettus leschenaultia, Desmodus rotundus murinus* (Wimsatt 1954; Bjorkman and Wimsatt 1968), *Glossophaga soricine* (Hamlett 1935), *Artibeus jamaicensis parvipes* (Wislocki and Fawcett 1941), *Myotis lucifugus lucifugus* (Wimsatt 1965; Enders and Wimsatt 1968), *Vesperugo leisleri* (Ramaswami 1933), *Scotophilus temmincki* (Gopalakrishna 1950a, b), *Pipistrellus* sp. (Phansalkar 1972; Gopalakrishna and Sapkal 1974; Gopalakrishna and Karim 1972), *Tadarida brasiliensis cynocephala* (Stephens 1962, 1969) and *Tadarida aegyptiaca* during the early development itself, the maternal endothelium is lost from the capillaries in the placenta, distinguishing it into a haemochorial placenta. In the haemochorial placenta of *Desmodus rotundus murinus*, *Myotis lucifugus lucifugus* (Enders and Wimsatt 1968), *Scotophilus heathi* (Ramakrishna and Madhavan 1977), *Pipistrellus ceylonicus*

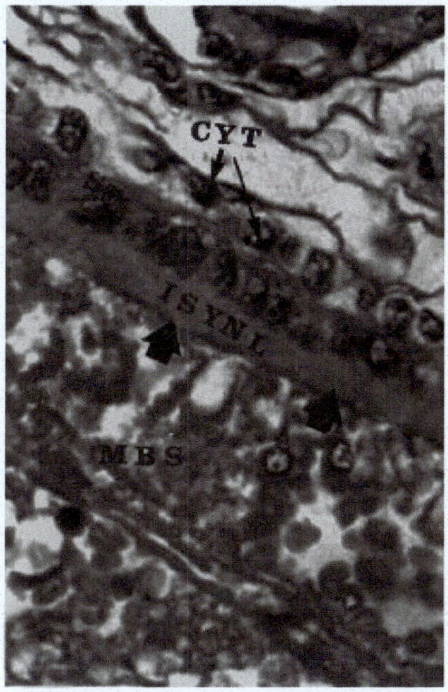

Fig. 4.11 Hemodichorial interhaemal membrane in *Rousettus leschenaulti* at mid-term of gestation. Note the maternal blood space (MBS) is bordered by the Intrasyncytial lamina (ISYNL), syncytiotrophoblast (SYN) and cytotrophoblast (CYT). Haematoxylin and Eosin stain [X640]. Bhiwgade et al. 2000

chrysothrix (Phansalkar 1972), *Pipistrellus mimus mimus* (Gopalakrishna and Karim 1972) and *Rousettus leschenaulti* (Bhiwgade et al. 2000) both syncytiotrophoblast and cytotrophoblast are seen and therefore, the placenta is classified as hemodichorial (Fig. 4.11). In the bats, *Tadarida brasiliensis cynocephala* (Stephens 1962, 1969), *Tadarida aegyptiaca* and *Pipistrellus dormer* (Gopalakrishna and Sapkal 1974), only one of the trophoblasts is present distinguishing the placenta as hemomonochorial.

G) Interstitial Membrane

A distinctive feature that is observed in bat placenta is the interstitial membrane which is situated between the endothelium and the syncytiotrophoblast in endotheliochorial placenta and is seen embedded within the cytoplasmic lamina around the maternal blood sinusoids in hemochorial placenta. It is discontinuous with a network-like appearance and its cytoplasm shows continuity with the syncytiotrophoblast or the cytoplasmic lamina (Karim et al. 1978). It was Wimsatt (1958) who first reported the presence of interstitial membrane in placental tubules of bats. According to him, the interstitial membrane arises either wholly or partly from the syncytiotrophoblast. There was also another opinion that the interstitial membrane could be remain of the endothelium that gets greatly reduced as the placenta develops. The interstitial membrane increases in thickness with the dilation of the maternal blood spaces as the placenta develops. Additionally, the interstitial membrane is always in close proximity with the syncytiotrophoblast. This could suggest the syncytiotrophoblast could be the origin of interstitial membrane. There is, however, another possibility that the basement membrane of the shrunken maternal endothelium could also get incorporated into the interstitial membrane. Thus, interstitial membrane could have a dual origin with the trophoblast being one of the contributors (Enders and Wimsatt 1968; Bjorkman and Wimsatt 1968).

In the bat, *Myotis lucifugus lucifugus,* it has been reported that the syncytiotrophoblast gives out cytoplasmic processes that penetrate the basement membrane of maternal blood vessels and spread below the maternal endothelium. This causes the maternal endothelium to loosen from its basement membrane and separates the endothelial cells, which are later lost. Thus, the basement membrane of the maternal blood vessels, now gets a covering of cytoplasmic extensions from the syncytiotrophoblast. The basement membrane becomes an intrasyncytial membrane with perforations at places through which the cytoplasmic extensions of the syncytiotrophoblast spreads over it like a continuous layer. Stephens (1969) reports that in *Tadarida brasiliensis cynocephala*, there is a discontinuous homogenous layer over which the trophoblastic cytoplasm forms a layer that is in direct contact with the maternal blood. This homogenous layer can also be considered as the interstitial membrane described by Wimsatt (1958), Bjorkman and Wimsatt (1968) and Enders and Wimsatt (1968).

H) Accessary Placental Structures

Different placental structures other than the placenta, arising from the chorioallantoic membrane have been described in bats. In *Myotis lucifugus lucifugus* (Wimsatt 1945a, b) and *Megaderma lyra* (Gopalakrishna and Khaparde 1978a, b), a villous-like syndesmochorial placenta situated on the maternal side of the main placenta has been reported. In hipposiderid and molossid bats, the parietal layers of the chorion are in direct contact with the endometrium. These structures could also be playing a vital role in the exchange of materials between the maternal and foetal environments. In *Miniopterus schreibersii fuliginosus,* the main placental disc shows a smaller disc inside, with a cavity filled with maternal blood. The labyrinth wall is made of trophoblastic cells that are large, spongy in nature, and is PAS positive in nature. The relationship of this inner disc with the main placenta is still to be explored. Gopalakrishna (1958) and Wimsatt and Gopalakrishna (1958) have reported the presence of a sac-like haematoma at the mesometrial side of the main placenta. This haematoma develops in the second half of the pregnancy. Numerous fold-like extensions from the chorion invade this haematoma. The chorionic cells are seen with maternal erythrocytes that are actively ingested from the maternal blood in the haematoma.

If one looks at the histogenesis in the placenta, it is evident that the change in the basic placental structure progressively evolves into the one that brings the foetal capillaries as close to the maternal blood as possible with progressive reduction in the number and type of intervening layers of different tissues. This could be viewed as evolutionary adaptations to sustain growth of the embryo within the mother who may be small in body size, especially like the insectivorous bats. One could link these adaptations with the gestation period of the species, the cyclicity of breeding, food sources and the seasonality of their availability and the body weight of mother. This progressive breakdown of placental barriers is enlisted in Table 4.1 where different placental types are arranged in order to emphasize this aspect.

Table 4.1 Minimal interhaemal barrier seen in some discoidal chorioallantoic placenta

Type of barrier in definitive placenta	Species	Structural components
Endotheliomonochorial	• *Sancus murinus* (Asian house shrew). (Bhiwgade et al. 2000)	Hypertrophied maternal endothelium forms the lining of placental labyrinth. Basal lamina is mesh-like and is penetrated, at several places, by processes of maternal endothelial cells and trophoblast. Trophoblast is multinucleated and they form the basal lamina. Foetal endothelial cells lack a basal lamina.
Endotheliomonochorial	• *Taphozous melanopogon.* Family: Emballonuridae (Rohatgi et al. 1992) • *Hipposideros Speoris.* Family: Hipposideridae (Kothari and Bhiwgade 1992) • Lesser mouse-tailed bat (*Rhinopoma hardwickei*) *Rhinopoma hardwickei hardwickei* Family: Rhinopomatidae (Panse, Mandal, Bhiwgade 1992). • Secondary placenta of *Miniopterus schreibersii fuliginosus.* Family: Miniopteridae	Maternal endothelium is present with a single layer of trophoblast. The hypertrophied maternal endothelium forms the lining of placental labyrinth. The basal lamina is thick (interstitial membrane). Syncytiotrophoblast is present along with basal lamina and foetal endothelium.
Endotheliodichorial	• *Rhinnopoma micropyllum*Family: Rhinopomatide (Panse, Mandal and Bhiwgade 1992). • *Megaderma lyra lyra.* Family: Megadermatide • *Rhinolophus rouxi.* Family: Rhinolophidae • *Hipposideros lankadiva.* Family: Hipposideridae (Kothari and Bhiwgade 1992) • *Hipposideros fulvs fulvs.* Family: Hipposideridae	Hypertrophied maternal endothelium forms the lining of placental labyrinth; basal lamina (interstitial membrane) is thick; syncytiotrophoblast and cytotrophoblast are present with single or double basal lamina and foetal endothelium.
Haemomonochorial	Human and Bonet monkey	Syncytiotrophoblast forms the lining of intervillous space; scattered cytotrophoblast cells are seen; basal lamina may be single or double with foetal endothelium.
Haemomonochorial	• *Tadarida brasiliensis cynocephala*Family: Molossidae (Stephens 1969). • *Molossus ater*Family: Molossidae (Rasweiler IV 1991a, b). • *Chaerephon plicata*Family: Molossidae.	Highly interdigitated cytotrophoblast with giant cells lining the placental labyrinth. Basal lamina is single or double with foetal endothelium.
Haemodichorial	• Short-tailed fruit bat. (*Carollia perspicillata*)	Syncytiotrophoblast forms the lining of placenta labyrinth as a discontinuous intrasyncytial lamina; cytotrophoblast is often shrunken with many interdigitating microvilli; basal lamina is single or double with foetal endothelium.

(continued)

Table 4.1 (continued)

Haemomonochorial	Human and Bonet monkey	Syncytiotrophoblast forms the lining of intervillous space; scattered cytotrophoblast cells are seen; basal lamina may be single or double with foetal endothelium.
Haemodichorial	• Rabbit (*Orcytolagus cuniculus*). • *Macrotus waterhousii.* Family: Phyllostomatidae (Bodley 1974) • *Desmodus rotundus murinus.* Family: Desmodontidae (Bjorkmman and Wimsatt 1968) • *Myotis lucifugus lucifugus.* Family: Vespertilionidae (Enders and Wimsatt 1968) • *Scotophilus heathi.* Family: Vespertilionidae • *Cynopterus sprinx.* Family: Pteropidae (Bhiwgade et al. 2000). • *Pteropus, giganteus.* (Karim and Bhatnagar 1996). • *Rousettus leschenaulti.* Family: Pteropidae Karim et al. 1978) • *Thryoptera tricolorspix.* Family: Thyropteridae (Wimsatt and Enders 1980) • *Tylonycterus pachypus* • Primary and tertiary placenta of *Miniopterus schreibersii fuliginosus* Family: Miniopteridae • *Rhinopoma hardwickei hardwickei* family: Rhinopomatidae. (Panse, Mandal, and Bhiwgade 1992) • Secondary placenta of *Miniopterus scheibersii fuliginosus.* Family: Miniopteridae	The placental labyrinth is lined with an outer layer of syncytiotrophoblast, with pores and sometimes highly reduced. The inner layer of trophoblast, however, varies from cellular to multinucleate or syncytial in nature. Basal lamina is formed by trophoblast with foetal endothelium.
Haemotrichorial	• Rat (*Rattus norvegicus*), mouse (*Mus musculus*), hamster (*Cricetus auratus*). Bandicoot (*Bandicoota bengalensis*)	The placental labyrinth is lined by cytotrophoblast layer which has pores. The middle layer is of syncytiotrophoblast while the inner layer of trophoblast appears syncytial. Basal laminae are of trophoblast with foetal endothelium.
Chorioallantoic placental barrier	• Australian bandicoots (*Isodon macrourus*, *Parameles nasuta*).	Maternal endothelium forms the basal lamina with thin endometrial stroma; shrunken portions of syncytial cells ('heterokaryons') are seen probably formed by the fusion of uterine epithelial and trophoblast cells or foetal endothelium.

References

Anderso JW, Wimsatt WA (1963) Placentation and foetal membranes of the central American noctilionid bat, *Noctilio labialis minor*. Am J Anat 112:181–202

Aplin JD (1989) Cellular biochemistry of the endometrium. In: Wynn RM, Jollie WP (eds) Diolngy of the uterus, 2nd edn. Plenum Medical Book Company, New York, pp 89–129

Badwaik NK, Rasweiler JJIV, Oliveira SF (1997) Formation of reticulated endoderm, Reichert's membrane, and amniogenesis in blastocysts of captive-bred, short-tailed fruit bats, *Carolliaperspicillata*. Anat Rec 247:85–101

Bhiwgade DA (1976) Observations on some early stages of development and implantation of the blastocyst of *Rhinolopusrouxi* (Temminck). Proc Indian Acad Sci B 84:201–209

Bhiwgade DA (1977) Development of the foetal membranes in the Indian horse-shoe bat, *Rhinolophus rouxi*(Temminck). Proc Indian Acad Sci B 86:61–72

Bhiwgade DA (1979) J Anat 128(2):349–364

Bhiwgade DA, Taskar AS, Bhatnagar KP (2000) Acta Chiropterologica 2(1):107–122

Bjorkman NH, Wimsatt WA (1968) The allantoic placenta of the vampire bat (*Desmodusrotundus murinus*): a reinterpretation of its structure based on electron microscopic observations. Anat Rec 162:8397

Bodley HD (1974) Ultrastructural development of the chorio-allantoic placental barrier in the bat, *Macrotuswaterhousii*. Anat Rec 180:351–368

Bonilla HD, Rasweiler JJ IV (1974) Anat Rec 179:385

Chakraborty I, Das SK, Wang J, Dey SK (1996) Developmental expression of the cyclo-oxygenase-1 and cyclo-oxygenase-2 genes in the peri-implantation mouse uterus and their differential regulation by the blastocyst and ovarian steroids. J Mol Endocrinol 16(2):107–122

Dickmann Z, Noyes RW (1960) The fate of ova transferred into the uterus of the rat. Reproduction 1(2): 197–212

Enders AC, Wimsatt WA (1968) Formation and structure of the hemodichorialchorioallantoic placenta of the bat, *myotis lucifuguslucifugus*. Am J Anat 122:453–489

Fukuda MN, Sato T, Nakayama J, Klier G, Mikami M, Aoki D, Nozawa S (1995) Trophinin and tastin, a novel cell adhesion molecule complex with potential involvement in embryo implantation. Genes Dev 9(10):1199–1210

Genbacev OD, Prakobphol A, Foulk RA, Krtolica AR, Ilic D, Singer MS, Yang Z-Q, Kiessling LL, Rosen SD, Fisher SJ (2003) Trophoblast L-selectin-mediated adhesion at the maternal-fetal interface. Science 299(5605):405–408

Gopalakrishna A (1949) Studies on the embryology of Microchiroptera, part IV: an analysis of implantation and early development in *Scotophiluswroughtoni* (Thomas). Proc Ind Acad Sci 30:226–242

Gopalakrishna A (1950a) Studies on the embryology of Microchiroptera, part V: placentation in the vespertilionid bat, *Scotophiluswroughtoni*(Thomas). Proc Ind Acad Sci 31:235–251

Gopalakrishna A (1950b) Studies on the embryology of Microchiroptera, part VI: structure of the placenta in the Indian vampire bat, Lyroderma lyra lyra (Geoffroy). (Megadermatidae). Proc Ind Acad Sci 16: 93–98

Gopalakrishna A (1958) Foetal membranes in some Indian Microchiroptera. J Morph 102:157–197

Gopalakrishna A, Bhiwgade DA (1974) Foetal membranes in the Indian horse-shoe bat, Rhinolophus rouxi (Temminck). Curr Sci 43:516

Gopalakrishna A, Karim KB (1972) Ibid. 41:144

Gopalakrishna A, Karim KB (1980) Female genital anatomy and the morphogenesis of fetal membranes of Chiroptera and their bearing on the phyloge-netic relationships of the group. Golden Jubilee volume of the National Academy of Sciences, Allahabad, pp 379–428

Gopalakrishna A, Karim KB, Rajgopal G (1974) Post-ovulatory changes in the eggs of some Indian bats. Curr Sci 43:454–455

Gopalakrishna A, Khaparde MS (1978a) Development of the foetal membranes and placentation in the Indian false vampire bat, *Megaderma lyralyra* (Geoffroy). Proc Ind Acad Sci 87B:179–194

Gopalakrishna A, Khaparde MS (1978b) Early development, implantation and amniogenesis in the Indian false vampire bat, *Megaderma lyra lyra* (Geoffroy). Proc Ind Acad Sci 87B:91–104

Gopalakrishna A, Moghe MA (1960) Developmenl of the foetal membranes in the Indian leaf nosed bat, *Hipposiderosbicolor pallidus*. Z Anat Entw Gesch 122:137–149

Gopalakrishna A, Sapkal VM (1974) Foetal membranes in the Indian pipistrellid, *Pipisirellusdormeri* (Dobson). J zool Soc India 26:1–9

Hamlett GWD (1935) Notes on the embryology of a phyllostomid bat. Am J Anat 56:327–353

Heideman PD (1989) Delayed development in Fischer's pygmy fruit bats, *Haplonycterisjischeri*, in the Philippines. J Reprod Fertil 85:363–382

Hertig AT, Rock J, Adams FC (1956) A description of 34 human ova within the first 17 days of development. Am J Anat 98(3):435–493

Jeschke U, Hutter S, Heublein S, Vrekoussis T, Andergassen U, Unverdorben L, Papadakis G, Makrigiannakis A (2013) Expression and function of galectins in the endometrium and at the human feto-maternal interface. Placenta 34(10):863–872

Kao LC, Tulac S, Lobo S, Imani B, Yang JP, Germeyer A, Osteen K, Taylor RN, Lessey BA, Giudice LC (2002) Global gene profiling in human endometrium during the window of implantation. Endocrinology 143(6): 2119–2138

Karim KB (1972a) Foetal membranes and placentation in the Indian leaf-nosed bat. *Hipposideros fulvus fulvus* (Gray). Proc Ind Acad Sci 76B:71–78

Karim KB (1972b) J Zool Soc India 24(2):135

Karim KB (1975) Early development of the embryo and implantation in the Indian vespertilionid bat, *pipistrellus mimusmimus* (Wroughton). J Zool Soc India 27:119–136
Karim KB (1976) Embryology of some Indian Chiroptera. DSc dissertation. Nagpur University, India
Karim KB, Bhatnagar KP (1996) Observations on the chorioallantoic placenta of the Indian flying fox, *Pteropus giganteus giganttus*. Ann Anat 178:523–530
Karim KB, Wimsatt WA, Gopalakrishna A (1978) Structure of the definitive placenta in the Indian bat, *Rousettus leschenaulti* (Pteropidae). Anat Rec 190:438
Koji Y (2018) Biol Reprod 99(1):175–195
Kothari A, Bhiwgade DA (1992) Ultrastructural studies of interhemal membrane in three species of hipposiderid bats. Proceedings of the Ninth International Bat Research Conference. Madurai, India, 54:134
Krutzsch PH, Crichton EG (1991) Fertilization in bats. In: Dunbar BS, O'Rand M (eds) A comparative overview of mammalian fertilization. Plenum Press, New York, pp 137–139
Li DD, Yue L, Yang ZQ, Zheng L-W, Guo B (2017) Evidence for Hmgn2 involvement in mouse embryo implantation and decidualization. Cell Physiol Biochem 44:1681–1695
Lim H, Paria BC, Das SK, Langenbach R, Trzaskos JM, Dey SK (1997) Multiple female reproductive failures in cyclooxygenase 2-deficient mice. Cell 91(2): 197–208
Martin L, Towers PA, McGuckin MA, Little L, Luckoff H, Blackshaw AW (1987) Reproductive biology of flying foxes (Chiroptera: Pteropodidae). Australian Mammalogy 10(115–18)
Monsivaias D, Clementi C, Peng J, Titus MM, Barrish JP, Creighton CJ, Lydon JP, DeMayo FJ, Matzuk MM (2015) Uterine ALK3 is essential during the window of implantation. Proc Natl Acad Sci U S A 112:E387–E395
Mori T, Uchida TA (1981) Ultrastructural observations of fertilization in the Japanese long-fingered bat *Miniopterusschreibersiifuliginosus*. J Reprod Fertil 63:231–235
Mossman HW (1937) Comparative morphogenesis of the fetal membranes and accessory uterine structures. Contrib Embryol, Carnegie Inst Washington 26:129–246
Nikas G (1999) Cell-surface morphological events relevant to human implantation. Hum Reprod 14(Suppl 2): 37–44
Phansalkar RB (1972) Early development and placentation in the vespertilionid bat, pipistrellus ceylonicuschrysothrix., Ph.D. thesis,. Nagpur University
Potts DM, Racey PA (1971) A light and electron microscope study of early development in the bat, *Pipistrellus pipistrellus*. Micron 2:322–348
Qivntero F, Rasweiler JJIV (1974) Ovulation and early embryonic development in the capt've vampire bat, *Desmcdusrotundus*. J Reprod Fert 41:265–273
Ramakrishna PA, Madhavan A (1977) Foetal membranes and placentation in the vespertilionid bat *Scotophilia heathi* (Horsefield). Proc Indian Acat Sci B 86:117–126
Ramaswami LS (1933) Some stages of the placentation in *Vesperugoleisleri* (Kuhl). Hatf-yrly Mysore Univ 7:1–41
Rasweiler JJ IV (1974) Reproduction in the long-tongued bat, Glossophagasoricina. II. Implantation and early embryonic development. Am J Anat 139:1–36
Rasweiler JJ IV (1977) Preimplantation development, fate of the zona pellucida, and observations on the glycogen-rich oviduct of the little bulldog bat, *Noctilioalbiventris*. Am J Anat 150:269–300
Rasweiler JJ IV (1979) Early embryonic development and implantation in bats. J Reprod Fertil 56:403–416
Rasweiler JJ IV (1991a) Spontaneous decidual reactions and menstruation in the black mastiff bat. *Molossus ater*. Am J Anat 191(1–22)
Rasweiler JJ IV (1991b) Development of the discoidal hemochorial placenta in the black mastiff bat, *Molossus ater*. Evidence for a role of maternal endothelial cells in the control of trophoblastic growth. Am J Anat 191:185–207
Rasweiler JJ IV (1993) Pregnancy in Chiroptera. J Exp Zool 266:495–513
Rasweiler JJ IV, Badwaik NK (1996a) Improved procedures for maintaining and breeding the short-tailed fruit bat (*Carolliaperspicillata*) in a laboratory setting. Lab Anim 30:171–181
Rasweiler JJ IV, Badwaik N (1996b) Unusual aspects of inner cell mass formation, endoderm differentiation. Reichert's membrane development, and amnio-genesis in the lesser bulldog bat. *Noctilioalbiventris*. Anatomical Record 246:293–304
Richardson EG (1977) The biology and evolution of the reproductive cycle of *Miniopterusschreibersii* and *M. australis* (Chiroptera: Vespertilionidae). J Zool 183:353–375
Roberts RM (1996) Interferon-τ and pregnancy. J Interf Cytokine Res 16(4):271–273
Rohatgi L, Bhiwgade DA, Menon SN (1992) Electron microscopic studies on the chorioallantoic placenta of the emballonurid bat, *Taphozousmelanopogon*. Bat Res News 34:39
Spencer TE (2014) Biological roles of uterine glands in pregnancy. Semin Reprod Med 32(05):346–357
Srivastava SC (1952) Placentation in the mouse tailed bat. RhinopomaKinneari. Proc Zool Soc Bengal 5:105
Stephens RJ (1962) Histology and histochemistry of the placenta and foetal membranes in the bat, *Tadaridabrasiliensiscynocephala*. Am J Anat 111: 259–286
Stephens RJ (1969) The development and the fine structure of the allantoic placental barrier in the bat, *Tadaridabrasiliensiscynocephala*. J Ultrastruct Res 28:371–398
Turco MY, Gardner L, Hughes J, Cindrova-Davies T, Gomez MJ, Farrell L, Holligshead M, Marsh SGE,

Brosens JJ, Critchley HO, Simons BD, Hemberger M et al (2017) Long-term, hormone-responsive organoid cultures of human endometrium in a chemically defined medium. Nat Cell Biol 19(5):568–577

Uchida T (1953) Studies on the embryology of the Japanese house bat, *Pipistrellus tralatitiusabramus* (Temminck) II. From the maturation of the ova to the fertilization, especially on the behaviour of the follicle cells at the period of fertilization. Science Bulletin, Faculty of Agriculture, Kyushu University, 14,153–168

Van der Stricht O (1909) La structure de l'oeuf des mammiferes (chauve-souris), Vesperugonoctula. 3epart.: L'oocyte de la fin du staded'accroissement, au stade de la maturation, au stade de la fecondation et au deT)ut de la segmentation. Memotres de I'academie royale de medicine de Belgique, Vol. 2

Van Mourik MSM, Macklon NS, Heijnen CJ (2009) Embryonic implantation: cytokines, adhesion molecules, and immune cells in establishing an implantation environment. J Leukoc Biol 85(1):4–19

Wimsatt WA (1944) An analysis of implantation in the bat, *myotis lucifuguslucifugus*. Am J Anat 74:355–411

Wimsatt WA (1944a) Further studies on the survival of spermatozoa in the female reproductive tract of the bat. Anat Rec 88:193–204

Wimsatt WA (1944b) An analysis of implantation in the bat. *Myotis lucifugus*. Am J Anat 74:355–411

Wimsatt WA (1945a) The placentation in the vespertilionid bat, myotis lucifuguslucifugus. Am J Anat 77:1–51

Wimsatt WA (1945b) Notes on breeding behavior, pregnancy and parturition in some vespertilionid bats of eastern United States. J Mammal 26:23–33

Wimsatt WA (1954) The fetal membranes and placentation of the tropical American vampire bat. Desmodusrotundus murinus. Acta Anat 21:285–341

Wimsatt WA (1958) The allantoic placental barrier in Chiroptera: a new concept of its organization and histochemistry. Acta Anat 32:141–186

Wimsatt WA (1965) Notes on breeding behaviour, pregnancy and parturition in some vespertilionid bats of the Eastern United States. J Mamm 26:23–33

Wimsatt WA (1975) Some comparative aspects of implantation. Biol Reprod 12:1–40

Wimsatt WA, Enders AC (1980) Structure and morphogenesis of the uterus. Placenta, and preplacental organs of the neotropical disc-winged bat. *Thyropteratricolorspix* (MicrochiropteraThyropteridae). Am J Anat 159:209–243

Wimsatt WA, Gopalakrishna A (1958) Occurrence of placental hematoma in the primitive sheathtailed bats (Emballonuridae), with observations on its structure, development and histochemistry. Am J Anat 103:35–68

Wislocki GB, Fawcett DW (1941) The placentation of the Jamaican bat. Artibeusjamaicensisparvipes. Anat Rec 81:307

Xie H, Wang H, Tranguch S, Iwamoto R, Mekada E, DeMayo FJ, Lydon JP, Das SK, Dey SK (2007) Maternal heparin-binding-EGF deficiency limits pregnancy success inmice. Proc Natl Acad Sci U S A 14(46): 18315–18320

Ultrastructure of Endotheliochorial Interhemal of the Indian Musk Shrew, *Suncus murinus*

5

The nature of the insectivoran chorioallantic placenta has been a controversial subject mainly because of the incorrect identification of trophoblast by Hubrecht (1894). He reported it as hemochorial. A more plausible interpretation of the placental nature of the Indian musk shrew, *Suncus murinus*, was presented by Owers (1960), who described it as endothelio-endothelial.

Here, the chorionic trophoblast was reported to be lost early in gestation with the vascularized allantois remaining to penetrate the maternal tissue. Three years later, Mossman and Owers (1963) noted the apparent loss of the trophoblast suggested that choriovitelline and chorionic or inverted placenta took over the trophoblastic function. The finding of an intermediate lamina in the placenta of *Blarina brevicauda* using electron microscopy led Wimsatt et al. (1973) to conclude that this lamina indeed was the syncytium (or symplasma). This conclusion was based on the consideration that the syncytial tissue is closely followed by a continuous basal lamina. Kiso et al. (1990) and Kiso and Nakagaw (1994) argued that, because both the maternal and foetal endothelium contact each other, the Suncus placenta is endothelial. To date, then, there is still no clear consensus on the nature of Suncus placenta. Utilizing ultra semithin sections for light microscopy we found an even arrangement of maternal villi and foetal villi giving a more complicated labyrinthine pattern (Fig. 5.1). So, we undertook to ultrastructurally characterize and correctly classify the interhemal membrane of the Indian musk shrew.

The four stages in the placentation of *Suncus murinus*—the early limb bud (CRL 6.6 mm), the late limb bud (CRL 8.2 mm), the advanced (CRL 12.8 mm), and the full-term pregnancy (CRL 16 mm) were investigated. The following description generally applies to all four stages. The interhemal membrane of the chorioallantoic placenta is composed of maternal endothelium syncytial trophoblast with the basal lamina, foetal mesenchyme, and foetal endothelium. Because of the persistence of the maternal endothelium and the trophoblast, the interhemal membrane is correctly characterized as endotheliochorial.

The maternal blood spaces are surrounded by a greatly hypertrophied endothelium throughout the gestation during which their cellular characters are retained. The luminal surface is covered with blunt microvilli and numerous pinocytotic vesicles (Figs. 5.2, 5.3 and 5.4). The nuclei show a moderate amount of heterochromatin. Majority of the mitochondria are dark, somewhat elongated, whereas others are ovoid. Much of the cytoplasm is occupied by rough endoplasmic reticulum with parallelly arranged, dilated cisternae provided with ribosomes (Fig. 5.3). A prominent and discontinuous fibrous zone of cytoplasmic filaments (tonofilaments) is seen associated with areas of greater densities, the demosomes (Figs. 5.2 and 5.4). The Golgi complex comprises three basic elements, the most characteristic being a stock of flattened sacs with

D. A. Bhiwgade, S. Menon, *Ultrastructural Investigations on the Pituitary-Gonadal Axis*,
https://doi.org/10.1007/978-981-99-3276-4_5

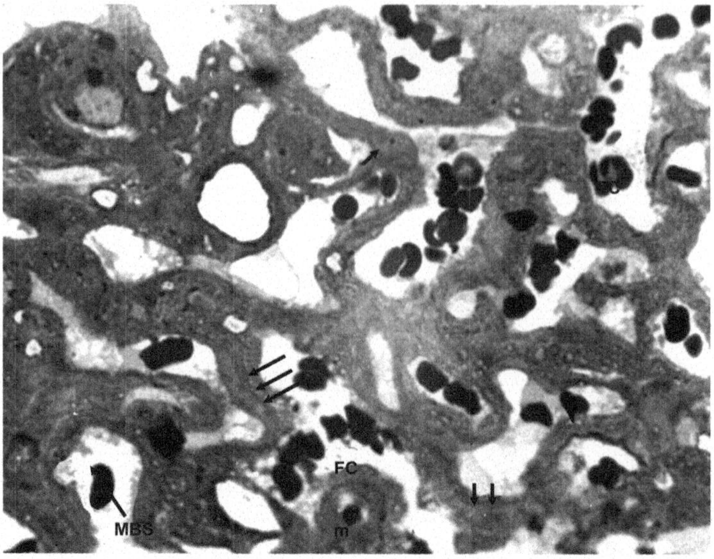

Fig. 5.1 Ultrasemithin sections of placenta of Suncus murinus at low power magnification showing placental tubules. Note the maternal blood space (MBS) along with maternal endothelium (arrows), foetal capillaries (FC), attenuated syncytiotrophoblastic layer (small arrows) and mesenchyme cells (m). (Toluidine blue stain) [X 1,000]. Unpublished photograph from Bhiwgade

often dilated ends. The cytoplasm is also characterized by numerous polyribosomes, dense bodies, micropinocytotic vesicles, coated vesicles, multivesicular bodies, moderate amounts of glycogen and junctional complexes (Figs. 5.3 and 5.4).

A trophoblastic layer, i.e. the 'intermediate cytoplasmic lamina' is observed beneath the basal laminar region of the maternal endothelium, and the same has been confirmed as the syncytial trophoblast (Figs. 5.2, 5.3, 5.4, 5.5 and 5.6). Due to its multiple infoldings, spaces are formed that open freely at both surfaces (Figs. 5.5, 5.6 and 5.7) giving this layer a sieve or sponge-like appearance. These suggest the existence of paracellular routes across the trophoblast which would be of enormous functional value physiologically and have been discussed in greater detail for the rabbit (Stulc et al. 1969), the guinea pig (Kaufmann et al. 1982, 1987), rat (Stulc and Stulcova 1986), (King 1992) and the human placenta (Kertschanska et al. 1994). To verify and test the continuity of such a paratrophoblastic route, lanthanum-hydroxide perfusion or section electron microscopy would be necessary.

The cytoplasm contains dense bodies, few lipid droplets, multivesicular bodies, tonofilaments and few coated vesicles. The trophoblastic lamina is very thin, rough endoplamic reticulum, mitochondria and Golgi complex appear to be fewer. However, during the limb bud and advanced pregnancy stages, a few mitochondria with prominent cristae and few short segments of the rough endoplasmic reticulum were observed in this region rich in cytoplasm. Though the trophoblastic layer shows varied thickness within the labyrinth it is never observed to be absent at any point throughout the placental development. The syncytial trophoblast in the labyrinth, however, appears as a finely perforated membrane. These various cytoplasmic organelles make the syncytial trophoblast synthetically active.

The basal lamina of the trophoblast is a discrete and continuous layer. However, on the opposite side (towards foetal endothelium) it is more irregular and is filled with flocculent material in the spaces between the cytoplasmic processes of the adjacent foetal endothelium.

The foetal mesenchymal cells greatly hypertrophied with a well-developed spherical nucleus with equally distributed heterochromatin (Fig. 5.8a). The cytoplasm contains well-developed ovoid mitochondria with prominent cristae (Fig. 5.9), in close association with a greatly developed rough endoplasmic reticulum with heavily studded ribosomes (Figs. 5.8 and 5.9). Associated whorled membranous bodies

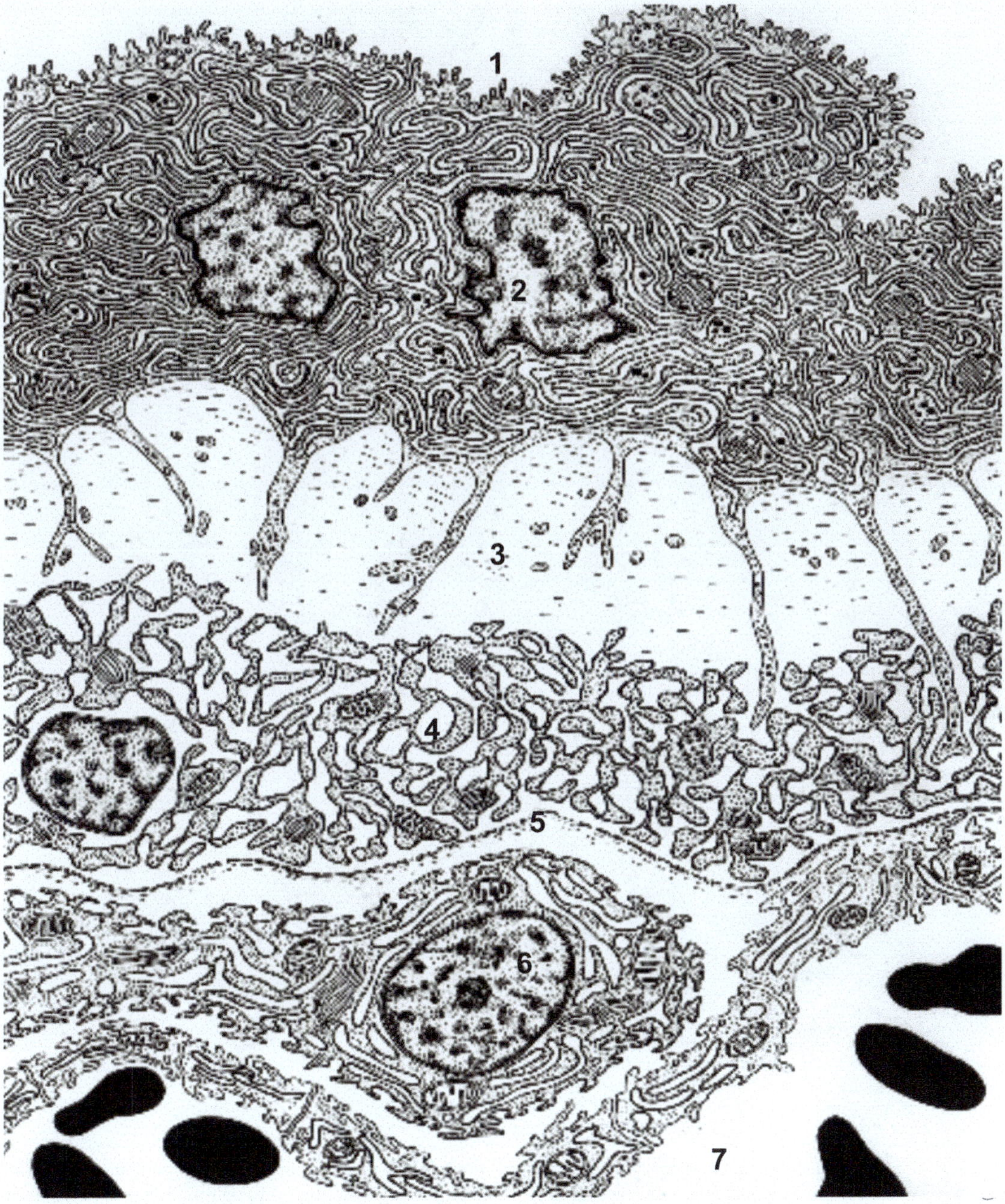

Fig. 5.2 Semi-schematic representation of the interhemal membrane of Suncus murinus to show the relationship between the maternal and foetal bloodstream. (1) Maternal blood space (2) Maternal endothelium (3) Basal laminar region (4) Syncytial trophoblast layer; (5) Basal lamina of trophoblast (6) Foetal mesenchymal cell (7) Foetal capillary. Published, Dr. Bhiwgade

are sometimes encountered. Some vesiculated rough endoplasmic reticulum with dilated cisternae are observed with their membranes studded with ribosomes (Fig. 5.9). The cytoplasm is characterized by numerous dense granules (probably secretory in nature), polyribosomes,

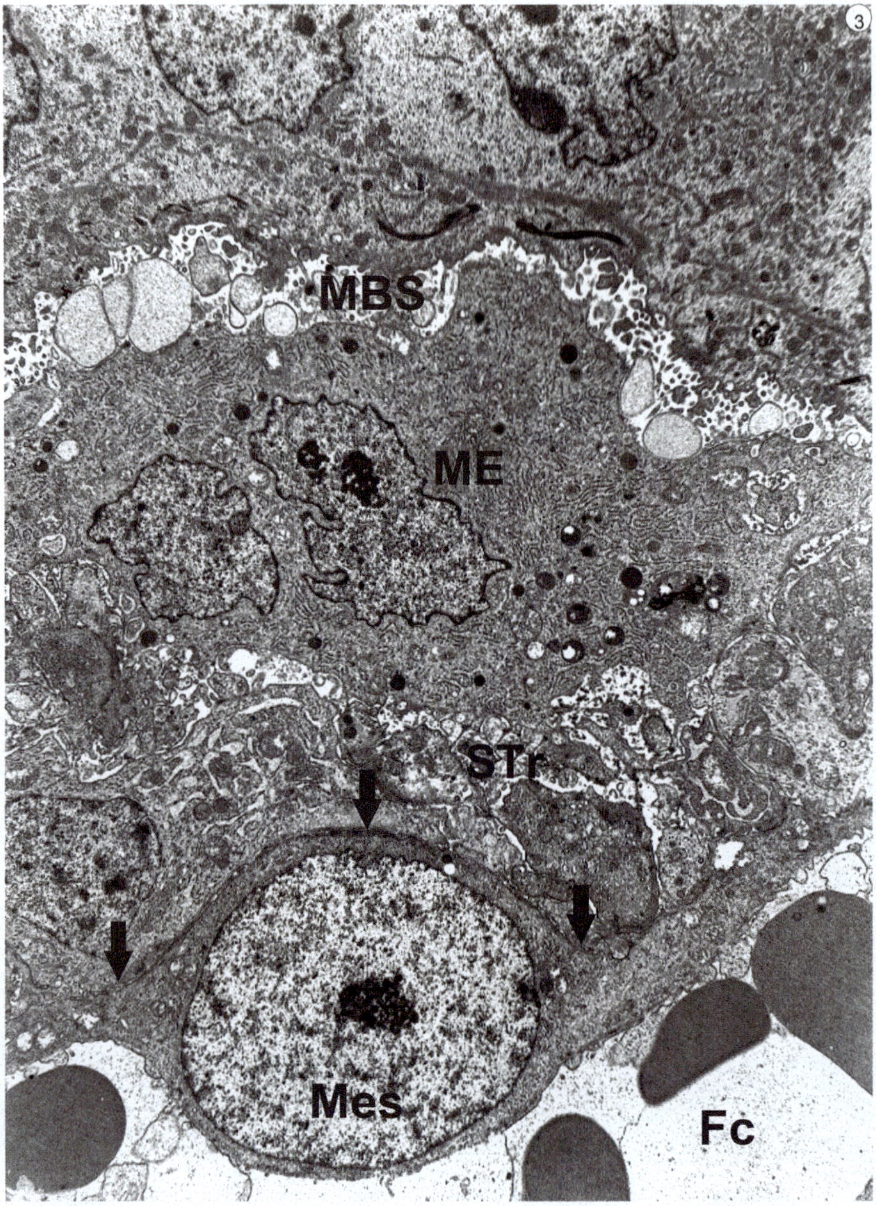

Fig. 5.3 Low-power electron micrograph of interhemal membrane in Suncus murinus during advanced pregnancy. Note the maternal blood space (MBS) surrounded by hypertrophied maternal endothelium which is irregular with a moderate amount of heterochromatin. Also note the well-developed rough endoplasmic reticulum, secretory granules and ovoid mitochondria in the maternal endothelium. Note the syncytiotrophoblast (STr) adjacent to it is the basal lamina of trophoblast (arrows). On the lower right is the foetal capillary (FC) surrounded by foetal endothelium and Mesenchyme (Mes) [X 3,000]. Published, Dr. Bhiwgade

coated vesicles and microtubules (Fig. 5.8a). Golgi zones are well developed in the juxtanuclear region. It consists of numerous associated vesicles and spherical microparticles (matrical lipid debris) are observed in small clusters (Fig. 5.8b). Numerous fine filaments are scattered throughout the cytoplasm.

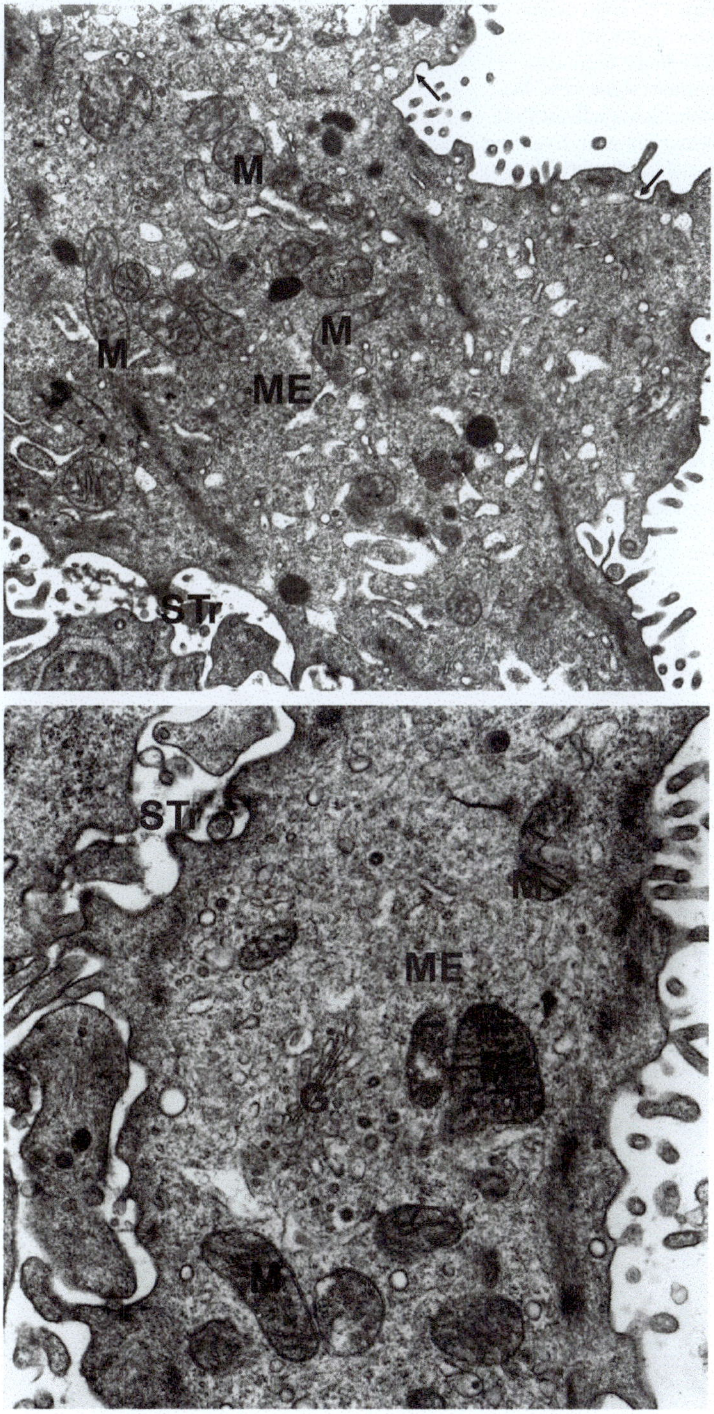

Fig. 5.4 High-power electron micrograph of internal membrane in Suncus murinus during limb bud embryo showing well-developed maternal endothelium (ME) and a thin layer of syncytiotrophoblast (STr). Note the mitochondria (M) and Golgi apparatus (G) in the maternal endothelium (ME). Also, note numerous coated vesicles and pinocytotic vesicles (arrows) facing towards the maternal surface [X 25,000] [X 13,000]. Published, Dr. Bhiwgade

The foetal capillary endothelium is surrounded by well-developed and numerous blunt cytoplasmic processes projecting into the foetal capillary and lacks a continuous basal lamina (Fig. 5.7). The cytoplasm of the foetal endothelial cells is characterized by a moderate number of oval

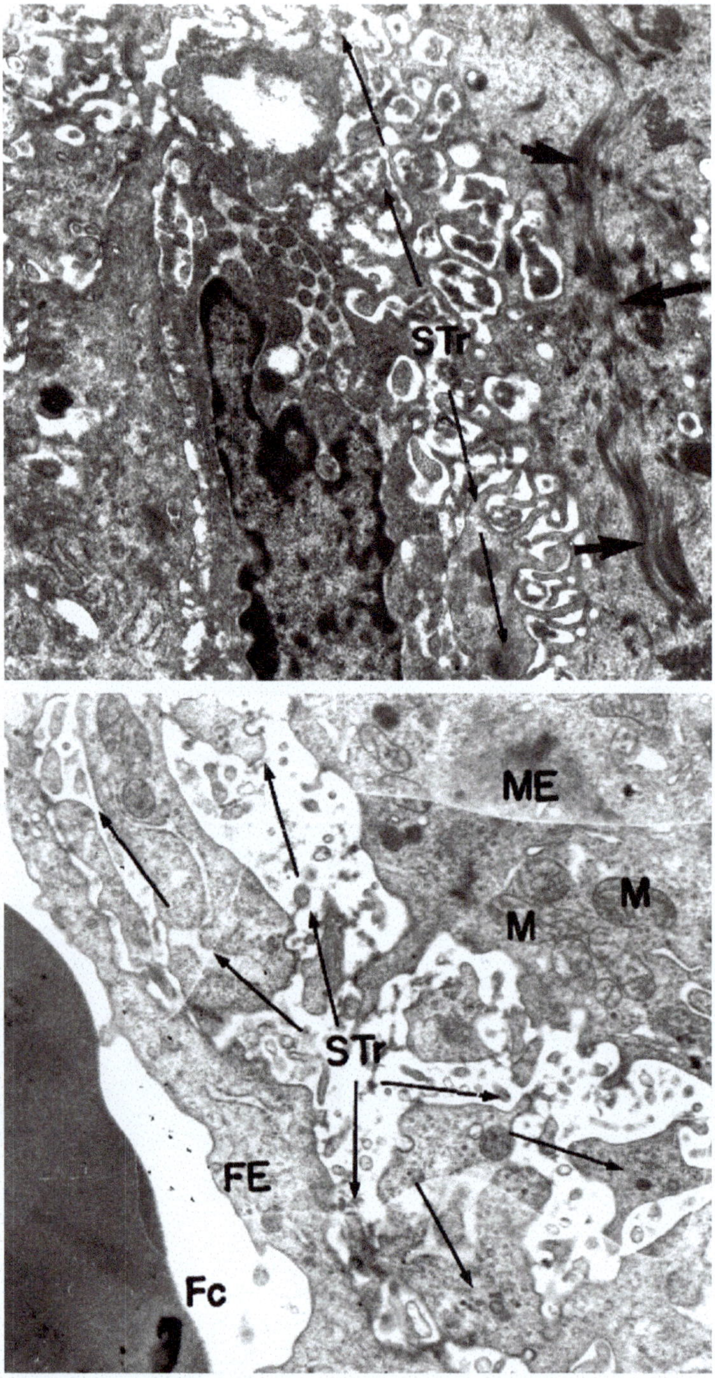

Fig. 5.5 High-power micrographs of internal membrane of Suncus murinus show sieve-like syncytiotrophoblast (STr—arros) and well-developed foetal capillary (FC) along with foetal endothelium (FE). Note the maternal endothelium (ME) with well-developed mitochondria (M) with few cristae and dense bodies. Also, note the discontinuous fibrous tract (arrows) near the syncytiotrophoblast (STr) [X 13,000] [X 13,000]. Published, Dr. Bhiwgade

mitochondria with prominent cristae, dilated cisternae of rough endoplasmic reticulum, numerous polyribosomes and coated vesicles.

As pointed out by Mossman (1987), it is rather difficult to conclusively characterize the mammalian interhemal membrane without electron microscopy. The present ultrastructural study of *Suncus murinus* placenta has clarified certain controversial issues concerning the organization of the interhemal membrane: (1) the exact identity

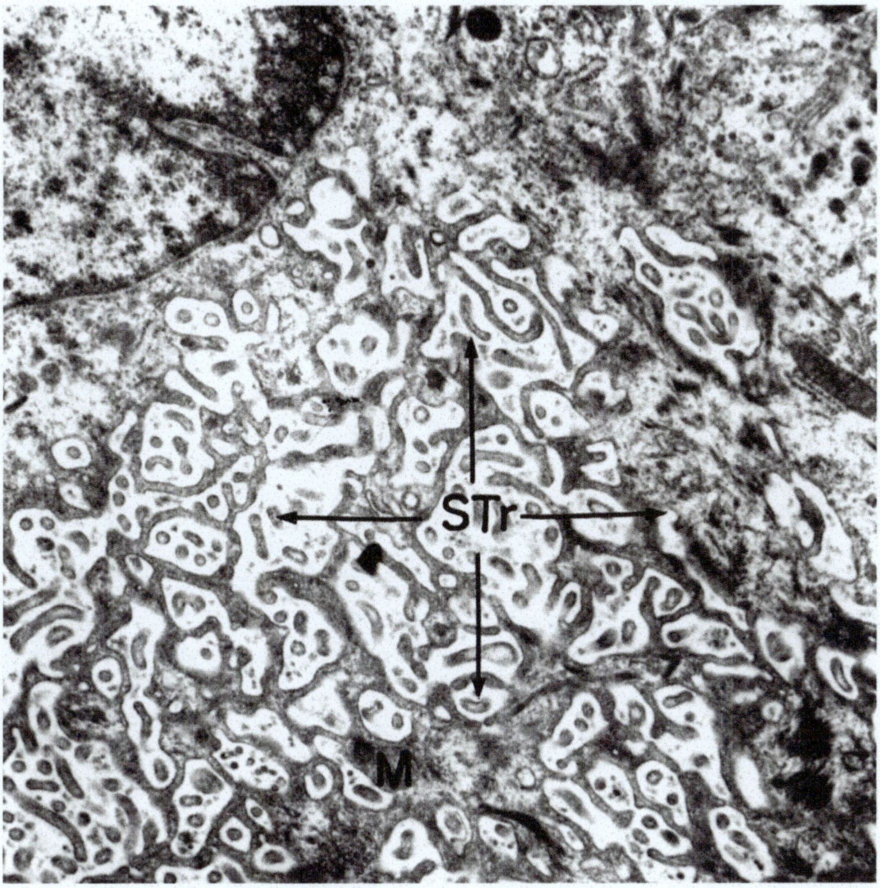

Fig. 5.6 High-power electron micrograph of interhemal membrane of Suncus murinus during full-term showing thin spongy nature of syncytiotrophoblast (STr). Note that the mitochondria (M) are generally larger than the intercellular spaces [X 16,000]. Published, Dr. Bhiwgade

of an attenuated cytoplasmic membrane which is situated between maternal endothelium and foetal constituents of the barrier membrane and, (2) the extent of the continuity of this layer throughout the placental labyrinth. Due to the retention of maternal endothelium and trophoblast throughout gestation, interhemal membrane is confirmed to retain an endotheliochorial nature throughout the gestation.

This is in accordance with the findings of Wimsatt et el. (1973) in *Blarina brevicauda.* Others who have examined the placenta of Blarina and Soerx (Wimsatt and Wislocki 1947; Wislocki and Wimsatt 1947), and *Suncus* (Owers 1960; Mossman and Owers 1963) with the light microscope, have emphasized the basic similarity of the placentae of these three shrew species. The continuity of the syncytial trophoblast is maintained in *Blarina brevicauda.* Other studies on the placenta of Blarina and Sorex have been the basis for reporting its nature as endotheliochorial (Wimsatt and Wislocki 1947).

Owers (1960) verified these interpretations and presented evidence that trophoblast (without specifying the layer) is lost in early gestation and therefore, the placenta of the Indian musk shrew is endothelio-endothelial, a view still maintained by Kiso and coworkers (see Kiso and Nakagaw 1994). Later, Mossman and Owers (1963) noted the absence of trophoblast and agreed with the earlier conclusion of an endothelio-endothelial placenta and suggested that the lamina being

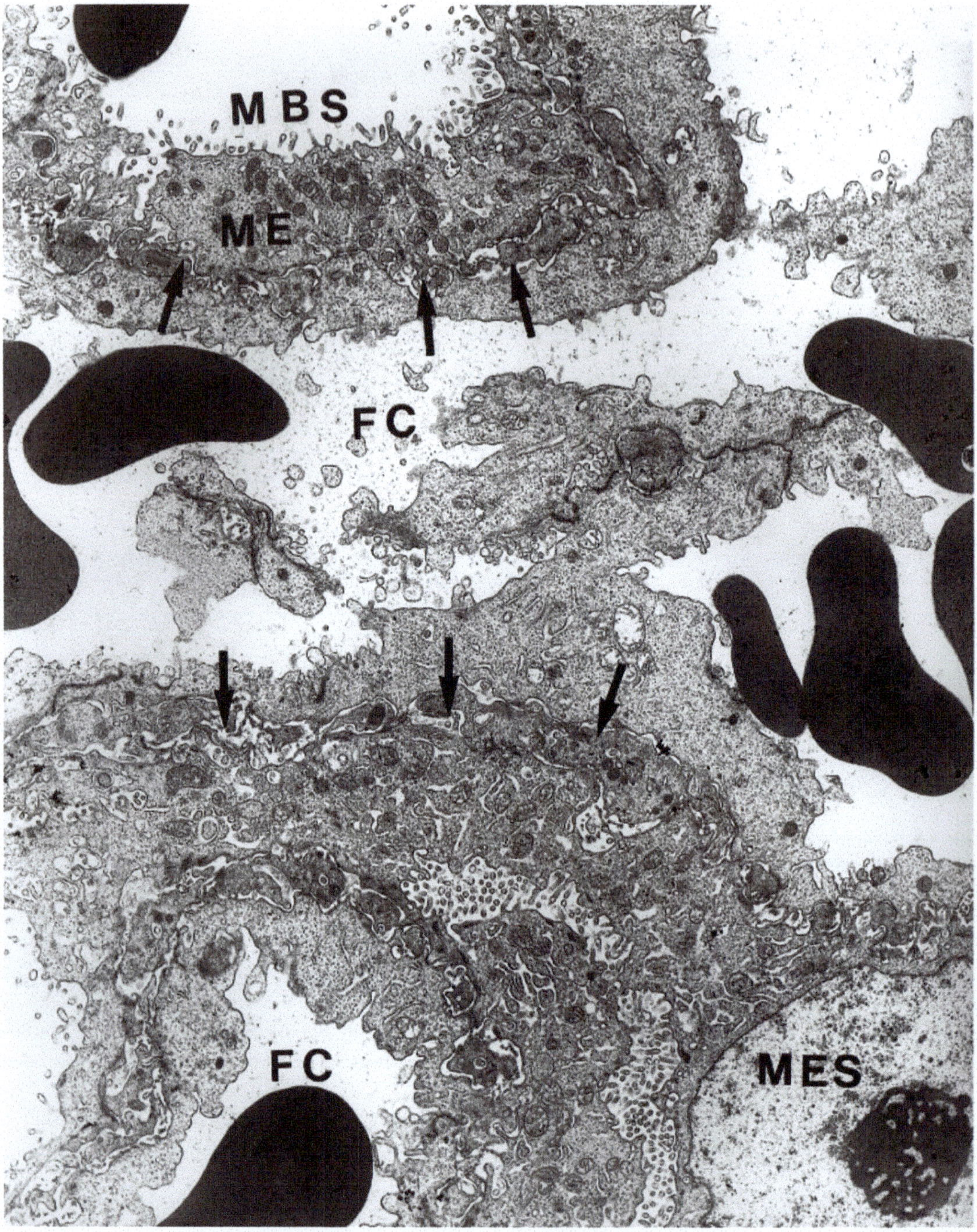

Fig. 5.7 Low-power electron micrograph of interhemal membrane in Suncus murinus during full-term pregnancy showing numerous blunt microvilli protruding into the maternal blood space (MBS). Each placental tubule consists of maternal endothelium (ME), a thin layer of syncytiotrophoblast (STr) and the foetal capillary (FC) surrounded by foetal endothelium [X 3,600]. Published, Dr. Bhiwgade

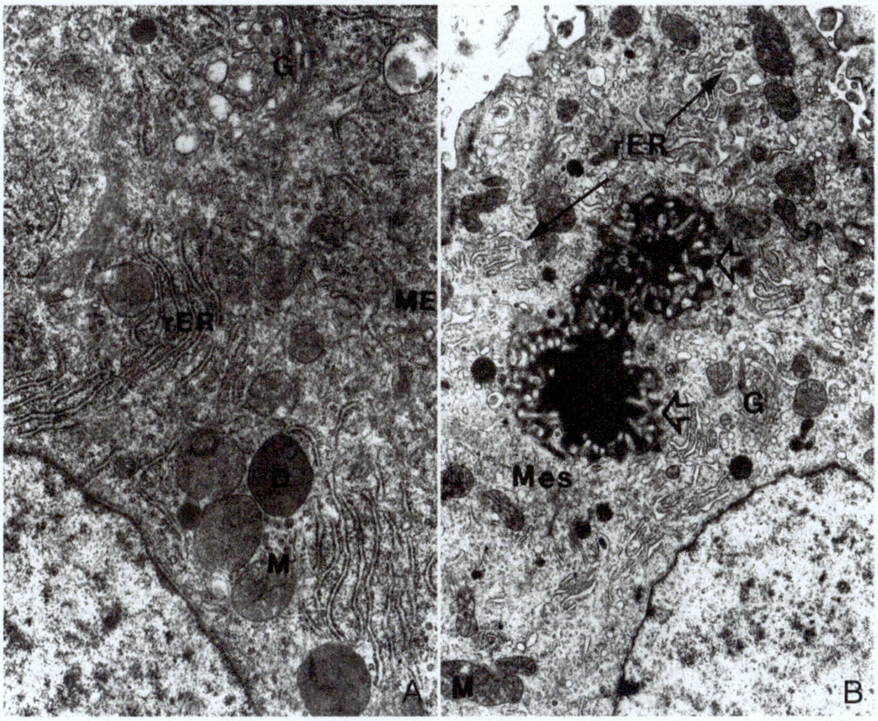

Fig. 5.8 (**a**, **b**). Magnified view of mesenchyme (Mes) cell of the internal membrane of Suncus murinus during early pregnancy to show vesiculated rough endoplasmic reticulum (V-rER). The polyribosomes have disaggregated to form solitary ribosomes, but are still attached to vesiculated rough endoplasmic reticulum (V-rER) [X 33,000]. Published, Dr. Bhiwgade

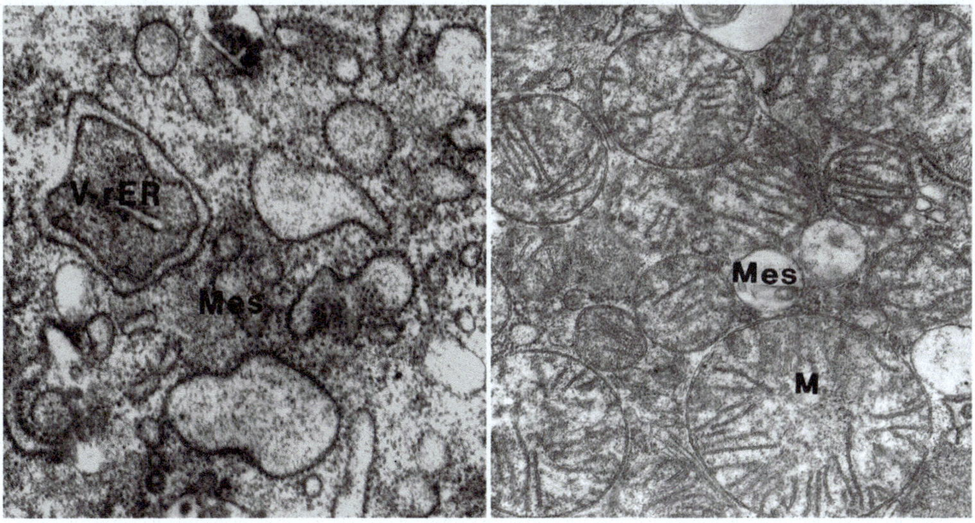

Fig. 5.9 High-power electron micrograph of Mesenchyme (Mes) cell of internal membrane of Suncus murinus during limb bud embryo. The cytoplasm contains well-developed mitochondria (M) with prominent cristae, dilated rough endoplasmic reticulum (rER), spherical microparticles (hollow arrow) in a small cluster and dense granules [X 10,000]. Published, Dr. Bhiwgade

maternal in origin, represents in their view an unusual form of symplasma derived from the modified maternal endothelium, and was discontinuous extensively over the labyrinth. Our observations correspond with those of Wimsatt et al. (1973), who reported the placenta of *Blarina brevicauda* as endotheliochorial at the ultrastructural level. However, our studies on Suncus have further delineated the intermediate layer as trophoblastic rather than belonging to the maternal symplasma as previously thought by Mossman and Owers (1963). This trophoblast is syncytial in character, extremely thin and sieve-like and it is an universal constituent of the interhemal membrane. Wimsatt et al. (1973) have described the presence of cytotrophoblast tissue in *Blarina brevicauda,* but it is absent from the interhemal membrane of *Suncus murinus.*

The maternal endothelium, although greatly hypertrophied, retains its cellular character, with the cytoplasm characterized by prominent discontinuous tonofibrillar zones in association with desmosomes and other cell organelle complements. Wimsatt and Wislocki (1947) have suggested that important synthetic activities, apart from the intrinsic synthesis of structural proteins needed to support hypertrophy and proliferation, may very well be maintained by the endothelium. Wimsatt et al. (1973) have reported that the tonofibril bundles (fibrous tracts) form the important cytoskeletal elements which help to maintain the cohesion and surface integrity in these highly distensible cells. It has also been suggested by them that the microvesicles and microvilli which are consistent constituents, greatly increase the functional surface area of thickened endothelial cells, and facilitate the passage between maternal and foetal circulation. The intimate interdigitations and numerous direct contacts between the basal endothelial cells and the cytoplasm of the subjacent trophoblastic layer are necessary for functions requiring close contact between the cytoplasmic constituents of the interhemal membrane.

The maternal endothelium in interhemal membrane of several species of bats shows the presence of microvesicles, caveolae and numerous blunt microvilli, which play an important role in active absorption of maternal blood (Bhiwgade 1990; Karim and Bhatnagar 1996). In our study, the close association of the mitochondria with the cisternae of rough endoplasmic reticulum and lipid droplets is suggestive of intimate relation with the active protein synthesis and lipid metabolism, as reported by Lawn and Chiquoine (1965) in Ferret placenta.

The syncytial trophoblast appears sieve-like within which the intercommunicating system of extracellular lacunae was observed which permitted numerous basal processes of the overlying maternal endothelial cells to penetrate deeply into the trophoblastic sponge and even pass through the trophoblast to its basal lamina. Besides the fact that the trophoblastic layer is continuous, it also meets the cytologic and developmental criteria for the trophoblast (Wimsatt et al. 1973), is also viable and synthetically active (Enders 1965; Bhiwgade 1990), and therefore it can be safely interpreted that it is the trophoblast. Kiso et al. (1990); Kiso and Nakagaw (1994) have presented arguments that after day 24 of gestation, the syncytiotrophoblast disappears from the interhemal membrane, a fact not borne out from their illustrations which depict the presence of this tissue. Moreover, in our preparations, syncytiotrophoblast clearly appears as a viable tissue confirming the interhemal membrane to be endotheliochorial rather than endothelio-endothelial.

The foetal endothelium lacks a continuous basal lamina, instead it bears numerous blunt processes that project into the foetal capillaries which, according to Wimsatt et al. (1973), is suggestive of the involvement of a large functional surface so formed for placental transport or in support of other functions as a whole.

The greatly hypertrophied cells (such as those of synovial cell membrane of the immobilized rabbit knee joint) display the loss of polyribosomal configuration by their disaggregation and those of the ribosomes (Ghadially 1988). It is believed that the polyribosomes break up and are not reformed thus giving rise to many solitary ribosomes. Similarly, disaggregation of polyribosomes without degranulation of the vesiculated rough endoplasmic reticulum is seen indicating impaired protein production for 'export'.

Disaggregation of polyribosomes lying in the cytoplasm is associated with depressed endogenous proteins. The spherical microparticles observed in small clusters and their massive accumulations have also been noted in the human placenta, near the trophoblastic epithelium (Liebhart and Janczewska 1973). These are suggested to be involved in neurosecretory function. Further comparative ultrastructural studies on the interhemal membrane of animals such as bats of diverse groups using tracer techniques are needed to clarify the functional role of this membrane in placental physiology.

References

Bhiwgade DA (1990) Comparative electron microscopy of chorio-allantoic placental barrier in some Indian Chiroptera. Acta Anat 138:302–317

Enders AC (1965) A comparative study of the fine structure of the trophoblast in several hemocho- rial placentas. Am J Anat 116:29–68

Ghadially FN (1988) Ultrastructural pathology of the cell and matrix, vol 2, 3rd edn. Butterworths, London

Hubrecht AAW (1894) Studies in mammalian embryology. III. The placentation of the shrew (Sorex vulgaris, L.). Quar J Micro Sci 35:481–537

Karim KB, Bhatnagar KP (1996) Observations on the chorioallantoic placenta of the Indian flying fox, Pteropus giganteus giganteus. Ann Anat 178:523–530

Kaufmann P, Schroeder H, Leichtweiss HP (1982) Fluid shift across the placenta: II. Fetomaternal transfer of horseradish peroxidase in the Guinea pig. Placenta 3: 339–348

Kaufmann P, Schroeder H, Leichtweiss HP, Winterhager (1987) Are there membrane-lined channels through the trophoblast? A study with lanthanum hydroxide. Tropho Res 2:557–571

Kertschanska S, Kosanke G, Kaufmann P (1994) Is there morphological evidence for the existence of transtrophoblastic channels in human placental villi? Tropho Res 8:581–596

King BF (1992) Ultrastructural evidence for transtrophoblastic channels in the hemomonochorial placenta of the degu (Octodon degus). Placenta 13: 35–41

Kiso Y, Nakagaw Y (1994) The laboratory shrew placenta: evidence for an endothelio-endothelial type. Endocr J 41(suppl):S57–S61

Kiso Y, Yasufuku K, Matsuda H, Yamauchi S (1990) Existence of an endothelioendothelial placenta in the insectivore, Suncus murinus. Cell Tissue Res 262:195–197

Lawn AM, Chiquoine AD (1965) The ultrastructure of the placental labyrinth of the ferret (Mustela putorius furo). J Anat (London) 99:47–69

Liebhart M, Janczewska E (1973) Ultrastructural changes in the placenta of a newborn with congenital diabetes mellitus. Path Europ 8:127–134

Mossman HW (1987) Vertebrate fetal membranes. Rutgers University Press, NJ

Mossman HW, Owers N (1963) The shrew placenta: evidence that it is endothelio-endothelial in type. Am J Anat 113:245–271

Owers NO (1960) The endothelio-endothelial placenta of the Indian musk shrew, Suncus murinus–a new intepretation. Am J Anat 106:1–26

Stulc J, Friedrich R, Jiricka Z (1969) Estimaion of the equivalent pore dimensions in the rabbit placenta. Life Scs 8:167–180

Stulc J, Stulcova B (1986) Transport of calcium by the placenta of the rat. J Physiol 371:1–16

Wimsatt WA, Enders AC, Mossman HW (1973) A reexamination of chorioallantoic placental membrane of a shrew, Blarina brevicauda: resolution of a controversy. Am J Anat 138:207–234

Wimsatt WA, Wislocki GB (1947) The placentation of the American shrews, Blarina brevicauda and Sorex fumeus. Am J Anat 80:361–436

Wislocki GB, Wimsatt WA (1947) Chemical cytology of the placenta of two north American shrews (Blarina brevicauda and Sorex fumeus). Am J Anat 81:269–308

6 Ultrastructure of Interhemal Membrane in some Bats

6.1 *Rousettus leschenaulti* and *Cynopterus sphinx* (Megachiroptera)

In the progression of placental development, it could be generally stated that the initial placenta encountered is endotheliochorial, which soon after changes into the hemochorial one. The term placenta of Rousettus (Bhiwgade 1990), with the hemochorial condition, is observed to be derived as a post-modification of the endotheliochorial condition, immediately after the trilaminar blastocyst stage, and not after midgestation as previously reported (Gopalakrishna and Karim 1980). We have examined six stages of Rousettus placenta from trilaminar blastocyst stage till the term to understand the better nature of the interhemal membrane.

Except for a short note (Shomita and Bhiwgade 1995) and some light microscopic observation (Wimsatt 1958), fine structural observations on the placenta of Cynopterus are lacking. Here, we also present the ultrastructure of the interhemal membrane of *Cynopterus sphinx* in the late limb-bud and full-term stages.

6.1.1 *Rousettus leschnaulti*

Proceeding from the maternal side to the foetal side, the placental membrane can be identified to be made up of the serosa, the myometrium, the junctional zone, the giant cell layer, the placental labyrinth which also shows a necrotic zone, the allantoic mesenchyme with the allantoic vessels and then the foetal capillaries (Fig. 6.1). It is in the centre of foetal side of the placental disc, that the umbilical cord is attached.

	Rousettus leschenaulti				*Cynopterus sphinx*		
Date	Stage	CRL	IMC	Date	Stage CRL		IMC
15 Apr.[a] 15 Apr.[a] 2 Nov.[b] 20 Jan.[b] 8 Feb.[b] 23 Feb.[b]	Trilaminar blastocyst neural groove early limb- bud Late limb- bud Mid- pregnancy Term pregnancy	4.00 6.90 8.95 10.00 14.00 18.00	* 1–6 1–6 1–6 1–6 1–6	26 Dec. 28 Jan.	Specimen not available Specimen not available Specimen not available Late limb-bud Specimen not available Advanced	8.85 13.90	1,3–6 1,9–6

[a]During second pregnancy; [b]During first pregnancy; [b]Maternal endothelial cells face the maternal blood space

The umbilical cord has five blood vessels. Two of them are the allantoic arteries, one is the allantoic vein, while the other is the vitelline artery. The fifth one is the vitelline vein. All these blood vessels are within the mucoid connection tissue.

The endometrial allantoic duct is still seen inside the umbilical cord. The yolk sac is seen adjacent to the foetal border of the placenta (Fig. 6.1). Both the endodermal and mesodermal elements undergo hypertrophy transforming the yolk sac into solid vascular, gland like.

D. A. Bhiwgade, S. Menon, *Ultrastructural Investigations on the Pituitary-Gonadal Axis*,
https://doi.org/10.1007/978-981-99-3276-4_6

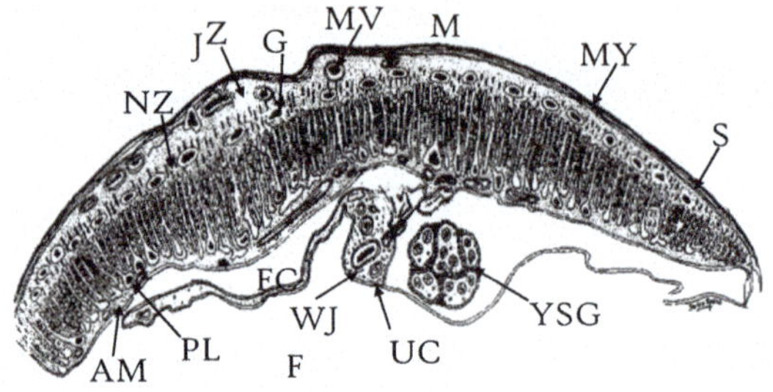

Fig. 6.1 This is a camera lucida drawing showing schematic relationship between different parts of the chorioallantoic placenta of *Rousettus leschenaulti* at full term. The various parts of the placenta can be discerned from the maternal (M) to the foetal side (F). Abbreviations: *AM* allantoic mesenchyme, *EAD* endodermal allantoic duct, *FC* allantoic vessels, *G* giant cell layer, *JZ* junctional zone, *MV* maternal efferent vessels, *MY* myometrium, *NZ* necrotic zone, *PL* placental labyrinth, *S* serosa, *UC* umbilical cord, *WJ* Wharton's jelly, *YSG* yolk sac gland. From Bhiwgade et al. 2000

6.1.2 Establishment of Placenta

The implantation of the blastocyst marks the beginning of the formation of placenta while the establishment of the trophoblastic placenta marks the end of the process. The placental disc becomes discernible externally after this. As the three layers of the placenta develop, each placental tubule consists of a maternal blood vessel which has a single cell lining made up of maternal endothelial cells (Fig. 6.2). The placental tubule rests atop an eosinophilic basement membrane and encapsulates the maternal blood space. It is situated adjacent to the syncytial trophoblastic layer. The interhemal membrane can be identified as hemodichorial from the neural groove stage and can be seen persisting through the limb-bud stage (Fig. 6.3), mid-term (Fig. 6.4), till the full-term stages (Fig. 6.5). In the hemodichorial condition, the maternal blood space is directly surrounded by the inner syncytial layer. The syncytium is followed by the outer cytotrophoblastic layer which is in close proximity to the foetal capillary endothelium. From the onset of the neural groove and in limb-bud stages the lack of maternal endothelium is very distinct.

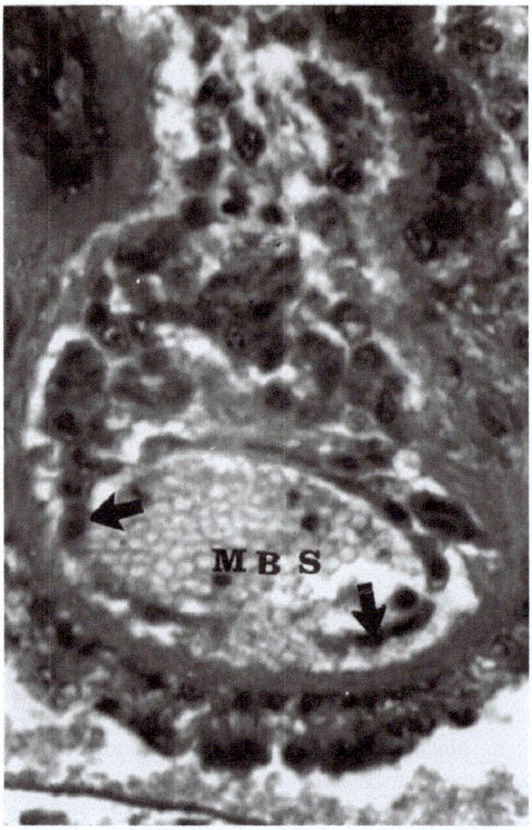

Fig. 6.2 Section of the placental tubule of *Rousettus leschenaulti* during the trilaminar blastocyst stage. The discontinuous layer of maternal endothelial cells (arrows) Haematoxylin and eosin stain (H & E). [X 150]. From Bhiwgade et al. 2000

The different layers of interhemal membrane, which remain constant throughout the last 5 of the 6 stages studied can be described as follows:

6.1.3 Maternal Blood Space (MBS)

The maternal blood space is seen lined by the ectoplasmic layer. During the trilaminar blastocyst stage, the ectoplasmic layer surrounding the lacunae remains undisturbed and its cytoplasm bears a few scattered vesicles but lacks major cell organelles (Fig. 6.2). Maternal endothelial cells (Fig. 6.2) form the perimeters of the MBS; therefore, this stage of placentation, unlike the stages following it, is endotheliochorial. From the neural groove stage onwards, the profile of

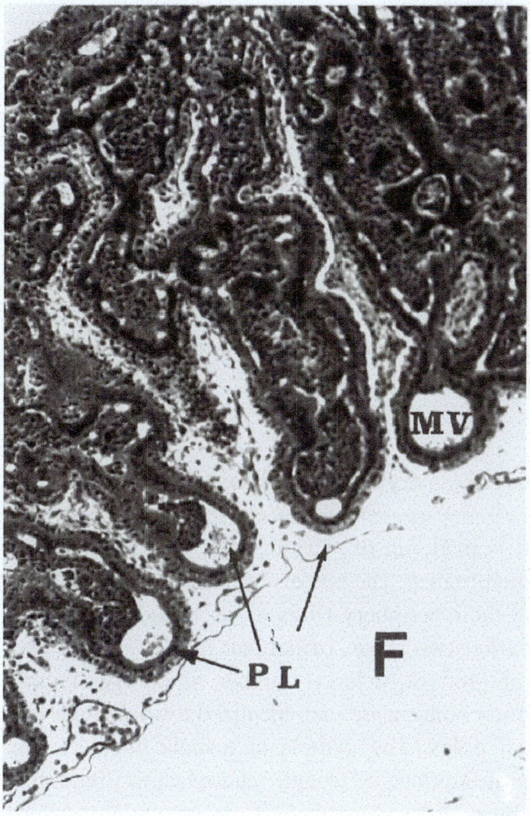

Fig. 6.3 Note the placental tubules (PL) 'hanging' from the uterine wall. Maternal vessels (MV) are large toward the foetal border (F). H & E. [X 240]. From Bhiwgade et al. 2000

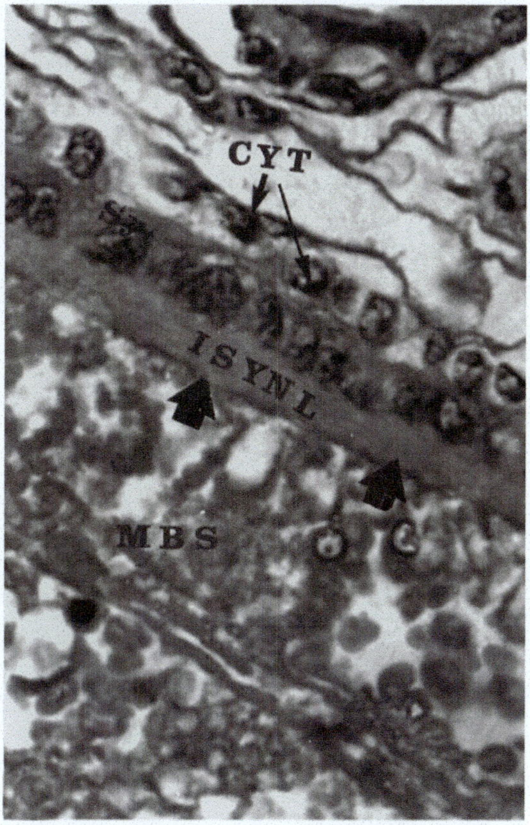

Fig. 6.4 Note the maternal blood space (MBS) bordered by the intrasyncytial lamina (ISYNL), syncytiotrophoblast (SYN) and the cytotrophoblast (CYT). H & E. [X 640]. From Bhiwgade et al. 2000

this thin, ectoplasmic, sycytiotrophoblastic layer is progressively transformed into increasing numbers of microvilli (Fig. 6.6) with a few lipid droplets. Multivesicular bodies and caveoli, some filled with electron-dense secretions, are also found within the cytoplasm as gestation advances.

6.1.4 Intrasyncytial Lamina

In all stages investigated except the trilaminar stage, the intrasyncytial lamina is seen as a constant feature. Intrasyncytial lamina comprises an ill-defined and discontinuous layer of homogeneous extracellular material. The cytoplasm is more electron dense than that of the syncytial trophoblast. Intrasyncytial lamina remains embedded in the syncytial trophoblast. The intrasyncytial lamina separates the outer ectoplasmic layer bordering the maternal blood space and the underlying syncytial mass. It shows blebbing in some areas. At term, the homogenesis material of the intrasyncytial lamina extends into the maternal blood space through the outer ectoplasmic layer by the desmosomally connected coarse syncytium (Figs. 6.6a, b and 6.7).

6.1.5 Syncytiotrophoblast

The syncytiotrophoblastic cytoplasm is dense and contains lipid droplets, secretory vesicles, Golgi complexes (Fig. 6.6a, b) and intracellular canalicular systems (coarse syncytium, Figs. 6.6a, b and 6.7). Nuclei display wrinkled contours and are

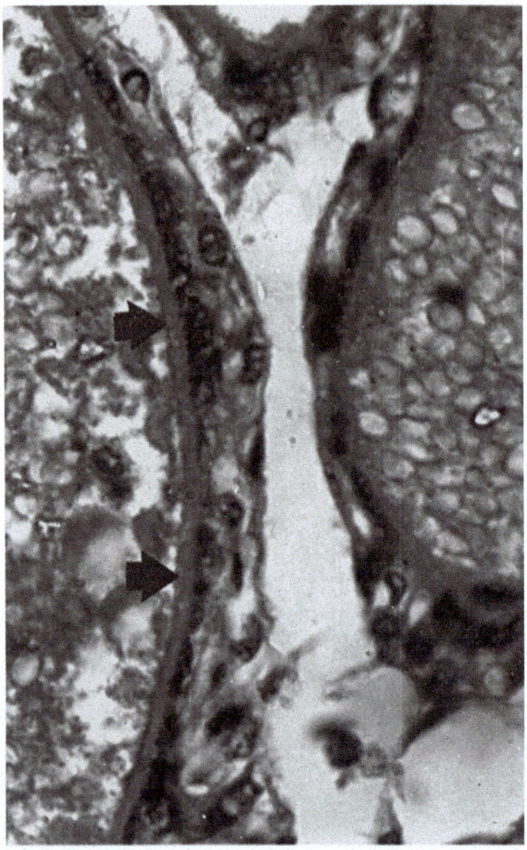

Fig. 6.5 Note the intrasyntial lamina (arrows), syncytiotrophoblast (SYN), and cytotrophoblast (CYT) Compare with Figs. 6.3 and 6.4. H & E. [X 640]. From Bhiwgade et al. 2000

embedded in the continuous cytoplasm uninterrupted by plasma membrane. From the neural groove stage onward, changes in the syncytium are noticeable. During this stage, the mitochondria, rather than the Golgi bodies predominate and tend to gravitate towards the cytotrophoblastic layer. By the limb-bud stage (Fig. 6.7), many absorption vacuoles, absent previously, appear in the syncytium. Fenestration and lacunae begin to appear in the syncytium and are conspicuous by mid-gestation. The syncytium is extensive at term, when the syncytial mass is reduced into long tubular stretches of maze-like lacunae (Figs. 6.6a, b and 6.7). Where present, the nuclei are pushed into localized pockets along the peripheral fringes of the layer. Strangely, desmosomes are present between the honeycombed syncytial mass which contains numbers of secretory vesicles and multivesicular bodies. Desmosomes within the syncytial layer (Fig. 6.6a, b) are probably unique to *Rousettus leschenaulti*. No other bat species, such desmosomes have been reported. An interdigitating membrane between the syncytiotrophoblast, is first encountered at the limb-bud stage, and becomes progressively more prominent with advancing development. It bears desmosomes and is most prominent at term (Fig. 6.6a, b).

6.1.6 Cytotrophoblast

It is made up of rows of cells with well-defined membranes. The nuclei show discontinuous halos at their periphery (Figs. 6.6a, b and 6.7). In the earlier two stages (trilaminar blastocyst and neural groove), a large number of darkly staining mitochondria are seen localized towards their apical poles. The cytoplasm reveals vacuoles and proliferation of rough endoplasmic reticulum (Figs. 6.6a, b and 6.7). Intercellular spacing between adjacent cells was present at all stages, interrupted by desmosomes and junctional complexes. At mid-term, the cytoplasm appears more electrons lucent and many coated vesicles adhere to it on the foetal side. The Golgi complex is inconspicuous. By mid-term, attenuation of the cytotrophoblastic layer commences until it is reduced to thin flanges. In the term placenta, cytotrophoblast is virtually absent in some regions. In these regions, the syncytium spans the space between the cytotrophoblast cells and rests directly on the basal lamina to face the foetal endothelium (Fig. 6.7). Thus, some regions of the interhemal membrane appear to be hemomonochorial.

6.1.7 Basal Lamina

Generally, at all stages, it is a continuous, double-layered structure of constant width between the cytotrophoblastic layer and the foetal endothelium. At mid-term, the cytotrophoblastic

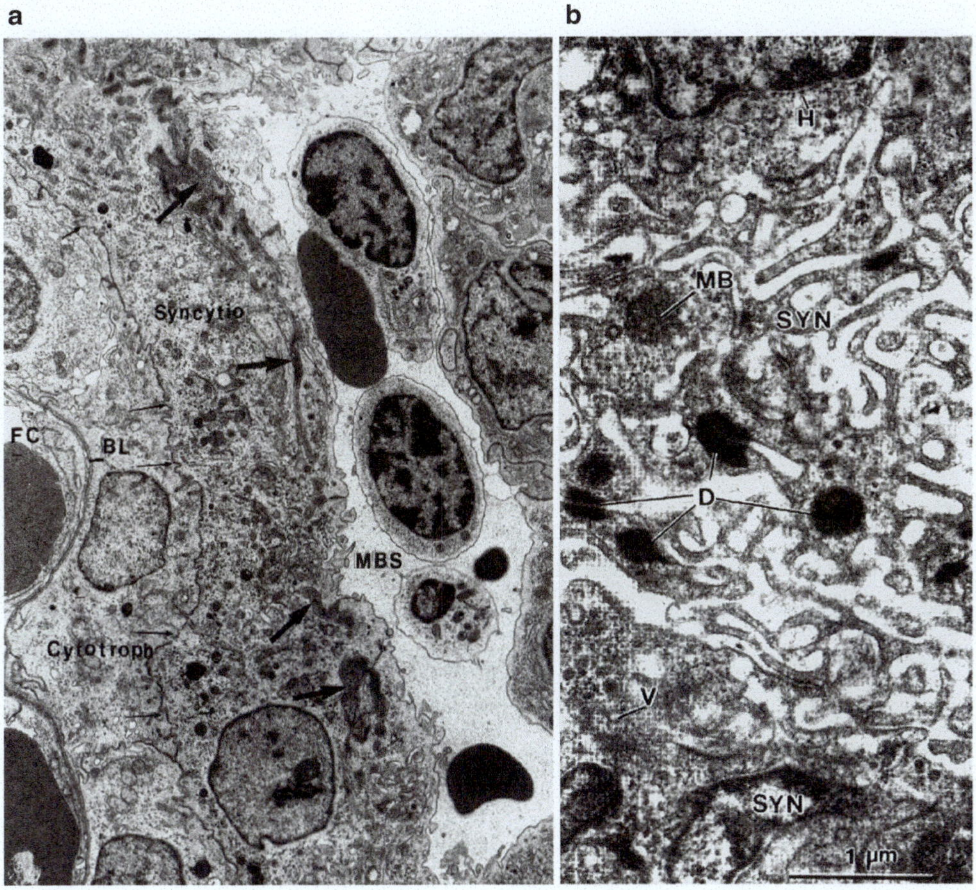

Fig. 6.6 (**a**) High-power E.M. of placenta *Rousettus leschenaulti* at the early limb-bud stage of gestation. Few isolated spindle-shaped cells floating free in cytotrophoblast can be seen. Interdigiting membrane is indicated by arrows [X 3500]. (**b**) Placenta of *Rousettus leschenaulti* at term. The spongy sieve-like syncytiotrophoblast (SYN) shows intrasyncytial desmosomes (D), nuclear halo (H), multivesicular body (MB) and vesicles (V). From Bhiwgade et al. 2000

cytoplasm appears more electron lucent and many coated vesicles adhere to its foetal side. In the term placenta, the cytotrophoblast attenuates bringing the syncytium in contact with the foetal endothelium at numerous places 5 B.

6.1.8 Foetal Endothelium Embedded in Mesenchyme

Foetal mesenchyme shows cytoplasm with abundant rough endoplasmic reticulum and mitochondria. Foetal capillaries are not fenestrated and occasionally abut directly against the trophoblastic basal lamina (Fig. 6.7). Pinocytotic vesicles and mitochondria were observed indicating active transfer of metabolites.

The presence of the maternal endothelium as a thin ectoplasmic layer, and both layers of trophoblast from the onset of placentation characterizes the placenta as hemodichorial. Fig. 6.7 is a schematic representation of different layers in the labyrinthine hemodichorial placenta of *Rousettus leschenaulti*.

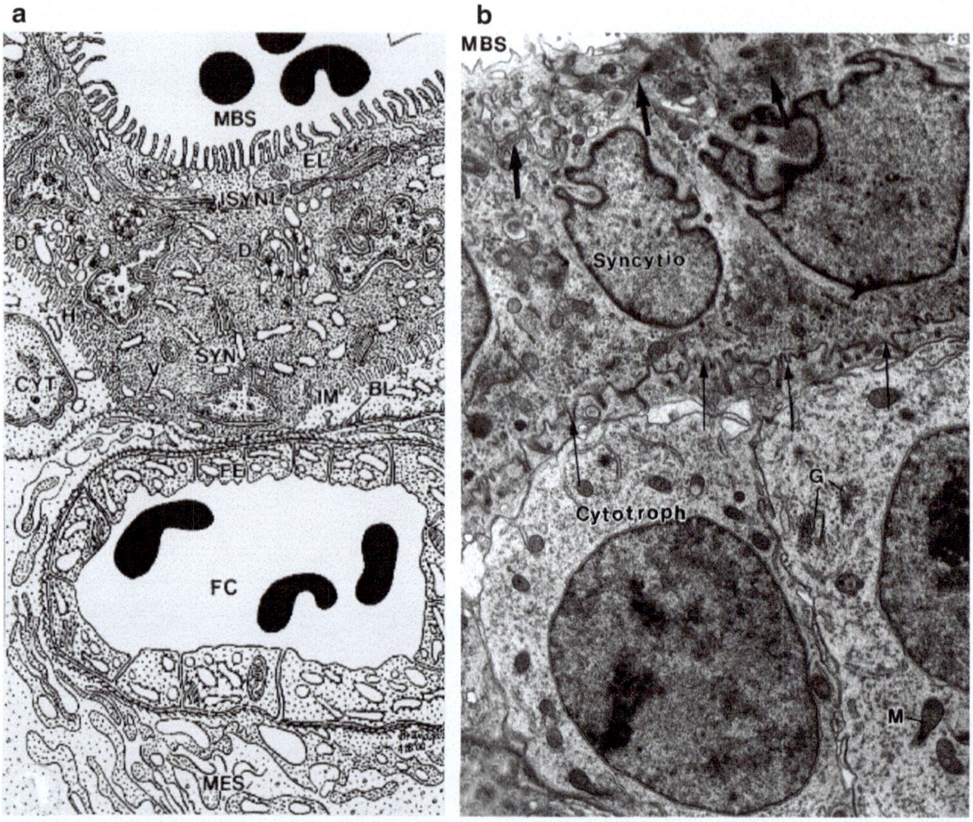

Fig. 6.7 (**a**) A is a schematic diagram showing various layers of the trophoblast seen in the labyrinthine hemodichorial placenta of *Rousettus leschenaulti*. Abbreviations: *BL* basal lamina, *CYT* cytotrophoblast, *D* desmosome, *EL* ectoplasmic layer, *FC* foetal capillary, *FE* foetal endothelium, *H* nuclear halo, *IM* interdigitating membrane, *ISYNL* intrasyncytial lamina, *MBS* maternal blood space, *MES* mesenchyme, *SYN* syncytiotrophoblast, *V* vesicles. (**b**) Electron Micrograph (high power) of the placenta in *Rousettus leschenaulti* at the early limb-bud stage showing two trophoblastic layers syncytiotrophoblast (Syt) and cytotrophoblast (Cyt). Golgi body (G), interdigitating membrane (arrows) and prominent mitochondria (M) are also prominent [X 6000]. From Bhiwgade et al. 2000

6.1.9 *Cynopterus sphinx*

Light microscopic evaluation of the placental disc of *Cynopterus sphinx* (Figs. 6.8, 6.9, 6.10, 6.11 and 6.12) shows that the placenta comprises the same components as in *Rousettus leschenaulti*. The umbilical cord and the yolk sac show similar structures. The special features that are observed in *Cynopterus sphinx* are the much higher ratio of syncytiotrophoblast to cytotrophoblast at the limb-bud stage of placental development and the ratio becoming much less at term. The term placenta, significantly, also shows the absence of maternal endothelial layer and the intrasyncytial lamina.

The interhemal membrane of Cynopterus comprises the following parts:

6.1.10 Maternal Blood Space

The maternal blood spaces are lined by the free surfaces of the syncytiotrophoblast since the maternal endothelium is absent. This surface shows long, slender, microvillar extensions protruding into the maternal blood space (Fig. 6.13).

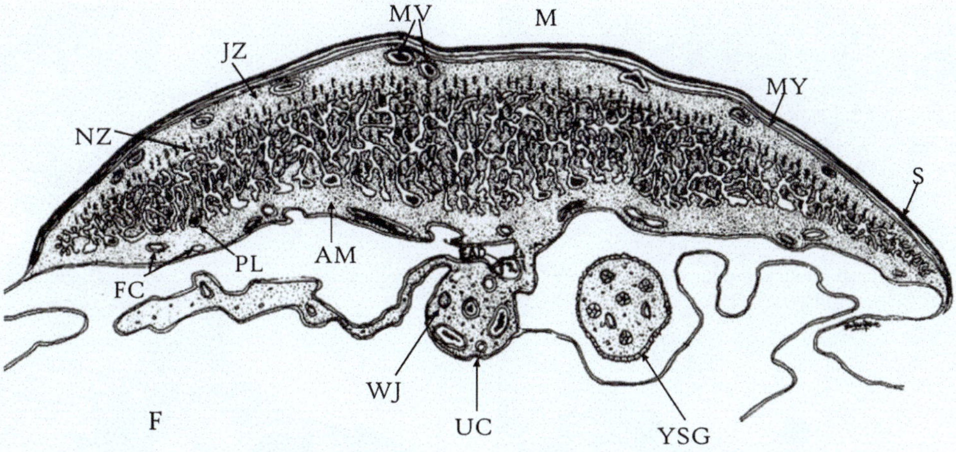

Fig. 6.8 The placental disc at full-term pregnancy (drawn with Camera lucida) shows various components of the chorioallantoic placenta in *Cynopterus sphinx gangeticus* starting from the maternal (M) side to the foetal (F) side. *AM* allantoic mesenchyme, *EAD* endodermal allantoic duct, *FC* allantoic vessels, *JZ* junctional zone, *MV* maternal efferent vessels, *MY* myometrium, *NZ* necrotic zone, *PL* placenta labyrinth, *S* serosa, *UC* umbilical cord, *WJ* Wharton's jelly (mucoid connective tissue), *YSG* yolk sac gland. From Bhiwgade et al. 2000

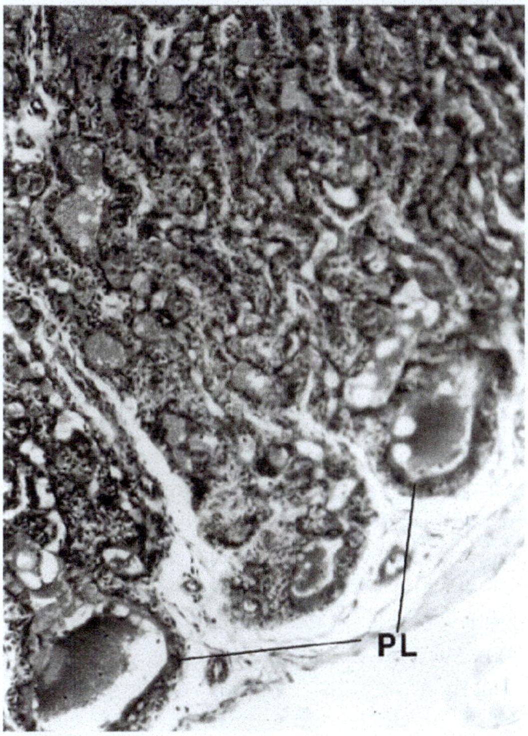

Fig. 6.9 Low-power light micrograph of the term placenta of *Cynopterus sphinx gangeticus* showing placental tubules (PL) arranged in a labyrinthine manner [X 150]. From Bhiwgade et al. 2000

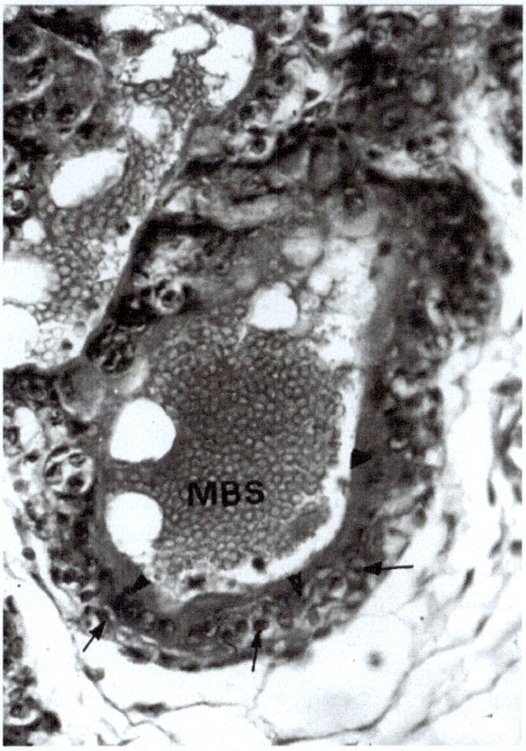

Fig. 6.10 Note the hemodichorial condition with maternal blood space (MBS), syncytiotrophoblast (arrow head) and cytotrophoblast (arrow) [X 340]. From Bhiwgade et al. 2000

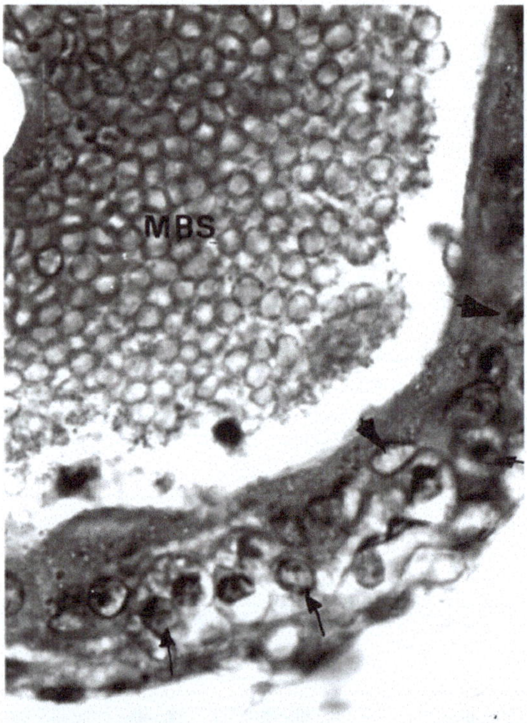

Fig. 6.11 Note the prominent hemodichorial condition with maternal blood space (MBS), syncytiotrophoblast (arrow head) and cytotrophoblast [X 640]. From Bhiwgade et al. 2000

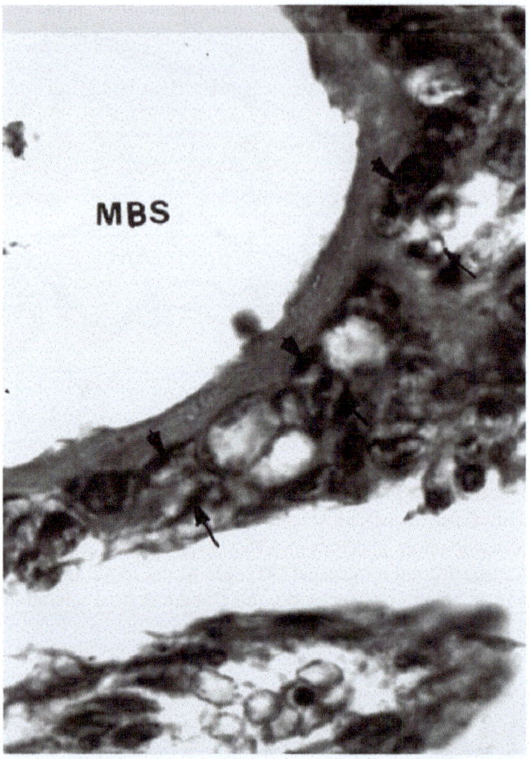

Fig. 6.12 Note the hemodichorial condition with the presence of a maternal blood space (MBS), syncytiotrophoblast (arrow head) and cytotrophoblast [X 640]. From Bhiwgade et al. 2000

6.1.11 Syncytiotrophoblast

This layer is deeply electron dense, compared to the underlying cytotrophoblast layer, which bears rows of irregularly shaped nuclei with darkly staining nucleoli. Replacing the intrasyncytial lamina, the space is packed with canalicular structures. They lie underneath the free surface of the syncytium at the edges of the maternal blood space. Due to this, these lacunae have a filigreed appearance and are aligned parallel to the long axis of the trophoblastic layer. During the late limb-bud stage, the syncytiotrophoblast comprises the bulk of the placental tissue. At several places, syncytiotrophoblast surrounds clumps of cytotrophoblastic cells so that the interhemal membrane appears to be hemomonochorial (Fig. 6.13). At term, on the other hand, the syncytial topography is extremely eroded, being reduced to thin flange at sites that are heavily invested with lipid droplets (Fig. 6.14). The organelle complement during the limb-bud stage includes numerous coated vesicles. Long intercellular canals bearing desmosomal connections meandering through the cytoplasm between the syncytium and the MBS (Fig. 6.13). The heavy deposits of lipid droplets are seen in this layer. Multivesicular bodies are also present. The organelles present in the syncytiotrophoblast at term are: well-defined intercellular channels with desmosomal connections, vesiculated and lamellar rough endoplasmic reticulum (rER), coated vesicles, Golgi complex, dense bodies and loosely aggregated glycogen granules (fig. 6.14). The syncytium and the apposed cytotrophoblastic layer together are thrown into distinct interlocking folds to form an interdigitating membrane which comprises brush-like fibrillar elements running along the length of the double membrane on both of its sides and embedded in

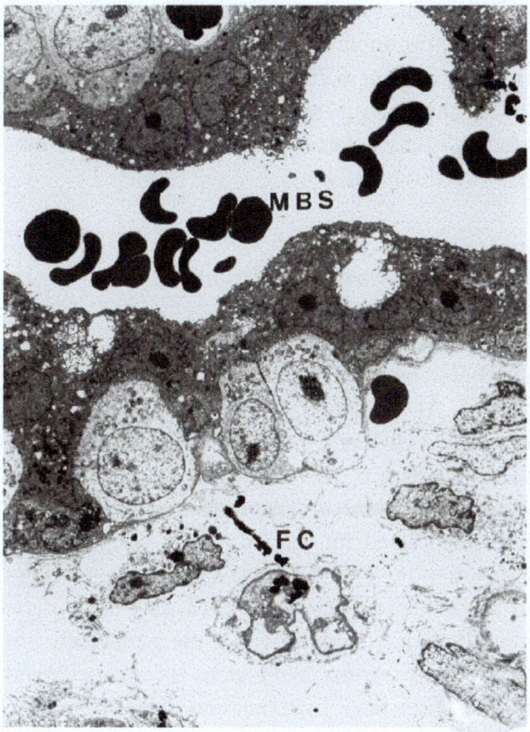

Fig. 6.13 Electron micrograph of the Interhemal membrane in *Cynopterus sphinx gangeticus* at late limb—bud stages of gestation (low power). The two trophoblastic layers, i.e. syncytiotrophoblast (SYN) and Cytotrophoblast (CYT), intervening between the maternal blood space (MBS) and the foetal capillary (FC) is clearly seen. The ratio of the syncytium to cytotrophoblast mass is roughly 2:1 with indentation of the syncytial layers resulting in isolated pools of cytotrophoblast (CYT). The mitochondria (M) are seen more in the cytotrophoblast region [X 2000]. From Bhiwgade et al. 2000

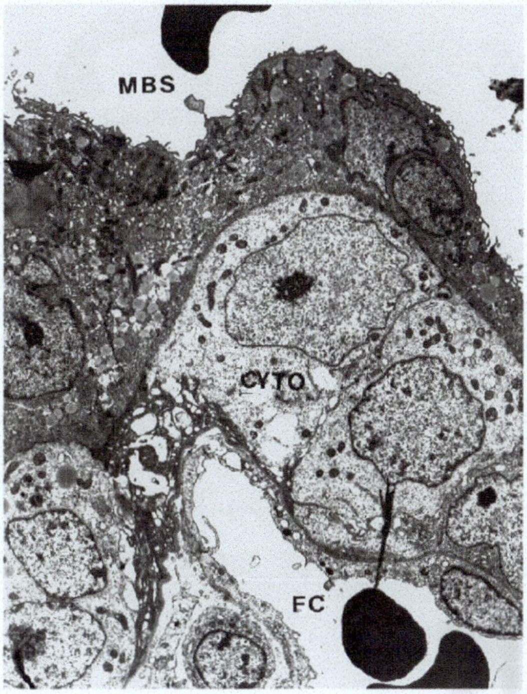

Fig. 6.14 Electron micrograph of the term placenta in *Cynopterus sphinx gangeticus*(low power) exhibiting a continuous syncytiotrophoblast (Syt) bordering the maternal blood space (MBS) on one side and the cytotrophoblast (Cyt) on the other. Basal lamina (BL) and foetal capillaries (FC) with endothelium (FE) are distinct. Syncytial layer bears many lipid droplets (L). Cytotrophs show intracellular space (ICS), mitochondria (M) along with osmophilli lipids droplets (*) in the binucleate Giant cells (B-Cyt). From Bhiwgade et al. 2000

the trophoblast. This configuration is more conspicuous at term than at the limb-bud stages and displays desmosomal and junctional complexes along its length (Figs. 6.14 and 6.15).

6.1.12 Cytotrophoblast

The cytotrophoblast, during the limb-bud stage, is restricted to isolated clumps of cells with predominant desmosomes, some junctional complexes and with the overlying syncytiotrophoblast (Fig. 6.13). It is characterized by large well-defined membrane-lined cells with large nuclei and prominent nucleoli. The cytoplasm is electron lucent and the cells are joined along their long axes into a single layer via desmosomal and junctional complexes. Intercellular spaces appear between adjacent cells at term.

The rough ER is prominent as numerous vesicles at the limb-bud stage and tends to change progressively into the labyrinthine and lamellated form by term. Ovoid or elongated mitochondria bearing lamellar cristae exist at both stages and are conspicuous toward the overlying syncytium. The Golgi complex is more prominent during the earlier stages and is represented by both lamellar, parallelly stacked cisternae and the vesicles. Also present are loosely aggregated glycogen granules, and a sparser population of dense bodies and lipid droplets (Fig. 6.14). A peculiarity of this layer, particularly at term, is the continuity of the

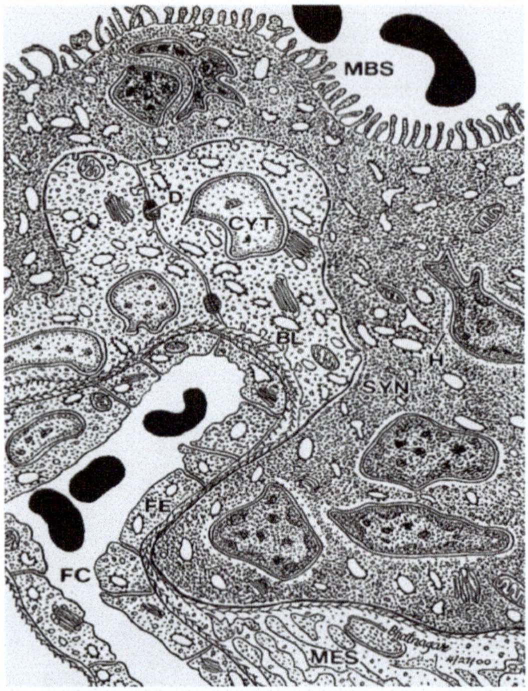

Fig. 6.15 Diagrammatic depiction of the trophoblast layers in the labyrinthine hemodichorial placenta of *Cynopterus sphinx gangeticus*. Note the absence of intrasyncytial lamina and the presence of desmosomes in the syncytium. From Bhiwgade et al. 2000

intercellular canalicular system from the interdigitating membrane to the basal lamina, which also bears desmosomal connections. The layers are thin at this stage, especially in the cytotrophoblast, whereas in some regions, it is greatly reduced compared to the syncytial layer. At other places, the invasive action of the syncytium isolates the cytotrophoblast cells breaking the continuity of the cytotrophoblast. The syncytium thus, comes in direct contact with the maternal blood on one side and is in contact with the foetal capillary on the other side (Fig. 6.29). In *Rousettus leschenaulti*, the minimal interhemal barrier is discernible as hemomonochorial at such regions.

6.1.13 Basal Lamina

It is thick, often double-layered structure of uniform width throughout its length, and is homogeneously electron dense. It lies between the cytotrophoblast and the foetal endothelium (Figs. 6.28 and 6.29). At term, granules apparently undergoing transportation are also seen along with associated vesicles that are in continuity with the membrane.

6.1.14 Foetal Endothelium

The foetal capillaries rest on the basal lamina and are lined by endothelial cells containing large nuclei. At limb-bud stage, the capillaries are embedded in foetal mesenchyme, while at term the cytoplasm is stretched into a thin ring lining the capillary lumen. There are labyrinthine eER with associated mitochondria and Golgi bodies. Erosion of the cytotrophoblast at this stage occurs to a greater extent (Figs. 6.13 and 6.14). The hemodichorial nature of the placenta in *Cynopterus sphinx gangeticus* is characterized by the absence of maternal endothelium and the presence of two layers of the trophoblast (Figs. 6.14 and 6.15).

The absence of the maternal endothelium and the presence of both the layers of trophoblast from the onset of placentation characterizes the placenta as hemochorial. The relationship of different layers in the labyrinthine hemodichorial placenta of *Cynopterus sphinx gangeticus* is schematically depicted in Fig. 6.15.

6.1.15 Discussion

The initial placenta is endotheliochorial which undergoes changes that transform the placenta into a final condition that is seen at term (Gopalakrishna and Karim 1980). The final configuration of placenta results from gradual erosion and consequent loss of placental layers. No mammal, so far, has been described with a placental condition in which a layer is added during progressive gestation.

The change from the initial endotheliochorial to hemochorial condition marks the transformation of interstitial lamina—a phenomenon well documented in *Macrotus waterhousii* (Bodley

1974), *Desmodus rotundus* (Bjorkman and Wimsatt 1968), *Hipposideros fulvus* (Bhiwgade 1990), *Tylonycteris pachypus* (Mahaiey et al. 1995), in the primary and tertiary placenta of *Miniopterus schreibersii* (Malassine 1970; Kimura and Uchida 1984; Mossman 1987; Bhiwgade et al. 1992) and *Pteropus giganteus* (Karim and Bhatnagar 1996). This change is also observed in *Rousettus leschenaulti*, but interestingly not in *Cynopterus sphinx gangeticus*.

The earlier stages in *Rousettus leschenaulti* have been identified as endotheliochorial (Gopalakrishna and Karim 1980), and it corroborates our own observations at the trilaminar stage. In fact, during the neural groove and early limb-bud stages, the spindle-shaped maternal endothelial cells do not form a continuous lining. These few floating spindle-shaped cells, according to Stephens (1969), are maternal in origin, but do not really account for an endotheliochorial condition. Therefore, according to him, such placenta can be regarded as being hemochorial right from the beginning. Wimsatt and Enders (1980) have reported a similar case in the neotropical disc-winged bat, *Thyroptera tricolor*, in which the maternal endothelium is lost much earlier, i.e. prior to the limb-bud stage. This led to the conclusion that the ultrastructural organization of the interhemal membrane is hemodichorial, but the maternal endothelium disappears relatively early, and the trophoblastic differentiation can be considered precocious. However, in our opinion, the very existence of the maternal endothelium during the trilaminar blastocyst stage, excludes the possibility of placenta in *Rousettus leschenaulti* being hemochorial from the very beginning. Instead, the placentation has been reported as endotheliochorial up to mid-gestation (Gopalakrishna and Karim 1980). According to observations, only the placenta at trilaminar blastocyst stage is endotheliochorial but all stages thereafter exhibit a hemochorial condition. The establishment of a hemochorial condition with the loss of the maternal endothelium indicates the transformation of the interstitial membrane into the intrasyncytial lamina. Similar description has been reported by Cukierski's (1987) based on observations on the formation of intrasyncytial lamina from the maternal endothelium in Chiroptera. Discontinuities in the syncytial layer permit the cytoplasm of the syncytiotrophoblast to flow through this layer, lining the maternal blood space directly adjacent to the intrasyncytial lamina (Enders and Wimsatt 1968; Bhiwgade et al. 1992). Thus, an ectoplasmic layer, comes into existence, that, defines the perimeter of maternal blood space.

The luminal syncytiotrophoblast surface edging the maternal blood space has irregular microvilli. These digitate extensions form a morphological modification for increased surface area and enhanced diffusion in both *Rousettus leschenaulti* and *Cynopterus sphinx gangeticus*. Similar observations have been made in *Desmodus rotundus* (Bjorkman and Wimsatt 1968), *Pteropus giganteus* (Karim and Bhatnagar 1996) and rat (*Rattus norvegicus*), rabbit (*Oryctolagus cuniculus*) and armadillo (*Dasypus novemcinctus*—Enders 1960, 1965).

The syncytiotrophoblast in *Rousettus leschenaulti*, from mid-gestation onward to the coarse syncytium, is a feature that shows a definite resemblance with the carnivore placenta (Enders 1960; Anderson 1969). This has been reported only in one other bat, *Tyloncycteris pachypus* (Mahaiey et al. 1995). The intrasyncytial desmosomes are a feature that has not been reported in any other bat species. These desmosomes probably support the syncytium and prevent it from collapsing. They also control the ports of entry and exit, as well as exchange between different layers of the tissue. They also provide continuity between the maternal blood space and the syncytium.

Similar, but not identical, observations on the coarse syncytium have been reported in rats (Enders 1965), ferrets (*Mustela putorius furo*—Lawn and Chiquoine 1965), and nine-banded armadillo (Enders 1960) where these structures help to increase surface area for better exchange of material (Gammal 1985; Bhiwgade et al. 1992). The spongiotrophoblast or coarse syncytial tissue is believed to drain the maternal blood from the placenta as in rodents (Wooding and Flint 1994), and as it is perfused by the maternal blood, it can introduce foetal hormones and other substances into the maternal circulation (Wooding and Flint 1994).

The interdigitating membrane at the syncytiotrophoblast and cytotrophoblast interface was also observed in the nine-banded armadillo (Enders 1960) and in most chiropteran placentae that have both the trophoblastic layers. In the syncytium of *Rousettus leschenaulti*, the adherent vesicles, according to Bjorkman and Wimsatt (1968), indicate a transport mechanism across the layers. Accordingly, in *Rousettus leschenaulti*, they would be involved in transport within the same syncytium.

At term, the disappearance of the cytotrophoblast suggests that after enough syncytiotrophoblast was produced, its role changed to minimal maintenance, as suggested by Midgley et al. (1963) in the monkeys. These authors stated that with the advancing gestation, the cytotrophoblast exhausts its synthetic machinery and transforms into syncytium. Because of this change, the cytotrophoblast layer becomes confined to isolated areas. By this modification of extreme attenuation, the syncytium is reduced into thin flanges in the term placenta of *Rousettus leschenaulti*. The diffusional distance is thus lessened (Anderson 1969). This is comparable to armadillo (Enders 1960) and human (Wislocki and Dempsey 1955) conditions. The few complexly folded and interlocked surfaces in close contact, and the numerous junctional zones between the cell membranes in *Rousettus leschenaulti* have also been reported in the nine-banded armadillo (Lawn and Chiquoine 1965), the golden hamster (*Cricetus auratus*—Carpenter 1972), and the marmoset (*Callitthrix jacchus*—Jollie 1973).

In *Cynopterus sphinx gangeticus*, the syncytial mass observed at term contains cytoplasmic complements that have been reported in the syncytia of the raccoon (*Procyon lotor*—Creed and Biggers 1963), rat, human, armadillo, rabbit, guinea pig (*Cavia porcellus*–Enders 1965) and dog placentae (Anderson 1969). According to Lawn and Chiquoine (1965) and Dempsey and Wislocki (1956), the abundance of granular endoplasmic reticulum is believed to be associated with protein synthesis. Lipid droplets have been reported in the syncytial trophoblast of both the nine-banded armadillo (Enders 1960) and humans (Wislocki and Dempsey 1955). Among bats, lipid droplets have been observed as syncytial inclusion in *Taphozous melanopogon* (Bhiwgade 1990), *Myotis lucifugus* (Enders and Wimsatt 1968), and in the yolk sac of *Tadarida brasiliensis* (Stephens 1969) as well as in the cytotrophoblast of *Rhinopomahar dwickei* (Bhiwgade 1990) and *Pteropus giganteus* (Karim and Bhatnagar 1996).

The interdigitating membrane reported in the nine-banded armadillo (Enders 1960) is also observed in *Cynopterus sphinx gangeticus*. In fact, this membrane seems to be an integral component of the placental layers in all mammals, including bats that have both the syncytiotrophoblast and cytotrophoblast layers. It is absent in the monochorial placenta. Attenuation is also noticed in the cytotrophoblast layer of *Cynopterus sphinx gangeticus* and its near disappearance can be seen in *Rousettus leschenaulti*, in which the syncytial mass and volume are enhanced at the cost of the cytotrophoblast (Midgley et al. 1963). This corroborates the view expressed by Enders (1965) and Body and Hamilton (1966). According to them, the cytotrophoblast is gradually eliminated from the interhemal membrane with the progression of gestation. The eliminated cells get incorporated into the outer syncytium, as observed by Jollie (1973) in the marmoset placenta.

The giant cells observed in our study are more typical of the bovine placenta and have also been noted in the Chinchilla placenta (King and Tibbitts 1976). Wooding and Flint (1994) reported the giant cells to be a characteristic of placentation in rodents. These cells are phagocytic and migratory in nature. They undergo endomitosis, i.e. DNA replicates without the subsequent cell division. This could be a probable reason for the existence of binucleate giant cells in *Cynopterus sphinx gangeticus* and *Pteropus giganteus* (Karim and Bhatnagar 1996). Along with the invasive syncytium, the phagocytosis by giant cells erode the maternal tissue (Wooding and Flint 1994). Deane et al. 1962 have suggested a probable role for steroid production for these giant cells. Wooding (1982a, b, 1984) reported that in sheep and goat, the foetal chorionic binucleate cells give rise to the syncytia.

Clearly, each layer of the interhemal membrane has a particular function. The main function of the interhemal membrane is enhancement the transport of various substances in both directions between maternal and foetal environments, so as to ensure foetal survival. The layers, therefore, display an ability to interact with each other. Placentologists have theorized that the morphological modifications seen in these layers like the coarse syncytia and the presence of indentations, help in improving efficiency of transplacental transport which ensure foetal survival. Through our observations, the interhemal membrane of *Rousettus leschenaulti* and that of *Cynopterus sphinx gangeticus* have been ascertained to be of hemodichorial nature, a feature that seems to be characteristic of the family Pteropodidae.

6.2 *Taphozous melanopogon*

The ultrastructural studies on *Taphozous melanopogon* placenta reveal the presence of different tiers of trophoblast, the maternal endothelial cells in chori-allantoic discoidal placenta, and the non-cellular barrier, the interstitial membrane that separates maternal and foetal environments.

During the limb-bud stages of development slight changes are observed in both the trophoblastic cells. In contrast to that of cytotrophoblast and syncytiotrophoblast of previous stage, the cytoplasm of both trophoblasts has extensive rough endoplasmic reticulum, hypertrophied mitochondria and a prominent Golgi zone. The most evident characteristic that distinguishes placental organization during the mid-pregnancy is the prominent presence of foetal capillaries in the region of the trophoblast. The mesenchyme also shows the adequate presence of foetal capillaries. A significant number of cytotrophoblast cells are seen in contact with the common basal membrane. Cytoplasm of these cells is quite less electron dense as compared to those of the syncytium during the mid-term.

The total absence of the cytotrophoblast during full term indicates that the placental barrier is of the endotheliomonochotial type. The placental barrier is characterized by four cellular layers that constitute the interhaemal membrane as seen in (Fig. 6.16a, b, c). A maternal endothelium, syncytial trophoblast, foetal endothelium and the mesenchyme.

In addition, the continuous interstitial membrane with a few perforations and the basal lamina are visible. Foetal capillaries distinctly invade the trophoblast layer at this stage. The trophoblast layer is seen in close relation to maternal blood capillaries. The maternal endothelium is greatly hypertrophied but it retains its cellular character (Fig. 6.16a, b, c). In some places, the maternal endothelium directly comes in contact with the foetal capillary and therefore the placental barrier appears to be endothelio-endotheliochorial in nature.

The maternal endothelium rests on an interstitial membrane. The zone of attachment is quite extensive and is distinctively fibrous. There are microvilli seen on the luminal surface of the endothelium (Fig. 6.16a, b, c). There is a single layer of syncytiotrophoblast which looks well organized with nuclei placed evenly. The mesenchyme shows the presence of foetal capillaries which places come to lie close to the trophoblast basal lamina (Fig. 6.16a, b, c). The foetal connective tissue space is reduced and is seen at the basal lamina of foetal capillary. The layers as observed in the labyrinthine endothelio monochorial placenta of the *Taphozous melanopogon* are schematically shown in Fig. 6.17. Under electron microscope, the placenta of *Taphozous melanopogon* shows a broad maternal endothelium, a continuous interstitial membrane and a single layer of syncytiotrophoblast. The placental barrier is endotheliomonochorial and not endotheliochorial (Sandhu 1986).

6.3 *Megaderma lyra lyra*

Under the electron microscope, the chorio-allantoic placental barrier at the full term of gestation in *Megaderma lyra lyra* shows sequential

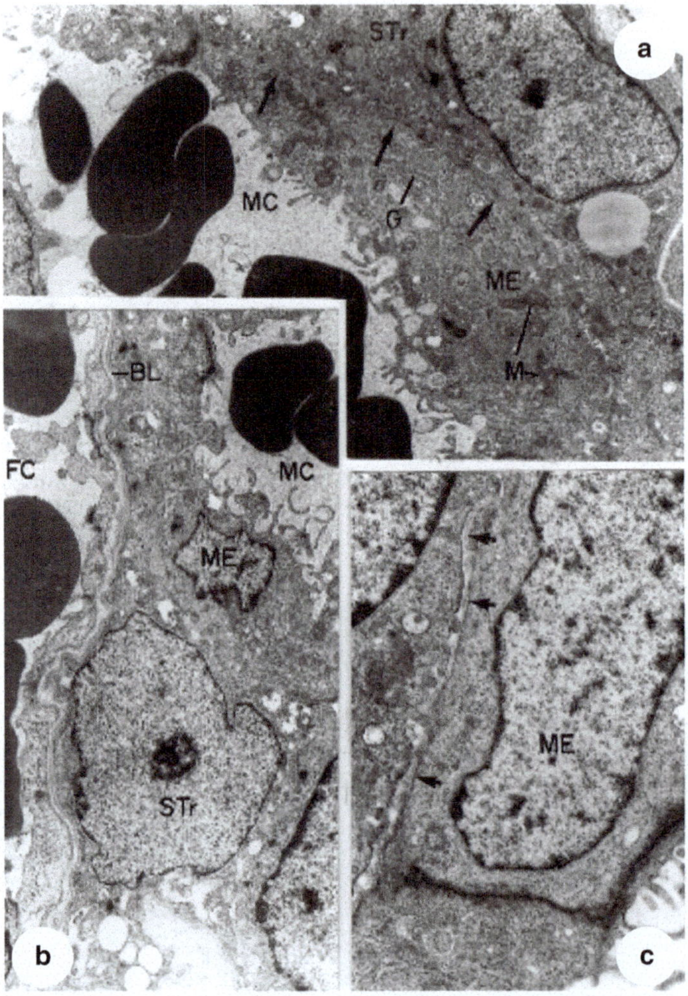

Fig. 6.16 (**a**) Transmission electron micrograph of the interhemal membrane of *Taphozous melanopogon* near term. Note the maternal blood capillary (MC) bordered by a distinct maternal endothelium (ME) resting on an interstitial membrane (arrows) and syncytiotrophoblast (STr). (**b**): The maternal endothelium contains mitochondria (M) and Golgi zones (G). The foetal blood capillary (FC) is separated from the maternal blood capillary (MC) by its endothelium and a basal lamina (BL). (**c**): The maternal endothelium (ME) and interstitial membrane (short arrows) are seen [X 6000] [X 6000] [X 9000]. From Bhiwgade 1990

layers from the maternal side. Viz. the maternal endothelium, an interstitial membrane which is discontinuous, a distinct syncytial trophoblast, cytotrophoblast as a thin layer, the foetal stroma and its capillaries with endothelium resting on a basal lamina (Fig. 6.18). The endothelial cells at the maternal end are hypertrophied showing a large spherical nucleus at the centre (Fig. 6.18). The mitochondria of these cells show the typical structure. The endothelial cells, however, show numerous endoplasmic reticulum which are granular in appearance. The Golgi complex is present in the juxtanuclear region and is comparatively smaller. The endothelium shows long microvilli at its luminal surface. These microvilli emanate in clusters from small indentations of the endothelial surface (Fig. 6.18). The interstitial membrane, unlike the maternal endothelium, is a homogeneous and discontinuous membrane. It shows gaps at many places adjacent to the syncytiotrophoblast. Through these gaps, the materials from the syncytiotrophoblast can percolate or diffuse. The interstitial membrane is present throughout the placental labyrinth like the trophoblast. In a few places, the maternal endothelium cells have undergone degenerative changes and have become rounded. At these sites, the maternal endothelial cells have lost their attachment to each other. In some places, they have separated from the interstitial membrane too. At these places, the maternal blood

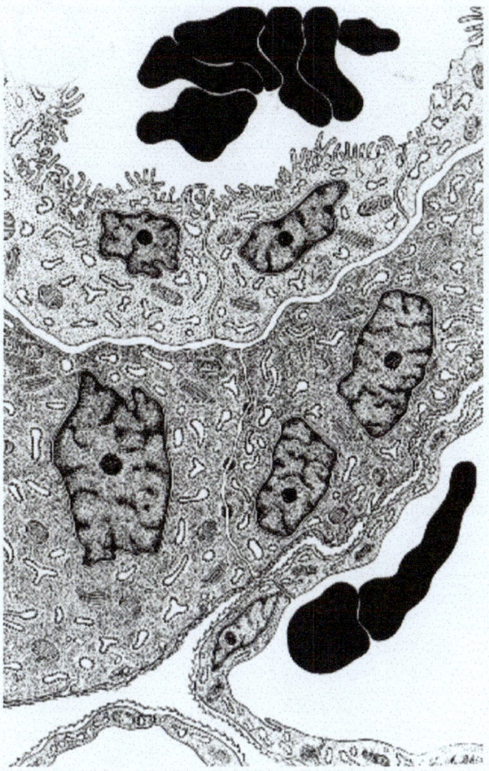

Fig. 6.17 A schematic representation of the layering of the trophoblast in the labyrinthine endothelio monocorial placenta of *Taphozous melanopogon*. The maternal blood capillary is on the top. A foetal capillary is seen in the lower right corner. Between the two bloodstreams there are maternal endothelium, a single layer of syncytial trophoblast, basal lamina (cross-hatched) and foetal endothelium. From Bhiwgade 1990

comes in direct contact with the interstitial membrane.

Within the placental barrier, the syncytiotrophoblast is a highly differentiated layer and shows no signs of the existence of cell boundaries. It appears as a continuous layer of syncytium but varies in thickness at places. The nuclei are irregularly scattered and at places are seen in groups within the syncytium. There is abundance of endoplasmic reticulum with cisternae that are partially dilated. Typical Golgi zones are also seen (Fig. 6.19a, b). Within the syncytium, mitochondria are scattered but are relatively more towards the endothelial border. In the mature placental labyrinth, cytotrophoblast cells are less and represent a minor percentage of the tissue. Cytotrophoblast cells are seen in immediate contact with the external surface of the syncytium and lie embedded beneath the basal lamina. The basal lamina separates the foetal connective tissue from the meshes of labyrinth. During this stage of gestation, the cytotrophoblast cells are not in contact with each other. At some places where cytotrophoblast cells adjoin the syncytium, the membranes of the cells and syncytium come in close contact to get complexly folded and interlocked (Figs. 6.18 and 6.20). Due to less number of organelles, the cytoplasm of cytotrophoblast cells is less dense than that of the syncytium. The cytoplasm also shows few mitochondria, endoplasmic reticulum, lipid bodies and some free ribosomes. The foetal connective tissue space is seen only at the basal lamina of the foetal capillary (Fig. 6.19). This nature of barrier appears to be sustained till parturition. The definitive chorio-allantoic placenta can thus be identified as endotheliodichorial (vasodichorial) (Figs. 6.20 and 6.21). The placenta of *Megaderma lyra lyra* consists of a typical endotheliochorial labyrinth with two layers of trophoblast and a discontinuous interstitial membrane. The placenta at term can be re-classified as endothelio–dichorial instead of endotheliomonochorial (Gopalakrishna and Khaparde 1978).

6.4 *Rhinolophus rouxi*

In the endotheliodichorial type of placenta of *Rhinolophus rouxi,* the elements that are encountered from the maternal blood space towards the foetal blood capillary are maternal endothelium, interstitial membrane, cytotrophoblast, basal lamina and foetal mesenchyme (Fig. 6.22). Throughout the gestation, maternal endothelium is seen surrounded by a continuous layer of syncytiotrophoblast with a discontinuous interstitial membrane in between (Fig. 6.23). The foetal blood capillary is enclosed by a foetal endothelium and has a distinct basal lamina. The cytotrophoblastic layer rests on the basal lamina (Fig. 6.23). The two trophoblastic layers are in close opposition and the membrane between them

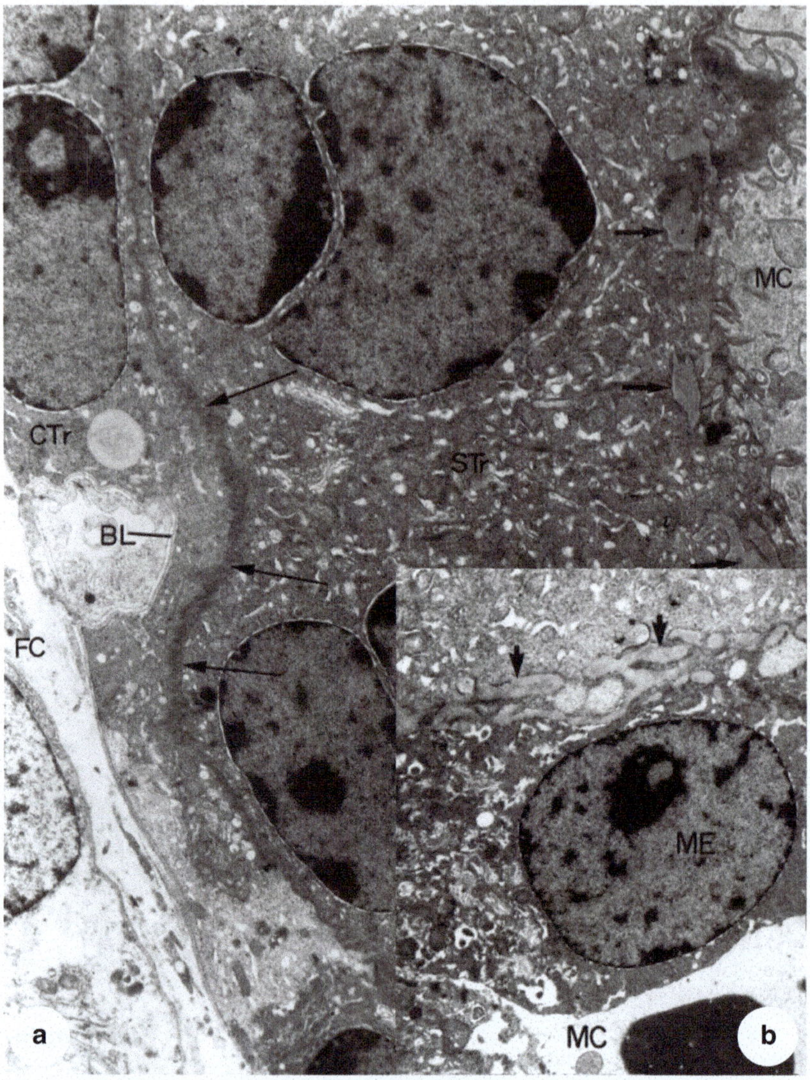

Fig. 6.18 (**a**) (inset). These figures show the placental labyrinth at term in *Megaderma lyra lyra* under the transmission electron microscope. The endothelium, a basal lamina (BL), a thin layer of cytotrophoblast (CTr), a layer of syncytiotrophoblast (STr) and the discontinuous interstitial membrane (large arrows) separate the foetal capillary (FC) from the maternal blood capillary (MC). Microvilli are seen on the luminal surface of the endothelium. The small arrows indicate an area where syncytial trophoblast and cytotrophoblast are in close contact and have interlocked. (**b**) (Inset) shows the maternal endothelium (ME) resting on the interstitial membrane (arrows) [X5,000] [X 6000]. From Bhiwgade 1990

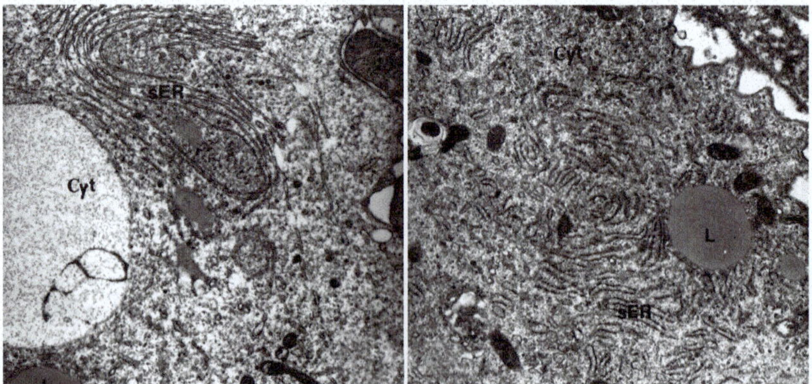

Fig. 6.19 These are high-power transmission electron micrographs of the placenta at the advanced stage of gestation in *Megaderma lyra lyra*. The figures show the cytotrophoblast (Cyt) with rough endoplasmic reticulum (rER) and lipid droplets (L) [X 16,500] [X 16,500]

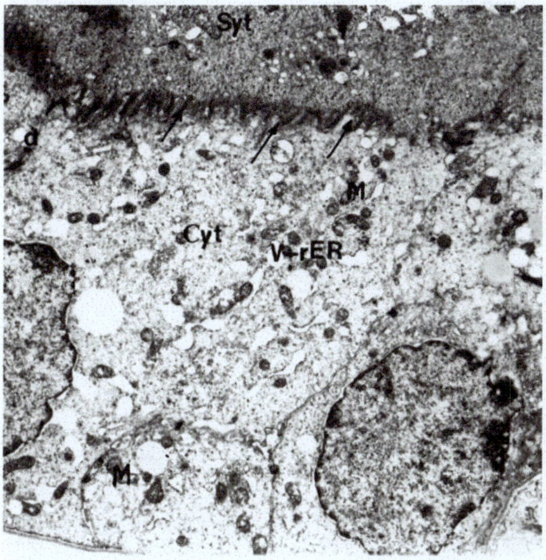

Fig. 6.20 Electron Micrograph of the term placenta (high power) of Megaderma lyra lyra showing apposition of the cytotrophoblastic (Cyt) layer with the Syncytiotrophoblast (Syt) along with intervening interdigitating membrane (arrow). Cytotrophoblast shows mitochondria (M) vesiculated VrER [X 8500]. Unpublished electron micrograph from Bhiwgade and Thakur 1992

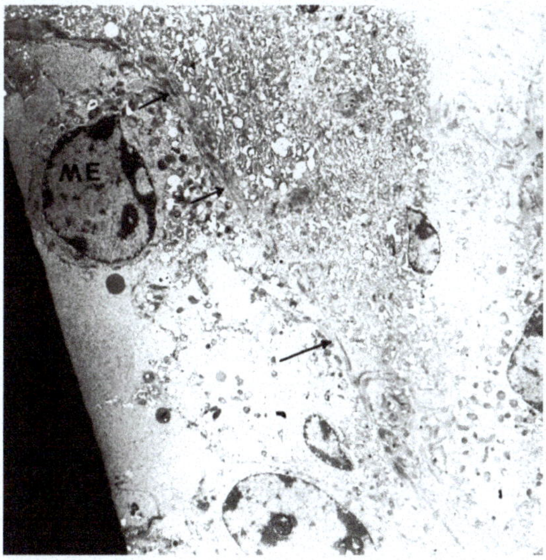

Fig. 6.21 The placenta at term of Megaderma lyra lyra. Maternal endothelial cells (ME) shows a large prominent nucleus and loose highly renovated cytoplasmic matrix bearing mitochondria (M). Interstitial membrane (arrows) is of same girth all along its length. Synlytium is highly vacuolated [X 3500]. From Bhiwgade and Thakur 1992

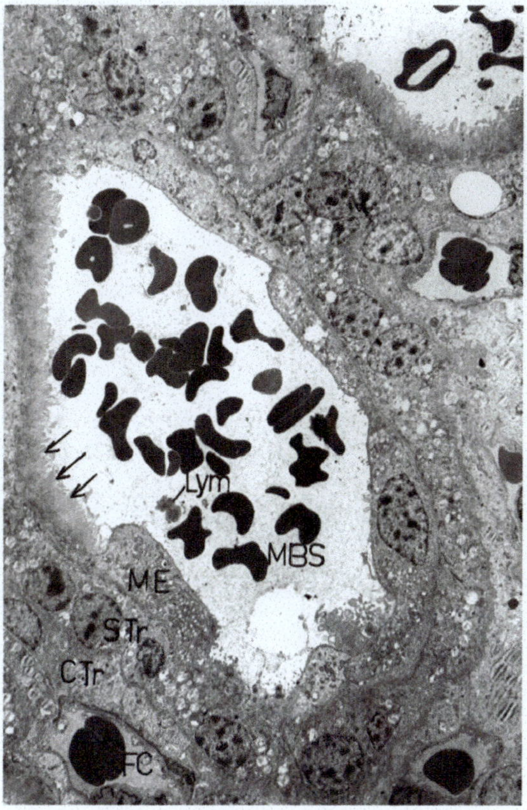

Fig. 6.22 This is a low-power electron micrograph of the placenta of *Rhinolophus rouxi* at term. Presence of lymphocytes (Lym) in the maternal blood space (MBS), the maternal endothelium (ME), syncytiotrophoblast (STr), cytotrophoblast (CTr), foetal capillary (FC) are distinct. Note the microvilli (arrows) extending from the maternal endothelium [X 1600]. From Bhiwgade 1990

shows interdigitations. Also, the two trophoblastic layers are found to be continuous and in some regions, the syncytiotrophoblast is thicker than the cytotrophoblast, whereas in other regions, it is vice versa (Figs. 6.22 and 6.23).

The maternal endothelium is well developed consisting of a single layer of large, elongated cells (Fig. 6.22). The maternal endothelial cells are in proximity to the neighbouring cells and the membrane between them shows the presence of tight junctions. The cytoplasm of this layer is light and granular. The nuclei are irregular in shape and have a double membrane with pores. Heterochromatin is scattered all over the nucleoplasm and is deposited marginally. Mitochondria are found in large numbers. They are present all

Fig. 6.23 This a Low-power electron micrograph of interhemal membrane of *Rhinoluphus rouxi* at full term. The foetal capillary (FC) can be seen surrounded by its thin endothelium (FE), basal lamina (arrow heads), cytotrophoblast (CTr), syncytiotrophoblast (STr), interstitial membrane (large arrows) and maternal endothelium. Numerous microvilli (mv) are seen extending into the maternal blood space (MBS). The interdigitation between the syncytiotrophoblast and cytotrophoblast is also seen (small arrows) [X 5000]. From Bhiwgade and Thakur 1992

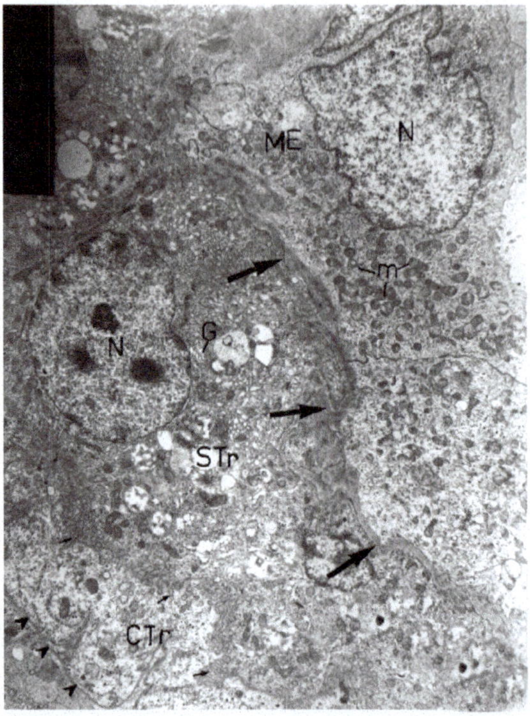

Fig. 6.24 Electron micrograph of the placenta of *Rhinolophus rouxi* at full term (low Power). The placenta shows discontinuous interstitial membrane (large arrows) between the maternal endothelium (ME) and syncytiotrophoblast (Str). Note the numerous mitochondria (m) and nucleus (N) in the maternal endothelium and Golgi zones (G) in the syncytiotrophoblast. Also seen are interdigitations (small arrows) between the syncytiotrophoblast and cytotrophoblast (CTr), which in turn rests on a basal lamina (arrow heads) [X 5000]. From Bhiwgade and Thakur 1992

over the cytoplasm but more in the cytoplasm towards the adjacent cell. Most of them are round, oval or elongated in shape. They have tubular cristae which extend across their width and their matrix is dense. Polyribosomes are found to be scattered all over the cytoplasm. During the full term stages, numerous coated vesicles and glycogen rosettes are present in the cytoplasm and their density is more towards the syncytiotrophoblast. On the side facing the discontinuous interstitial membrane, the cytoplasm also contains continuous pinocytotic vesicles. During the full-term stage, the unicellular nature of this layer is lost in some regions, wherein the cytoplasmic content decreases marginally and towards the maternal blood space it forms fine finger-like microvilli. The stubby microvilli which are few in number and uniform in length and thickness are seen on the free surface of maternal endothelium (Fig. 6.24).

The syncytiotrophoblast is well developed with irregular nuclei. Each nucleus has a double membrane with pores. The nucleoplasm contains a distinct nucleolus with few chromatin centres. Heterochromatin is also found to be deposited marginally (Figs. 6.22 and 6.23). They are small or large in size while in shape they are round, oval

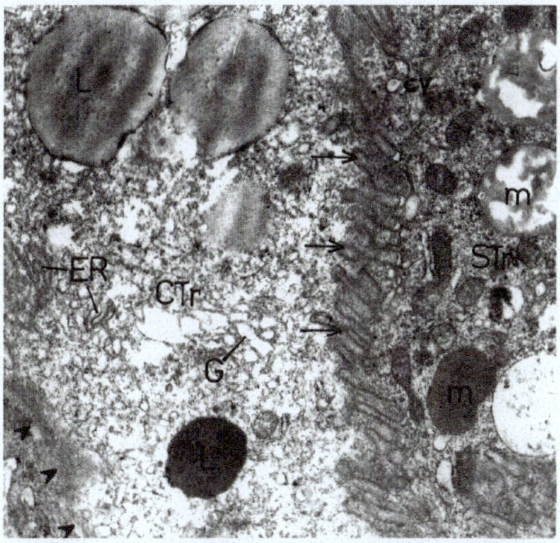

Fig. 6.25 The placenta of *Rhinolophus rouxi* at full term under the high power of transmission electron microscope. It shows the interdigitation (arrows) between the cytotrophoblast (CTr) and syncytiotrophoblast (STr). Large lipid droplets (L), smooth endoplasmic reticulum (ER) and Golgi zones (G) in the cytoplasm of cytotrophoblast can be seen. Numerous mitochondria (m) in the cytoplasm of syncytiotrophoblast are seen. Also seen is the basal lamina (arrow heads) towards the left side bottom [X 48,000]. From Bhiwgade and Thakur 1992

or elongated. They have tubular cristae which usually extend across their breadth and their matrix is dense. Some of them show hypertrophy where the matrix is less dense and sometimes very little. This is mainly seen in the late stage of gestation. Rough endoplasmic reticulum is found to be present in the cytoplasm around the nucleus. They form parallel cisternae and may be present in small segments or form concentric lamellae. They are studded with ribosomes at regular intervals. During full-term pregnancy, rough endoplasmic reticulum shows extensive dilation and vesiculation. The Golgi zones are few in number. They are present all over the cytoplasm but more towards the cytotrophoblast. On the side facing the interstitial membrane, tiny, coated vesicles are present in large numbers. Their number increases marginally during advanced and full-term stages. Occasionally, few lipid droplets are present during the advanced stages. They are small in size and contain lipids of medium density (Fig. 6.24).

The cytotrophoblast is a unicellular layer which is continuous with cells that are in close association. The membranes between the adjacent cells show the presence of desmosomes at regular intervals. The cytoplasm is characterized by few mitochondria which are scattered all over the cytoplasm. The mitochondria are small in size. They are round, oval or elongated in shape with tubular cristae, some of which extend across their width. Their matrix is light and granular. Some of them show hypertrophy (Fig. 6.25) and some have concentric cristae. Short rough endoplasmic reticulum is present mainly towards the foetal mesenchyme and are seen arranged along the plane of the cytotrophoblast.

The foetal endothelium from midterm to full term is seen as a thin layer made up of a single layer of cells (Fig. 6.25). Its thickness is not uniform. In regions where nuclei are present, it is quite thick and in such regions, it occupies a major area of the foetal blood space. The cytoplasm is light and granular, with very few cell organelles like mitochondria, coated vesicles and ribosomes. The nuclei are few in number and are irregular and elongated. They have a double nuclear membrane with pores. Nucleoplasm is usually without a nucleolus, but heterochromatin is present which is deposited marginally. Mitochondria are small and round or oval in shape. They have a dense matrix due to which cristae are not clearly visible. Most of them show hypertrophy. Many coated vesicles and ribosomes are scattered all over the cytoplasm.

The mature placenta in *Rhinolophus rouxi,* resembles that of *Magaderma lyra lyra.* The placental labyrinth has large maternal capillaries with continuous layer of maternal endothelial cells. There is an interstitial membrane which is continuous. There are two layers of trophoblast. The placenta in *Rhinolophus rouxi,* therefore, can be designated as endotheliochorial (Bhiwgade 1977). At the luminal surface, there are microvesicles, caveolae, coated vesicles and numerous blunt microvilli which indicates that the endothelium plays an important role in active absorption of materials from the maternal blood.

6.5 Hhipposiderid Bats

A *Hipposideros lankadiva.*
B *Hipposideros speoris.*
C *Hipposideros fulvus.*

6.5.1 *Hipposideros lankadiva*

The interhemal membrane in *Hipposideros lankadiva* at term is endotheliodichorial. Starting from the maternal to foetal blood space, the layers seen are maternal endothelium, syncytiotrophoblast, cytotrophoblast, mesenchyme and foetal endothelium (Fig. 6.26). The maternal blood space is seen surrounded by maternal endothelium, which in turn is enclosed by the syncytiotrophoblast. The syncytiotrophoblast rests on interstitial membrane. Similarly, the foetal capillaries are enclosed by its endothelium which in turn is surrounded by the cytotrophoblast. The cytotrophoblast rests on the basal lamina. The two trophoblastic layers are in close opposition throughout and the membranes between them show interdigitations of medium dimension. Although, the two layers appear uniform, their thickness varies. The cytotrophoblast appears to be discontinuous in some regions and the syncytiotrophoblast comes directly in contact with the basal lamina of cytotrophoblast. The cytoplasm of the syncytiotrophoblast is darker and denser than the cytoplasm of the cytotrophoblast. The maternal endothelium is well developed except in some regions where its cellular nature is lost, and presence of syncytial blocks is noticed.

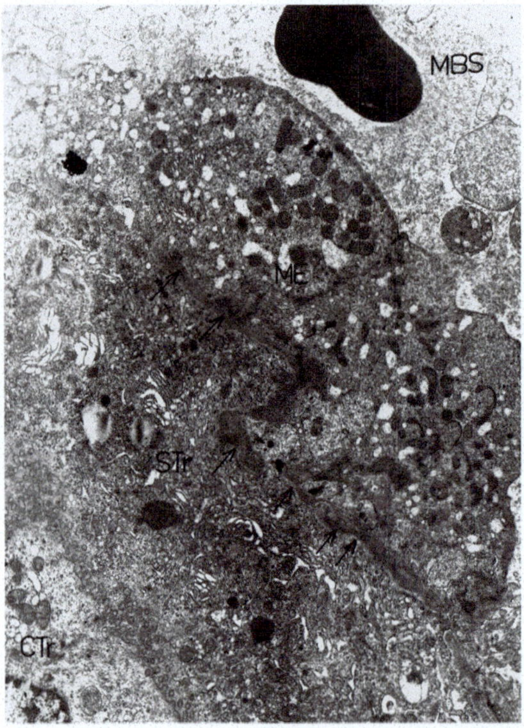

Fig. 6.26 Low-power micrograph of placenta in Hipposideros lankadiva showing maternal blood space (MBS) with syncytial blocks (Sb), maternal endothelium (ME), discontinuous interstitial membrane (arrows), syncytiotrophoblast (STr) and cytotrophoblast (CTr) [X 6600]. Unpublished electron micrograph from Kothari

Numerous microvilli can be seen emanating from the maternal endothelium along the side facing the maternal blood space. The cytoplasm of this layer has a granular matrix of high density with many vacuolated regions and shows uneven thickness. In regions where it is thick, the cell organelles like mitochondria, Golgi zones and coated vesicles are concentrated and in regions where it is thin or discontinuous, syncytial blocks are present. The nuclei of this layer are large, elongated and irregular in shape. There is deposition of heterochromatin towards their margin and the nucleolus is also marginally placed. Many chromatin centres are found scattered in the nucleoplasm. Mitochondria are numerous and are of various sizes. Most of them are round, but quite a few are elongated. They have a dense granular matrix with tubular cristae, which extends across their width. The Golgi zones are few and each one comprises 3–4 cisternae that appear dilated. The orientation of the Golgi zones is not uniform. The coated vesicles are tiny in size, few in number and present all over the cytoplasm (Fig. 6.26).

The syncytiotrophoblast is dark, well developed and rich in cell organelles. The cell organelles are aggregated more towards the maternal endothelium and around the nucleus. There are many caveolae present near the discontinuous interstitial membrane. The cytoplasm of this layer becomes continuous with the cytoplasm of maternal endothelium. The nuclei of this layer are quite large and elongated. Their margin is irregular and with prominent pores in the nuclear membrane. Around the nucleus endoplasmic reticulum with parallel cisternae can be seen. They are abundant and appear granular due to uniformly spaced ribosomes. Their cisternae are arranged in all planes and in many places they are partly dilated. Free ribosomes are also found in the vicinity of endoplasmic reticulum. Mitochondria are large in size as compared to those present in the other layers. They have a dense granular matrix and the granules are finer than their surrounding cytoplasm. Most of them have tubular cristae, although some have lamellar cristae. The mitochondria are largely arranged parallel to the plane of the syncytiotrophoblast, but their cristae are perpendicular to this plane. Those which are situated near the interstitial membrane show hypertrophy and appear swollen or balloon like with scanty matrix. Occasionally, in their matrix are present dark myelin bodies (Fig. 6.26). Numerous Golgi zones are seen distributed all over the cytoplasm of this layer. They are very conspicuous with four to five cisternae and few vesicles. In some, all the cisternae often appear as vesicles. Many lipid droplets, which are a characteristic feature of this layer are seen. Each droplet is enclosed by a unit membrane and forms a liposome. They are seen accumulated in small or large groups. They are of various size, but most of them are large. Even their density varies from low to very high and many of them show areas of different density. This gives rise to bands of half-light and half-dark regions, and collectively they resemble a typical zebra pattern. In addition to these cell organelles, the cytoplasm also shows the presence of desmosomes, tight junctions, coated vesicles and glycogen granules.

The cytotrophoblast is of unicellular layer which is made up of cells and in many places it appears discontinuous. The cytoplasm is light and granular but less electron dense than the syncytiotrophoblast and maternal endothelium. Its thickness is not uniform. In regions where the layer is thick, each cell contains a single large nucleus and its cytoplasm contains cell organelles like mitochondria, coated vesicles and polyribosomes. Golgi zones and endoplasmic reticulum are totally absent. The nucleus of each cell is usually central in position and round in shape. Occasionally, it is elongated. Each nucleus has a double nuclear membrane with pores. The nucleoplasm contains few chromatin centres and a prominent nucleolus. Inter-chromatin granules are also seen in the nucleoplasm. Small, medium and large-size mitochondria with oval shape are present and have a granular matrix of high density. Those which are present towards the syncytiotrophoblast have tubular or lamellar cristae whereas those present towards the foetal side show hypertrophy. The matrix of the hypertrophied mitochondria is of lesser electron density. Many, small and large, coated vesicles and micropinocytotic vesicles are present towards the syncytiotrophoblast. Small aggregates of glycogen, free and polyribosomes are also present all over the cytoplasm (Figs. 6.27 and 6.28).

The foetal endothelium is poorly developed. It is generally in the form of a thin layer. Only in regions where the nucleus is present, this layer becomes thick but its cytoplasmic content remains the same. Cell organelles are sparse. The nuclei are elongated and somewhat rectangular in shape. There is marginal heterochromatin and the nucleus is also situated towards the margin of the layer. There are interchromatin granules scattered in the nucleoplasm. The nuclear membrane shows presence of prominent pores. The cytoplasm of this layer also shows presence of free and clustered glycogen granules, polyribosomes and a few lipid droplets. Oval lipid droplets which are tiny in size are seen. Their contents are of low electron density.

The foetal mesenchyme is uniformly granular and of low density. Around the foetal capillaries are found small cytoplasmic islands surrounded by a distinct cell membrane and with very few cell organelles. In some regions, they show presence of an irregular nucleus, dense bodies and glycogen granules. Glycogen granules which are present in small clusters, often form rosettes. Fibroblasts are totally absent.

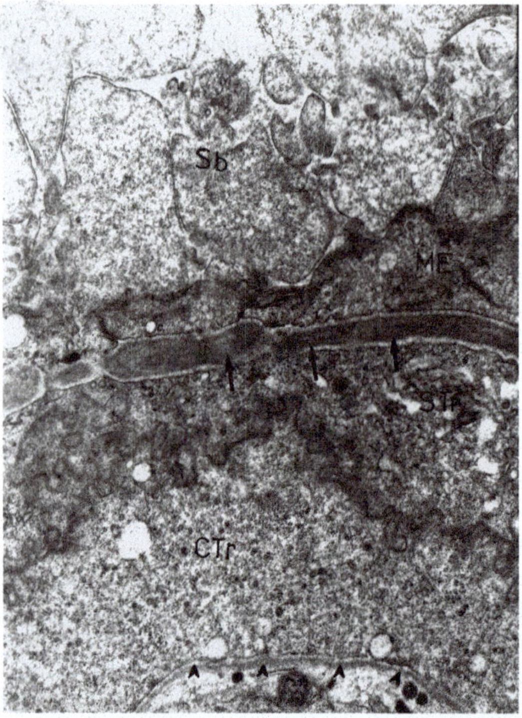

Fig. 6.27 High-power electron micrograph of placenta in Hipposideros lankadiva at term. Note the syncytial blocks (Sb) above the maternal endothelium (ME), discontinuous interstitial membrane (arrows), syncytiotrophoblast (STr) and cytotrophoblast (CTr) resting on a distinct basal lamina (arrow heads) [X 16,000]. Unpublished electron micrograph from Kothari

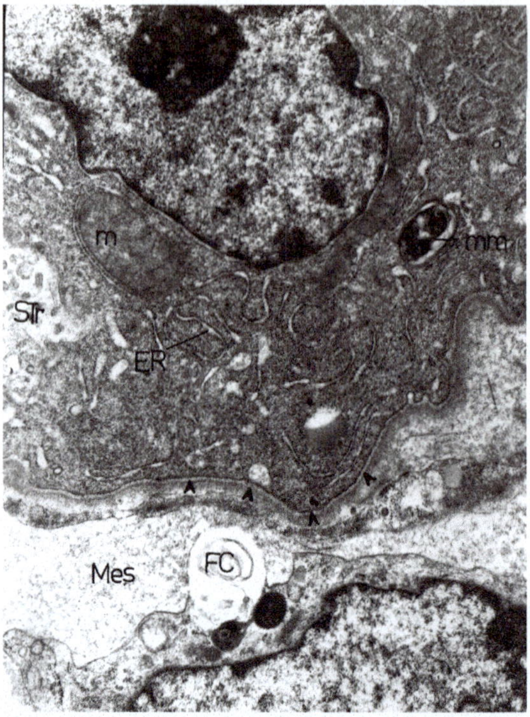

Fig. 6.28 High-power electron micrograph of Hipposideros lankadiva term placenta showing the syncytiotrophoblast. Note the syncytiotrophoblast (STr) resting on the basal lamina (arrow heads). Also note the large mitochondria (m), some with myelin inclusion (mm) and dilated parallel cisternae of rough endoplasmic reticulum (rER). Also seen is mesenchyme (mes) and foetal capillaries (FC) [X 13,000]. Unpublished electron micrograph from Kothari

6.5.2 *Hipposideros speoris*

The interhemal membrane of *Hipposideros speoris,* at term, is endotheliomonochorial. As seen from the maternal side to the foetal blood space; maternal endothelium, syncytiotrophoblast, mesenchyme and foetal endothelium are seen in sequence. The cytotrophoblast layer is completely absent (Fig. 6.29). The maternal blood space is surrounded by the maternal endothelium which in turn is encircled by syncytiotrophoblast. The syncytiotrophoblast rests on discontinuous interstitial membrane which lies towards the foetal side and is bound by a distinct trophoblastic basal lamina. The foetal mesenchyme, on the other hand is surrounded by the foetal endothelium which in turn is bound by its own basal lamina. Between the two basal lamina, mesenchyme is present. In some regions, however, the two basal lamina are so close to each other that they appear as a single unit (Fig. 6.29).

The maternal endothelium is continuous, unicellular and well developed. It consists of a single layer of cells (Figs. 6.29). Each cell has a nucleus situated more towards the syncytiotrophoblast. It is surrounded by cytoplasm rich in mitochondria and Golgi zones. Each nucleus is large in size and oval in shape. There is a double nuclear membrane with distinct pores. Nucleolus is usually absent, but few chromatin centres are present. The nucleoplasm is granular, and heterochromatin is deposited marginally. Mitochondria are numerous, of various sizes and are large and hypertrophied. Most of them are round, except a

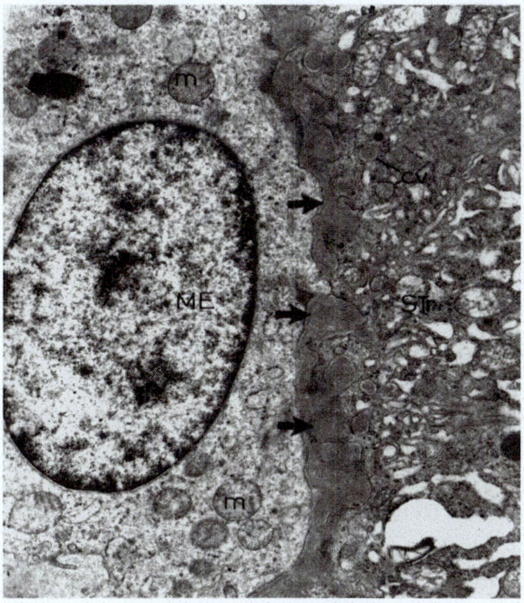

Fig. 6.29 High-power electron micrograph of placenta in Hipposideros Speoris during full-term stage. Numerous mitochondria (M) of the maternal endothelium (ME) are concentrated near the interstitial membrane (arrows). Numerous coated vesicles (CV) are seen in the syncytiotrophoblast (Str) [X 10,000]. Unpublished electron micrograph from Kothari

few which are oval. They have double outer membrane and a dense granular matrix inside. Their cristae are short and tubular. The Golgi zones are many and scattered in the cytoplasm, more towards the nucleus. They are made up of few cisternae and vesicles, all of which appear dilated. The cytoplasm also contains sparse endoplasmic reticulum, few polyribosomes, free ribosomes and rosettes of glycogen granules.

The interstitial membrane, although discontinuous, is quite thick and well developed. However, its thickness is not uniform even in those regions where it is continuous (Fig. 6.29).

The syncytiotrophoblast is thick, continuous and well developed. Its cytoplasm is dark and dense. The nucleoplasm usually does not contain a distinct nucleolus but few chromatin granules are present. Heterochromatin is also present which is deposited marginally. The cytoplasm all around the nucleus shows a dense network of cisternae of rough endoplasmic reticulum arranged in parallel. Their concentration is more towards the foetal side and their membranes are uniformly studded with ribosomes. In many places, the cisternae are dilated due to which it appears vesiculated and they are also found to enclose cell organelles like mitochondria, Golgi zones, coated vesicles and lipid droplets. The mitochondria are found to be scattered all over the cytoplasm oval or round in shape and large in size. They have a double outer membrane with tubular cristae and a light granular matrix. In some cases, the cristae are quite long and extend across the width but most of them are hypertrophied. Many Golgi zones are presently scattered in the cytoplasm, but most of them are enclosed within the rough endoplasmic reticulum. Their orientation is not uniform. Each Golgi zone is made up of few cisternae. In most of the cases, they appear dilated to a large extent. Few lipid droplets are found in the cytoplasm. They are medium and large size droplets and their lipid contents are of medium density. Also present in the cytoplasm are numerous ribosomes, dense bodies, coated vesicles and rosettes of glycogen granules.

The mesenchyme is bound by the basal lamina of the trophoblast on one side and on the other by that of the foetal endothelium. It is in the form of a thin layer with sparse cell organelles. Scattered all over the mesenchyme are numerous fibrous inclusions, pinocytotic vesicles and vesicles containing granules.

The foetal endothelium is thin and continuous layer enclosing the foetal blood space and capillaries. The foetal capillaries are enclosed in association with its endothelium. The cytoplasm of this layer is dark and dense but very few cell organelles are present. Mitochondria are large, less in number, round or oval in shape but many are hypertrophied. Few small lipid droplets are present, the contents of which are of high density. Also present in the cytoplasm are numerous polyribosomes, free ribosomes, coated vesicles and membrane stacks.

6.5.3 *Hipposideros fulvus*

The ultrastructural observation of interhemal membrane in *Hipposideros fulvus fulvus* at term reveals that the placental labyrinth is endotheliodichorial (Fig. 6.30). Starting from the maternal side to the foetal blood space; maternal endothelium, syncytiotrophoblast, cytotrophoblast and foetal endothelium are seen in order (Fig. 6.31). The maternal blood space is enclosed by the syncytiotrophoblast. The interstitial membrane is present this membrane is very close to the maternal endothelium and is discontinuous in nature (Fig. 6.31). Numerous stubby microvilli are seen emerging from the luminal surface of the interstitial membrane. The syncytiotrophoblast is of varying thickness and its cytoplasm is darker than that of the cytotrophoblast. The cytotrophoblast is well defined, unicellular layer and rests on a distinct basal lamina.

In some regions cytotrophoblast appears to be thicker than the syncytiotrophoblast. The maternal blood space shows presence of syncytial blocks which are spherical or elongated in shape. They have a light granular matrix and are devoid of any cell organelles (Fig. 6.32a, b, c).

The syncytiotrophoblast is a thick, dark and continuous layer. The cytoplasm towards the maternal blood space shows the presence of distinct desmosomes and cell membranes (Fig. 6.32a, b, c). The nuclei in this layer are few in number and smaller in size as compared to the nuclei of the cytotrophoblast. They are irregular in shape, differ in their sizes and are placed randomly in the cytoplasm. Each nucleus has a double nuclear membrane with pores. A distinct nucleolus along with a few chromatin centres are present in the nucleoplasm. In some cases, the nucleolus is of the open type. However, a thin layer of heterochromatin is present all along their margin. In some nuclei, deep invaginations are seen. Golgi zones, mitochondria and lipid droplets are present in large numbers (Fig. 6.32a, b, c). Endoplasmic reticulum is sparse. They are of smooth type. Mitochondria are small in size and few in number. They have a double outer membrane and a dense matrix. Most of them show hypertrophy. Numerous Golgi zones are present around the nucleus and all over the cytoplasm. They are mainly made up of cisternae and the vesicles are few in number. Their orientation is not uniform but many of them are arranged in the plane perpendicular to the plane of the syncytiotrophoblast. They form at sites slightly towards the cytotrophoblast whereas their maturing face is towards the maternal side. Most of the cisternae are dilated (Fig. 6.32a, b, c). Abundant large-sized lipid droplets are present around the nucleus and all over the cytoplasm.

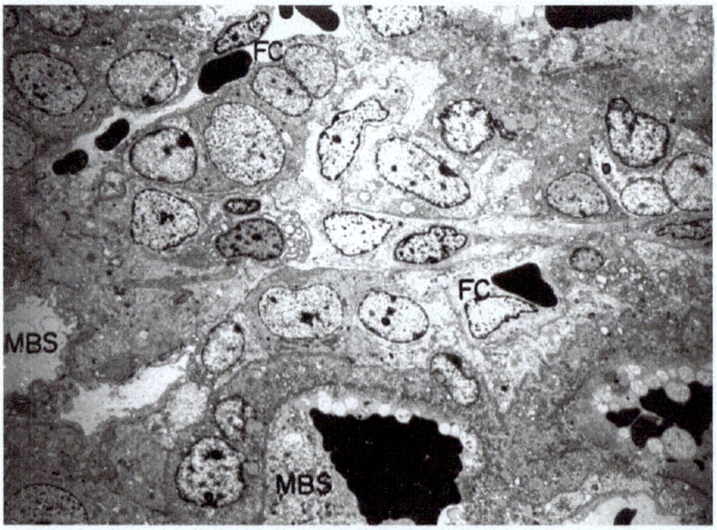

Fig. 6.30 Electron micrograph of placental labyrinth of Hipposideros fulvus fulvus at term showing its general organization. Note the syncytiotrophoblast (STr) and cytotrophoblast (CTr) between the maternal blood space (MBS) and foetal capillary (FC) [X 1500]. From Bhiwgade 1990

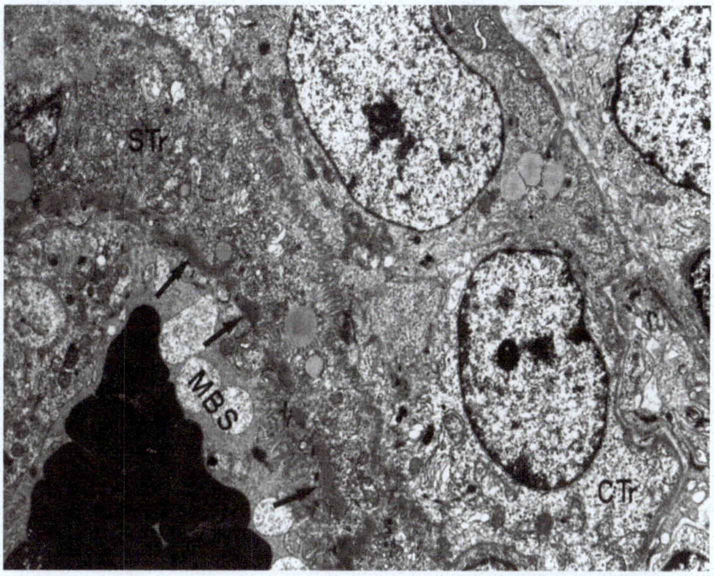

Fig. 6.31 Electron micrograph of placental labyrinth of Hipposideros fulvus fulvus at term showing the maternal blood space (MBS), maternal endothelium (ME), syncytiotrophoblast (STr) and cytotrophoblast (CTr). Note the discontinuous interstitial membrane (arrows) and the distinct basal lamina (arrow heads) [X 4000]. From Bhiwgade 1990

Most of them are round, however, some of them are oval in shape. They contain lipids of medium and low density. Many of them are bound by a distinct membrane forming liposomes. Throughout the cytoplasm are present, ribosomes and rosettes of glycogen granules in large numbers. There are numerous coated vesicles and pinocytotic vesicles all over the cytoplasm but more towards the intrasyncytial lamina. Also present towards the maternal side are fine tubular microfilaments with numerous desmosomes.

The cytotrophoblast is unicellular, continuous and is seen as a thick layer. The adjacent cells of this layer are in close association and their closeness is marked by the presence of desmosomes between them. The cytoplasm of this layer is light and granular. Each cell has a single nucleus which is centrally placed and occupies a major area of the cell (Fig. 6.32a, b, c). It is large in size and elongated in shape. It has a double nuclear membrane with pores. A distinct nucleolus is present which is generally of the open type. In the nucleoplasm, there are few chromatin centres and there is deposition of heterochromatin towards the margin. The cytoplasm around the nucleus is rich in cell organelles. Mitochondria are abundant and are scattered in the cytoplasm. They are large and appear round or elongated. They have a double outer membrane with sharp tubular cristae and light granular matrix. Some of them, however, show hypertrophy. The cisternae of rough endoplasmic reticulum are parallel, well developed and embedded with ribosomes at regular intervals. Sometimes they are found to enclose cell organelles like mitochondria and lipid droplets. Numerous Golgi zones are present all over the cytoplasm but more towards the syncytiotrophoblast. They are oriented in different planes and are made up of few cisternae along with few vesicles. Most of the cisternae appear dilated. Lipid droplets are abundant, round or oval in shape and contain lipid of high and medium electron density. Scattered uniformly in the cytoplasm are numerous ribosomes, glycogen rosettes and coated vesicles.

The foetal endothelium is poorly developed. It is unicellular in nature and made up of a thin layer of cytoplasm. In regions where the nucleus is seen, the endothelium is thick with light cytoplasm and with very less cell organelles. The nuclei are large, elongated in shape with a double nuclear membrane that has pores. A distinct nucleolus is present in the nucleoplasm with very few chromatin centres. The cytoplasm around the nucleus contains a few coated vesicles, free ribosomes and polyribosomes. The cytoplasm also contains hypertrophied mitochondria

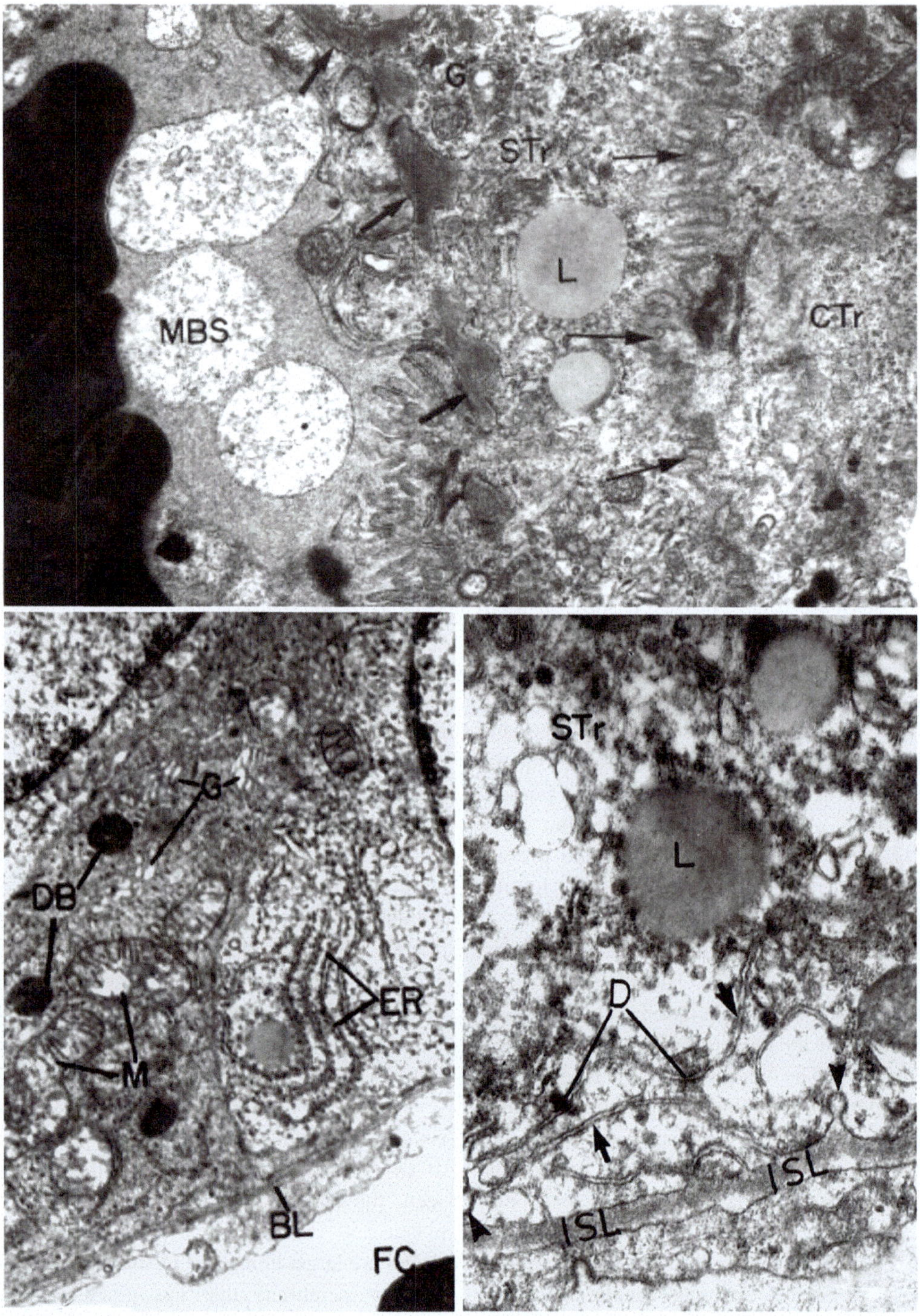

Fig. 6.32 Electron micrograph of placenta from *Hipposideros fulvus fulvus* at full term. The syncytial trophoblast (STr) has lipid droplets (L) and Golgi zones (G). The interstitial membrane (arrows) is discontinuous with finger-like interdigitations (arrow heads). The cytotrophoblast (CTr) is visible. Stubby microvilli towards the maternal blood apace (MBS) is seen [X 8000]

which are large and round. They have a double outer membrane, and their cristae are tubular.

The comparison of ultrastructural characteristics of interhemal membranes from megadermatid and three species of Hipposiderid bats indicates that the endotheliochorial condition is widely distributed among these species. Our study also reveals the heterogenicity in ultrastructure of trophoblastic layers, along with the changes and differences in *Rhinolophus rouxi, Hipposideros lankadiva, Hipposideros spepris* and *Hipposideros fulvus fulvus.*

For correctly defining a placental barrier the light microscopic study has to be confirmed by ultrastructural observations. Enders and Wimsatt (1968), and Bjorkman and Wimsatt (1968), for instance, had established that the placenta of *Myotis lucifugus* and *Desmodus routunds murinus* is hemodichorial, which revised the earlier report that it was endotheliodochorial (Wimsatt 1945a, b, 1954).

Gopalakrishna and Karim (1980), while reviewing the placentation in Chiroptera, have opined that light microscopic studies are not enough to distinguish syncytiotrophoblast and cytotrophoblast. In our electron microscopic study, we have observed that the maternal endothelium, a discontinuous interstitial membrane, syncytiotrophoblast and cytotrophoblast are present throughout gestation, in the placenta of *Megaderma lyra, lyra, Rhinolophus rouxi, Hipposideros lankadiva* and *Hipposideros fulvus fulvus.* The placental barrier in these bats is endotheliodichorial throughout gestation. Both in general organization and fine structure, they resemble the placenta of Megaderma (Bhiwgade 1989) and *Rhinopoma microphyllum* (Mandal 1991). The placenta of *Hipposideros speoris* is endotheliomonochorial due to the presence of a thick maternal endothelium, a discontinuous interstitial membrane and a single layer of syncytiotrophoblast.

The ultrastructural analysis of the chiropteran placenta belonging to the primitive families brings out the observation that these placentae consist of three layers of trophoblast, a non-cellular barrier 'interstitial membrane' and the maternal endothelium that separates maternal and foetal components.

6.5.4 Interstitial Membrane

The interstitial membrane has been reported in the ferret (Lawn and Chiquoine 1965), the cat (Wynn and Bjorkman 1968), the dog (Anderson 1969) and recently in *Taphozous melanopogon, Rhinopoma hardwickie hardwickie* (Bhiwgade 1990) and *Rhinopoma microphyllum* (Mandal 1991). An interstitial membrane seen under light microscopy in the endotheliochorial placentae of bats, as reported by Wimsatt (1958), is histologically very similar. In the out study, we have observed the presence of discontinuous acellular membrane in the placentae of *Megaderma lyra lyra, Rhinolophus rouxi, Hipposideros lankadiva, Hipposideros fulvus fulvus* and *Hipposideros speoris.* This membrane has been reported to be physiologically important (Cukierski 1987) and we feel that the membrane assists in the exchange of materials across the interstitial membrane either in one direction only or in both directions.

6.5.5 Syncytiotrophoblast

In the discoidal placentae of all families of chiroptera, the trophoblast surrounding the maternal channels has been described as syncytiotrophoblast (Wimsatt 1958), except in *Tadarida* (Stephens 1969), *Mollossus ater* (Rasweiler 1991) and *Rhinopoma hardwickie hardwickie* (Panse and Bhiwgade 1991) where a cytotrophoblast persist throughout the gestation. In our study, the main features of syncytiotrophoblast in different bat species studied, are the presence of its cell organelles, i.e. rough endoplasmic reticulum, mitochondria and Golgi zones. Wynn and Davies (1965) have described a similar observation in hyena. Bhiwgade (1989, 1990) and Bhiwgade et al. (1992) observed similar characteristics in several species of bats, whereas Enders (1965) has shown similar characteristics in many other species of mammals. Wislocki and Wimsatt (1947) on the basis of histochemical evidence have suggested that the presence of organelles indicates that the layer is very active in synthesis. Additionally in our observations, we have reported the presence of coated vesicles, pinocytotic vesicles, numerous vacuoles and multivesicular bodies which are situated near the trophoblastic surface that borders the maternal blood. This is an evidence of the fact that the syncytiotrophoblast is active in the transport of macromolecules both into and from the maternal blood.

6.5.6 Cytotrophoblast

Ultrastructural observation by us of placental trophoblast in four species of bat, except *Hipposideros speoris,* shows the presence of numerous mitochondria, distinctive granular endoplasmic reticulum, numerous Golgi zones, lipid droplets etc. which are clear evidence of active synthesis as reported in other species of mammals (Enders 1965; Anderson 1969; Wynn and Davies 1965; Wimsatt et al. 1973). Under the electron microscope, trophoblastic area shows the presence of coated vesicles and caveolae. These structures extend from the free syncytial surface to foetal capillary endothelium and help in active transport through the entire interhaemal membrane. Wimsatt (1945a, b) after observing mitotic activity in these cells, has suggested that the cytotrophoblast cells, act as stem cells. When need arises, they give rise to the syncytial trophoblast (Rasweiler 1991). In the placenta of *Hipposideros speoris,* however, there is total absence of the cytotrophoblast. This could be an adaptation to minimize the distance between the foetal and maternal environments so as to make efficient exchange of materials to provide for the demands of the developing foetus.

6.6 *Rhinopoma microphyllum* and *Rhinopoma hardwickei hardwickei*

The endotheliochorial condition with some unique features seen in *Rhinopoma hardwickei* with a cytotrophoblastic interhemal barrier is an oddity. The loss of syncytiotrophoblast leading to only cytotrophoblast forming the interhemal barrier is a midway transition during morphogenesis. This is unlike the more usual trend of cytotrophoblast giving rise to syncytiotrophoblast in other endotheliomonochorial conditions (Bhiwgade 1990). Our study has shown for the first time, a unique ultrastructural morphology of final placenta which is not seen in any other family of bats. Another interesting feature encountered, in this bat, is the presence of both rough and smooth endoplasmic reticulum being seen at the same time in the same cell. In *Rhinopoma microphyllum*, interestingly, a microtubule channel system is seen in the syncytio—and—cytotrophoblast throughout the placental development.

By the early and late limb-bud stages of embryonic development, in both *Rhinopoma hardwickei* and *Rhinopoma microphyllum,* the interhemal barrier separating the maternal and foetal blood streams consists of following layers: (Mandal 1991) (i) Maternal endothelium. (ii) A continuous layer of syncytiotrophoblast. (iii) A well-defined cytotrophoblast. (iv) Basal lamina and (v) Foetal capillary along with foetal mesenchyme (Figs. 6.33, 6.34 and 6.35).

This condition prevails till term pregnancy in *Rhinopoma microphyllum* and is designated as endotheliodichorial. However, in *Rhinopoma hardwickei* during the late limb-bud stage, the syncytiocytotrophoblast appears to be irregular and varying in thickness. It is thick where the nuclei are situated and very thin and sometimes virtually absent at other places. The most visible change during this stage is that at some places the cytotrophoblast comes in close contact with interstitial membrane.

By mid-gestation, there is complete disappearance of the syncytiotrophoblast layer thereby reducing the effective placental barrier layers to maternal endothelium, interstitial membrane, cytotrophoblast, basal lamina and foetal mesenchyme with foetal capillaries. This condition is endotheliomonochorial.

The near-term placenta resembles the preceding stage, i.e. it is endotheliomonochorial in nature with a cellular trophoblast layer in *Rhinopoma hardwickei.*

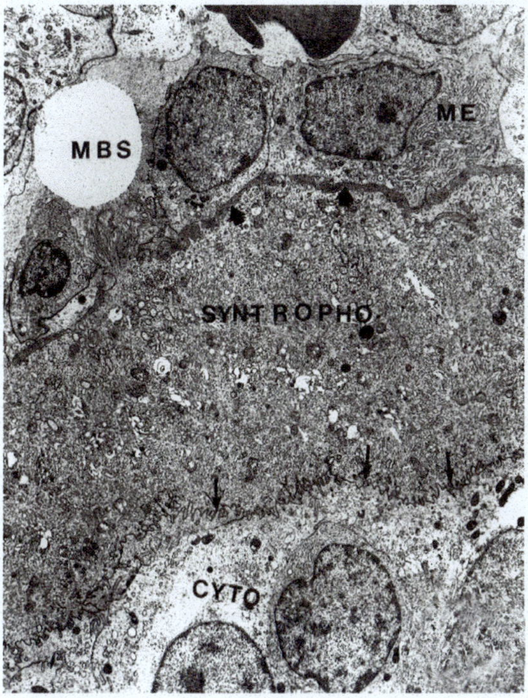

Fig. 6.33 Ultrastructure of the placenta during full term. Discontinuous interstitial membrane (arrows) syncytiotrophoblast. Cytotrophoblast and small arrow interdigital membrane (small arrows) are seen [X 3500]. From Bhiwgade 1990

A well differentiated and continuous layer of syncytiotrophoblast devoid of any cell boundaries with evenly placed nuclei is observed in the early limb-bud stage in *Rhinopoma hardwickei* but, it is a constant feature throughout gestation in *Rhinopoma microphyllum* (Figs. 6.34 and 6.35). The cytoplasm is dense, characterized by spherical or ovoid mitochondria, conspicuous Golgi apparatus, numerous associated vesicles, granular endoplasmic, polyribosomes, dense bodies, dark and light multivesicular bodies, coated vesicles and lipid droplets. The well-formed microtubular system separates the syncytiotrophoblast from the cytotrophoblast and often displays numerous desmosomes which unite these epithelia at points in *Rhinopoma microphyllum* (Figs. 6.34 and 6.35).

In *Rhinopoma hardwickei*, the syncytiotrophoblast becomes progressively irregular during the late limb-bud stages, loses its characteristic number of organelles and becomes thinner at some places or virtually disappears by mid-pregnancy (Figs. 6.36 and 6.37).

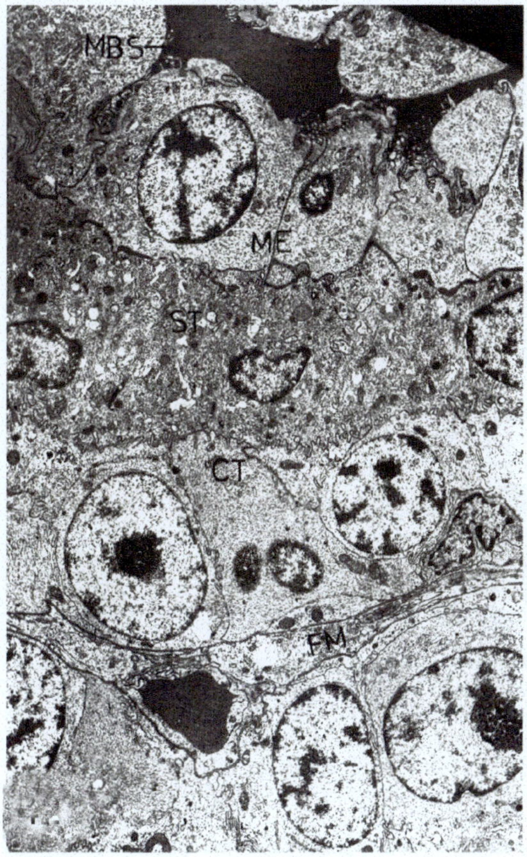

Fig. 6.34 Low-power electron micrograph of placental labyrinth of *Rhinopoma microphylum* during mid pregnancy showing maternal blood space (MBS), maternal endothelium (ME), cytotrophoblast (cyt) and basal lamina BL. On lower right is foetal capillary (FC) [X 5000]. From Mandal 1991

The maternal endothelium appears hypertrophied throughout the embryonic development. The nucleus is invaginated with considerable aggregation of chromatin material near the nuclear envelope. The cytoplasm is characterized by numerous elongated and ovoid mitochondria with fairly prominent cristae. Well-developed rough and smooth endoplasmic reticulum, Golgi apparatus, ribosomes, numerous polyribosomes, few coated vesicles, multivesicular bodies and micropinocytotic vesicles are also seen (Figs. 6.37 and 6.38a, b).

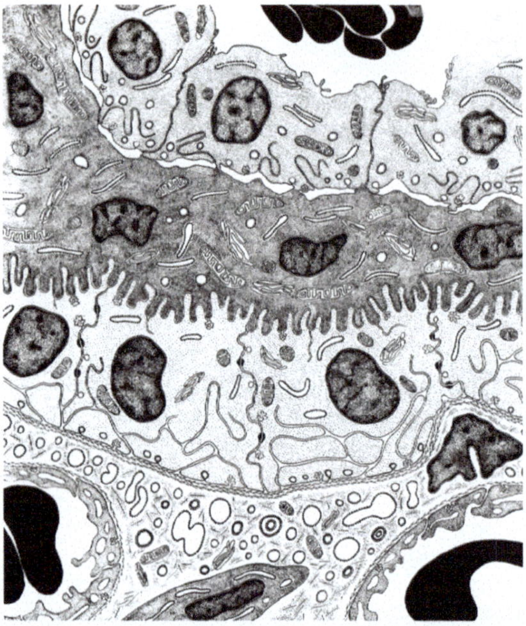

Fig. 6.35 A schematic representation of the layer of the trophoblast in the chorio-allantoic placenta of *Rhinopoma microphylum*. The maternal blood capillary is across the top. Foetal capillaries are seen in the lower right and left corners. Note that in between the two bloodstreams there are maternal endothelium, interstitial membrane, well defined two layers of Trophoblast inner syncytiotrophoblast and outer cytotrophoblast foetal lamina and foetal endothelium. From Mandal 1991

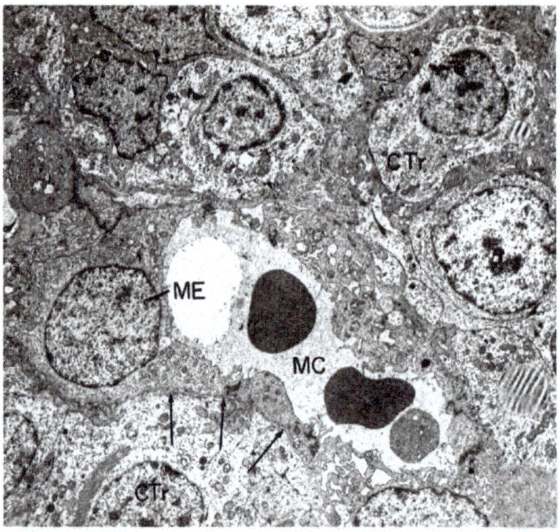

Fig. 6.36 Low-power electron micrograph of interhemal membrane during mid-pregnancy of *Rhinopoma hardwickei*. The placental barrier between maternal blood space (MBS) and Foetal Capillary (Fc) consists of maternal endothelium (ME) along with cell junctions (JC), discontinuous interstitial membrane (arrows), Cytotrophoblast (cyt) and basal lamina (BL). Note a well-developed Golgi complex (G) in the cytotrophoblast. Also note the micropinocytotic vesicles (arrows) in maternal endothelium facing the MBS [X 5000]. From Bhiwgade 1990

The maternal endothelium sits on a distinct acellular interstitial membrane which is continuous during early embryonic development (Fig. 6.34) but shows some gaps at irregular intervals in later stages (Fig. 6.37). Coated vesicles, caveolae and micro pinocytic vesicles are seen on either side of interstitial membrane.

This layer is a constant feature of the barrier throughout gestation in both *Rhinopoma microphyllum* and *Rhinopoma hardwickei* and its cellular character is indicated by the presence of well-developed cell membrane provided with desmosomes. The cytoplasm is less dense than the syncytiotrophoblast and is rich in mitochondria which are elongated or ovoid and show tubular cristae. Endoplasmic reticulum is well developed and is tubular in shape with slightly expanded cisternae which have high electron density. Additionally, ribosomes, Golgi complex with numerous associated vesicles are also

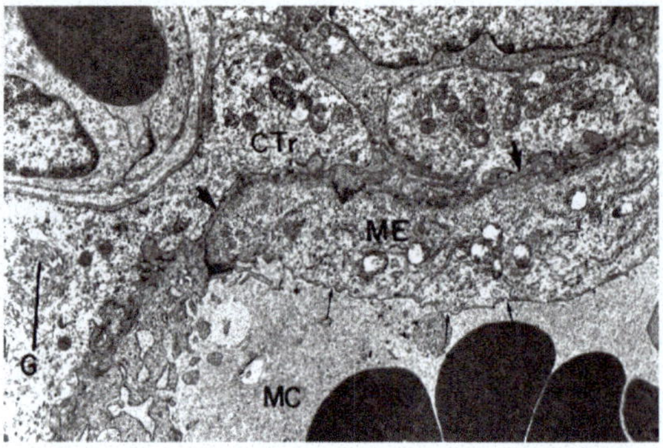

Fig. 6.37 High-power electron micrographs of placenta in *Rhinopoma hardwickei* during mid-pregnancy. The maternal blood space (MBS) and the maternal endothelium with abundant cell organelles (ME) are seen. Ovoid mitochondria (M), Golgi apparatus (G), multivesicular bodies (MVB), coated vesicles and polyribosomes. Just below the maternal endothelium is seen the discontinuous interstitial membrane (arrows). Well-developed Golgi apparatus (G) and Lipid droplets (L) are visible in the cytotrophoblast (cyt) [X 6000]. From Bhiwgade 1990

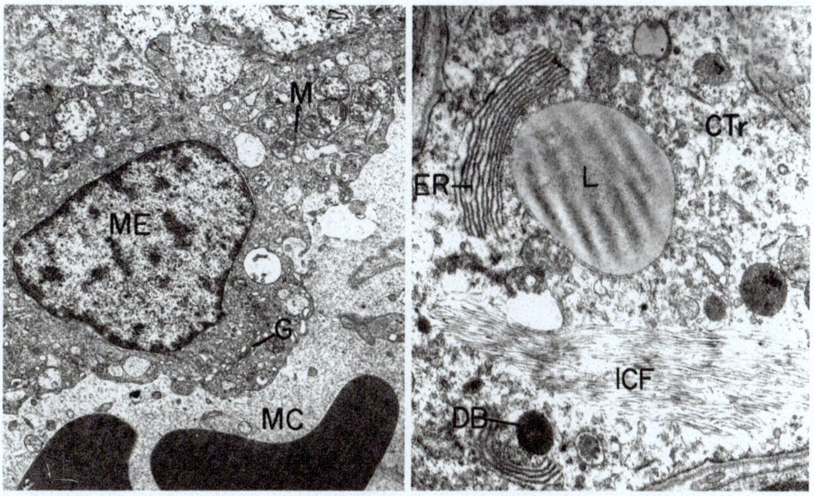

Fig. 6.38 High-power electron micrographs showing cytotrophoblast (cyt) in placenta of *Rhinopoma hardwickei* during mid-pregnancy. Note the oval nuclei, well-developed Golgi apparatus (G) and closely packed lamellae of smooth endoplasmic reticulum (sER). Mitochondria (M), intracytoplasmic filaments (ICF), Lipid droplets (L), dense bodies (DB) and Basal lamina (BL) are also seen [X 10,000] [X 10,000]. From Bhiwgade 1990

observed. In *Rhinopoma hardwickei* the cytoplasm is also characterized by the presence of closely packed lamellae of smooth endoplasmic reticulum along with rough ER—a feature not reported in any species of bats (Fig. 6.38a, b). In *Rhinopoma microphyllum*, during mid-pregnancy a large population of microtubular channel systems are seen to extend from free syncytial surface through the cytotrophoblast to the foetal endothelium (Figs. 6.39 and 6.40a, b). On the other hand, in *Rhinopoma hardwickei* during mid- and term pregnancy there is infolding of the trophoblastic basal lamina at several places, between lateral cell borders. Foetal capillaries are seen at the sites of these indentations (Figs. 6.39 and 6.41), all of which reduce the thickness of the interhemal barrier in these regions.

Underlying the cytotrophoblast the basal lamina is a double membraned structure bearing lumen and micropinocytotic and coated vesicles often discharging their contents into the irregularly shaped lumen (Fig. 6.40a, b).

The mesenchymal cells are characterized by ovoid mitochondria, tubular endoplasmic reticulum, pinocytotic vesicles, coated vesicles, dense bodies and microfilaments. In *Rhinopoma hardwickei*, the foetal capillaries are in direct contact with maternal blood in some places (Figs. 6.39 and 6.40a, b).

The observations on placenta in *Rhinopoma hardwickei* and *Rhinopoma microphyllum*

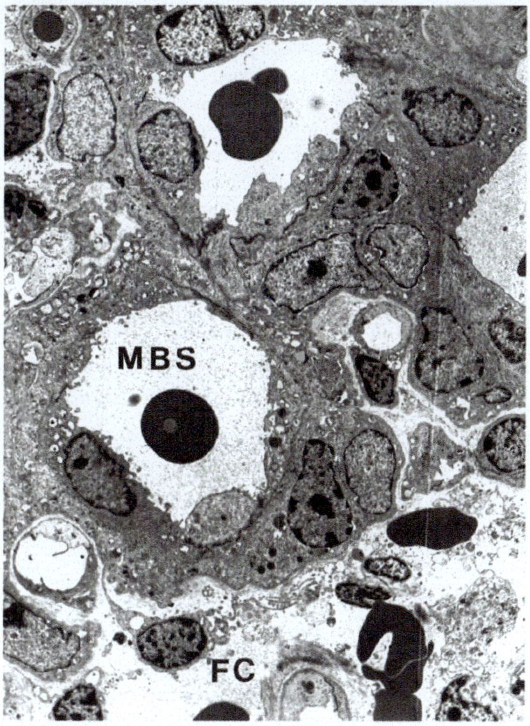

Fig. 6.39 Low-power electron micrograph of the interhemal membrane during full term in *Rhinopoma hardwikei* showing the maternal blood space (MBS) surrounded by maternal endothelium (ME). Note the frequent infolding of basal lamina (BL) and foetal capillaries (FC) occupied between the indentation of cytotrophoblast (cyt). Also note the foetal capillaries are distributed in the mesenchyme (Mes) [X 5000]. From Bhiwgade 1990

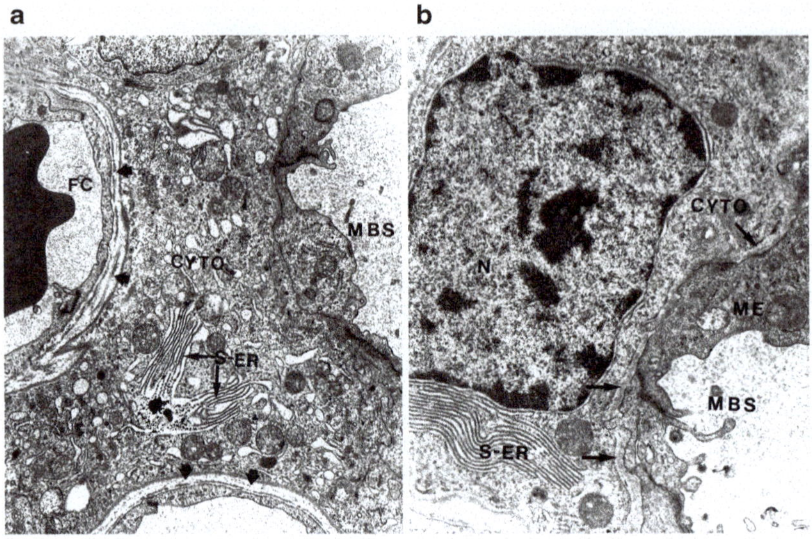

Fig. 6.40 (**a**, **b**) High-power electron micrograph of the interhemal membrane of *Rhinopoma hardwikei* during full term showing maternal blood space (MBS) surrounded by the maternal endothelium (ME). Maternal endothelium is seen with rough endoplasmic reticulum (rER) and mitochondria (M). Distinct cell junctions (JC) are seen between the endothelial cells. The maternal endothelium sits on the discontinuous interstitial membrane (arrows). Junctional complex with numerous desmosomes (D) is seen between two cytotrophoblast cells. The cytoplasm contains numerous ovoid mitochondria (M), well-developed Golgi apparatus (G), Vesicular rough endoplasmic reticulum (rER) and closely packed lamellae of smooth endoplasmic reticulum (sER). Cytotrophoblast rests on a distinct basal lamina (BL). Also note the foetal capillary (FC) and its endothelium (FE) [X 8000] [X 13,000]. From Bhiwgade 1990

indicate that the placental barrier is endotheliodichorial during early and late limb-bud stages. This continues till term in *Rhinopoma microphllum*. In *Rhinopoma hardwikei,* however, from mid-pregnancy onwards a striking ultrastructural change is witnessed with the disappearance of the syncytial layer of the trophoblast from barrier. This outstanding feature revealed in our study i.e. the presence of cytotrophoblast alone as a representative of the trophoblastic layer in the definitive interhemal barrier has been observed only in the secondary placenta of one another bat i.e. *Miniopterus schreibersii fuliginosus* (Bhiwgade et al. 1992). Unlike that in *Rhinopoma hardwickei*, the endotheliomonochorial condition is present right from the beginning and remains throughout the gestation in *Miniopterus schreibersii fuliginosus* (Bhiwgade et al. 1992).

The maternal endothelium remains greatly hypertrophied throughout the embryonic development with cells bearing rough endoplasmic reticulum, numerous polyribosomes, abundant mitochondria, well development Golgi apparatus. Few coated vesicles, caveolae, multivesicular bodies and micropinocytotic vesicles are also seen (Bhiwgade 1990). The presence of micropinocytic vesicles, caveolae and coated vesicles in the maternal endothelium indicates its active role in the absorption of materials from maternal blood.

The presence and orientation (Dempsey and Wislocki 1956 for cat) of rough endoplasmic reticulum suggest that the synthesized proteins, pass through the interstitial membrane. This is also reported in carnivore placentas of ferret (Lawn and Chiquoine 1965), dog (Anderson 1969) and the insectivore, shrew (Wimsatt et al. 1973). The presence of both rough and smooth endoplasmic reticulum in a single cell is a feature unique to the cytotrophoblast cells of *Rhinopoma hardwickei* and has not been described in any other chiropteran species to date. This feature

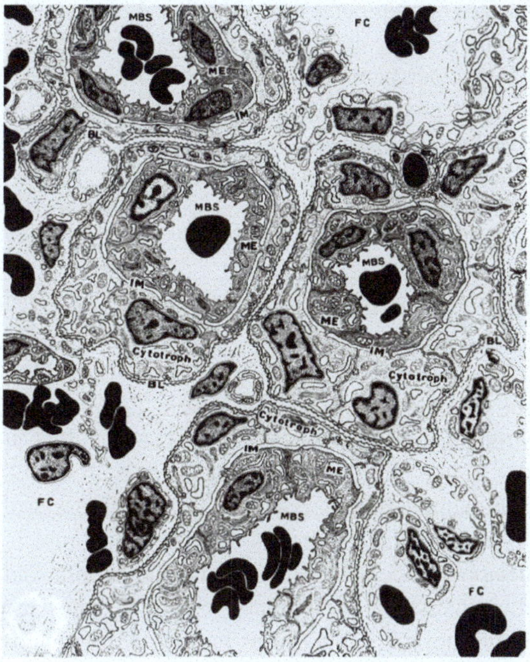

Fig. 6.41 A semi-schematic representation of the layering of trophoblast in the interhemal membrane of *Rhinopoma hardwikei* at full term. The maternal blood space (MBS) with maternal endothelium (ME) is seen. Maternal endothelium rests on the discontinuous interstitial membrane (cyt) and basal lamina (BL). Foetal capillaries (FC) along with mesenchyme are also seen. From Bhiwgade 1990

could be of some physiological importance that needs in-depth studies prior to any conclusive comment. Luckett (1970) described the fine structure of *Macaca mulatta* placenta and encountered both types of ER, i.e. rough and smooth in Langhan cells that were often closely related to the mitochondria. No significant comments, however, were given on the presence of this unusual dual occurrence. According to Ghadially (1988) cell types that have abundant rough endoplasmic reticulum usually have little smooth endoplasmic reticulum and vice versa with the exception of hepatocytes where both forms of endoplasmic reticulum are fairly well represented.

The acellular membrane is similar to the interstitial membrane that has been described in ferret (Lawn and Chiquoine 1965), pinniped (Harrison and Young 1961), dog (Anderson 1969), shrew (Wimsatt et al. 1973) and recently among chiropterans, i.e. *Taphozous melanopogon*, *Rhinopoma hardwickei*, *Rhinolophous rouxi* and *Megaderma lyra lyra* (Bhiwgade 1990) and *Miniopterus schreibersii fuliginosus* (Bhiwgade et al. 1992). Caveolae, coated vesicles and micropinocytotic vesicles seen on either side of interstitial membrane indicate the transport of metabolites bi-directionally across the membrane (Bhiwgade et al. 1992).

In *Rhinopoma microphyllum*, a usually well-developed system of channels extending from the foetal to the maternal side is observed and is very similar to these observed in the macaque placenta. Gammal (1985) reported the presence of tubular channel-like structures and microcrypts in the rhesus monkey. Oduor-Okelo et al. (1983) have described these channels in the placental villi of the spotted hyena to be reminiscent of the extracellular lacunae that may serve to provide a larger absorptive surface area and improved efficiency of both intra and intercellular transport within the intrasyncytial matrix (Gammal 1985; Bhiwgade 1991). In Armadillo, infoldings of the basal cytoplasmic membranes give rise to extensive canaliculi that apparently criss-cross the syncytia and increase the surface area available for the exchange of material between the syncytium and the intervillous space (Enders 1960).

In *Rhinopoma hardwickei hardwickei* with advancing gestational age, the syncytiotrophoblast becomes less dense until it disappears completely by mid-pregnancy. This is an unusual trend in placental morphogenesis among mammals with endotheliochorial placenta. Generally, endotheliomonochorial placentation possesses a single layer of syncytiotrophoblast instead of cytotrophoblast as encountered in *Rhinopoma hardwickei*. Such a condition has been noticed to be more in interhemal membranes that have hemochorial placentation as in molossid bats, *Tadarida brlasilensis cyanocephala* (Stephens 1969), *Molossus ater* (Rasweiler 1991) and *Chaerephon plicata*, (Gopalakrishna et al. 1989). It is also the case in jumping mouse (Zapus) and probably the Jeroba-*Jaculus Jaculus* (King and Mossman 1974) and the rock hyrax *Heterohyra brucei* (Oduor-Okelo et al. 1983) among other mammals. Hence, this unusual cytotrophoblastic definite endotheliochorial placenta of *Rhinopoma hardwickei hardwickei* needs much more extensive studies to understand its role in placental morphogenesis.

The cellular organelle components of the cytotrophoblast like Golgi apparatus, well-developed endoplasmic reticulum, mitochondria, micropinocytotic vesicles and coated vesicles may be involved in the transport and segregation of macromolecules in the interhemal membrane (Enders and Wimsatt 1968; Bhiwgade 1990). Additionally, active protein synthesis by the trophoblast is reported (Bhiwgade 1990). Similar cytoplasmic features have been reported in other species of mammals (Enders 1965; Anderson 1969; Wynn and Davies 1965). The cytotrophoblast appears undifferentiated with its major cell organelles in various species (Anderson 1969; Bjorkman and Wimsatt 1968; Enders 1965; Enders and Wimsatt 1968; Stephens and Cabral 1970), a condition seen when a syncytium is present between the maternal tissue and cellular trophoblast. In *Rhinopoma hardwickei hardwickei,* the nature of organelles present in the cytotrophoblast prior to mid-pregnancy supports this point but post mid-pregnancy, however, the presence of well-developed major cell organelles indicates a more decisive role and capability of the cytotrophoblast. Cytotrophoblast at this stage, probably plays a greater role in the physiological, endocrinological as well as metabolic day-to-day activity of the trophoblast. It could also quite possibly mean a major takeover of the trophoblastic endocrine function from the syncytium that has now been reduced to an incomplete layer confined to isolated patches prior to its complete disappearance. This, however, is yet to be ascertained.

The frequent infolding of trophoblastic basal lamina between the lateral cell borders and the foetal capillaries together with the indentation in the cytotrophoblast serve to reduce the thickness of the barrier membrane in such regions. Enders and Wimsatt (1968) have suggested that in Myotis, thinning of the trophoblast in later stages of pregnancy, cytologically less active cytotrophoblast, and infoldings that increase the surface on the foetal side of the basal layer are all features that the chorioallantoic placenta of Myotis bat shares with the placenta of other species. Stephens (1969) and Rasweiler (1991) have further suggested that such 'thinning' characteristics facilitate expansion of tubules and help in more efficient transplacental transfer.

6.7 *Scotophilus heathi*

The Horsefield bat, *Scotophilus heathi*, belongs to the family Vespertilionidae and its placental structure and histogenesis have been studied at the light microscope level. In this bat, the endotheliodichorial condition lasts till the late Limb-bud stage. Afterwards, there is complete disappearance of the maternal endothelium and the interstitial membrane converts into the intrasyncytial lamina. The progressive development of the ectoplasmic layer from mid-term to full term leads to the establishment of a hemodiochorial placenta.

Like the other vespertilionid bat, *Myotis lucifugus lucifugus* (Enders and Wimsatt 1968), the interhemal barrier of *Scotophilus heathi* is endotheliodichorial initially. It changes into the hemodichorial condition at term with the conversion of the interstitial membrane into the intrasyncytial lamina.

At the early limb-bud stage, the following main layers are seen in the placenta sequentially from the maternal side:

(1) Maternal endothelium, (2) Syncytiotrophoblast, (3) Cytotrophoblast, (4) Basal lamina and (5) Foetal capillary, foetal endothelium along with mesenchyme.

The maternal endothelium is seen only in the early limb-bud stage (Fig. 6.57) which immediately undergoes disintegration. By the late limb-bud stage, it completely disappeared (Fig. 6.58). The interstitial membrane was indistinct, with gaps at few places and thinning at few regions. The layer appeared to be on the verge of being converted to intrasyncytial lamina. The syncytiotrophoblast showed varying thickness and the ratio of syncytiotrophoblast: cytotrophoblast appeared to be more or less

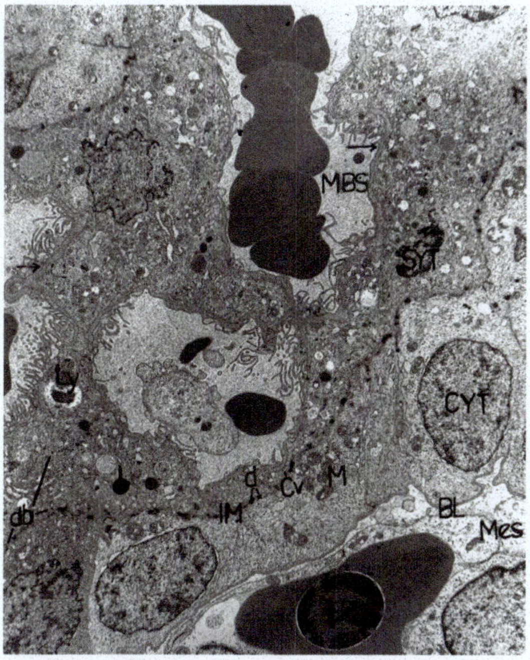

Fig. 6.42 Electron micrograph of the interhemal membrane of *Scotophilus heathi* at the early limb-bud stage. Both the trophoblastic layers, i.e. syncytiotrophoblast (SYT) and cytotrophoblast (CYT) are separated by the interdigitating membrane (IM) and show desmosomal connections (d) along its length. Lipids (L) and lysosomes (Lys) are noticed in the syncytium. The basal lamina (BL) is seen separating the cytotrophoblast and the foetal mesenchyme (Mes) which has the foetal endothelium (FE) and foetal blood vessel (FBV). The maternal blood space (MBS) is filled with its blood capillaries. Coated vesicles (Cv), dense bodies (bd) and polyribosomes are seen in the cytoplasm. The interstitial membrane (thin arrow) is continuous and adjacent to the maternal blood space (MBS) [X 2500]. Unpublished electron micrographs from Bhiwgade and Deshbratar

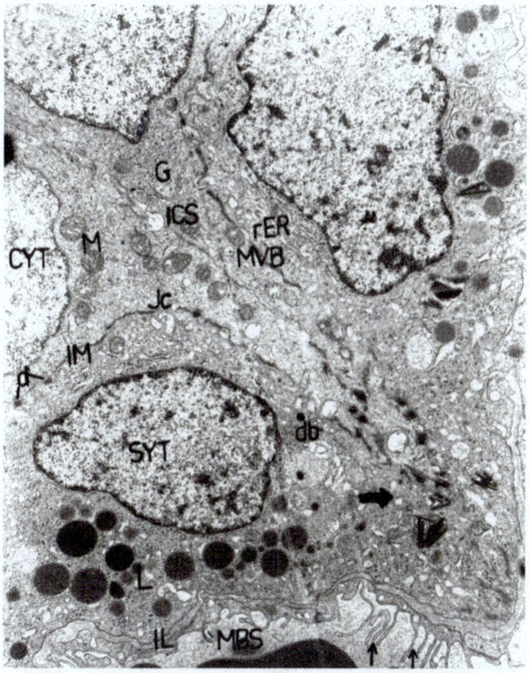

Fig. 6.43 Electron micrograph of the interhemal membrane of *Scotophilus heathi* at the late limb-bud stage. Syncytiotrophoblast (SYT) with lipid deposition (L), dense bodies (db) and rough endoplasmic reticulum (rER) is seen. Note the intercanalicular channel system (thick arrow). The interdigitating membrane (IM) shows desmosomes (d), Junctional complex (Jc) and Inter Cellular Spaces (ICR). The cytotrophoblast (CYT) shows numerous mitochondria (M), multivesicular bodies (MVB), rough endoplasmic reticulum (r-ER), Golgi (G), and polyribosomes. Note the intrasyncytial lamina (IL) with microvillous extensions projecting into the maternal blood space (MBS) and making direct contact with its capillaries (thin arrow) [X 5000]. Unpublished electron micrographs from Bhiwgade and Deshbratar

equal. The two layers of trophoblast, i.e. syncytiotrophoblast and cytotrophoblast are separated by interdigitating membrane with desmosomal connections (Fig. 6.57). The cytotrophoblast sits on a distinct basal lamina, followed by the foetal capillary, with its foetal endothelium and mesenchyme (Fig. 6.42).

By mid-pregnancy, the most important change is the complete disappearance of the maternal endothelium. The interstitial membrane is converted into the intra-syncytial lamina. There is progressive development of the ectoplasmic layer at the late limb-bud stage till mid-pregnancy (Figs. 6.43, 6.44 and 6.45).

The maternal endothelium is distinctly absent from the late limb-bud stage of development and undergoes complete disintegration. The irregularly shaped lacunae of the maternal blood space now comes in direct contact with the cytoplasmic components of the syncytiotrophoblast, i.e. ectoplasmic layer (Figs. 6.43 and 6.44). This ectoplasmic layer is formed with the complete loss of the maternal endothelium. The loss of maternal endothelium is mainly due to the

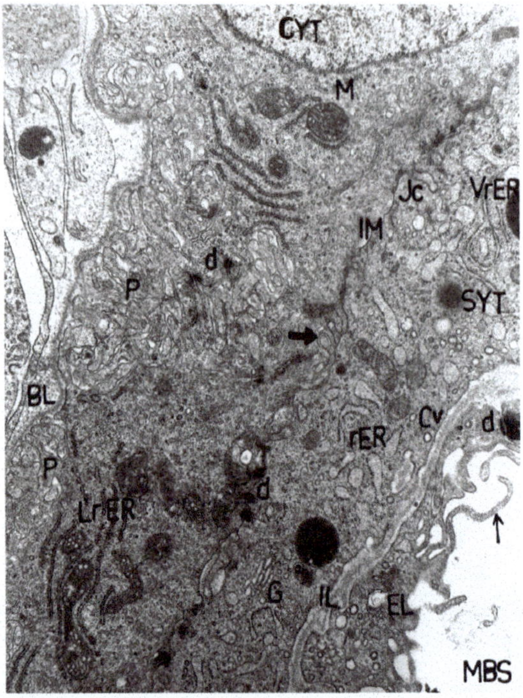

Fig. 6.44 Electron micrograph of placenta in *Scotophilus heathi* at the mid-term stage under high power. The podocytic modification (P) of the basal lamina (BL) with tubular continuity across the cytotrophoblast (CYT) is seen. Desmosomes (d) are quite distinct at these sites. The rough endoplasmic reticulum is seen as lamellae (L-rER). The mitochondria (M) are electron lucent. Numerous vesiculated rough endoplasmic reticulum (VrER), and coated vesicles (Cv) are observed in the syncytiotrophoblast (SYT). Golgi (G) with associated vesicles, lipid (L) and mitochondria (M) are observed in the syncytium. A double layered, uneven intrasyncytial lamina (IL) with ectoplasmic layer (EL) its microvillous processes (thin arrow), desmosomes (d) and Junctional complexes (Jc) are seen [X 8000]. Unpublished electron micrographs from Bhiwgade and Deshbratar

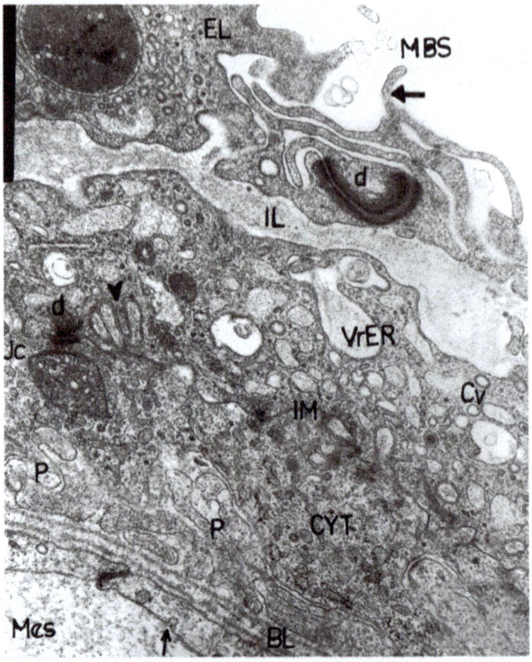

Fig. 6.45 Electron micrograph of the syncytiotrophoblast from the placenta of *Scotophilus heathi* at the mid-term stage (high power). Interdigitating membrane (IM) thrown in loops (arrow heads) along its length with desmosomes (d) can be seen. The podocytic modifications (P) of the basal lamina (BL) are noticed in the cytotrophoblastic margin (CYT). The foetal mesenchyme (Mes) adjoining the basal lamina exhibits pinocytosis (thin arrow). The syncytiotrophoblast looks fenestrated because of the vesiculated rough endoplasmic reticulum (VrER). The intrasyncytial lamina (IL) is uneven with homogenous matter within it. Ectoplasmic layer (EL) shows coated vesicles (Cv) and desmosomes (d). The maternal blood space (MBS) shows microvillous extensions (thick arrows) from the ectoplasmic layer [X 15000]. Unpublished electron micrographs from Bhiwgade and Deshbratar

proliferative activity of the syncytiotrophoblastic cytoplasm which loosens the maternal endothelium from its base on the interstitial membrane. This proliferating cytoplasm of the syncytiotrophoblast pushes through the uneven width of the interstitial membrane at different places and makes gaps and discontinuities in its length. The cytoplasm of the syncytiotrophoblast percolates through these gaps to line the maternal blood space as an ectoplasmic layer. Through these gaps, the ectoplasmic layer extends as a number of thin cytoplasmic filamentous structures at the luminal surface. These extensions are not uniform in width and length but are highly proliferative at limb-bud stage and mid-pregnancy (Figs. 6.43 and 6.44). The ectoplasmic layer with well-developed microvillous projections is noticed throughout the stages of development from late-limb bud to full-term stage (Figs. 6.43, 6.44, 6.45 and 6.46).

By late limb-bud stage and advancing to full-term pregnancy cell organelles like multivesicular bodies, coated vesicles, desmosomes, dense

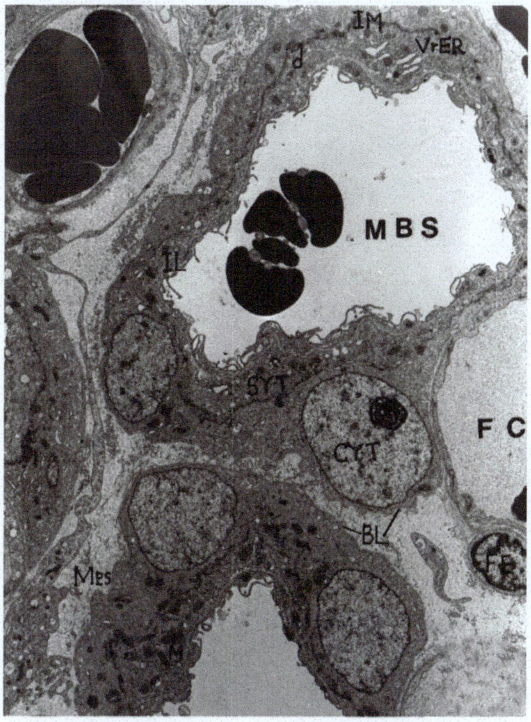

Fig. 6.46 Low power electron micrograph of the interhemal membrane of *Scotophilus heathi* at full-term. The foetal capillary (FC) is in close proximity with maternal blood space (MBS) and the maternal blood capillaries indicating a reduction in the thickness of the syncytiotrophoblast (SYT) and cytotrophoblastic (CYT) layers. The interdigitating membrane (IM) separating the trophoblastic layers bears desmosomal connections (d) in its continuity. Note the vesiculated rough endoplasmic reticulum (VrER) and mitochondria (M). The basal lamina (BL) is distinct, double layered membrane followed by foetal mesenchyme (Mes) showing lose ground matrix and granulated structures. The foetal endothelium (FE) is embedded in the mesenchyme. The interasyncytial lamina (IL) appears to be continuous, double layered with microvillous processes along its luminal surface protruding into the maternal blood space (MBS) and thus giving an irregular profile to the same. [X 4000]. Unpublished electron micrographs from Bhiwgade and Deshbratar

bodies and mitochondria are added into the ectoplasmic layer (Figs. 6.44 and 6.45).

The syncytiotrophoblast is fairly electron dense, much more than the underlying cytotrophoblast and both are fairly equal in ratio at the early limb-bud stage. Irregular and uneven wrinkled nuclei embedded in a common ground matrix devoid of any cell boundaries. The width of syncytiotrophoblast layer varies during the different stages of development and is thicker at sites with nuclei and thinner at sites without nuclei (Fig. 6.42). At limb-bud stage, the cytoplasm is still fairly electron dense along its entire length with few nuclei grouped together (Fig. 6.43). At mid- and full-term stage the interhemal layer is quite attenuated and appears as a thinner mass as compared to that at the limb-bud stage except at few places. Syncytiotrophoblast cytotrophoblast ratio appears to be in equal proportion of cytotrophoblast (Fig. 6.43).

The rough endoplasmic reticulum of the syncytial cytoplasm at early limb-bud stage is both tubular as well as some dilated in form with amorphous material within it (Fig. 6.42). At early limb-bud stage itself and sometimes by late limb-bud stage onwards the tubular rough endoplasmic reticulum dilates and become more and more vesiculated type indicating the initiation of a coarse syncytium. The rough endoplasmic reticulum gains prominence during late limb-bud stage wherein it is a dominant feature with both tubular and vesiculated type (Fig. 6.43). The vesiculation of rough endoplasmic reticulum peaks at term whereupon the coarse syncytial structure too becomes more extensive. The syncytiotrophoblast is the most electron dense at term and it resembles a coarse syncytium with distinct spongy porous mass due to the presence of unusually high concentration of extensively modified vesiculated rough endoplasmic reticulum (Fig. 6.43). Such a type of modification of the syncytiotrophoblastic layer is commonly referred to as 'inter-lobular', 'coarse syncytium' or 'spongiotrophoblast', i.e. trophospongium.

The darkly stained ovoid and or elongated mitochondria bound by a double membrane with lamelliform or shelf-like cisternae are noticed in the cytoplasm of syncytiotrophoblast. They are in more proximity to the maternal face at early limb-bud stage, whereas mitochondria are generally present in groups throughout the cytoplasmic mass in the preceding stages. Mitochondria are extremely hypertrophied and attain different sizes and shapes at full-term stage (Fig. 6.46). Few of the mitochondria are swollen and show cavitations of the matrix and distorted cisternae.

The Golgi complex becomes more prominent and distinct in the advancing stages of pregnancy. At the early limb-bud stage the Golgi is indistinct. During late limb-bud stage the Golgi shows stacked cisternae with associated secretory vesicles. At mid-term the Golgi gains prominence and appear to be parallelly stacked with more proximity towards the cytotrophoblastic layer. The Golgi seemed to be associated with the other cellular inclusions like dense bodies, vesiculated r-ER and mitochondria (Fig. 6.44).

At the confluence of the basal portion of the syncytiotrophoblast and the underlying cytotrophoblast layer the interdigitating membrane is formed. This membrane is observed throughout the stages of development right from the limb-bud stage to the full-term stage. It is ill-defined during early development but shows presence of a number of desmosomes (Figs. 6.44 and 6.45).

Blocks of cytoplasmic membrane or loops like convoluted folds as well as invading fissure-like canals and crevices reach far into the cytotrophoblast that is observed beneath their membrane limits at the mid-term and full-term stage. These may be the rudiments of microtubular system (Fig. 6.46). These loops or folds serve to separate the two trophoblastic layers. At localized sites, it is in continuity with the trophoblastic basal laminar through a canal system and also bears desmosomal connections as well as cell spaces between the adjoining cytotrophoblast (Fig. 6.47).

In contrast to the syncytiotrophoblast, the nuclei in the cytotrophoblast layer show well-defined membrane that borders the surrounding cytoplasm.

A moderate number of ovoid to elongated or hypertrophied mitochondria with lamellar cisternae are noticed preferentially towards the apices of the cells, immediately underlying the syncytial phase. Few of the mitochondria exhibit cavitations with disoriented cisternae.

Endoplasmic reticulum is observed at all stages of development but is most prominent during the mid-term and full-term stages (Fig. 6.47). Strands of lamellar endoplasmic

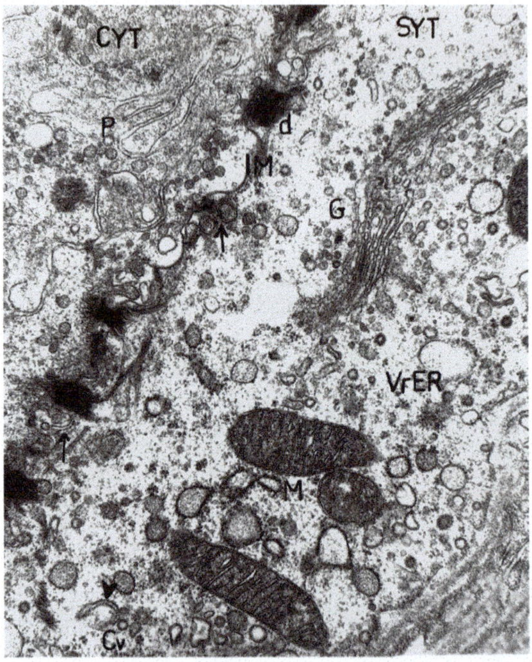

Fig. 6.47 Ultrastructure of the placenta in *Scotophilus heathi* at full term under high-power electron microscopy. The podocytic modifications (P) of basal lamina (BL) that penetrate deep into the cytotrophoblast (CYT) is distinct. The interdigitating membrane (IM) shows desmosomes (d) along its length and is folded into loops (thin arrow) Hypertrophied mitochondria (M) with strands of regular cisternae, vesiculated rough endoplasmic reticulum (VrER), coated vesicles (Cv) and the microtubular intracanalicular channels (arrow head) are observed in the syncytiotrophoblast (SYT). The Golgi complex is arranged in stacks of 4–5 and is associated with vesicles [X 15000]. Unpublished electron micrographs from Bhiwgade and Deshbratar

reticulum were observed in close association with mitochondria at the mid-term and full-term stage (Figs. 6.60 and 6.62).

The Golgi complex appears to be concentrically arranged at late limb-bud stage and stacked in parallel during the full-term stage (Figs. 6.48 and 6.49).

The other cellular organelles observed in the cytoplasm include sparse, dense bodies, multivesicular bodies, coated vesicles and lysosomes. The cytotrophoblastic matrix is less electron dense as compared to the syncytiotrophoblast and as such comparatively fewer cell organelles are present.

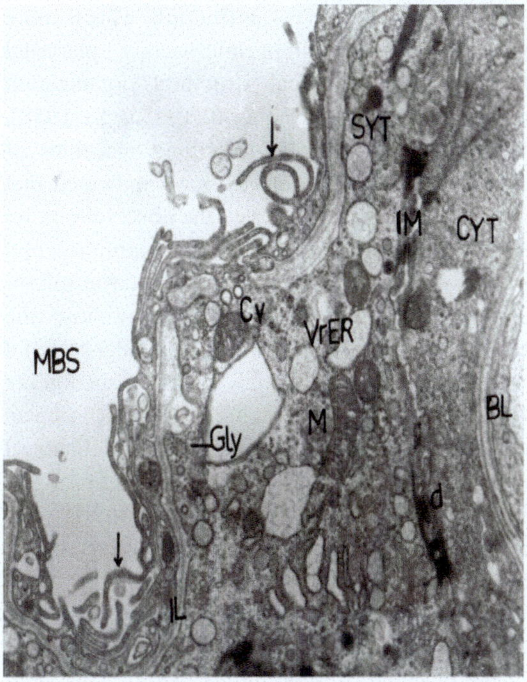

Fig. 6.48 Magnified view of the syncytiotrophoblast (SYT) in *Scotophilus heathi*. It looks extremely fenestrated and porous. It shows coated vesicles (Cv), vesiculated rough endoplasmic reticulum (VrER), mitochondria (M), and sparse glycogen bodies (Gly) scattered in the cytoplasm. The intrasyncytial lamina (IL) shows uneven width. It is discontinuous and the ectoplasmic processes are thread-like processes (thin arrow) which form loops and extends into the maternal blood space (MBS). The interdigitating membrane (IM) with desmosomes (d) is followed by cytotrophoblast (CYT) and then the basal lamina (BL) [X 12,000]. Unpublished electron micrographs from Bhiwgade and Deshbratar

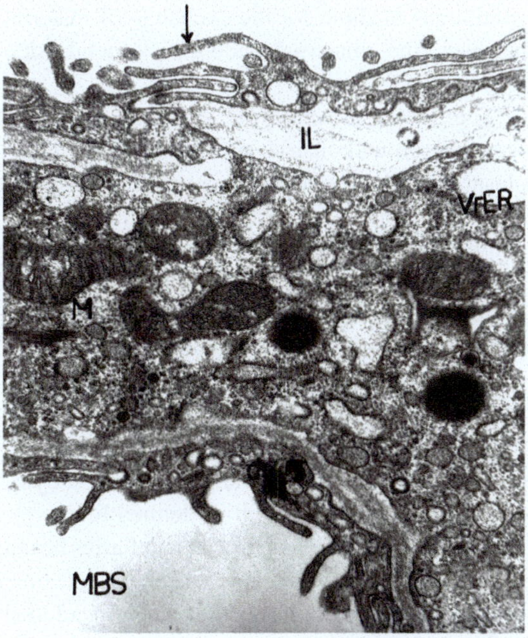

Fig. 6.49 Ultrastructure of the syncytiotrophoblast in *Scotophilus heathi* at full-term stage (high power). It shows spongy, and fenestrated appearance. Numerous hypertrophied mitochondria (M), vesiculated rough endoplasmic reticulum (VrER) along with and lipids (L) are seen in the syncytium. The intrasyncytial lamina (IL) is of uneven width, double-layered and discontinuous. The ectoplasmic layer is sculptured with microvillus extensions (Thin arrow). These microvilli expel their contents into the maternal blood space through. Desmosomes are seen at the junctions of microtubular canals (curved arrow) [X 15,000]. Unpublished electron micrographs from Bhiwgade and Deshbratar

A specific characteristic feature during full-term pregnancy is the podocytic protrusion that appears like microtubular channels, which enclose small amounts of cytoplasm with their folds and loops with extensive coiling at sites. This feature is prominent from mid-term stage (Fig. 6.45), but is most prominent at full-term and shows tubular continuity along with a few desmosomal connections (Fig. 6.48). These protrusions originate from the basal lamina and outer membrane of the cytotrophoblast that run along its length is transformed into conspicuous vesicles and blocks. They probably enable transportation within their lumen. As compared to earlier stages, the cytotrophoblast shows attenuation into thin phalanges at term. At full-term stage, these reductions are sometimes accompanied by highly reduced cytotrophoblast layer at some sites (Fig. 6.47).

The basal lamina is a very well-developed continuous, double-membraned layer, which runs along the proximal length of the cytotrophoblast, in an even width. The contents are homogenous and of a cellular nature enclosed within the membrane limits that bear podocytic vesicles (Fig. 6.46) and also podocytic blocks at the cytotrophoblastic side during the full-term stage. The basal lamina is so positioned that it separates the identities of the cytotrophoblastic layer and foetal endothelium that lie on either side of its membrane. It thus provides a continuous medium for the exchange of material between the maternal side and foetal side (Fig. 6.47).

The foetal mesenchyme exhibits a loosely scattered ground matrix surrounded with cytoplasm, that has mitochondria, coated vesicles, tubules of rough endoplasmic reticulum, lipids, dense bodies, glandular structure, lysosomes, microvillous canals as well as free and bound ribosomes during different stages of development. These are more pronounced with advanced stages of gestation (Fig. 6.46).

The foetal capillaries are embedded in the mesenchymatous mass and often the mesenchyme displays an extremely fenestrated cytoplasmic ground matrix with ill-defined limits.

The endothelial cells of foetal capillaries bear a loose matrix at mid-pregnancy. The cytoplasmic organelles seen during development included darkly stained ovoid mitochondria, Golgi complex, coated vesicles and dense bodies at the full-term stage (Fig. 6.46).

Our electron microscopic studies indicate that observations on the definitive placental barrier in *Scotophilus heathi,* reported earlier, need some modifications. The earlier studies were based on light microscopic observations and histochemical studies which were also under the light microscope. What has been earlier reported as an 'enucleate maternal endothelium' that adjoins the maternal blood space is in fact the 'ectoplasmic' layer of syncytiotrophoblast which shows abundant microvilli under the electron microscope. Therefore, barring the early post-implantation stages, the placenta is actually haemochorial rather than endotheliochorial, as reported earlier. It was earlier reported that the cytotrophoblast layer in *Scotophilus heathi* is lost during the latter half of pregnancy. In our observations, the cytotrophoblast cells become greatly reduced into a syncytium in such a way that the two layers of the trophoblast appear as a single smooth unit. Under electron microscope, the definitive chorio-allantoic placenta of *Scotophilus heathi* shows two distinct layers of trophoblast till term. Therefore, the placenta should be classified as hemodichorial. Thus, our findings in *Scotophilus heathi* suggest that, probably, similar conditions may be seen in other vespertilionids. This is more so because, under light microscopy, placental barrier shows identical structural organization as seen in *Myotis* (Wimsatt 1945a, b, 1958). Wimsatt (1958), while describing placentas of several Phyllosomatide bats, have indicated that at the light microscopic level, the observations are similar to those reported in *Scotophilus and Myotis.* It is plausible that under electron microscope, these placentae would also show the unusual intrasyncytial lamina of acellular material which serves as the initial basement membrane for the maternal endothelium. These placentae could also then be classified as hemodichorial characterized by PAS-positive 'interstitial membrane'. As seen in vespertilionids, the intrasyncytial lamina could be a characteristic of the placenta in phyllostomids too.

The continuous intrasyncytial lacuna system containing the lamina is the most striking feature of the placenta of *Scotophilus heathi.* This lacunae system is consistent and uniform in its distribution and position. The intrasyncytial lamina develops from the maternal tissue, but all through the gestation, it is seen as completely embedded within the syncytium. Even during the displacement of the maternal endothelium and the subsequent enlargement of the labyrinth, the intrasyncytial lamina remains intact. The enlargement of the labyrinth that is observed, with the elongation of the maternal blood space, indicates that during this stage, the lamina must be formed from the material synthesized by the syncytium, at least, in part.

The continuous intrasyncytial lacuna system seen in *Scotophilus heathi* placenta is not entirely unique, since in *Eutamias* (Chipmunk), a subsurface lacuna system has been reported within the syncytium that contains an extracellular material. This system is also overlaid by an ectoplasmic layer (Enders 1965). The analogous arrangement as seen in two divergent animals indicates the probability that the arrangement could have distinct functional advantages. It is interesting to note that only an ectoplasmic layer overlays the intrasyncytial lamina. This acellular lamina, therefore, separates the maternal blood from the majority of foetal tissues, the endoplasmic reticulum and other synthetic apparatus of the syncytium. It could be suggested that this separation could serve as an immunological mechanism by which antigenic substances are prevented from entering the maternal side from the foetal tissues. Kirby et al. (1964) have recently shown evidence in the mouse that, acelluler maternal layer may facilitate immunological isolation of ectopic implants from maternal tissues.

Several cytological features observed in the placenta of *Scotophilus heathi* are also seen in the hemodichorial placenta of rabbit and in other hemochorial placentas as seen in *Myotis*. Thick syncytial layer with its granular endoplasmic reticulum and the large Golgi zone oriented towards the foetal side of the syncytium are features that are also reported in the placenta of rabbit, human, armadillo and guinea pigs. The basal layer of trophoblast (Cytotrophoblast) which is cytologically less active in appearance than the overlaying syncytium, thinning of the trophoblast layer seen at later stages of pregnancy, infoldings of the basal layer seen towards the foetal side are all features that the chorioallantoic placenta of *Scotophilus heathi* shares with the placenta of other species.

6.8 *Tylonycteris pachypus*

Tylonycteris pachypus, the Flat headed bat, belongs to the family Thyropteridae (Koopman 1993) and its placental structure and histogenesis have not been studied to date neither at the light microscopic level nor at the electron microscopic level. We undertook a study to establish the familial similarities if any between the *Tylonycteris pachypus* and T*hyroptern tricolor spix* and at the same time characterize the specific peculiarities of placenta in this species of bat. We studied the placenta of *Tylonycteris pachypus*, under an electron microscope through different developmental stages.

The placenta of *Tylonycteris pachypus* shows the presence of the characteristic coarse syncytium in its placenta which is also seen in the pteropid bat, *Rousettus leschenaulti* (Bhiwgade et al. 2000). The placenta also shows podocytic modifications of the basal lamina which is akin to those found in ungulate placenta, macaque, and human (Wynn and Davies 1965). Our observation reveals that, at the late neural groove stage, no particular layering of the trophoblast is seen. There is distinct maternal endothelium along with a compactly arranged cytotrophoblast layer followed by the syncytiotrophoblast of varying thickness. The cytotrophoblast layer at this stage of gestation shows many vesiculated rough endoplasmic reticulum and several well-developed concentrically arranged Golgi complexes.

The mid and term placenta, however, reveals the absence of the maternal endothelium. The maternal blood space comes in direct contact with the cytoplasmic mass of the syncytiotrophoblast, i.e. ectoplasmic layer. This is followed by discontinuous intrasyncytial lamina within the syncytiotrophoblast. The syncytiotrophoblast shows vesiculated rough endoplasmic reticulum that has a spongy appearance. The cytotrophoblastic basal that maintains distinct continuity across the limits of both layers shows extensively modified podocytic extensions, with some desmosomal connections. Glycogen rosettes are seen at the peripheral limits of the two trophoblastic layers. At some places the layers become significantly reduced to thin phalanges, whereby, the maternal blood space and foetal capillary come very close to each other. The thickness of the interhemal barrier at these places is thus, greatly reduced.

Since the maternal endothelium is absent and since both trophoblastic layers are seen at term, the definitive placenta in *Tylonycteris pachypus* can be identified as hemodichorial. Till the late limb-bud stage, the placenta is endotheliodichorial but after that the complete disappearance of the maternal endothelium and the transformation of interstitial membrane into the intrasyncytial lamina leads to the establishment of a hemodichorial placenta. This is similar to that reported in another thyropterid bat, *thyroptera tricolor spix* (Wimsatt and Enders 1980).

For our study, *Tylonycteris pachypus* (Temminck) were obtained from Kerala, the southern part of India from early January to middle of April. The breeding habits of *Tylonycteris pachypus* have been reported by Midway L. (1972). We studied the placental structure both under the light microscope and under the electron microscope during the following stages of gestation:

1. Late neural groove stage
2. Early limb-bud stage
3. Late limb-bud stage
4. Mid--regnancy
5. Full term

6.8.1 Light Microscopic Observations

At the late neural groove stage, the allantoic diverticulum arises before the closure of the neural groove but does not establish contact with the placenta until a little later in development. The placenta is cup shaped with its base towards the antimesometrial side of the uterus. The placenta shows a zone of syncytiotrophoblast, which extends deep into the endometrium and forms a thick mantle of syncytiotrophoblast. It occupies almost half the thickness of the endometrium. At the border of the placenta, several large maternal vessels are seen surrounded by a layer of syncytiotrophoblast and a layer of cytotrophoblast (Fig. 6.50).

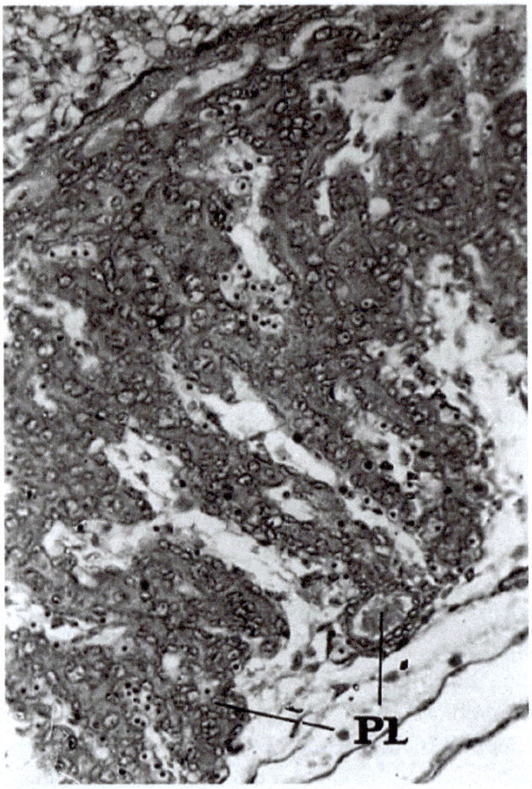

Fig. 6.50 Light micrograph of the placenta in *Tylonycteris pachypus* during early limb-bud stage. Note the placental tubules (PL) hanging from the uterine wall (Haematoxylin and Eosin stain) [X75; X 150]. Unpublished light micrograph from Bhiwgade and Mahaley

The syncytiotrophoblast is seen enclosing the maternal blood capillaries, in which hypertrophied and cuboidal endothelial cells are seen. Many large maternal blood vessels with similar endothelium, are also seen just below the foetal margins of the placenta.

In some places, particularly in the region of the chorionic placenta, a distinct lining of cytotrophoblast occurs on the foetal margins of the placenta, that is, the inner surface of the placenta cup. The cytotrophoblast layer pushes into the syncytiotrophoblast layer in the form of solid columns of cells. In vertical section, these columns appear to be made up of two rows of cytotrophoblast cells. Some of these are hollow, into which the chorionic mesoderm is seen to have penetrated. These are the beginnings of the formation of chorionic villi. The foetal margin of the chorionic placenta, appear to be indented, due to these hollow villi.

During the early limb-bud stage in the chorio-allantoic placenta (Fig. 6.50), the cytotrophoblastic villi which were seen in the previous stage, are seen to have extended deeper into the syncytiotrophoblast zone. The foetal mesenchyme and allantoic blood capillaries are seen within the hollows of the villi. In many places the villi are expanded, and the maternal vascular channels surrounded by the syncytiotrophoblast become compressed into tubules. These tubules appear to be hanging from the uterine wall. Each placental tubule in the transverse section consists of a central maternal capillary surrounded by a layer of syncytiotrophoblast and the cytotrophoblast on the outside. The endothelial cells of the maternal capillary inside the tubule

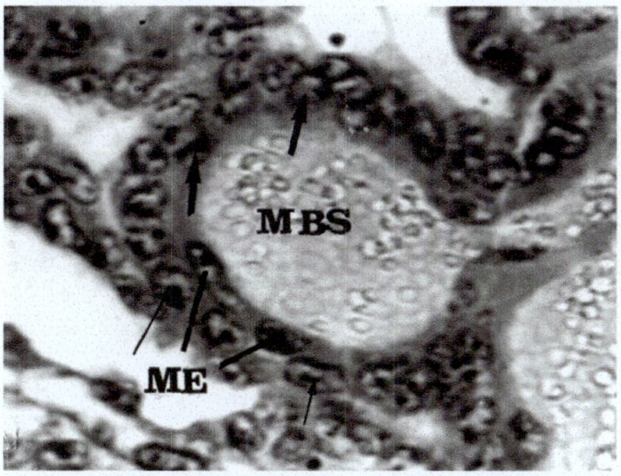

Fig. 6.51 Light micrograph of placenta in in *Tylonycteris pachypus* showing the Maternal Blood Space (MBS) within the tubule with few maternal endothelial cells (ME) (Haematoxylin and Eosin stain) [X 240]. Unpublished light micrographs from Bhiwgade and Mahaley

are rare and if seen, are flat with fusiform nuclei (Fig. 6.51, 6.52, and 6.53).

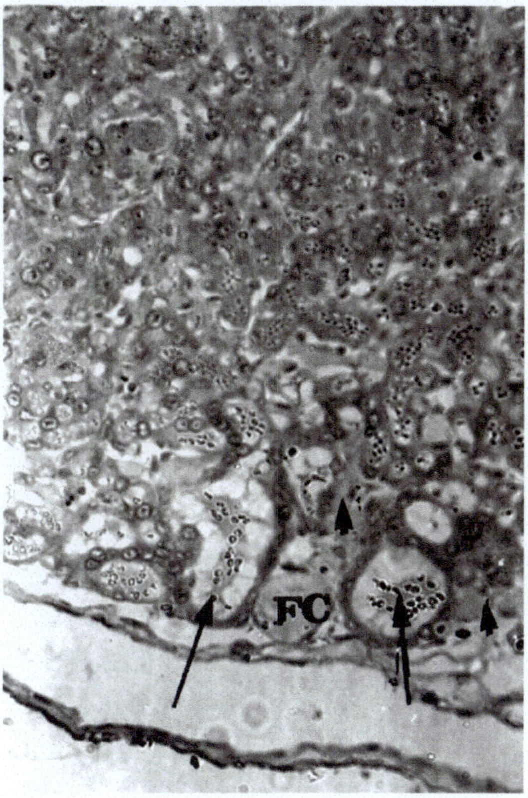

Fig. 6.52 Light micrograph of term placenta in *Tylonycteris pachypus*. The placental tubules are seen as more compactly arranged. At most of the places foetal capillaries (FC) are very much close to the maternal blood space (thin arrows). Note the foetal mesenchyme (arrow heads). (Haematoxylin and Eosin stain) [X 150]. Unpublished light micrographs from Bhiwgade and Mahaley

6.8.2 Electron Microscopic Observations

The histogenesis of the placenta in *Tylonycteris pachypus* follows the pattern as in other bats. During the late neural groove stage, no particular layering system of trophoblast is observed. The trophoblastic placenta, at this stage has a multi-layered cellular character with well-developed cell membranes. Desmosomes are seen to be connecting the borders of adjoining cells to form a thick tissue layer. The cell organelles include cisternae of concentrically arranged Golgi complex and darkly stained ovoid mitochondria. Foetal capillaries, however, are not distinct during this stage.

The trophoblastic placental organization gives way to the characteristic layering as gestation advances from limb-bud stage to full-term stage. The maternal vasculature is well defined by large, irregularly shaped blood lacunae that are bounded by a unicellular layer of large endothelial cells. The endothelial cells have large darkly stained nuclei and nucleoli during the early and late limb-bud stages (Fig. 6.54). The contour of the nucleus is wrinkled. The cytoplasmic matrix is characterized by numerous predominantly spherical mitochondria with shelf-like cristae. A distinct feature of the cytoplasm at this stage is the presence of a large number of vesiculated rough endoplasmic reticulum. Other inclusions like dense bodies, coated vesicles and polyribosomes are also seen. Junctional complexes and distinct desmosomes connect adjacent cells into a single layer that rests on a conspicuous continuous interstitial membrane. Interstitial membrane is well developed with acellular, homogeneous, matrix limited by a double membrane and is of uneven width. The membrane attenuates at some regions as the gestation progresses. Gaps start forming in

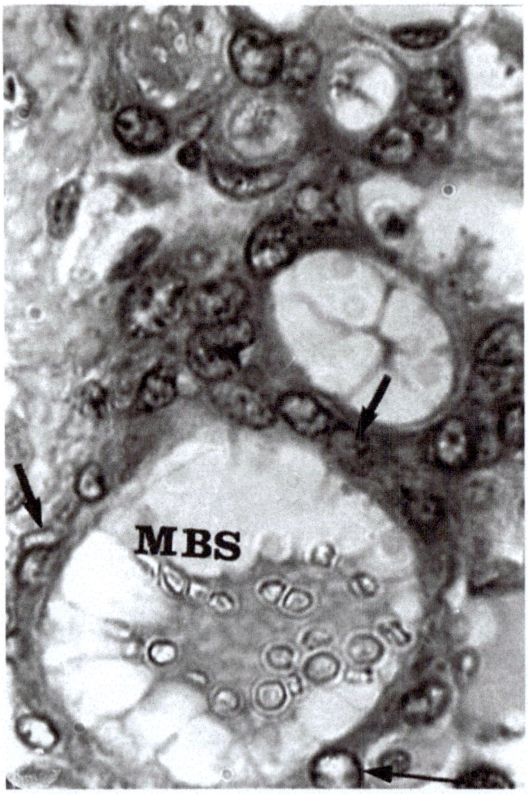

Fig. 6.53 Light microscopic photograph of chotioallantoic placenta in *Tylonycteris pachypus* at full term. The maternal blood space (MBS) is surrounded by inner layer of syncytiotrophoblast (arrow) followed by few cellular cytotrophoblast cells (thin arrow) (Haematoxylin and Eosin stain) [X 240]. Unpublished light micrographs from Bhiwgade and Mahaley

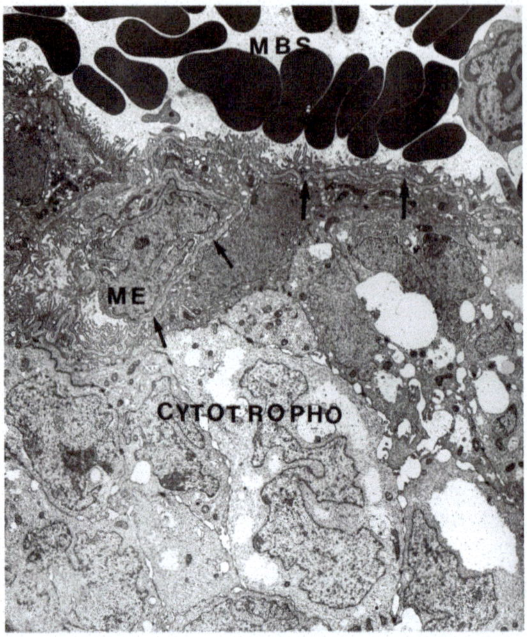

Fig. 6.54 Electron micrograph of interhemal membrane of *Tylonycteris pachypus* at late limb bud stage. The maternal blood space (MBS), a distinct maternal endothelium (ME), with large nucleus (N) a continuous acellular interstitial membrane (arrows) are distinct. At the luminal surface of maternal endothelium microvillous projections are visible. The apposition of the cytotrophoblast layer with the syncytiotrophoblast along with the intervening membrane (arrow heads) is seen [X 7000]. Unpublished electron micrographs from Bhiwgade and Mahaley

this structure by late limb-bud stage. The maternal endothelial layer is, however, conspicuously absent in the subsequent stages of mid and term pregnancy.

The irregularly shaped lacunae within the maternal blood space come in direct contact with the cytoplasmic component of the syncytiotrophoblast, i.e. the ectoplasmic layer (Fig. 6.55). This layer is formed as the maternal endothelium is lost. The disappearance of maternal endothelium is caused by the proliferative activity of the cytoplasm of syncytiotrophoblast which loosens the maternal endothelium from its base of the interstitial layer. The proliferating cytoplasm pushes through the uneven width of the interstitial membrane at sites where it is thin enough to break off. At these sites, the cytoplasm pushes through creating gaps and discontinuities in the interstitial membrane (Fig. 6.55). The cytoplasm of the syncytia without most of the cytoplasmic organelle, except mitochondria flows through these gaps to line the maternal blood space as an ectoplasmic layer. This protruding syncytium is produced into thin cytoplasmic fingers at the luminal surface (Fig. 6.55). It is not uniform in width and is highly proliferative at mid-pregnancy. At term multivesicular and dense bodies are added to the population of cytoplasmic organelles in the ectoplasmic layer (Figs. 6.56 and 6.57).

This morphogenetic development leads to the incorporation of the interstitial membrane, in parts into the syncytial ground matrix to form the ‘intrasyncytial lamina’. Therefore, the

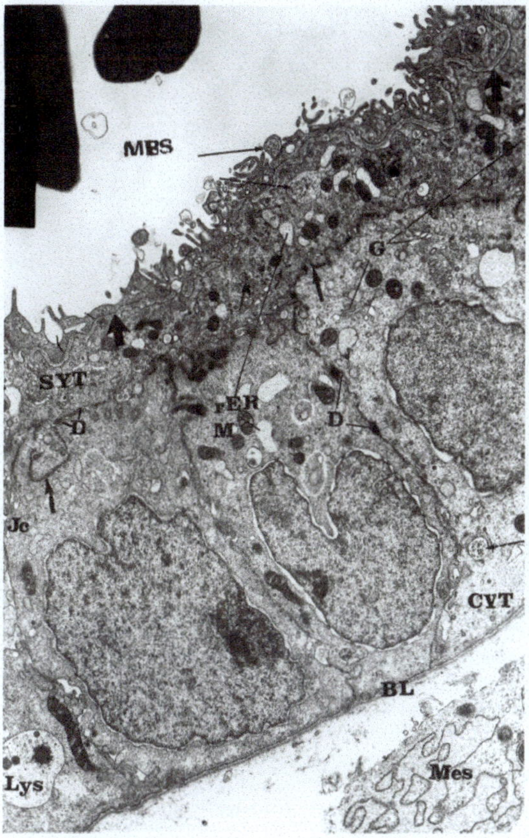

Fig. 6.55 Ultrastructure of interhemal membrane in *Tylonycteris pachypus* at the late limb-bud stage. The cellular organization as well as difference in the ratio of thickness of the trophoblast are distinct. A discontinuous interstitial membrane (thick arrows) at irregular intervals underneath the maternal blood space (MBS) separates the thin zone of syncytiotrophoblast (SYT) from a wider zone of cellular cytotrophoblast (CYT). The cytotrophoblast lies between the upper desmosomally connected (D) interdigitating membrane (arrows) and the lower basal lamina (BL). Cytotrophoblastic membrane bears desmosomes (D), junctional complex (Jc) that unite the adjacent cells in a unicellular layer. The thick layer of cytotrophoblast is a characteristic feature of the early stage placenta. A moderate number of mitochondria (M), rough endoplasmic reticulum (rER), light multivesicular bodies (thin arrows) and Golgi complex (G) are noted in both the trophoblast. Note the lysosome (Lys) in the cytotrophoblast. Foetal mesenchyme (Mes) is at the bottom right-hand corner [X 7000]. Unpublished electron micrographs from Bhiwgade and Mahaley

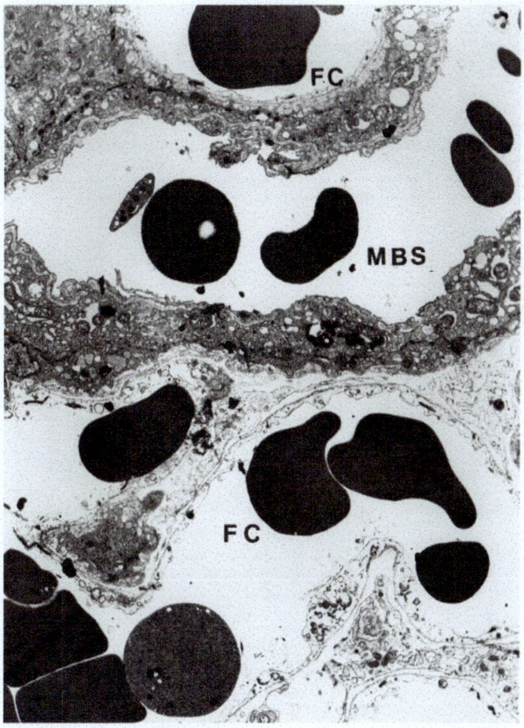

Fig. 6.56 Electron micrograph of fully established placental membrane of *Tylonycteris pachypus* at full term. It shows two thin layers of the trophoblast i.e. the syncytiotrophoblast and the cytotrophoblast between the maternal blood space (MBS) and the foetal capillary (FC). Cytotrophoblast is isolated from the syncytiotrophoblast by the invasion of the interdigitating membrane (arrow) [X 3500]. Unpublished electron micrographs from Bhiwgade and Mahaley

intrasyncytial lamina established by mid-pregnancy is lined by an ectoplasmic layer (Fig. 6.55) at the luminal edge and the syncytial ground matrix at the trophoblastic face. It is characterized by its discontinuous, homogeneous acellular (Fig. 6.57) electron-dense matrix limited by a double membrane which shapes it into a thin lining of more or less even thickness. At sites, both during mid and term pregnancy, thin tubular folds and convoluted looped channels with desmosomal connections are given out, to open into the maternal blood space directly, thereby providing a continuity with the same. At full-term, many coated vesicles are seen, in continuation with the membrane of intrasyncytial lamina (Fig. 6.57) along with some pinocytotic vesicles in the process of exuding their contents into the same. The intrasyncytial lamina and the ectoplasmic layer are both features that are seen consistently in the placenta of this bat.

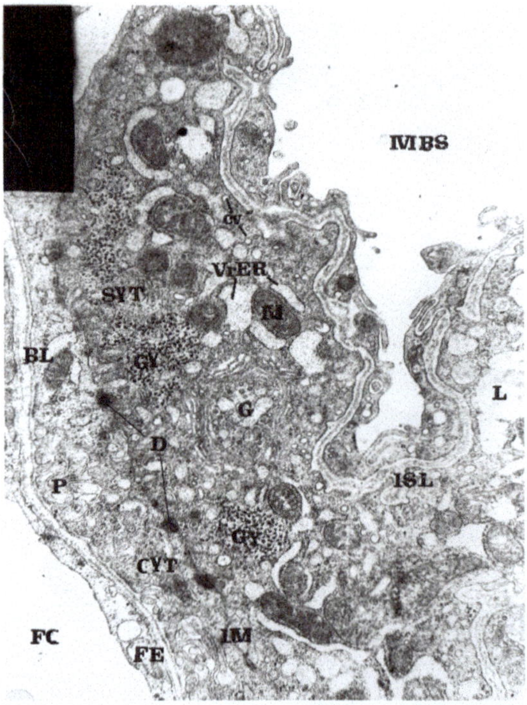

Fig. 6.57 Magnified part of interhaemal membrane of *Tylonycteris pachypus* at full term stage. The organization of placental layers as well as cell organelles within the different layers are seen. Between the foetal capillary (FC) and maternal blood space (MBS), is the discontinuous double membrane layer, the intrasyncytial lamina can be seen. Also seen are the syncytiotrophoblast with well-developed electron dense organelle, followed by the cytotrophoblast, with the desmosomally connected (D) as the intervening layer. Cytotrophoblast is resting upon basal lamina (BL) and is followed by foetal endothelium (FE). Hypertrophied mitochondria (M) are surrounded by vesiculated rough endoplasmic reticulum (VrER), concentrically arranged Nukber of Golgi complex (G), Glycogen bodies (GY) in groups, cluster of coated vesicles (cv) and lipid droplets are seen within the syncytiotrophoblastic mass. A number of tubular channel system, the podocytes (p) arising from the basal lamina covering most of the cytotrophoblastic mass are also noted [X 13,000]. Unpublished electron micrographs from Bhiwgade and Mahaley

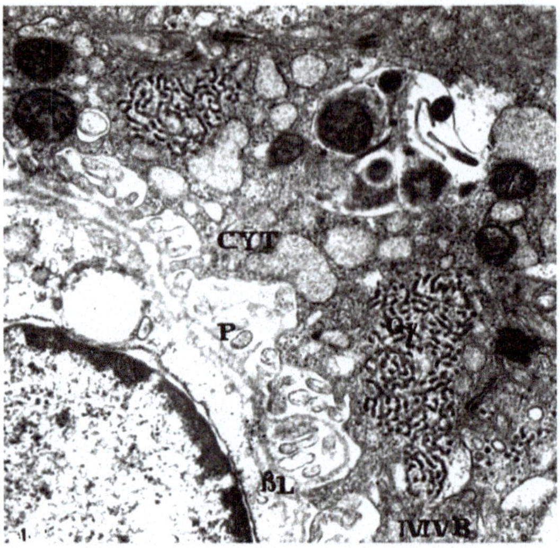

Fig. 6.58 Enlarged view of cytotrophoblast (CYT) in *Tylonycteris pachypus*. A specific characteristic feature is the presence of numerous convoluted loops-the podocytes (P) arising from basal lamina (BL). Note the glycogen granules (Gy) in rosettes and multivesicular body (MVB) [X 25,000]. Unpublished light micrographs from Bhiwgade and Mahaley

The syncytiotrophoblast is a highly differentiated layer with unevenly spaced nuclei embedded in a common ground matrix of cell boundaries throughout its entire mass. Its width varies and is thicker at sites of nuclei location and thinner elsewhere. During early and late limb-bud stage, the cytoplasm is extremely electron dense with nuclei grouped together in 2's or 3's that exhibit intracellular spacing around its perinuclear limits. At mid-term and full-term the layer is quite attenuated and presents a thinner mass as compared to that of the cytotrophoblast (Figs. 6.55, 6.56 and 6.57). The nuclei are irregular with conspicuous nucleoli. The syncytiotrophoblast, however, is most electron dense at full-term and it resembles a 'coarse syncytium' with distinct spongy porous mass due to the presence of an unusually high concentration of extensively, modified vesiculated rough endoplasmic reticulum (Fig. 6.58). The expanded vesicles are distributed randomly, but in close cluster throughout the syncytium and bears no particular pattern or design. The syncytial mass bears an extremely fenestrated appearance and these percolations are characterized by vesiculated expansions of rough endoplasmic reticulum. The rough endoplasmic reticulum gains prominence during mid-pregnancy wherein it is a dominant feature of both tubular and vesiculated types. As the development advances, the tubular rough endoplasmic reticulum undergoes dilation and becomes more and more

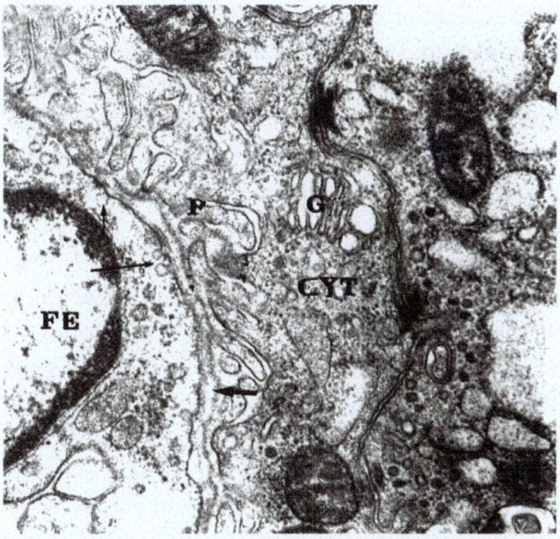

Fig. 6.59 Enlarged view of cytotrophoblast in *Tylonycteris pachypus*. Podocytic (P) modification of basal lamina (arrows) into the block-like structure. This may help to increase the absorptive area. Micropinocytotic vesicles (thin arrows) are seen along the length of the basement membrane and Golgi complex (G) within the cytotrophoblast (CYT). Note the foetal endothelium (FE) in the bottom left-hand corner [X 33,000]. Unpublished light micrographs from Bhiwgade and Mahaley

vesiculated indicating the beginnings of a coarse syncytium. The vesiculization of rough endoplasmic reticulum peaks at full term whereupon the 'coarse syncytial' structure too becomes most extensive.

Among the cytoplasmic inclusions, polyribosomes are abundant in the early limb-bud stage. They are much more in number than the dense bodies, light multivesicular bodies, coated vesicles and lipid droplets that are otherwise seen. Polyribosomes, thus gain prominence during the mid-pregnancy along with a large number of coated vesicles. A peculiar feature observed in the syncytium at mid-pregnancy is the presence of many thin convoluted channels provided with desmosomes that seemingly traverse the entire width of the syncytia and at some places extend from interdigitating membrane to the intrasyncytial lamina (Fig. 6.59). Whereas, at full-term, for the first time during the entire embryonic developmental process, glycogen rosettes and clusters of glycogen bodies are seen concentrated towards the cytotrophoblastic face in the syncytium (Fig. 6.57). During this stage, the syncytium also shows extreme attenuation.

The syncytium is a double-membraned structure first observed during late limb bud as a distinctly electron dense and very slightly undulated structure at the confluence of the two trophoblastic layers, i.e. the overlying syncytiotrophoblast and the closely apposing underlying cytotrophoblast. It is poorly defined during this stage but bears a large number of desmosomes and is a constant feature throughout development. The syncytium runs along the entire stretch of the two trophoblastic layers. It is of more or less uniform width and becomes progressively prominent and distinct with advancing gestation. It is most prominent at term and is thrown into convoluted folds and separates the two trophoblast layers.

The cellular characteristics of the cytotrophoblast are indicated by the presence of well-developed limiting membranes provided with desmosomes that unite the individual cytotrophoblast cells length-wise in a unicellular layer. A number of junctional complexes are present prominently during the limb-bud stages. It forms a continuous well-defined layer presenting a thinner mass than the overlying syncytia during mid-pregnancy. The shape of the nucleus is uniformly smooth during the early limb-bud stage but becomes irregular during the later stages and is surrounded by intracellular spacings. The cytoplasm of the cytotrophoblast during all the stages of development is more electron lucent than dense as compared to the ground matrix in the syncytia (Fig. 6.59).

The peculiar characteristic feature during term is the podocytic protrusions that appear as microtubular channels entrapping small amounts of cytoplasm within their folds and loops. These protrusions originate from the basal lamina in the form of conspicuous blocks some also possess vesicles en route transportation within their lumen.

The basal lamina is a very well-developed double membraned layer (Fig. 6.59). It stretches continuously along the proximal length of the cytotrophoblast in an even width. Its contents are of homogeneous, acellular nature, enclosed within membrane limits that bear podocytic vesicles in continuation with the same at the foetal face during term, and podocytic blocks at the cytotrophoblastic face (Figs. 6.58 and 6.59). It serves to separate the identities of the trophoblastic and foetal endothelium of the interhemal barrier.

By the limb-bud stage, foetal capillaries are seen to be embedded in loose patches of mesenchyme and bear distinct foetal endothelium (fig. 6.55) with prominent nuclei. These endothelial cells bear a dense matrix during mid-pregnancy and the cytoplasmic contents during development include darkly staining ovoid mitochondria, coated vesicles, microvesicles and dense bodies (Fig. 6.57).

The definitive discoidal placenta of the bat, *Tylonycteris pachypus*, at term, is hemodichorial due to the absence of maternal endothelium and presence of both layers of syncytiotrophoblast and cytotrophoblast.

Our study shows the conversion of the endotheliodichorial condition of the placenta during earlier stages of gestation into the hemodichorial condition at term with the concomitant conversion of the interstitial membrane into the intrasyncytial lamina along with the establishment of 'coarse syncytium' which is a modification of basal lamina (Figs. 6.60a, b and 6.61).

The presence of organelle inclusions like microvesicles, caveolae and coated vesicles in the maternal endothelium indicates absorption of materials from the maternal blood (Bhiwgade 1990). Numerous pinocytotic invaginations and coated vesicles indicate heightened endocytotic activity (Wynn and Davies 1965, Enders 1965, and Luckett 1970). The presence of rough endoplasmic reticulum suggests that proteins synthesized pass through the interstitial membrane. This is also true in case of carnivore placentas of ferret (Lawn and Chiquoine 1965), dog (Anderson 1969) and in case of insectivore placentas of shrew (Wimsatt et al. 1973).

The presence of well-developed endoplasmic reticulum and Golgi zones in maternal endothelium in *Tylonycteris pachypus* is a feature shared

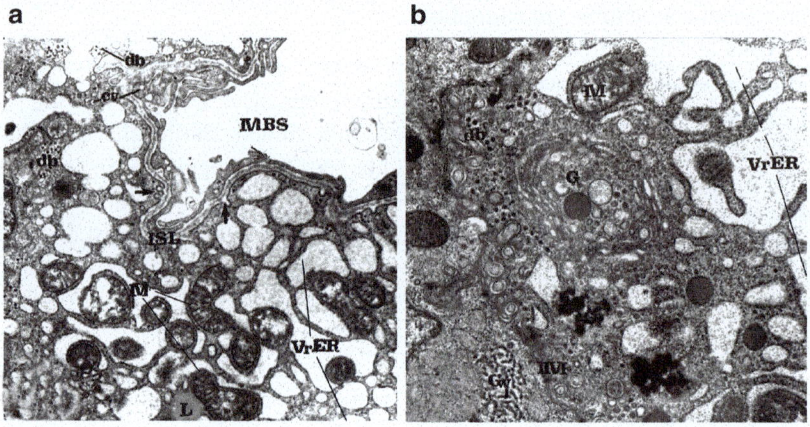

Fig. 6.60 (**a**) Enlarged view of syncytiotrophoblast in *Tylonycteris pachypus* showing spongy appearance due to highly vesiculated rough endoplasmic reticulum (VrER) with enclosed hypertrophied mitochondria (M). A number of coated vesicles (cv), dense bodies (db) and lipid droplets (L) are seen within the matrix of syncytiotrophoblast. Micropinocytotic vesicles (arrows) are seen along the edge and some are in the process of expelling their content into the intrasyncytial lamina (ISL). Note the maternal blood space (MBS) in the top right-hand corner [X 17,000]. (**b**) Enlarged view of syncytiotrophoblast showing in *Tylonycteris pachypus* showing the rough endoplasmic reticulum (VrER) with a sac-like appearance due to extreme dilation of cisternae with enclosed hypertrophied mitochondria (M). A number of glycogen bodies (Gy) and dense bodies (db) are noted along the convoluted folds of interdigitating membrane (IM). Also note the concentrically arranged cisternae of Golgi Complex (G) [X 25,000]. Unpublished electron micrographs from Bhiwgade and Mahaley

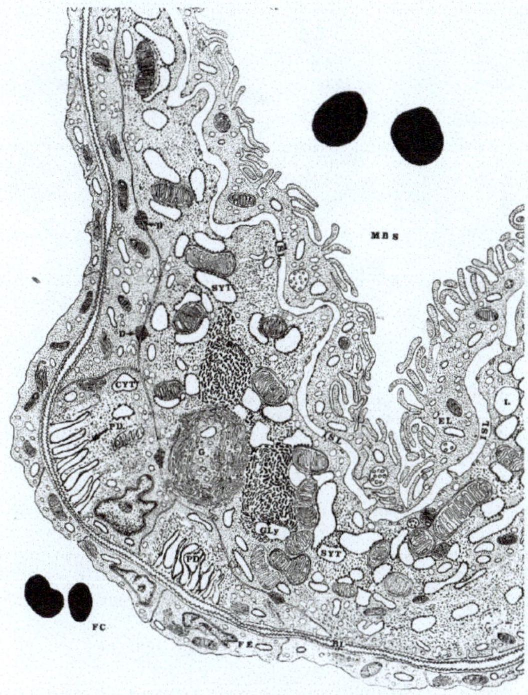

Fig. 6.61 The maternal and foetal elements of the placenta at full term in *Tylonycteris pachypus* are represented in in this diagram. Maternal blood space (MBS) is seen on the upper side followed by the continuous ectoplasmic layer (EL), which is completely separated from the main layer of syncytiotrophoblast (SYT) by the discontinuous layer of intrasyncytial lamina (ISL). The layer of cytotrophoblast (CYT) is below the syncytiotrophoblast, and is seen resting on the basal lamina (BL—single cross harch). A foetal capillary (FC) is in lower bottom. Single cross hatch is the basal membrane of the foetal endothelium. Glycogen bodies (GY), lipid droplets (L) and concentrically arranged Golgi complex (G) are seen within the syncytiotrophoblastic mass. Note the podocytic protrusions (PD) along the length of the cytotrophoblast. Also note the desmosomes (D) along the interdigitating membrane. From Mahaley 1998

with *Megaderma lyra lyra*; *Rhinopoma hardwickei* and *Taphozous melanopogon* (Bhiwgade 1990).

6.9 *Chaerephon plicata*

Earlier investigations have shown that the family Molossidae exhibits a decidual reaction, i.e. endometrial stromal changes in the absence of pregnancy as typified by *Molossus ater,* the Black mastiff bat *(*Rasweiler 1991). The bat, *Chaerephon plicata*, is also an insectivorous bat belonging to the family molossidae. The ultrastructure of placenta of the other two species of this family, i.e. *Tadarida brasiliensis cynocephala* (Stephens 1969) and *Molossus ater* (Rasweiler 1991) have been described. The chorioallantoic placenta of the Indian molossid bat has already been described, at the light microscopic level, by Gopalakrishna and Karim (1980). Our investigation attempts to establish the nature of placentation in *Chaerephon plicata* at the ultrastructural level, and draw similarities if any, with the placentation of the other two members of the family.

Cytologically, at the electron microscopic level, the term placenta shows a foetal endothelium made up of irregular, elongated, more or less, squamous cells. It shows the usual features of the endothelium including the micropinocytotic vesicles, abundantly distributed throughout the matrix, especially on the outer surfaces. The endothelium is relatively thin in places and does not seem to possess any pores. The mesenchyme too does not exhibit any unusual features. However, the term placenta exhibits relatively closely spaced foetal capillaries along the perimeter of each maternal blood space (Figs. 6.62 and 6.63).

A sparse, but distinctly visible population of the trophoblastic cells, i.e. the cytotrophoblast are seen interspersed between the foetal blood capillaries. In fact, the foetal blood capillaries are observed to be occupying extremely indented portions of the cytotrophoblast layer thereby reducing the interhaemal membrane to a thin layer between the maternal blood space and the foetal blood vessel (Figs. 6.62 and 6.63). The foetal capillary is a bed of a close network of capillaries that are oriented parallel to the maternal blood space. They surround the maternal blood and so when the maternal channels (MBS) are cut in cross section, the foetal capillaries too get cut in cross section.

Under the electron microscope a discontinuous layer of 'homogeneous material' is seen to overlay large parts of the apical portions of the cytotrophoblasts. The cytoplasm of the

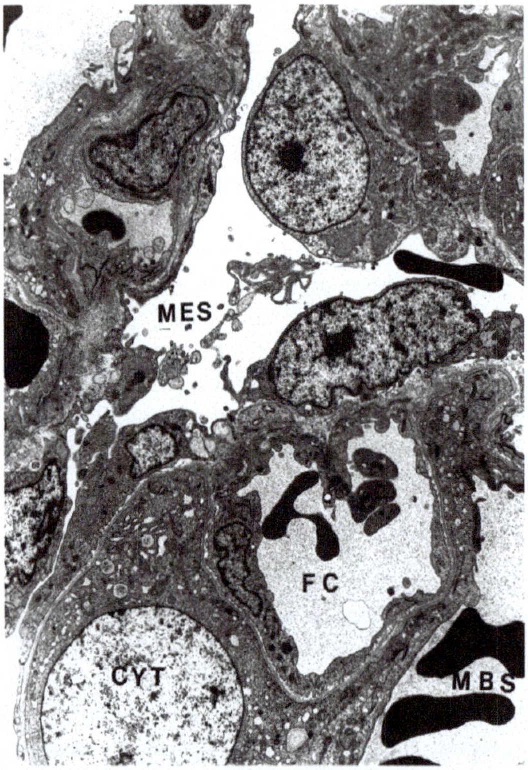

Fig. 6.62 Electron micrograph (low power) of the placenta of *Chaerephon plicata* at full term. Note that the basal lamina is continuous, underlying the cytotrophoblastic layer (CYT). Also note the foetal capillary (FC) along with foetal endothelium (FE) embedded in foetal mesenchyme (MES) [X 5000]. From Bhiwgade 1990

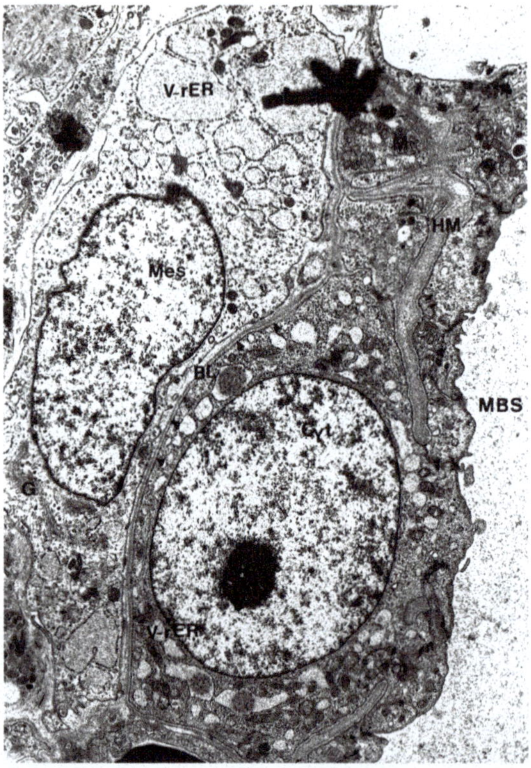

Fig. 6.63 Electron micrograph of the placenta of *Chaerephon plicata* at full term. Note maternal blood space (MBS), cytotrophoblast (Cyt) and foetal Mesenchyme (Mes) with vesiculated rough endoplasmic reticulum (VrER). Homogeneous material (HM) is seen to flow into cytotrophoblastic layer to underlie the free laminal edge of the same towards the maternal blood space from its point of connection with the basal lamina (BL). Golgi bodies (G) are also prominent [X 8000]. From Bhiwgade 1990

cytotrophoblast protrudes through the discontinuities in the 'homogeneous material' (HM), to line the maternal blood space directly (Fig. 6.63). The discontinuous layer of 'homogeneous material' at the apical portions of the cytotrophoblast, continues with the basal lamina which underlay the trophoblast. It encloses the cytoplasm and establishes direct contact with the maternal blood space (MBS). The layer exhibits fair amount of desmosomal connections towards the maternal blood space. In some places, the cytoplasm of the cytotrophoblast extends laterally under the 'homogeneous material' and protrudes out, little away, through the discontinuities, to establish direct contact with the maternal blood (Fig. 6.63).

The cytotrophoblast has a single large nucleus with 1–2 nucleoli, displaced basally in the cell. The organelle of the cytoplasm includes small round / ovoid mitochondria, of which a few are a typical dumbbell and rod shaped, all bearing lamelliform cristae. They exhibit a random distribution in the cytoplasm. The Golgi complexes seen at the apical cytoplasm above the nucleus are relatively smaller. Rough endoplasmic reticulum is lamellar but mostly vesiculated. They are in various stages of dilation, abundant in the cytoplasm, and in most parts, are mainly near the apical end. Numerous coated vesicles are seen in the cytoplasm of the cytotrophoblast often in close association with the 'homogeneous material' layer (Fig. 6.63). Some are even seen in continuation with the 'homogeneous material' and basal lamina. Free ribosomes too are located throughout the cytoplasm. Numerous characteristic smooth membraned, micropinocytotic vesicles and a few microtubular structures are also observed, mostly in association with the foetal endothelial portion of the barrier, and also in the apical portions of the cytotrophoblast. Occasionally, multi-vesicular bodies containing smaller vesicles are also seen in both, cytotrophoblasts and foetal mesenchyme. Rare occurrence of dense bodies is noted mostly in the foetal endothelium (Fig. 6.63).

The placental barrier is hemomonochorial. The trophoblast is cellular rather than syncytial. Intercellular junctions are abundant in the trophoblast layer. Another unusual feature of the barrier is that at many places, the trophoblast basal lamina shows infoldings, especially at the lateral cell borders.

The definitive placenta of the molossid bat, *Chaerephon plicata*, shows the absence of both maternal endothelium and the syncytiotrophoblast. The cytotrophoblastic layer is seen in confluence with the maternal blood space and the basal lamina borders the foetal endothelium on either side. These features could be used to designate the placental barrier as hemomonochorial as per Ender's terminology (Enders 1965). The endotheliochorial condition is the precursor to all other types of placental organizations since they can be considered as derived as a post-modification of the endotheliochorial condition (Gopalakrishna and Karim 1980). The establishment of hemochorial condition generally brings in changes in the conventional linings of the endotheliochorial placenta. The absence of the maternal endothelium itself initiates the conversion of the 'interstitial membrane' into the 'intrasyncytial lamina'. In *Chaerephon plicata*, however, a 'homogeneous material' similar to that observed in *Tadarida brasiliensis cynocephala* (Stephen,1969) seems to exist in place of the intrasyncytial lamina.

According to Stephens (1969), the 'homogenous material' could serve to control the invasiveness of the trophoblast or could prevent the immune rejection of the foetus by the mother (Kirby et al. 1964). It could also contribute significantly to the nutritional requirements of the developing foetus (Gulamhussein and Beck 1975).

Indentation of the trophoblastic layers of foetal capillaries is a peculiar feature most notably exhibited by the molossid family of chiroptera, represented by *Tadarida brasiliensis cynocephala* (Stephens 1969), *Molossus ater* (Rasweiler 1991) and also *Chaerephon plicata*. These capillaries have a fairly good distribution and are in large numbers in the term placentae of *Chaerephon plicata*. It was previously noted (Stephens 1969) and further confirmed, in our study, that in places where capillaries are well developed and established, the trophoblastic barrier is reduced to extremely thin phalanges. These could serve to reduce the distance for diffusion and reduce the time of diffusion. It could also increase the rate of bidirectional transport and improve the efficiency of exchange between maternal and foetal blood circulations.

A similar hypothesis was postulated by Rhodin and Terzakis (1962) and Wynn and Davies (1965) in the human placenta. They suggested that the close approximation of the foetal and maternal bloodstreams, as gestation advances, can facilitate rapid diffusion of carbon dioxide and oxygen.

In conclusion, the characteristic features observed at the ultrastructural level, in *Chaerephon plicata* bear a definite familial resemblance to those reported for *Tadarida brasileinsis cynocephala* (Stephens 1969) and *Molossus ater* (Rasweiler 1991).

The absence of the maternal endothelium clearly signals the conversion of the 'interstitial membrane' into the 'intrasyncytial lamina'. This conversion follows the incorporation of the maternal endothelium, either in part or in whole into the apical region of the syncytiotrophoblast. This has been supported by Cukierski's (1987) observation of the formation of the intrasyncytial lamina from the basal membrane of the maternal endothelium during the development of chiropteran placental labyrinth.

The intrasyncytial lamina functions as an immunological barrier to prevent antigenic substances from the foetal tissue entering the maternal environment (Enders and Wimsatt 1968; Bodley 1974). Cukierski (1987) suggested that the intrasyncytial lamina can serve as a selective filter and increase the surface area of the apical plasmalemma. It also provides structural support and maintains cell polarity as in the placenta of *Myotis lucifugus lucifugus*.

6.10 *Miniopterus schreibersii fuliginosus* (Hodgson)

In *Miniopterus schreibersii fuliginosus*, the chorioallantoic placental membrane shows three main types of placentae which develop in sequence: primary placenta, secondary placenta and tertiary placenta. During the neural groove and limb-bud stages, the primary placenta consists of a continuous ectoplasmic layer, intrasyncytial lamina, syncytiotrophoblast, cytotrophoblast, basal lamina, mesenchyme and then foetal endothelium. By full term, the primary placenta degenerates and only remains as a thin syncytiotrophoblastic layer with its basal lamina. Foetal capillaries are not seen in the mesenchyme. At early limb-bud stage, the placenta is in the form of secondary placenta. This placenta is of the endotheliomonochorial type when observed under the electron microscope. It shows a maternal endothelium, a cellular trophoblast surrounding the maternal blood tubule, basal lamina, mesenchyme and the foetal endothelium. The trophoblast is not syncytial as previously reported and the plasma membrane on the apical part of these trophoblastic cells is in direct connection with the discontinuous interstitial membrane. At term, a tertiary placenta is formed, which is hemodichorial in nature. It shows a thin ectoplasmic layer, a thick intrasyncytial lamina, syncytiotrophoblast, cytotrophoblast, basal lamina, mesenchyme and the foetal endothelium. The definitive placenta in this bat is very different from what has been reported earlier by Chari and Gopalakrishna based on their light-microscopic observations. They had reported the absence of maternal endothelium in the primary placenta from the neural groove and early limb-bud embryos, and the existence of only cellular trophoblast in the secondary placenta throughout the gestation. They had identified the tertiary placenta as hemodichorial.

Based on morphological comparisons, the genus Miniopterus is included in the family Vespertilionidae of Microchiropera. Embryological development of *Miniopterus schreibersii fuliginosus*, however, is different from all vespertilionid bats. The bat shows distinct embryological characteristics in the nature of implantation of the blastocyst, the development of the amnion, the morphogenesis and final structure of the chorioallantoic placenta (Gopalakrishna and Chari 1983). According to them, convergent evolution has resulted in the morphological resemblance between Miniopterus and other vespertilionid bats. Due to this, they have suggested a new family, Miniopteridae to accommodate this genus.

Branca (1927), Grosser (1927), and Kempcrmann (1929) have described the placenta during mid-pregnancy in the bat *Miniopterus schreibersii* of Europe. The Light-microscopic descriptions by them did not bring out the uniqueness of the development of the placenta of this bat. Chari and Gopalakrishna (1984), on the other hand, have proposed that the chorioallantoic placenta develops in three stages and results in the formation of three different types of placentae, that develop in a chronological sequence. They therefore designated these placentae as primary, secondary and tertiary placentae.

This difference in description regarding the nature of the three types of placentae, arises due to lack of ultrastructural studies. Our study, therefore, focuses on the development and fine structure of the interhemal membrane of each of the three types of placentation in this bat.

To make it easier to understand the morphological arrangement of the primary, secondary and tertiary placentae at term is described so that the ultrastructural studies could be understood better. Figure 6.64 is an illustrative diagram explaining the arrangement of the primary,

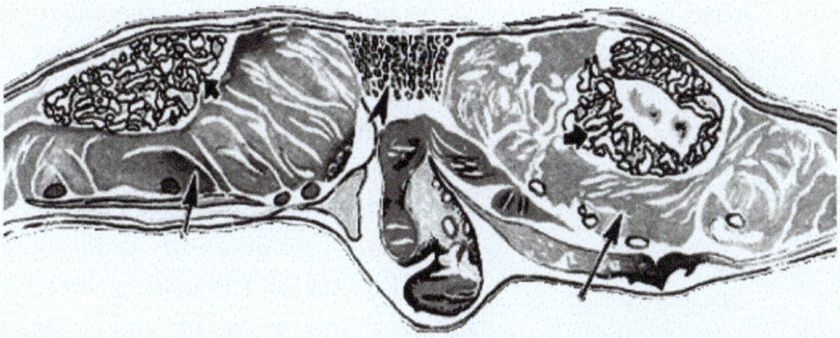

Fig. 6.64 Diagrammatic representation of the dissected uterus showing the morphological position of chorio-allantoic placenta. The antimesometric portion of the uterus, in sagittal section, at full term shows the relative positions of primary (arrow head), secondary (broad arrows) and tertiary placentae (Short arrows). The umbilical cord is connected at the centre of the primary placenta. The degenerative primary placenta is visible with two secondary placental discs on either side of it. The tertiary placenta is embedded deep within the secondary placental disc near its myometrial border

secondary and tertiary placentae at full term. Few placental tubules of the primary placenta at different stages of degeneration can be seen. The secondary placental disc consists of a network of closely arranged tubules in three dimensions. The tertiary placenta is embedded deep within each of the secondary placental disc near its myometrial border. It consists of a labyrinth of large, interconnected lacunae filled with maternal blood.

We studied the primary and secondary placentae during the neural groove, early limb-bud, mid-pregnancy and at term. The tertiary placenta was studied at full term.

6.10.1 Primary Placenta

Neural Groove and Early Limb-Bud Embryo

In the early stages of gestation, the interhemal membrane consists of syncytiotrophoblast, cytotrophoblast, basal lamina and foetal capillary endothelium (Fig. 6.65). In this placenta, identified as the primary placenta, the maternal endothelium is totally absent at the neural groove stage. The interstitial membrane is in the form of an intrasyncytial lamina with gaps (Fig. 6.65 inset). Syncytial cytoplasm protrudes as tongues through these gaps and spread along the endothelium, forming nearly a continuous ectoplasmic

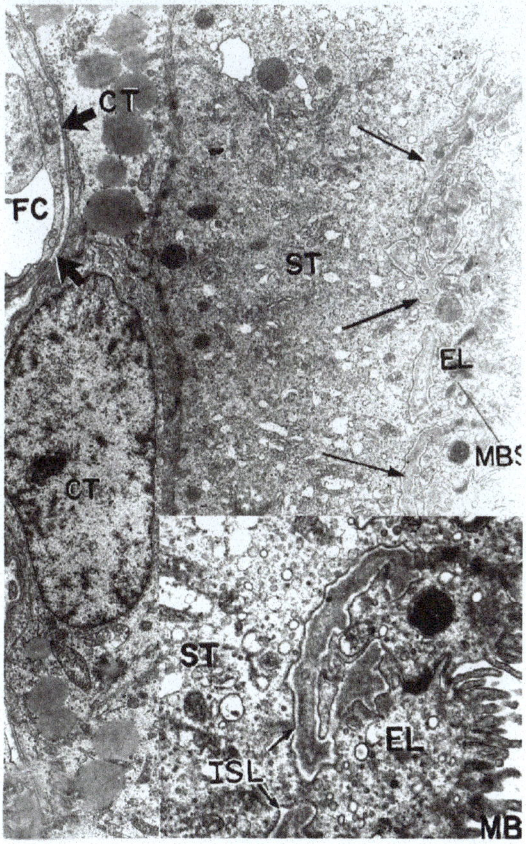

Fig. 6.65 Electron micrograph of the primary placenta of *Miniopterus schreibersii fuliginosus* during the early limb-bud stage. The maternal blood space (MBS), ectoplasmic layer (EL), discontinuous Intrasyncytial lamina (arrows), syncytiotrophoblast (ST), cytotrophoblast (CT), basal lamina (Broad Arrows) and foetal capillary (FC) are seen in sequence [X 5000]. Inset shows the Intrasyncytial Lamina (ISL) in detail. From Bhiwgade et al. 1992

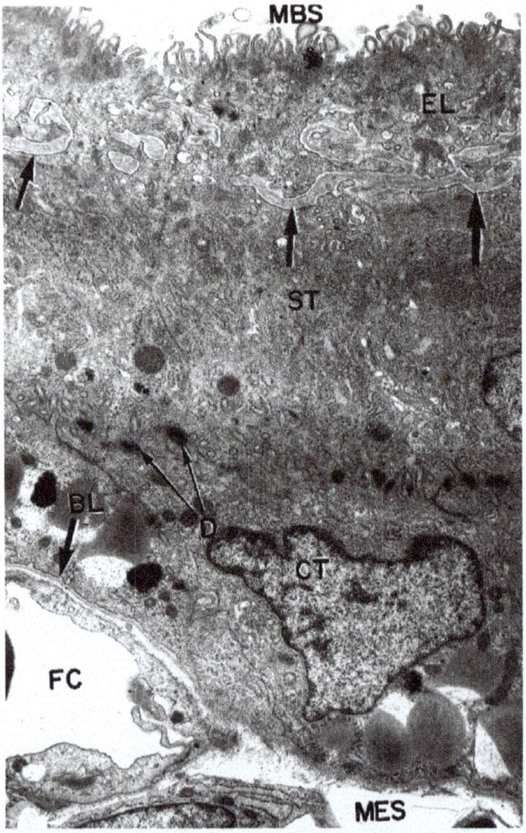

Fig. 6.66 Electron micrograph of the primary placenta in *Miniopterus schreibersii fuliginosus*, during the mid pregnancy showing the different elements in the placental labyrinth; viz.; Maternal blood space (MBS), Ectoplasmic layer (EL), Intrasyncytial lamina (arrows). Syncytiotrophoblast (St), Cytotrophoblast (CT), resting on basal lamina (arrow heads) and Foetal capillary (FC) and Mesenchyme (MES). Note the desmosomes (D) connecting the two trophoblastic layers. Also note the large lipid droplets in the cytotrophoblast [X 7000]. From Bhiwgade et al. 1992

lining. Except for the pinocytotic vesicles, dense bodies numerous multivesicular bodies, and microtubules, this ectoplasmic layer is devoid of most cellular organelles. Desmosomes are seen clearly along cell borders. The cell border also shows numerous elongated microvilli (Fig. 6.65 inset).

The syncytiotrophoblast, adjacent to the maternal blood space, shows numerous round or rod-shaped mitochondria with prominent parallelly arranged cristae. Polyribosomes are also seen in plenty. Golgi zones are with 3–4 lamellae and are associated with many Golgi vesicles. The distinct feature of syncytiotrophoblast layer is the presence of many strands of granular endoplasmic reticulum, large number of coated vesicles, dense bodies, and pinocytotic vesicles. This layer also shows some infoldings of its basal plasma membrane (Fig. 6.65). The cytotrophoblast is in the form of a single layer of cells that lies close to the overlying syncytiotrophoblast. The two trophoblastic layers come in close contact with each other at several places at regular intervals. At these sites, desmosomes are often seen connecting the two trophoblastic layers. The cytotrophoblast shows abundant polyribosomes, a few large rod-shaped mitochondria, Golgi zones with Golgi vesicles and a few strands of rough endoplasmic reticulum. The main feature of the cytotrophoblast at this stage of gestation is the presence of numerous large lipid droplets. These cytotrophoblast cells are in contact with a continuous basal lamina. Foetal capillaries are distributed within the mesenchyme and show typical squamous endothelium (Fig. 6.65).

Mid Pregnancy

During mid-pregnancy, the structure of the primary placenta is similar to the one already described except for the changes illustrated in Fig. 6.66.

The density of the cytoplasm of syncytiotrophoblast cells is not uniform but shows three zones of different densities. The first zone, near the intrasyncytial lamina, is similar to the ectoplasmic layer, with pinocytotic vesicles, microtubules, and microvilli (Fig. 6.66). The second zone shows the presence of nuclei and other cell organelles. The third zone shows many infoldings of the basal plasma membrane. This zone also shows many dense bodies, coated vesicles, pigment granules and multivesicular bodies. This layer also has many desmosomes with filamentous structures. The desmosomes are arranged in a parallel fashion. The cytotrophoblast appears to be reduced.. The primary placenta at mid-pregnancy is, therefore, hemodichorial. Fig. 6.67 schematically depicts

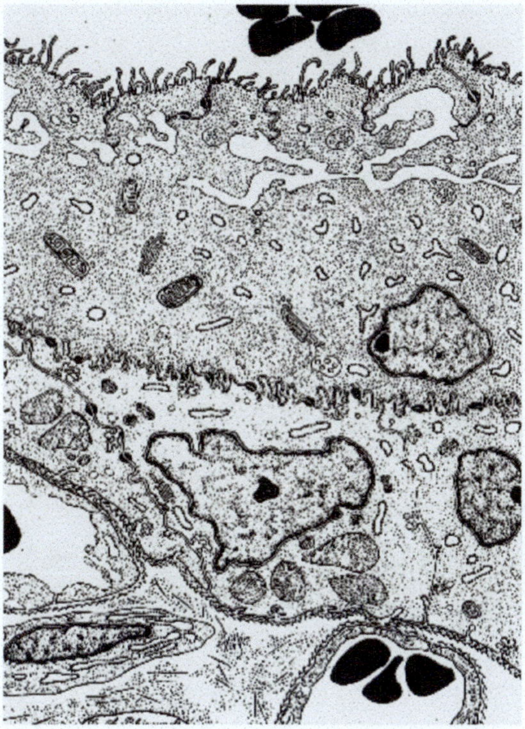

Fig. 6.67 A schematic diagram of maternal and foetal elements of the primary placenta of *Miniopterus schreibersii fuliginosus* at mid-pregnancy stage. Maternal blood space is seen on the upper side, followed by the ectoplasmic layer, The ectoplasmic layer is separated from syncytiotrophoblast by the intrasyncytial lamina (double cross-hatched). Below the syncytiotrophoblast is cytotrophoblast, resting on the basal lamina (cross-hatched). On the lower left and right corner, foetal capillaries can be seen lying in the mesenchyme

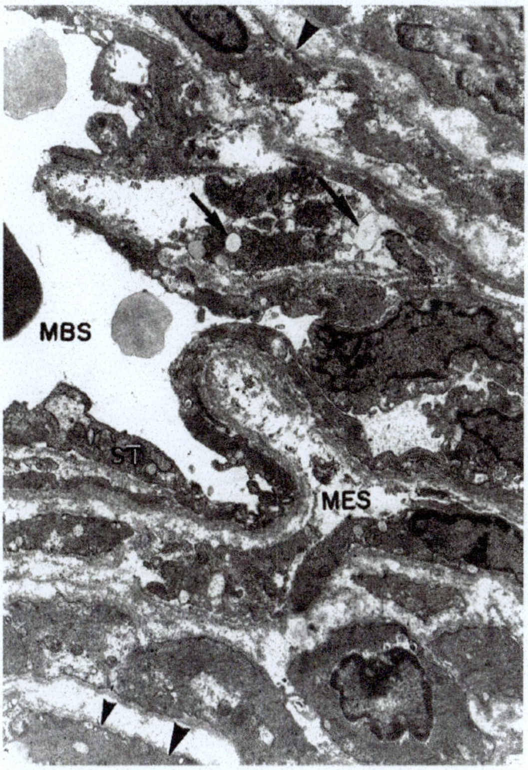

Fig. 6.68 The low-power electron micrograph of the primary placenta in *Miniopterus schreibersii fuliginosus* at term. Highly compressed degenerating placental tubules are seen projecting into the maternal blood space (MBS). Each placental tubule consists of thin layer of syncytiotrophoblast (ST) with their highly shrunken nuclei. Syncytiotrophoblast rests on basal lamina (arrows). Note the absence of foetal capillaries in the mesenchyme (MES) [X7,000]. From Bhiwgade et al. 1992

the relationship of the placental layers in the hemodichorial placenta during this stage.

Full-Term

At full term, the primary placenta becomes greatly reduced. The placental tubules, therefore, appear to be in a highly degenerative state (Fig. 6.68). The syncytiotrophoblast shows few shrunken nuclei and few cell organelles. Occasionally dense bodies, lysosomes and vacuoles are noticed. The syncytiotrophoblast directly rests on a basal lamina since the cytotrophoblast is totally absent. Foetal capillaries are absent from the placental labyrinth, but few stromal cells are seen.

6.10.2 Secondary Placenta

Neural Groove and Early Limb-Bud Embryos

The secondary placenta is endotheliochorial right from its formation till the full term. It shows well-developed maternal endothelium resting on the interstitial membrane. This in turn is surrounded by a single layer of trophoblast cells. The cytotrophoblast rests on a distinct basal lamina. Foetal capillaries with foetal endothelium are seen in the mesenchyme (Fig. 6.69). The maternal endothelial cells show hypertrophy and are wedge-shaped with a central nucleus. Within the cells, numerous circular mitochondria and well-

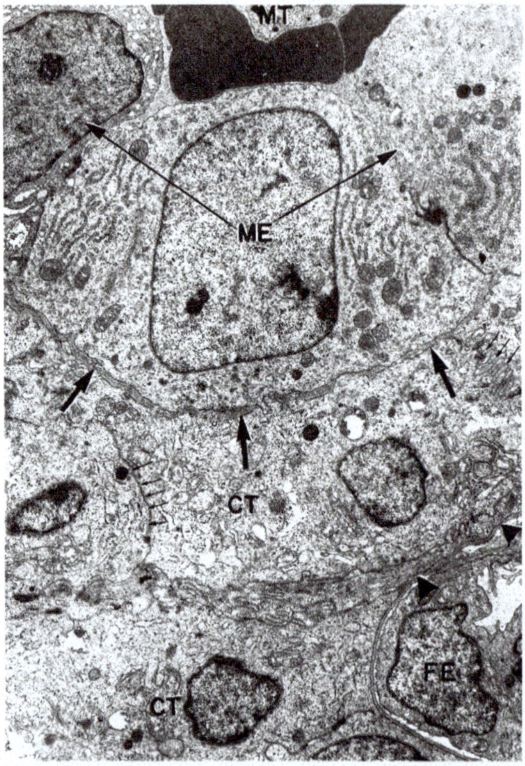

Fig. 6.69 Secondary placental membrane in *Miniopterus schreibersii fuliginosus* at early limb-bud stage. On the upper side of the image is the maternal capillary (MC) and immediately below it is the interstitial membrane (arrows). This is followed by two trophoblastic layers, upper syncytiotrophoblast (ST) and lower cytotrophoblast (CT) Cytotrophoblast is seen resting on basal lamina (arrow heads). On the lower right side is the foetal capillary with well-developed foetal endothelium (FE). The hypertrophic condition of the maternal endothelium (ME) can be noted [X 5000]. From Bhiwgade et al. 1992

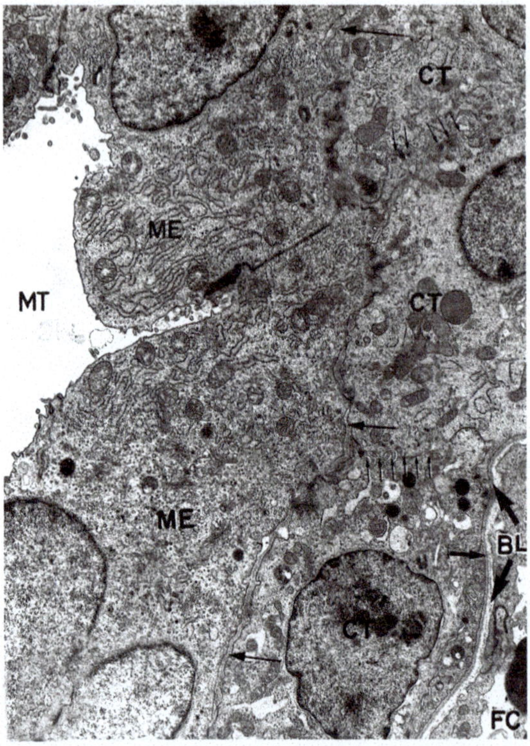

Fig. 6.70 Secondary placental membrane in *Miniopterus schreibersii fuliginosus* at mid-pregnancy. The barrier appears reduced with the maternal endothelium (ME) showing abundant cell organelles Viz., Rough endoplasmic reticulum, free polyribosomes and mitochondria. Syncytiotrophoblast (ST) is seen resting on the basal lamina (BL). Foetal Capillary (FC) is seen on the lower left side [X5,000]. From Bhiwgade et al. 1992

developed rough endoplasmic reticulum are seen. Rough endoplasmic reticulum is oriented to the base. Few lipid droplets, caveolae and a few micropinocytotic vesicles are also seen. Desmosomes are seen between adjacent endothelial cells. The cytotrophoblast shows abundant polyribosomes, and numerous small mitochondria. At the perinuclear side, Golgi bodies together with a few rough endoplasmic reticulum, few dense bodies, caveolae and coated vesicles are seen. The most unique feature of the cytotrophoblast cell in this placenta is the presence of cell junctions with desmosomes. Foetal capillaries are seen distributed within the mesenchyme. Foetal endothelial cells have abundant ribosomes due to which the cytoplasm appears very dense. The foetal endothelial cells have few mitochondria, dense bodies, pinocytotic vesicles and a few strands of rough endoplasmic reticulum.

Mid-Pregnancy

The placenta, at this stage, shows a well-developed maternal endothelium resting on the discontinuous interstitial membrane. Cytotrophoblast, basal lamina and foetal endothelium are also observed (Fig. 6.70). Maternal endothelial cells are hypertrophied with tubular rough endoplasmic reticulum and polyribosomes (Fig. 6.70). Since interstitial membrane is absent

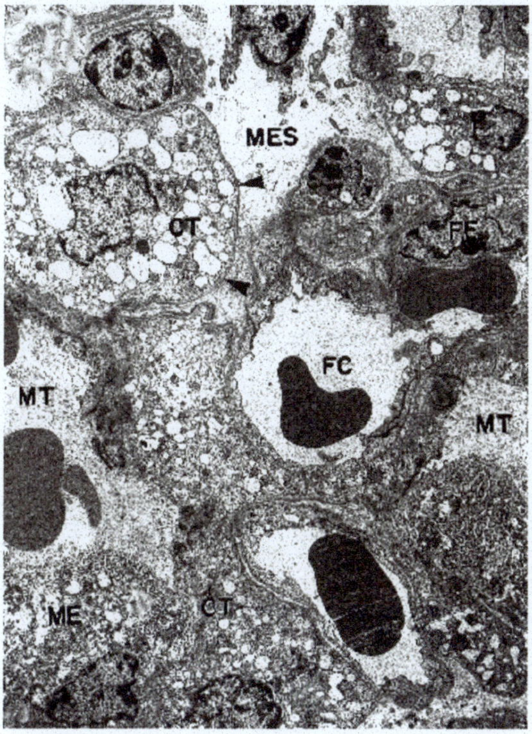

Fig. 6.71 High-power electron micrograph of secondary placenta in *Miniopterus schreibersii fuliginosus* at full term showing few foetal capillaries in close relation to a maternal blood space. The barrier consists of thin well-developed maternal endothelium (ME), Syncytiotrophoblast (ST) with well-developed ER and foetal endothelium (FE). Note that the maternal endothelium is absent at some places. The interstitial membrane ruptures at few points and the syncytium comes in direct contact with maternal blood (arrow) [X 5000]. From Bhiwgade et al. 1992

at some places, the maternal endothelial cells are in direct contact with cytotrophoblast. Hypertrophied cell organelles like mitochondria and rough endoplasmic reticulum are seen in the cytotrophoblast cells. In the cytoplasm outside the interstitial membrane mitochondria are relatively numerous. Golgi zones are small. Cell junctions with well-formed desmosomes are seen. Foetal capillaries greatly intend into the trophoblastic layer, resulting in a reduction in thickness of the interhemal membrane. The foetal capillaries show very thin endothelium.

Full-Term

At full-term, the placenta consists of a rich network of foetal capillaries seen around the circumference of each maternal tubule. Few trophoblastic cells are seen located between the foetal capillaries. The interstitial membrane forms a discontinuous layer over the trophoblastic cells and is seen continuous with the basal lamina that underlies the trophoblast (Fig. 6.71). As the placenta matures and foetal capillaries push in between the trophoblastic cells by separating the lateral membranes of the trophoblast. Due to this, the interstitial membrane of 'homogenous material' is brought in close proximity to the basal lamina which underlies the trophoblastic cells. In this stage, the interstitial membrane frequently comes into direct contact with the basal lamina of the trophoblast. The interhemal membrane, therefore, comprises a hypertrophied maternal endothelium, a discontinuous interstitial membrane, layer of trophoblast, a basal lamina common to both the trophoblast layers and the foetal endothelium. The cytoplasm of maternal endothelium is rich in rough endoplasmic reticulum, free ribosomes and hypertrophied mitochondria. Its thickness is, however, reduced and in some places it is absent. In such regions, where maternal endothelium is absent, the cytotrophoblast comes in contact with the maternal blood. The cytotrophoblast shows the characteristic presence of numerous highly dilated cisternae of the endoplasmic reticulum (Fig. 6.72). The cytotrophoblast rests on a distinct trophoblastic basal lamina. Foetal capillaries are distributed in the foetal stroma. Because of the presence of numerous indented foetal capillaries, the thickness of the interhemal membrane is reduced and in some places the foetal capillaries are in direct contact with the maternal blood. The relationship of the various layers in this placenta is depicted schematically in Fig. 6.73.

6.10.3 Tertiary Placenta

At full-term, the tertiary placental membrane consists of an intrasyncytial lamina which is

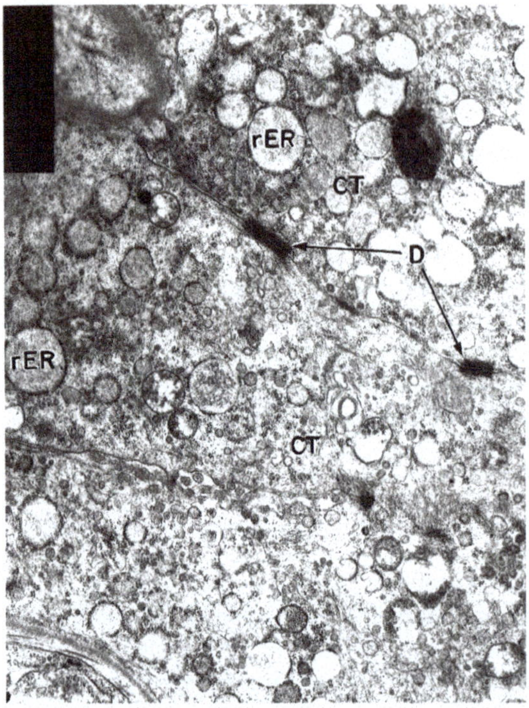

Fig. 6.72 Secondary placenta in *Miniopterus schreibersii fuliginosus* at term showing the magnified view of rough endoplasmic reticulum (rER) of syncytiotrophoblast (ST). Note the position of the desmosomes (D) [X 13,000]. From Bhiwgade et al. 1992

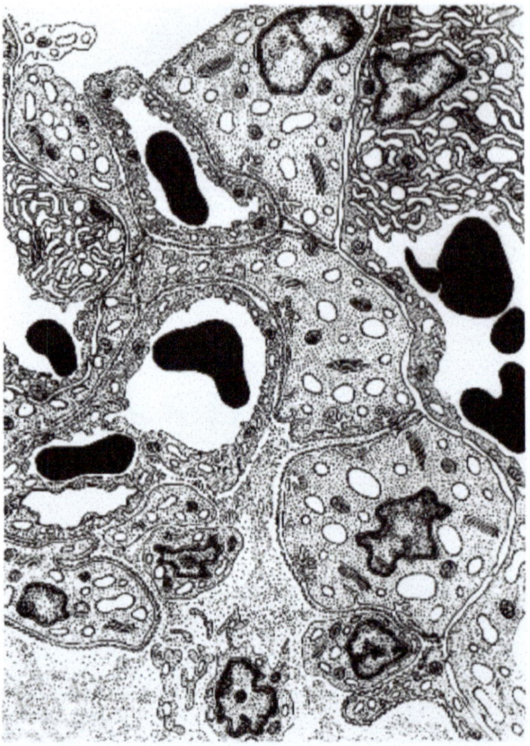

Fig. 6.73 A semi-schematic representation of layers of trophoblast, in secondary placenta of *Miniopterus schreibersii fuliginosus* at full term. The maternal capillary with maternal endothelium on the left side and lower right half resting on interstitial membrane can be seen. Syncytiotrophoblast is seen resting on basal lamina (cross-hatched). Upper right half shows mesenchyme with foetal capillaries

perforated at irregular gaps. The syncytial cytoplasm protruded through these gaps and spread out near its upper surface. This protruding cytoplasm forms a nearly continuous ectoplasmic layer lining the trophoblastic tubule (Figs. 6.74 and 6.75). This ectoplasmic layer is devoid of cell organelles except for numerous filaments associated with desmosomes and particulate complexes (Fig. 6.74).

The syncytiotrophoblast is a thick layer that shows darkly stained spherical nuclei. The cytoplasm of the syncytiotrophoblast appears hypertrophied with abundant dilated endoplasmic reticulum. Golgi zones, dense bodies, coated vesicles and a few mitochondria are also present.

Golgi complexes and mitochondria in cytotrophoblast are relatively less but contain large lipid droplets. The nuclei are large with a prominent nucleolus. Foetal connective tissue consists of only the basal lamina of the foetal capillary and occasional foetal mesenchyme cells. The definitive tertiary chorioallantoic placenta which is a unique feature in *Miniopterus schreibersii fuliginosus* is thus hemodichorial. Various layers in this placenta are depicted schematically in Fig. 6.75.

6.10.4 Discussion

The placenta of *Miniopterus schreibersii* of Europe at early stages has been described by several authors (Branca 1927; Grosser 1927; Kempcrmann 1929; Malassine 1970). These authors classified the secondary placenta as a mid-gestation 'Haupt Placenta'. Malassine (1970) called it 'Disque placenta'. None of

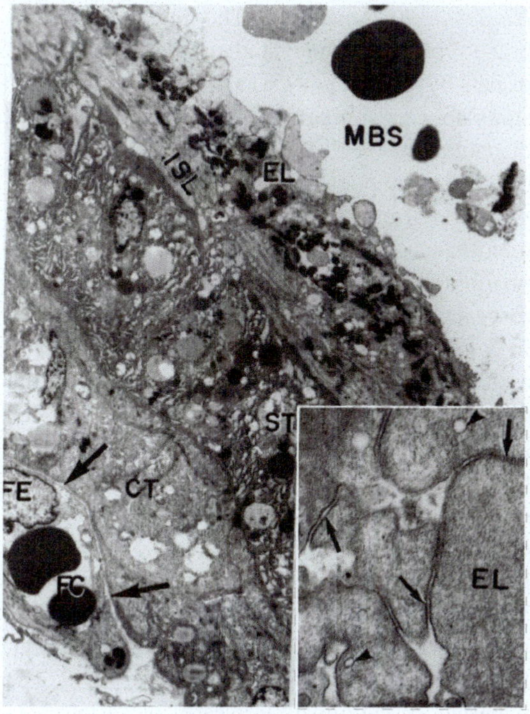

Fig. 6.74 Electron micrograph (low power) of the tertiary placenta in *Miniopterus schreibersii fuliginosus* at term. It consists of thin ectoplasmic layer (EL) surrounding the maternal blood space (MBS), thick intrasyncytial lamina (ISL), syncytiotrophoblast (ST) and cytotrophoblast (CT) [X 4000]. Inset shows desmosomes (short arrows) and particulate complexes (arrow heads). From Bhiwgade et al. 1992

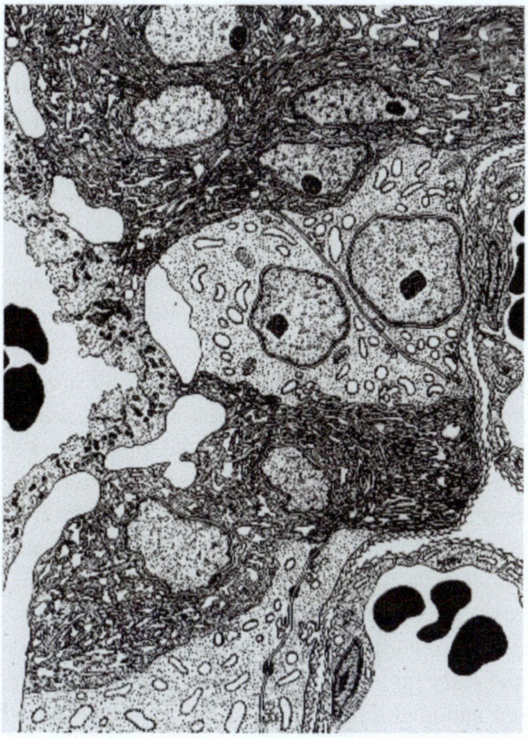

Fig. 6.75 A diagrammatic representation of the tertiary placenta in *Miniopterus schreibersii fuliginosus* at full term. Notice the layers of trophoblast. Maternal blood space is seen at the left centre, surrounded with the endoplasmic layer. It is separated from syncytiotrophoblast by thick intrasyncytial lamina. Cytotrophoblast is seen on the upper left half and in the centre. Foetal capillaries are seen on the lower left half and right centre

them, however, described the full-term secondary placenta or the development of the tertiary placenta. Later, Chari and Gopalakrishna (1984) described the developmental stages of placenta in *Miniopterus schreibersii fuliginosus* at light microscopic level. They considered the early stages of development of placenta in this animal as similar to those in other bats (Wimsatt 1945a, b, 1954, 1958; Gopalakrishna and Moghe 1960; Gopalakrishna and Karim 1979). According to Chari and Gopalakrishna (1984), the secondary and tertiary placenta are formed in a chronological sequence. The secondary placenta was categorized as endotheliochorial, while the tertiary placenta was considered to be hemomonochorial in nature. The nature of the trophoblast in bats has, thus remained obscure.

In our studies using electron microscopy, the ultrastructure of primary, secondary and tertiary definitive chorioallantoic placental barrier in *Miniopterus schreibersii fuliginosus* have been found to very unique, not observed in any other bat species or in any other mammal. The unique features are listed below:

6.10.5 Hemochorial Condition in Primary Placenta

Placentae in *Miniopterus schreibersii fuliginosus,* studied at neural groove and early limb-bud stages show hemochorial condition in the primary placenta. Due to the initial dilation of the maternal channels, the maternal blood is in direct contact with the trophoblast. This observation is contrary to what had been reported by earlier investigators who studied the placenta under the light microscope. They described the discoidal placenta at early limb-bud stage as endotheliochorial on the basis of a few spindle-shaped cells seen floating in the maternal blood space and not forming a capillary structure (Chari and Gopalakrishna 1984). In three species of bats, *Rousettus leschenaulti* (Bhiwgade et al. 2000), *Thyroptera tricolor spix* (Wimsatt and Enders 1980) and the early-formed placental pad of *Tadarida brasiliensis cynocephala* (Stephens 1969), maternal endothelium has been reported to be absent at the neural groove and early limb-bud embryos. Few spindle-shaped cells have been reported to be seen floating in the maternal blood space. These cells were called the maternal endothelial cells. According to Stephens, (1969) such a placenta is regarded as hemochorial at neural groove and early limb-bud stages because the free maternal endothelial cells seen in the maternal blood space do not form a lining of the maternal blood space.

As seen in our observations with *Miniopterus schreibersii fuliginosus,* the primary placenta in *Miniopterus shreibersii* of Europe, has been identified as hemochorial during the neural groove and early limb-bud stages. This is very similar to the placenta of *Tadarida* (Stephens 1969), Thyroptera (Wimsatt and Enders 1980) and *Rousettus* (Bhiwgade et al. 2000).

6.10.6 Interstitial Membrane

Interstitial membrane is an acellular component of endotheliochorial placentae and many authors have reported ultrastructural studies of the same. The interstitial membrane has been reported in the ferret (Lawn and Chiquoine 1965), the cat (Wynn and Bjorkman 1968), the pinniped (Harrison and Young 1961), dog (Anderson 1969), the shrew (Wimsatt et al. 1973) and recently in *Taphozous melanopogon, Rhinopoma hardwickei hardwickei, Rhinopoma microphyllum* (Mandal 1991), *Rhinolophus rouxi, Megaderma lyra lyra* (Bhiwgade 1990) and in our own study in *Hipposideros lankadiva.* The interstitial membrane reported in the endotheliochorial placentae of bats by Wimsatt (1958) after light microscopic study appears similar in histology. In our study, a discontinuous acellular membrane is observed in the secondary placenta of *Miniopterus shreibersii fuliginosus* and is comparable to the interstitial membrane described in the above species. This membrane has been attributed with some importance in placental physiology (Cukierski 1987). In our opinion, the membrane is dynamic and assists in the bi-directional exchange of material across the interhemal membrane. It is reported that interstitial membrane first appears under the maternal endothelium in the hemochorial placenta of vampire bats, *Demodus rotundus mernus* (Bjorkman and Wimsatt 1968), in the little brown bat, *Myotis lucifugus lucifugus* (Enders and Wimsatt 1968) in phyllostomid bats, *Mactrotus waterhousii* (Bodley 1974) and in leaf-nosed bat, H*ipposideros fulvus fulvus* (Bhiwgade 1990). The maternal epithelium is lost early in gestation in *Tadarida brasiliensis cynocephala* (Stephens 1969), *Thyroptera tricolor spix* (Wimsatt and Enders 1980) and, *Rousettus leschenaulti* (Bhiwgade 1991). In our studies on the primary placenta of *Miniopterus schreibersii fuliginosus*, the maternal endothelium also appears to be lost quite early in gestation. In studies on late-implanted bats, corresponding to amniogenesis, early neural groove and early limb-bud stages, the maternal endothelium is observed to have completely disappeared. Hence, in *Miniopterus schreibersii fuliginosus*, although the basic pattern of development was similar to that seen in other bats, the hemochorial condition is probably achieved much more precociously. In all the four species of bat mentioned earlier, the placenta at the ultrastructural level, showed no visible remnants of the maternal endothelium. With the exception of Tadarida, in other three bat species, the tongue-like protrusions of the syncytial cytoplasm through the gaps in endothelial basement membrane are seen incorporated into the syncytium as a clearly recognizable intrasyncytial lamina (Stephens 1969). The 'homogenous material' remains in cytotrophoblast throughout the gestation period. A significant and constant component of the definitive primary and tertiary placenta in *Miniopterus schreibersii fuliginosus* is the discontinuous intrasyncytial lamina made up of acellular material. It is, irregularly thickened, homogenous, and appears electron dense. On comparing the structure of the placental tubules of the primary and tertiary placentae of *Miniopterus schreibersii fuliginosus* with that of *Rousettus leshenaulti* (Bhiwgade 1991), Hipposideros *fulvus fulvus* (Bhiwgade 1991), *Mactrotus waterhousii* (Bodley 1974), *Desmodus rotundus murinus* (Bjorkman and Wimsatt 1968), *Myotis lucifugus lucifugus* (Enders and Wimsat, 1968) and *Thyroptera tricolor spix* (Wimsatt and Enders 1980), it becomes evident that the placental barrier in these bats has a prominent intrasyncytial lamina over which there is a layer of cytoplasm on the maternal border. This cytoplasmic layer could only have been derived from the syncytiotrophoblast. This cytoplasmic layer with almost no cell organelles is called the ectoplasmic layer (Enders and Wimsatt 1968). The fact that the intrasyncytial lamina is discontinuous, supports the hypothesis that the cytoplasm from the syncytiotrophoblast has percolated through the discontinuities towards the lumen of the placental tubule to form an almost continuous layer of cytoplasm bordering the lumen of the tubule.

The intrasyncytial lamina may have an immunological functional advantage, by which antigenic substances are prevented from entering the maternal environment from the foetal tissues (Enders and Wimsatt 1968; Bodley 1974). In our study, some structures like numerous caveolae, coated vesicles, thin-walled pinocytotic vesicles and microtubules are seen to be associated with intrasyncytial lamina and the ectoplasmic layer. This suggests that the transport of the metabolites could be possible in either direction. The large number of microvilli that project into the maternal blood from the free surface of the ectoplasmic layer of the primary placenta help to increase the surface area for the absorption. The microfilaments with desmosomes are the prominent features of the ectoplasmic layer in both the primary and the tertiary placenta of *Miniopterus schreibersii fuliginosus*. Bodley (1974) has identified these structures as the particulate complex and may be indicative of a membrane specialization of the syncytial trophoblast to function as a unique barrier.

6.10.7 Syncytiotrophoblast

The main features of syncytiotrophoblast in primary and tertiary placentae during the definitive condition are the presence of its cell organelles, like mitochondria, Golgi complex and rough endoplasmic reticulum. Similar characteristics of syncytium have been reported in several other species of mammals (Enders 1965). The pinocytotic vesicles, coated vesicles and multivesicular bodies near the trophoblastic border towards the maternal blood support the role of syncytiotrophoblast in the transport of macromolecules.

6.10.8 Cytotrophoblast

There are very few cell organelles seen in the cytoplasm of cytotrophoblast cells. Golgi complexes, mitochondria and endoplasmic reticulum are seen. Cytotrophoblast layer, therefore, may be less involved in synthetic activity. Wimsatt (1945a, b) has suggested that these cytotrophoblastic cells may act as the stem cells to give rise to the syncytiotrophoblast cells when needed. In the primary placenta, the cytotrophoblast cells disappear completely by term. The disappearance of the cytotrophoblast from the primary placenta may be because the entire primary placenta degenerates by term and does not have any foetal capillaries. The primary placenta, therefore, reaches a senescent stage by term.

The development of secondary placenta within the primary placenta in *Miniopterus schreibersii fuliginosus* is obviously a unique feature. The secondary placenta is first recognizable during the early limb-bud embryo (Chari and Gopalakrishna 1984). The earlier studies of secondary placenta were under the light microscope where, it was described as being composed of a mixed population of cellular and syncytial trophoblast (Chari and Gopalakrishna 1984). In our ultrastructural studies, it is now clear, that the secondary placenta is wholly made up of cellular trophoblast and it remains so throughout the gestation period. This is contrary to the previously reported acellular structure (Chari and Gopalakrishna 1984). This is an important observation in our study, The secondary placenta in *Miniopterus schreibersii fuliginosus* may thus be considered unique in this respect since the trophoblastic cords are cellular throughout their existence and it is very similar to that of Tadarida (Stephens 1969).

6.11 Ultrastructural Changes in the Corpus Luteum of Two Bats, during the Reproductive Cycle

6.11.1 Introduction

Corpus luteum is the fastest-growing endocrine structure but being a transient endocrine organ it also shows periodic changes of growth and regression at different stages of reproductive cycle. Regnier de Graff, is the pioneer in observing and studying corpus luteum as early as in 1672 (Short 1977). The term corpus luteum was coined much later in 1903 by Malphigi, when he noted the presence of 'yellow body' in the cow ovary. The significance of corpus luteum in maintenance of early pregnancy was established by Short in 1977. His study further demonstrated its significance not only in maintenance of pregnancy but also in the normal functioning of the reproductive cycle. Defective function of corpus luteum is a cause of reproductive failure (Jones 1949; Horta et al. 1977). Due to this, corpus luteum has become a topic of immense interest to the scientists working in the field of mammalian reproduction. Allen and Wintersteiner (1934), after performing various experiments, established the role of corpus luteum as a steroidogenic organ. Csapo et al. (1972) later confirmed it as a progesterone-secreting organ. The steroidogenic activity of corpus luteum is governed by an interplay of gonadotrophic hormones from the pituitary gland. The steroidogenic function of corpus luteum as progesterone secreting organ correlates with the typical cyclic cytological changes seen under the light microscope. Substantial work has been done in light microscopy of corpus luteum in a number of animals. Ultrastructural observations of corpus luteum also support the view that the ultrastructural changes in the organ correlate with the changes in the level of progesterone during various phases of reproductive cycle.

6.11.2 Ultrastructure Studies on Corpus Luteum

Corpus luteum is formed from the wall of mature follicle after ovulation and functions as an ovarian endocrine gland (Danell 1987; O'Shea 1987; Niswender and Nett 1988; Brar et al. 1994a). It consists of luteal cells, endothelial cells, pericytes, smooth muscle cells, fibrocytes, macrophages, leucocytes and plasma cells (Guraya 1971a, 1972, 1978; Niswender and Nett 1988). The luteal cells are steroidogenic cells responsible for secretion of hormones. There are two types of luteal cells; the small luteal cells and the large luteal cells. These two cell types originate from theca cells and granulosa cells, hence are also known as theca lutein cells and granulosa lutein cells (O'Shea et al. 1979). Sinha et al. (1971a) has mentioned the presence of three types of luteal cells in corpus luteum of raccoon, of which only one type increases during pregnancy. Thwaites and Edey (1969) have reported five types of luteal cells in the corpus luteum of ewe.

Luteal cells have been reported to be derived from theca granulosa cells of the ovarian follicles (Christensen and Gillim 1969; Enders 1973; Koering 1974; Crisp et al. 1970; Paavola 1977). Recent studies, however, suggest that a population of stem cells in the corpus luteum differentiates into small steroidogenic cells that later differentiate into large steroidogenic cells (Alila and Hansel 1984; Niswender et al. 1983; Schwall et al. 1986; Niswender and Nett 1988 and

1994, Farin et al. 1989 and Fields 1991). Small luteal cells are stellate in shape, containing abundant smooth endoplasmic reticulum, mitochondria and lipid droplets (O'Shea et al. 1979; Fitz et al. 1982 and Rodgers, et al., 1984). Large luteal cells are polyhedral in shape containing numerous mitochondria, lipid droplets and membrane bound granules (Mc.Clellan et al. 1975 and O'Shea et al. 1979).

The steroidogenic cells undergo a number of cytological transformations which are specific to a particular phase of reproductive cycle. These transformations are common to many species, while some are species specific (Christensen and Gillim 1969; Guraya 1971b). Commonly observed cytoplasmic organelles in an active corpus luteum are, moderate rough endoplasmic reticulum, extensively developed smooth endoplasmic reticulum, numerous pleomorphic mitochondria, Golgi complex, and secretory granules of varying sizes and densities. (Enders and Lyons 1964; Green and Maqueo 1965; Blanchette 1966; Pedersen and Larsen 1968; Adams and Hertig 1969a; Crisp et al. 1970). During the inactive phase or during regression the steroidogenic cells show crenated nucleus, dominant lipid droplets, autophagic vacuoles, moderate smooth endoplasmic reticulum, vesiculated rough endoplasmic reticulum and variedly shaped rarefied mitochondria (Cavazos et al. 1968; Priedkalns and Weber 1968b; Belt et al. 1970; Crisp et al. 1970; Parry et al. 1980; Fields 1991; Meyer 1991; Brar et al. 1994a).

Progesterone is the primary secretory product of the corpus luteum secretes and its level varies during various reproductive phases, which in turn correlates with changes in its cytological structures (Paavola 1977; Koering et al. 1973; Rothchild 1981; Rodgers, et al., 1984 and Guraya 1978). Corpus luteum along with progesterone, secretes prostaglandins, oestradiol-17beta and a variety of protein and peptide hormones such as oxytocin, oxytocin-related neurophysin I, relaxin, vasopressin and inhibin (Stacy et al. 1976; Bronstein et al. 1984; Fields 1991; Fields et al. 1992).

The functional life of corpus luteum varies greatly among different species of mammals and depends upon the mode of reproduction, and also several environmental and physiological factors. (Short and Buss 1965; Guraya 1968, 1971a, 1972; Rothchild 1981; Garverick and Smith 1986; Hansel and Dowed 1986; Lauderdale 1986; Gasrchiawinder, et al., 1988; Niswender and Nett 1988; Hillier 1994). Maintenance of corpus luteum has been studied in rabbit (Keyes and Nalbandov 1968; Spies et al. 1967, 1968 and Fuller and Hansel 1971); in ewe (Thwaites and Edey 1969; Karsch et al. 1971); in hamster, rat, mice and rabbit (Hilliard 1973); in guinea pig (Paavola 1977) and in primates (Richardson et al. 1985).

Regression of corpus luteum is yet another important phenomenon and is a responsible factor for successive pregnancies in the female. Autophagy plays an important role in the regression of luteal cells in the corpus luteum. Phagocytosis by macrophages also helps in the breakdown of the corpus luteum. Ultrastructure of such corpus luteum is characterized by presence of numerous dense bodies or lysosomes, lipids and a variety of autophagic vacuoles (Paavola 1979). Bats with multiple reproductive cycles and long nonbreeding periods in between are interesting animals to study corpus luteum since some of them show persistent corpus luteum till the next pregnancy.

In our study, we have studied the ultrastructural characteristics of corpus luteum of *Rousettus leschenaulti* and *Scotophilus heathi* with special reference to the various phases of their reproductive cycles.

6.11.3 *Rousettus leschenaulti*

(i) *Persisted corpus luteum of second pregnancy after lactation and during anestrous stage.*

Corpus luteum in Rousettus leschenaulti shows large sized steroidogenic cells along with small luteal cells, pericytes and vascular elements. The small luteal cells are irregular while the large cells are round. They are compressed in some regions due to their compact arrangement (Fig. 6.76). Numerous round to oval shaped mitochondria

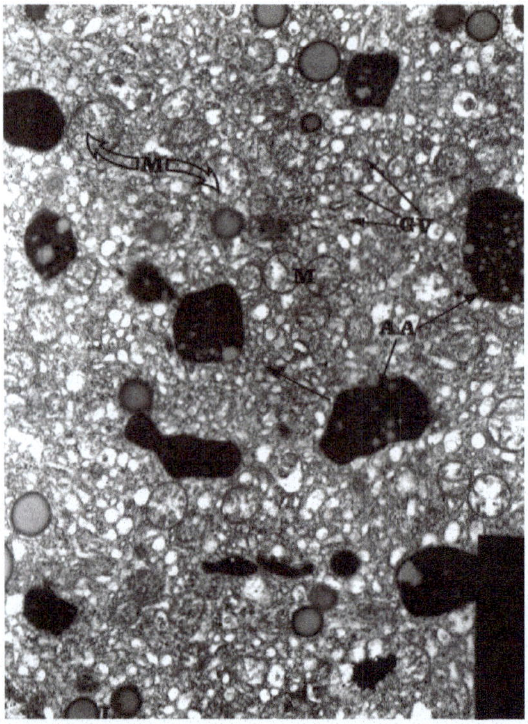

Fig. 6.76 Persisted corpus luteum of *Rousettus leschenaulti* of second pregnancy during anoestrous stage. The steroidogenic cell illustrates the stage of regression with abundant autophagocytes (AA) and lipid droplets. Mitochondria (m) and Golgi vesicles (GV) are seen. Note that the central region of the lipid droplet is electron lucent while its periphery is electron dense [X 16,000]. Unpublished electron micrographs from Bhiwgade and Savant

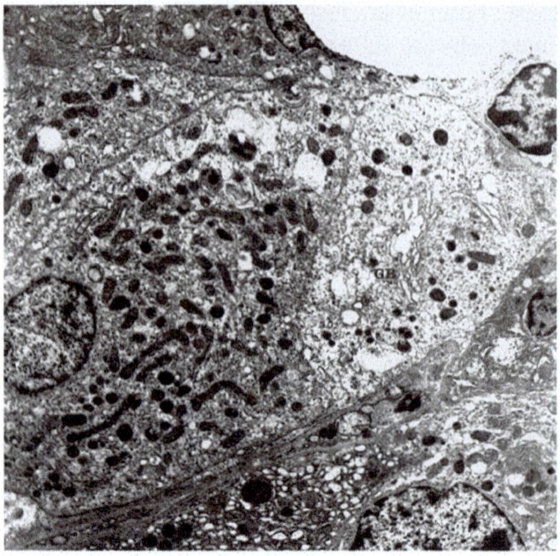

Fig. 6.77 Corpus luteum of *Rousettus leschenaulti* during first mid pregnancy. The steroidogenic cell shows numerous pleomorphic mitochondria (M) and dense bodies. Note the Sacculated Golgi bodies (GB) in the adjacent cell [X 7000]. Unpublished electron micrographs from Bhiwgade and Savant

dominate the cytoplasmic matrix. Many of the mitochondria appear vacuolated due to sparse cristae. Many mitochondria show dense woolly inclusions of varying sizes.

Endoplasmic reticulum is not a characteristic feature in this stage, but smooth endoplasmic reticulum is seen dispersed in the cytoplasm. Isolated regions of vesiculated rough endoplasmic reticulum are also recorded. The main feature of these steroidogenic cells is the abundant zones of Golgi complex associated with Golgi vesicles which are spread throughout the cytoplasm. Lipid droplets of varying sizes are seen throughout the cytoplasm. The central region of the lipid droplet is electron lucent as compared to its periphery. Some lipid droplets are in different stages of autophagocytosis. Significantly high number of autophagocytic vacuoles of varying shapes and sizes is a distinctive feature of this phase. Autophagocytosis takes away accumulated lipid droplets to mark the beginning of regression. In some regions autophagic vacuoles are seen near Golgi vesicles.

(ii) *Corpus Luteum of First Mid-pregnancy*

The steroidogenic cells of this stage show number of polymorphic forms in *Rousettus leschenaulti*. Some cells are small, irregular in shape and with large nuclei (Fig. 6.77). The nucleus of such a cell occupies almost whole of the cell leaving very little space for cytoplasm. Such nuclei are distorted, heterochromatic and without any nucleolus. The nucleus shows distinct margination of chromatin. The prominent cytoplasmic organelle in such cells is mitochondria with sparse cristae. Other types of cell population consist of slightly larger irregularly shaped cells with prominent nuclei. The nucleus of these cells shows patches of chromatin and inclusions. The cytoplasmic organelles seen in these cells are mitochondria and Golgi saccules. Dominant cells in the corpus luteum at this stage correlate with the characteristic features of steroid secretory cells. These secretory cells are large in size with oval shape. Their nucleus is round without any indentations. This euchromatic nucleus is placed eccentrically and is devoid of nucleolus. The most characteristic feature observed at this stage is the presence of varied inclusions within the nucleus. The nuclear matrix shows perichromatin as well as interchromatin granules. The intranuclear inclusions are in the form of myelinations, multivesicular bodies, condensed whorls of smooth endoplasmic reticulum and dense bodies. These cells are loaded with numerous pleomorphic mitochondria. Mitochondria are elongated with tubular or lamellar cristae but some of them are oval in shape with tubular cristae.

Vesiculated rough endoplasmic reticulum is seen in some regions while smooth endoplasmic reticulum in tubular or vesicular form is seen scattered in the cytoplasm. A distinct feature of these cells is abundance of Golgi in different forms and in different stages of development. Dilated Golgi saccules of various shapes and sizes are seen as a characteristics feature. Numerous coated vesicular are seen close to Golgi sacs. Dense bodies of various sizes and densities are seen along with Golgi in the cell periphery. Large dense bodies are seen associated with vesiculated rough endoplasmic reticulum, while coated vesicles are seen alongside extremely sacculated Golgi zone. Another distinct feature of cells at this stage is the complete absence of lipid droplets in all cell types.

(iii) *Corpus Luteum of First Full-Term Pregnancy*

The steroidogenic cells at term show characteristic features both of active cells (Fig. 6.78) and of regressive cells. The cells are basically oval in shape with a prominent large nucleus. These cells dominate the corpus luteum during this stage while the other polymorphic form observed during the previous phase are less. The smooth endoplasmic reticulum is arranged in a compartmentalized pattern in the periphery of

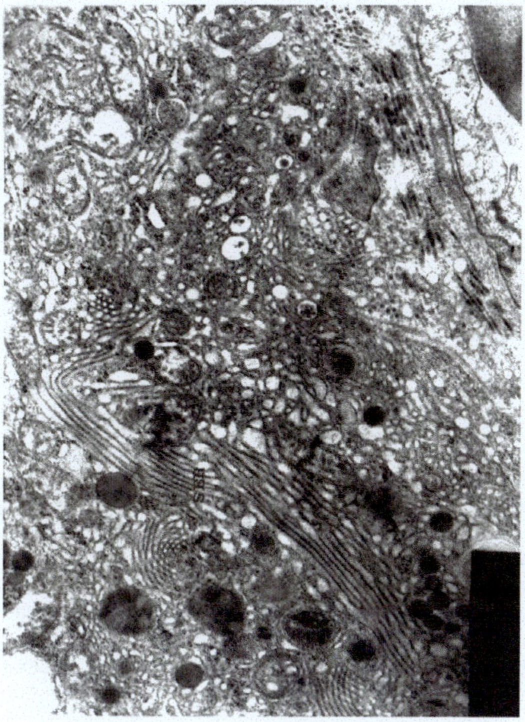

Fig. 6.78 Corpus luteum of *Rousettus leschenaulti* during the first full-term pregnancy. The steroidogenic cell shows abundant cisternae of smooth endoplasmic reticulum (SER), filaments and multi vesicular bodies. Note the honeycomb pattern of smooth endoplasmic reticulum in some regions [X 13,000]. Unpublished electron micrographs from Bhiwgade and Savant

cells. The mitochondria and Golgi complex are seen in the centre. There is a relative decrease in the number of mitochondria as compared to the previous phase while a greater number of pleomorphic mitochondria with tubular cristae are seen. They are also localized in the centre of the cells.

Endoplasmic reticulum is well developed and is seen in good amount. Rough endoplasmic reticulum is arranged in parallel stack in some regions (Fig. 6.79), while it is dilated around the nucleus. Free ribosomes are seen throughout the cytoplasmic matrix and are arranged in rosette pattern in some regions. Smooth endoplasmic reticulum is a prominent feature during this stage and is seen in different patterns with a compartmentalized arrangement. Elongated parallel cisternal arrangement of smooth endoplasmic reticulum is very

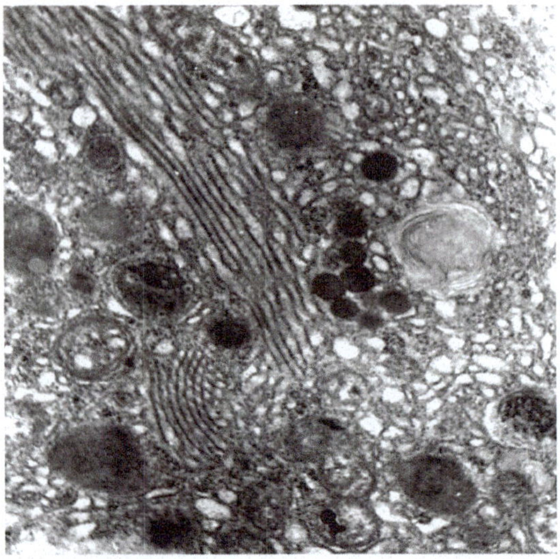

Fig. 6.79 Corpus luteum of *Rousettus leschenaulti* during the first full-term pregnancy. A region of steroidogenic cell shows parallel cisternae of smooth endoplasmic reticulum, dense bodies containing vesicles and whorled membranous structure [X 25,000]. Unpublished electron micrographs from Bhiwgade and Savant

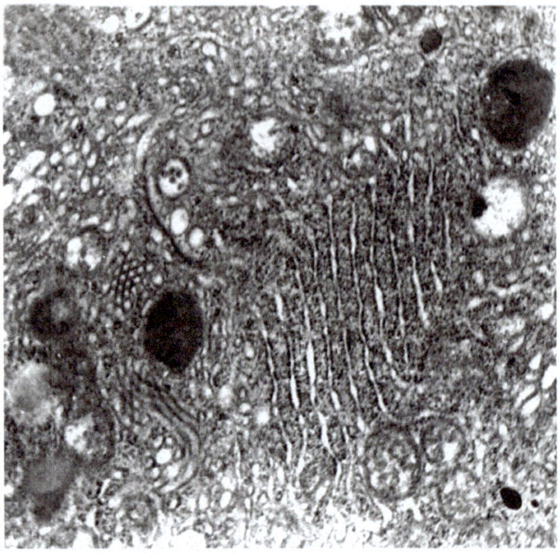

Fig. 6.80 Corpus luteum of *Rousettus leschenaulti* during first full-term pregnancy. A steroidogenic cell shows well developed parallel cisternae of rER, Golgi complex multivesicular bodies, and dense bodies. Note the continuity between rER and smooth ER cisternae [X 25,000]. Unpublished electron micrographs from Bhiwgade and Savant

common as compared to the whorled arrangement. Some regions show hexagonal honeycomb-like pattern since smooth endoplasmic reticulum in these regions gets cut in cross-section. In some regions, a close association with continuity between rough endoplasmic reticulum and smooth endoplasmic reticulum is seen. Few lipid droplets with high electron density are seen in some regions (Fig. 6.80).

Thin flat Golgi sacs arranged in 3–4 stacks are seen near rough endoplasmic reticulum in the periphery of the cell. Golgi with dilated saccules is seen in many regions along with Golgi vesicles. Numerous autophagic vacuoles are seen dispersed in both rough endoplasmic reticulum and the smooth endoplasmic reticulum. Multivesicular bodies of varying sizes with inclusions are prominently seen around Golgi zone. Bundle of filaments are seen in the periphery of the cell.

(iv) *Corpus Luteum of First Lactation*

During this stage, there is a drastic change in the cytoarchitecture of steroidogenic cells as compared to the precious phase. The partially degenerative cells seen during the earlier phase, seem to revive during this phase. There are, however, some cells that have reached a state of degeneration from which they cannot be revived. The active steroidogenic cells are oval in shape and compactly arranged. They form the main cell population of corpus luteum (Fig. 6.81). Numerous oval or round mitochondria are found randomly distributed in the cytoplasm. The cristae of these mitochondria are sparse either tubular or lamellar in form. Few elongated mitochondria with longitudinal cristae are also found. The mitochondrial matrix shows deposits of glycogen granules.

Cytoplasmic matrix shows presence of both types of endoplasmic reticulum, however, quantitatively, rough endoplasmic reticulum is less as compared to smooth endoplasmic reticulum. Rough endoplasmic is seen with irregularly arranged tubular cisternae with prominent ribosomes on the membranes. Scattered all over in the cytoplasm, are ribosomes either in singly or in groups. Ribosomes are arranged in rosette

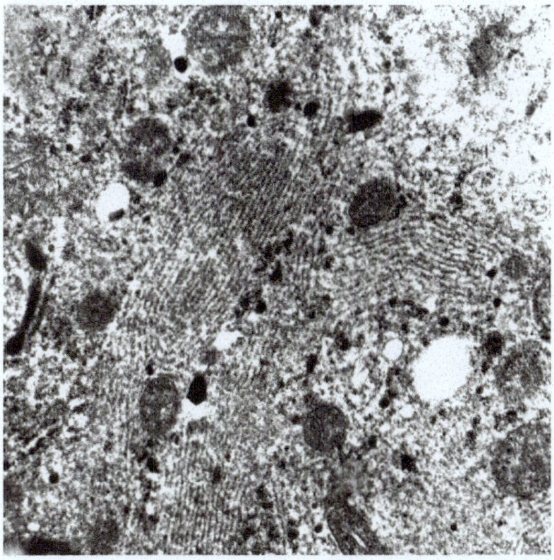

Fig. 6.81 Corpus luteum of *Rousettus leschenaulti* during the first lactation. The cell shows mitochondria with glycogen granules. Smooth endoplasmic reticulum is seen in prominent fingerprint pattern [X25,000]. Unpublished electron micrographs from Bhiwgade and Savant

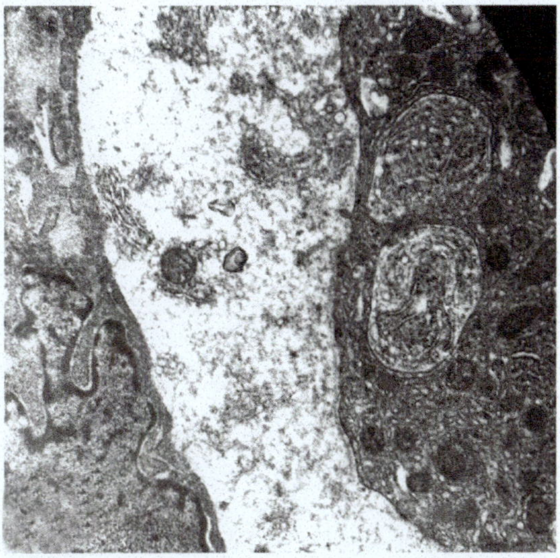

Fig. 6.82 Corpus luteum of *Rousettus leschenaulti* during second early pregnancy (during first lactation). The cell showing irregularly shaped nucleus. The autophagic vesicles are of single membrane containing endoplasmic autophagic vacuoles (arrow) are seen [X 16,000]. Unpublished electron micrographs from Bhiwgade and Savant

pattern. Hypertrophy of smooth endoplasmic reticulum is a distinct feature of this phase. In the peripheral region of the cell, smooth endoplasmic reticulum is seen prominently arranged in small whorls in a fingerprint pattern. This is a common feature of these cells. Smooth endoplasmic reticulum is long, tubular arranged in parallel and associated with dense bodies. In some places, distinct autophagic vacuoles are seen in the smooth endoplasmic reticulum region. Very few dense bodies of varying sizes are seen in the region of smooth endoplasmic reticulum or in the periphery of the cell.

(v) *Corpus Luteum of Second Early Pregnancy and During Lactation*

The steroidogenic cells of this stage appear roughly polyhedral in shape. The cells are relatively less compact and the cell membrane between the adjacent cells show modifications (Fig. 6.82). Nucleus is seen towards the cell periphery. It is round and heterochromatic with a distinct nucleolus. In some areas, the nucleus appears crenated and the chromatin is pushed to the margins. Perichromatin and interchromatin granules are seen in the nucleus of this stage. Many pleomorphic mitochondria are seen scattered throughout the cytoplasm. Some of the mitochondria that are abundant are oval in shape. Those which are tubular are with lamellar or tubular cristae. In some of the regions, the mitochondria are highly elongated and show longitudinal cristae (Fig. 6.83).

The steroidogenic cells show both types of endoplasmic reticulum. Well-developed rough endoplasmic reticulum arranged in 5–6 parallel tubular cisternae or in whorl pattern are seen. In some regions, rough endoplasmic reticulum shows vesiculation near smooth endoplasmic reticulum tubules or around mitochondria. Free ribosomes are scattered and arranged in rosette pattern. Smooth endoplasmic reticulum is made up of tubular and branched cisternae, that are seen as vesicles in cross-section. When in close association, rough endoplasmic reticulum and smooth endoplasmic reticulum show continuity between them (Fig. 6.83). Golgi zone consists of dilated saccules and vacuoles of varying size. Coated vesicles of varying sizes and densities are also

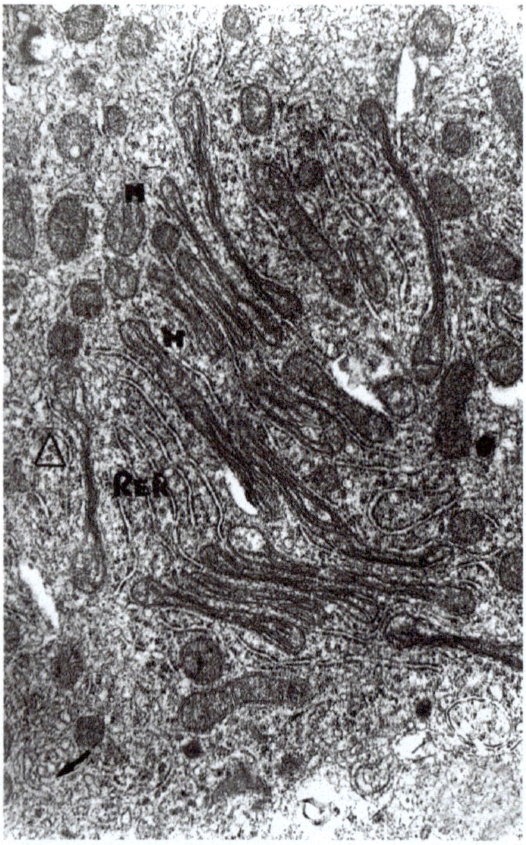

Fig. 6.83 Corpus luteum of *Rousettus leschenaulti* during second early pregnancy (during first lactation). Abundant pleomorphic mitochondria with tubular and lamellar cisternae are seen. Numerous cisternae of rER and free ribosomes arranged in rosette pattern (in Δ) are seen [X 25,000]. Unpublished electron micrographs from Bhiwgade and Savant

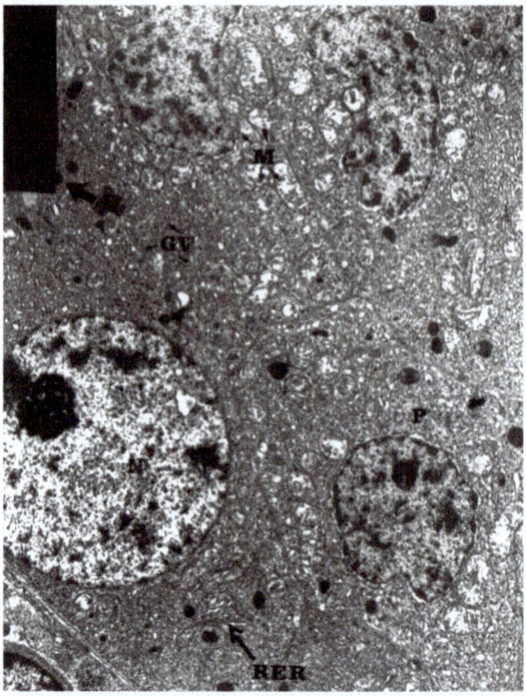

Fig. 6.84 Corpus luteum of *Rousettus leschenaulti* during second full-term pregnancy. The steroidogenic cells show pleomorphic nuclei (N), one of which shows eccentrically placed nucleolus (N). Pleomorphic mitochondria (m), rough endoplasmic reticulum, peripherally placed smooth endoplasmic reticulum and autophagic vacuoles can be seen [X 7000]. Unpublished electron micrographs from Bhiwgade and Savant

seen. Some of these are present near the cell membrane. Extremely large autophagic vacuoles or macrophages are seen prominently. Some of these show different types of inclusions such as rough endoplasmic reticulum or myelin structures (Fig. 6.83).

(v) *Corpus Luteum of Second Full-Term Pregnancy*

The steroidogenic cells exhibit polymorphism due to various shapes of nuclei seen in the cells. The cells are oval in shape with a nucleus showing a distinct nucleolus. The cells are compactly arranged, and the intercellular spaces are decreased (Fig. 6.84). The cells show compartmentalized arrangement of smooth endoplasmic reticulum. They also show sparse mitochondria, in the periphery and abundant mitochondria in centre. Presence of polymorphic nuclei is the main reason for polymorphism seen in the steroidogenic cells. The nuclei of active steroidogenic cells are round with a prominent nucleolus placed eccentrically. The nucleus shows pars amorpha of reticulate type. Condensed chromatin is seen randomly distributed in the nucleus. Some of the cells possess highly distorted euchromatic nuclei without nucleolus. The nuclear matrix shows condensed chromatin patches of perichromatin and interchromatin granules. Numerous pleomorphic mitochondria are distributed in an orderly pattern in the cytoplasm. Amongst the various forms, oval or round mitochondria dominate the total

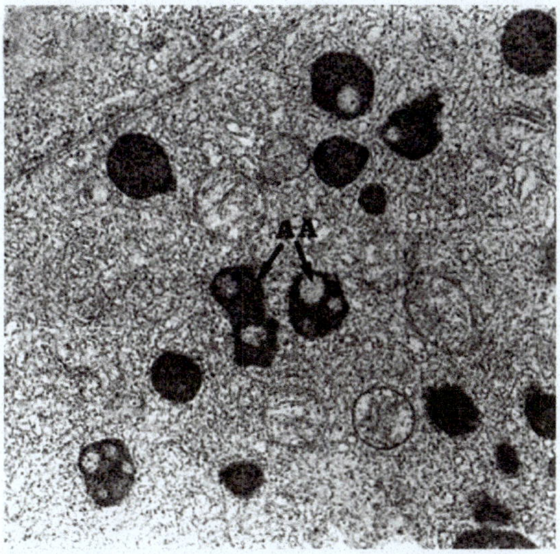

Fig. 6.85 Corpus luteum of *Rousettus leschenaulti* during second full term pregnancy. Region of steroidogenic cell showing abundant autophagic vacuoles and smooth endoplasmic reticulum, few mitochondria and vesiculated rough endoplasmic reticulum are also present [X 25,000]. Unpublished electron micrographs from Bhiwgade and Savant

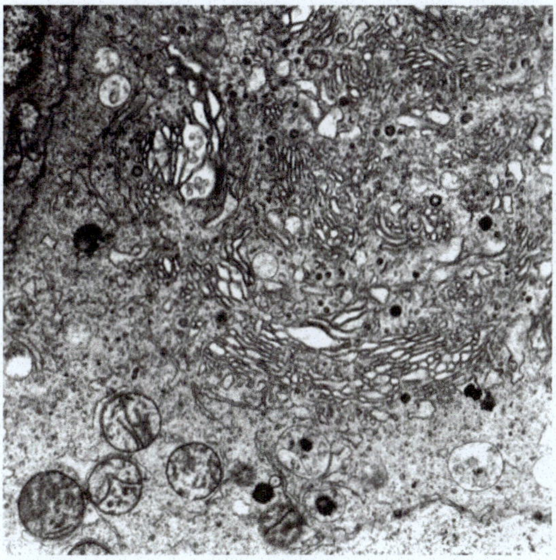

Fig. 6.86 Corpus luteum of *Rousettus leschenaulti* during second lactation. Note the numerous Golgi zones and large mitochondria with sparse cristae. Well-developed rough endoplasmic reticulum and dense bodies are also seen [X 13,000]. Unpublished electron micrographs from Bhiwgade and Savant

mitochondrial population. Though rare, elongated mitochondria with longitudinal cristae are present (Fig. 6.85).

Endoplasmic reticulum of both types is seen abundantly. Rough endoplasmic reticulum exhibits its typical pattern of parallel tubular cisternae with prominent ribosomes attached to its membrane. Rough endoplasmic reticulum is in vesiculated form in some regions. Rough endoplasmic reticulum is closely associated with smooth endoplasmic reticulum while mitochondria are in the cell periphery (Fig. 6.84). Smooth endoplasmic reticulum is distinctly seen as abundant in the peripheral region of the cell. It consists of small-branched tubules, which appear as small vesicles, where they are cut in cross-sections. Whorls of smooth endoplasmic reticulum are also seen in some regions. Isolated regions with Golgi complex are seen around rough endoplasmic reticulum along with Golgi vesicles. Dense lipid droplets are seen in moderate amount at the cell periphery. They are seen to be associated with smooth endoplasmic reticulum. Some of them show autophagy which is very common for this stage (Fig. 6.85). Numerous autophagic vacuoles are seen indicating active process of degeneration. The shape, size and density of these vacuoles vary greatly (Fig. 6.85).

(vi) *Corpus Luteum of Second Lactation*

The steroidogenic cells of this stage show many degenerative changes that are indicative of regression of corpus luteum. The cells are compactly arranged and hence show an irregular oval shape (Fig. 6.86). Numerous round to oval mitochondria with sparse tubular cristae are seen. Tubular mitochondria with extremely sparse tubular cristae are seen occasionally. Some of the mitochondria show myelination and complete loss of cristae (Fig. 6.86). Endoplasmic reticulum abundant and is seen in different forms. Rough endoplasmic reticulum is arranged in parallel and is with tubular cisternae and prominent ribosomes. Ribosomes are also seen scattered throughout the cytoplasmic matrix. Rough endoplasmic reticulum with various degrees of vesiculation is seen in different cells. Smooth

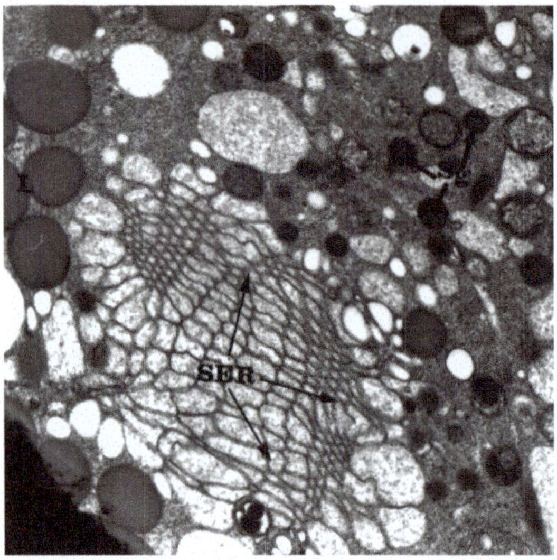

Fig. 6.87 Corpus luteum of *Rousettus leschenaulti* during second lactation. A steroidogenic cell showing large, accumulated lipid droplets, vesiculated rER and dilated smooth endoplasmic reticulum (SER) which are characteristic feature of regressing cell [X16,000]. Unpublished electron micrographs from Bhiwgade and Savant

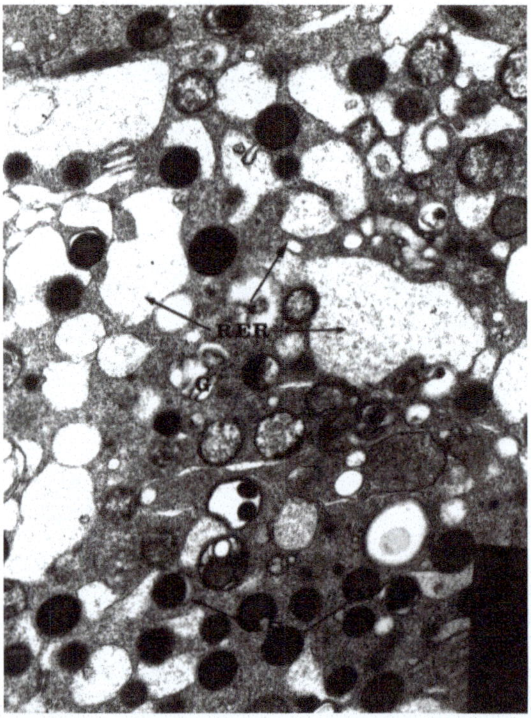

Fig. 6.88 Corpus luteum of *Rousettus leschenaulti* showing vesiculated rough endoplasmic reticulum (arrows), surrounded by dense bodies (Db) in steroidogenic cell [X16,000]. Unpublished electron micrographs from Bhiwgade and Savant

endoplasmic reticulum is present with regular tubular cisternae. In some regions, smooth endoplasmic reticulum shows a honeycomb pattern and is hypertrophied with dilated vesicles (Figs. 6.87 and 6.88). Cells show hypertrophy and dilation of Golgi complex. Numerous coated vesicles are seen around the Golgi zone. Numerous lipid droplets are also seen around smooth endoplasmic reticulum, rough endoplasmic reticulum and nucleus. Dense bodies of various sizes are distributed around the smooth endoplasmic reticulum and Golgi complex (Figs. 6.88 and 6.89).

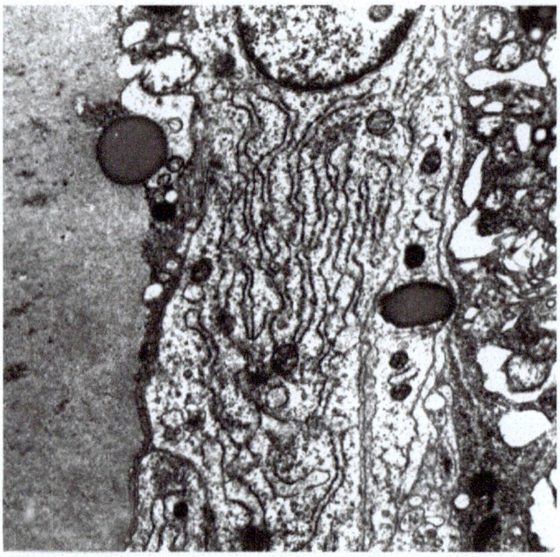

Fig. 6.89 Corpus luteum of *Rousettus leschenaulti* during second lactation. Note the region of steroidogenic cell showing distended cisternae of rough endoplasmic reticulum in juxta999 nuclear position. The adjacent cell shows Golgi elements and autophagic vacuoles [X 13,000]. Unpublished electron micrographs from Bhiwgade and Savant

6.11.4 *Scotophilus heathi*

i. *Early Pregnancy*

The corpus luteum in *Scotophilus heathi*, during early pregnancy, predominantly consists of one type of active steroidogenic cells in addition to the endothelial cells and vascular elements. The steroidogenic luteal cells are the largest of all cell types, closely arranged and separated by narrow

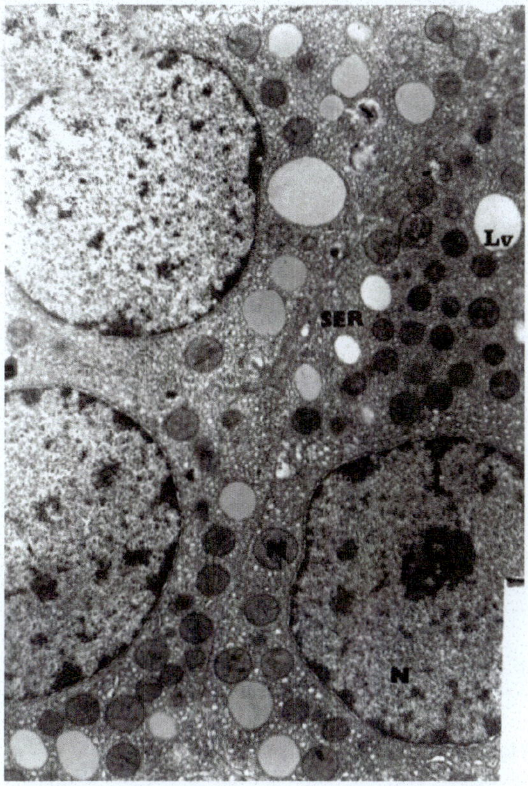

Fig. 6.90 Corpus luteum of *Scotophilus heathi* during early pregnancy. The compactly arranged steroidogenic cell shows round nuclei (n) with margination of chromatin (arrow). Note the *pars amarpha* in one of the nucleoli. Mitochondria and smooth endoplasmic reticulum (SER) are present throughout the cell. Large lipid vacuoles (Lv) are present [X 8000]. Unpublished electron micrographs from Bhiwgade and Savant

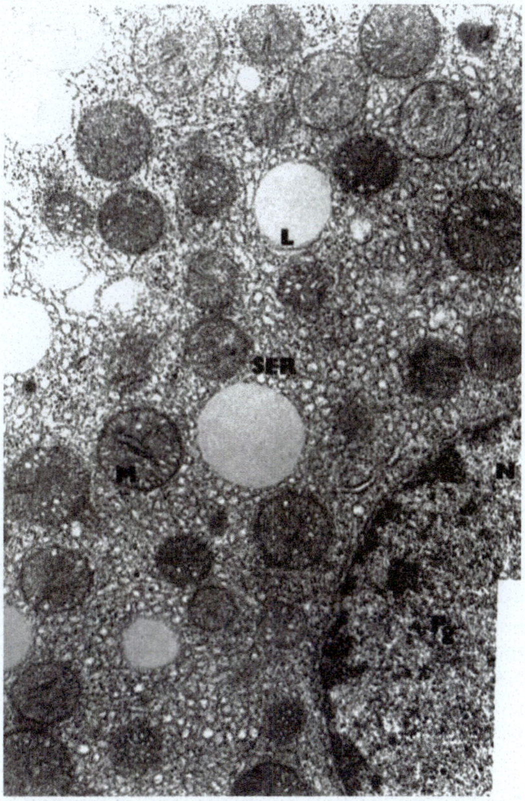

Fig. 6.91 Corpus luteum of *Scotophilus heathi* during early pregnancy. Magnified view of the steroidogenic cell cytoplasm showing abundant mitochondria (m) and smooth endoplasmic reticulum (SER) as the major cell organelles confirming it being an active steroid secretors. A few vacuolated large lipid (L) droplets are present [X 16,000]. Unpublished electron micrographs from Bhiwgade and Savant

spaces between them. The cells are irregular but oval in shape and compactly arranged (Fig. 6.90). Nucleus is round in shape with homogenous matrix of low electron density. The nuclear envelope is smooth without deep indentations. Sparse heterochromatin is confined to the nuclear periphery. Perichromatin granules form the distinct feature of the nuclear matrix but sparse interchromatin granules are also found. A single prominent eccentrically placed nucleolus is seen in some of the cells while it is absent in others. The nucleolus when present shows vacuolation due to the presence of pars amorpha. The nuclear-cytoplasmic ratio is high due to the large nucleus.

Numerous round or oval mitochondria of varying sizes are dispersed throughout the cytoplasm (Fig. 6.91). Occasional elongated or cup-shaped mitochondria are also observed. Mitochondrial cristae are mainly tubular and rarely vesicular or lamellar. Some of the elongated mitochondria show longitudinal cristae. In many regions, mitochondria are seen closely associated with Golgi region. Mitochondrial matrix is denser as compared to the background cytoplasm. The abundance of mitochondria and their distinctive features distinguishes the luteal cell at this stage. Smooth rough endoplasmic reticulum which is tubular and branched, dominate the luteal cell cytoplasm as compared to rough endoplasmic reticulum (Fig. 6.91). Smooth endoplasmic reticulum is the major cell inclusion seen at this stage.

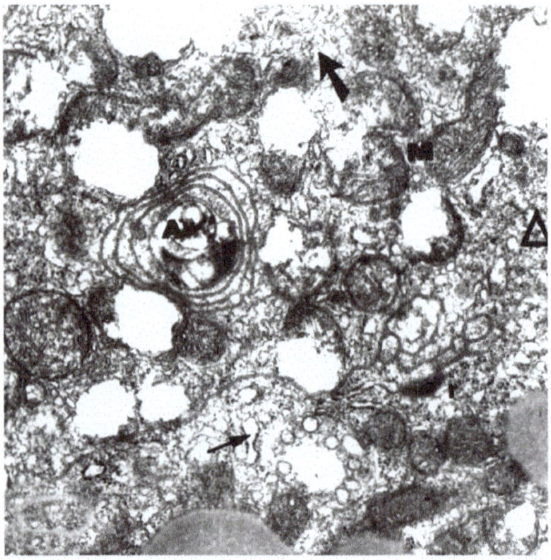

Fig. 6.92 Corpus luteum of *Scotophilus heathi* during early pregnancy. The steroidogenic cell showing abundant mitochondria (m) with tubular cristae vesiculated rER (thin arrow), free ribosomes arranged in rosette pattern. Note the presence of three layered autophagic vacuole (AV) which is unusual for this phase [X 25,000]. Unpublished electron micrographs from Bhiwgade and Savant

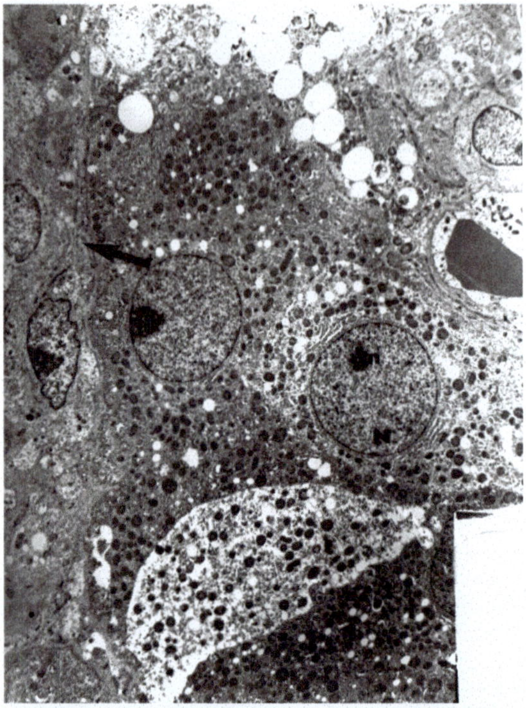

Fig. 6.93 Corpus luteum of *Scotophilus heathi* during limb-bud stage of development. Note the irregularly shaped active steroidogenic cell with increased intercellular spaces. The eunchomatic nucleus (N) with prominent nucleolus is seen [X 3500]. Unpublished electron micrographs from Bhiwgade and Savant

Rough endoplasmic reticulum is present moderately in the form of one to two short tubular cisternae around the nucleus and in a few places in the cytoplasm. Ribonucleoprotein granules or ribosomes are seen attached at irregular intervals along the membranes of the endoplasmic reticulum. Ribosomes are seen either dispersed or arranged in a rosette pattern at few areas in the cytoplasm. Vesiculated rough endoplasmic reticulum with ribosomes are also seen in some areas (Fig. 6.92). Golgi zone is distinct in some regions and is marked by the presence of typical Golgi elements. In other regions, Golgi complex consists of stacked, flattened, curved, membranous sacs. That are dilated at the ends. Outer convex immature side show vesicles of varying size and density, whereas inner concave mature side show a number of secretory granules or vacuoles. The Golgi region is found to be surrounded by elongated mitochondria or by numerous oval mitochondria. Varying size lipid droplets are seen abundantly along with the lipid vacuoles (Fig. 6.90). Some of the distinct but unusual inclusions observed in early pregnancy are, multivesicular bodies, autophagic vacuoles, autophagic vacuoles which are occasionally seen with three-layered membrane enclosing multivesicular bodies and densely developed microvilli surrounded by various-sized multivesicular bodies (Fig. 6.91).

ii. *Limb-Bud Stage of Embryo*

The corpus luteum in limb-bud stage is seen dominated by the active steroidogenic cells as in earlier stage. The cells show roughly oval or irregular cell boundaries. The intercellular spaces appear to have increased slightly as compared to the previous phase. The cells distinctly show the presence of numerous secretory granules (Fig. 6.93) and an extreme abundance of lipid vacuoles. Nucleus is round in shape and with homogenous matrix. The nuclear matrix has an electron density which is similar to that of the cytoplasm (Fig. 6.93). The nucleus is euchromatic with a single nucleolus which is placed eccentrically and more towards the periphery. The nucleus shows presence of perichromatin granules and single nucleolus. Nucleolus shows distinct pars amorpha (Fig. 6.93) with nucleoplasmic spaces due to the reticulate arrangement. The nuclear-cytoplasmic ratio is more or less equal. In one of the regions fragmentation of nucleus is seen which is very unusual for this phase. Numerous dense secretory granules are seen around the fragmenting nucleus.

The mitochondrial number has greatly reduced as compared to the previous phase. Extremely few elongated mitochondria are seen around the nucleus. Mitochondrial cristae show change in their structure and are replaced by single membrane intramitochondrial glycogen inclusions. The glycogen content in some regions is seen as condensed globular structures in the form of glycogenosomes. Unlike the previous phase, smooth endoplasmic reticulum is no more a prominent feature of the cell at this stage. It is limited to a few isolated areas and appears as tubular condensed mass. Rough endoplasmic reticulum shows progressive development as it shows increased tubular cisternal stacks. Rough endoplasmic reticulum is localized around the nucleus as 3–4 layered parallel cisternae (Fig. 6.93). Luteal cells at this stage are characterized by the presence of vacuolated lipid droplets. Some of the lipid droplets are in contact with the endoplasmic reticulum.

Numerous secretory granules of varying sizes and densities are observed. Some granules are less electron dense whereas some are extremely electron dense. Unusual structures or cell inclusions are seen near electron-dense secretory granules. Highly granular secretory granules are seen near smooth endoplasmic reticulum and vacuolated lipid droplets (Fig. 6.94). Different types of cell surface specializations are observed in different regions. Non-fenestrated cisternae with tubular extensions are seen protruding from the cell periphery. Many of these tubules are in association with secretory granules, lipid droplets or mitochondria. The cell membrane in some regions shows intracellular projections.

iii. *Mid-Pregnancy*

Corpus luteum of mid-pregnancy in *Scotophilus heathi* is dominated by the luteal cells, however, its cytoarchitecture shows a number of changes as compared to the previous stage. The steroidogenic cells show an irregularly oval shape with dense cytoplasmic matrix. Nucleus is round in shape and heterochromatic with chromatin dispersed in the nuclear matrix. The nuclear envelope exhibits smooth contour without any indentations (Fig. 6.95). Masses of chromatin are gathered towards the periphery of the nucleus with a few isolated regions in the centre. Very few perichromatinic granules are seen due to the margination of the chromatin (Fig. 6.95). The nuclear-cytoplasmic ratio is more or less equal. Abundant pleomorphic mitochondria are seen randomly distributed in the cytoplasmic matrix under low magnification. Many mitochondria show oval to circular boundaries, while few mitochondria show elongated or tubular and

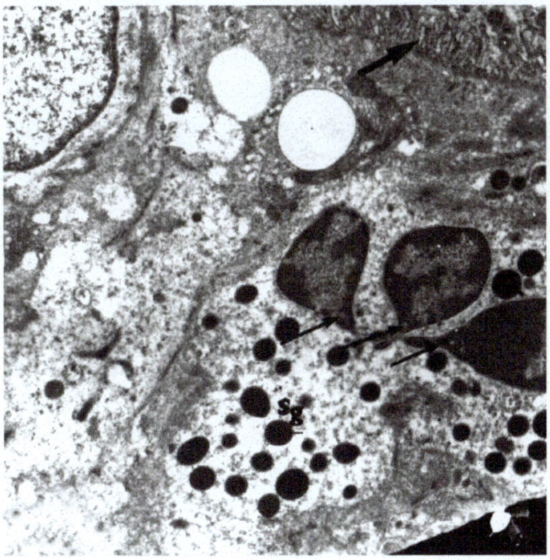

Fig. 6.94 Corpus luteum of *Scotophilus heathi* during limb-bud stage. The steroidogenic cell showing dense secretory granules (Sg) fragmenting nuclei (arrows) and non-fenestrated tubular cisternae (Bold arrow) [X 10,000]. Unpublished electron micrographs from Bhiwgade and Savant

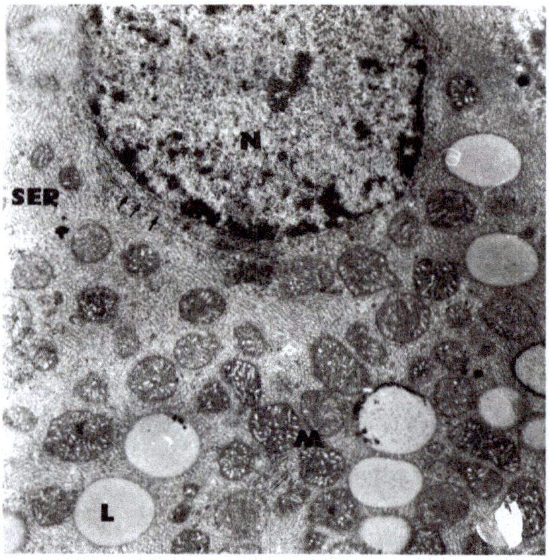

Fig. 6.95 Corpus luteum of *Scotophilus heathi* during mid-pregnancy. The steroidogenic cell shows the nucleus surrounded by rough endoplasmic reticulum (thin arrows) abundant smooth endoplasmic reticulum (SER) and pleomorphic mitochondria (m) with tubular cristae. Note the presence of lipid droplet (L) [X13,000]. Unpublished electron micrographs from Bhiwgade and Savant

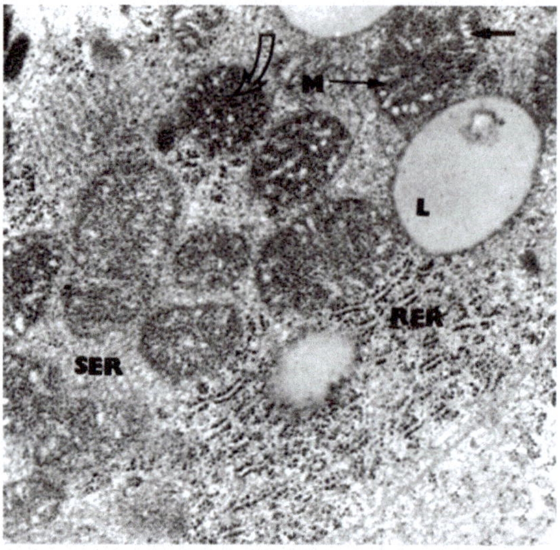

Fig. 6.96 Corpus luteum of *Scotophilus heathi* during mid-pregnancy. The cell shows intramitochondrial glycogen deposits (arrow). Abundant smooth endoplasmic reticulum (SER) few lipid droplets (L) and randomly arranged rER are also seen. [X 33,000]. Unpublished electron micrographs from Bhiwgade and Savant

irregular in shape (Fig. 6.95). Distinct tubular, vesicular or lamellar mitochondrial cristae are seen in the homogenous matrix (Fig. 6.96). Modifications in mitochondria with formation of vesicular cristae and intramitochondrial glycogen deposits are seen. Some mitochondria show less cristae due to intense single membrane bound glycogen accumulation in the mitochondrial matrix Extreme deposition of glycogen granules has totally changed the appearance of mitochondria (Fig. 6.96).

Once again, endoplasmic reticulum can be considered as the prominent feature of the luteal cells at this stage. Both types of endoplasmic reticulum are seen abundantly. Rough endoplasmic reticulum is well arranged in 3–4 tubular, parallel cisternae around the nucleus and in some other regions of the cytoplasm. Those seen around the nucleus are closely associated with lipid droplets and mitochondria. Smooth endoplasmic reticulum is seen as extremely hypertrophied with an extensive branched tubular pattern throughout the cytoplasm (Fig. 6.97). The tubular cisternae of smooth endoplasmic reticulum are well developed and arranged in whorls. This is a distinct feature of the luteal cells during mid-pregnancy. The cross-sectional

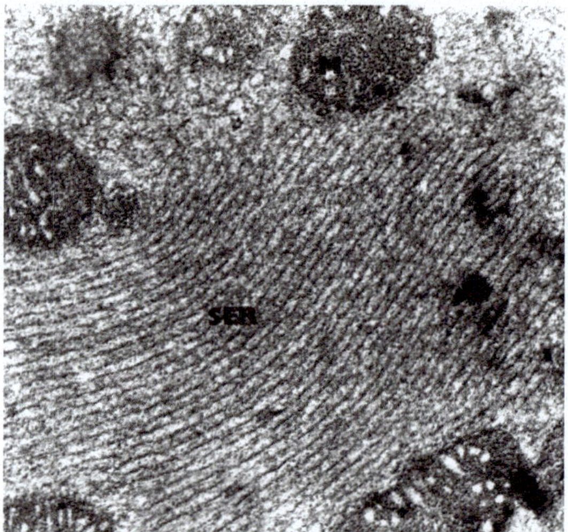

Fig. 6.97 Corpus luteum of *Scotophilus heathi* during mid pregnancy. The steroidogenic cell showing hypertrophy of smooth endoplasmic reticulum (SER) which is arranged in parallel tubular cisternae. It is seen surrounded by mitochondria with glycogen deposits [X 40,000]. Unpublished electron micrographs from Bhiwgade and Savant

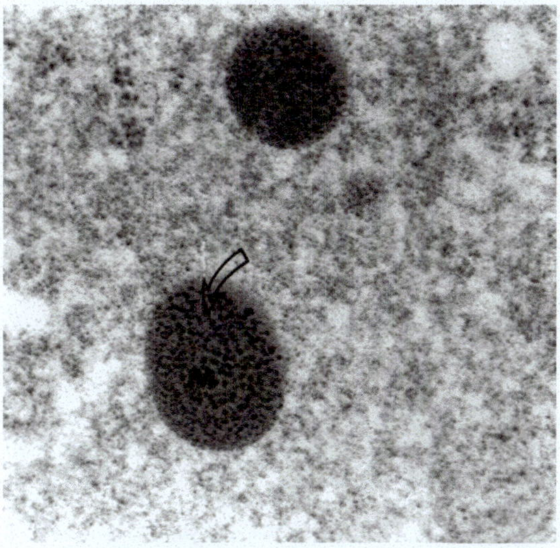

Fig. 6.98 Corpus luteum of *Scotophilus heathi* during mid-pregnancy. Extremely modified mitochondria (m) with heavy deposits of glycogen (arrow) as seen under high magnification [X 80,000]. Unpublished electron micrographs from Bhiwgade and Savant

Fig. 6.99 Corpus luteum of *Scotophilus heathi* during mid-pregnancy. A region of steroidogenic cell showing hypertrophy of smooth endoplasmic reticulum (SER) and honeycomb structure of tubules cut in cross section. Unpublished electron micrographs from Bhiwgade and Savant

view of these whorls show a typical honeycomb pattern which also is a characteristic feature of mid-pregnancy luteal cells (Fig. 6.99). Lipid droplets of varying sizes ranging from small, large to irregular are seen in aggregated pattern. They are associated with rough endoplasmic reticulum and mitochondria (Fig. 6.98). Vacuolated lipid droplets are also seen in some regions (Figs. 6.98 and 6.99).

iv. *Late Pregnancy*

Corpus luteum of late pregnancy shows distinct luteal cytoarchitecture that makes it distinct and easily distinguishable from the earlier stages. The luteal cells are shrunken with irregular cell boundaries. The cytoplasmic matrix is also shrunken, displaying an increased nuclear-cytoplasmic ratio. The nuclei are greatly distorted and a glance would distinguish them as nuclei of regressing cells (Fig. 6.100). Nuclei are highly distorted from their original round shape and show a variety of shapes. Some of the nuclei are oval in shape with a number of crenations and invaginations in the nuclear envelope. Some of the nuclei have become highly flattened elongated and tubular whereas some of them show a distinct semilunar shape (Fig. 6.101). Nucleolus is not present, and nucleus shows margination of chromatin.

Mitochondrial numbers seem to have decreased due to their aggregation in some regions and complete absence in other. They show changes in their morphology displaying a variety of shapes. Some of the common shapes of mitochondria are circular, oval, elongated and irregular. The mitochondrial matrix is lucid showing clear structure of their cristae. Many of the mitochondria exhibit tubular and lamellar cristae, while few of the mitochondria are rarefied and located around the smooth endoplasmic reticulum (Fig. 6.101).

The prominent cytoplasmic organelle seen in the luteal cells at term pregnancy is endoplasmic reticulum. The dominance of smooth endoplasmic reticulum over rough endoplasmic reticulum is distinctive (Fig. 6.101). Rough endoplasmic reticulum is seen with dilated and vesiculated cisternae and it is present around the nucleus. Vesiculated rough endoplasmic reticulum with

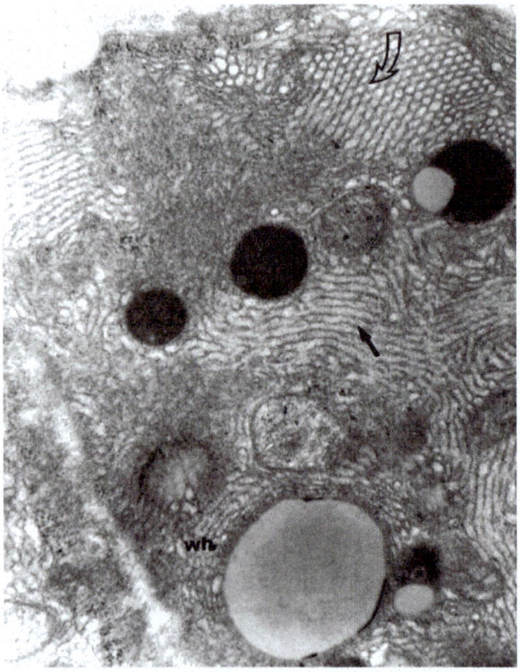

Fig. 6.100 Corpus luteum of *Scotophilus heathi* during late pregnancy showing characteristic whorls of smooth endoplasmic reticulum and lipid vacuoles of a typical large luteal cell. Note various profiles of smooth endoplasmic reticulum. Long cisternae of smooth endoplasmic reticulum tubules aligned parallel to each other (arrow) and closely packed whorled arrangement (Wh) are very common. Note the hexagonal honey comb pattern of smooth ER, (hollow arrow) which forms the prominent filament of this stage [X 25,000]. Unpublished electron micrographs from Bhiwgade and Savant

scattered ribosomes are seen scattered throughout the cytoplasm.

The abundance of smooth endoplasmic reticulum in varied patterns is seen in the cytoplasm of luteal cell at full-term pregnancy. The regular tubular pattern of elongated, parallel cisternae is prominent all throughout the cytoplasm (Fig. 6.101). Some of these cisternae show ribosomes around them. These cisternae in cross-sections are seen as hexagonal arrays with distinct honeycomb pattern. Smooth endoplasmic reticulum is seen as concentric membranous whorls of fenestrated cisternae, which are connected amongst themselves by tubules. The condensed whorls of smooth endoplasmic reticulum are surrounded by rarefied mitochondria.

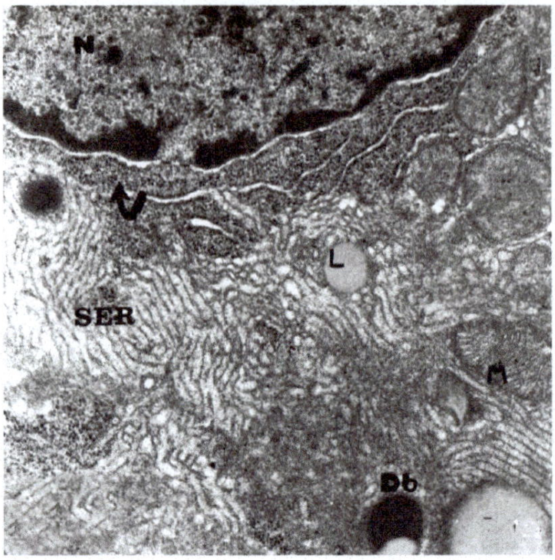

Fig. 6.101 Corpus luteum of *Scotophilus heathi* during full-term pregnancy. The steroidogenic cell shows part of nucleus (N) with marginal chromatin surrounded by distended cisternae of rough endoplasmic reticulum (curved arrow). Well-developed smooth endoplasmic reticulum (SER) few mitochondria (m); few lipid droplets (L) and dense bodies (Db) are also present [X 25,000]. Unpublished electron micrographs from Bhiwgade and Savant

Lipid droplets of varying sizes and electron densities are common. In many regions, lipid droplets have fused together to form large-sized droplets. Lipid droplets are generally seen surrounded by smooth endoplasmic reticulum, dense bodies and mitochondria. Numerous electron-dense bodies surround the mitochondria and smooth endoplasmic reticulum. Some of these bodies show vacuolation (Figs. 6.100 and 6.101).

6.12 Discussion

The changes in the fine structure of corpus luteum of *Rousettus leschenaulti* during various phases of reproductive cycle correlate well with the changes in progesterone level for the corresponding phases as recorded. The distinctive presence of two types of steroidogenic cells in the corpus luteum of *Ronsettus leschenanlti* is in accordance with the observation of Corner

(1921) in sow; Belt et al. (1970) in pig; Crisp et al. (1970) in human; Sinha et al. (1971b) in white-tailed deer and O'Shea (1987) in sheep, goat, cow and buffalo. The larger round steroidogenic cells though present in large numbers, show (atrophic) regressive changes, during anoestrous phase. This is in agreement with the observations of Azmi et al. (1984), in regressed corpus luteum of guinea pig. These large cells which are probably the progesterone secretors are steroidogenic cells as reported by Corner, (1921), in sow; Belt et al. (1970), in pig; Crisp et al. (1970), in human and Sinha et al. (1971b), in white-tailed deer.

The distorted heterochromatic nucleus of anestrous phase indicating the degenerative status of the cell, is similar to the observations of necrotic chondrocytes of rabbit by Ghadially and Roy (1969). These degenerative cells, seen during anestrous phase, in corpus luteum of *Rousettus leschenaulti*, undergo further regression in the next phase during which, a new corpus luteum is formed in the contralateral ovary.

The fall in the oestrogen level might be due to the exhaustion of cells owing to regression, whereas the cells showing features of active cells could be responsible for the rise in progesterone level (Fields 1991; Brar et al. 1994, b; Niswender et al. 1994).

The differentiation of steroidogenic luteal cells into different forms based on their nuclear morphology is seen during mid-pregnancy. This could be attributed to the variation in the degree of development of these cells according to the requirement of the progesterone (Paavola 1977; Guraya, 1971a and 1991). The presence of these cells correlates with the observation of Gemmell et al. (1976), in regressed corpus luteum of ewe. The presence of active cells during this phase, however, does not correlate with the low progesterone levels, as recorded by Date (1996). It is possible that the steroidogenic cells develop in batches under the influence of gonadotrophins released from the pituitary gland (Horvath and Kovacs 1988; Saeger 1992). This could explain the increased level of progesterone during this phase and also, the presence of regressive as well as active cells being seen simultaneously.

Presence of intranuclear inclusions during this phase may be an indicator of exhaustion of cells due to hyperactivity as suggested by Adams and Hertig (1969b) during mid-pregnancy in humans. Weber, et al. (1964) suggested that spheroidal bodies in the nucleus may represent a cell that has been subject to prolonged stimulation. Weber, et al. (1964), also showed variable evidence of secretory exhaustion, in the form of electron opaque, dehydrated appearing stellate cells with a pycnotic nucleus. Similarly, presence of intranuclear inclusions and their variation in human diseases have been recorded by Leduc and Wilson (1959) and Bouteille et al. (1967). They also concluded that nuclear bodies are related to cellular hyperactivity, the cause of which may be physiological, hormonal, drug induced, viral or tumour induced.

The luteal cells during the next phase of first lactation, show characteristic features of active cells, may be probably due to their revival. Such revival of the cells may be under the action of luteinizing hormone, which has the capacity to differentiate the cells of corpus luteum into the small and the large luteal cells (Donaldson and Hansel 1965). Based on our ultrastructural observations of corpus luteum during first lactation, we suggest that the revival could be induced steroidogenically, which is in accordance with the observation of Knobil (1973) during his study on the regulation of the primate corpus luteum during various phases of reproductive cycle.

Parallelly during the existence of a fully active corpus luteum of first lactation, a new corpus luteum is formed in the contralateral ovary due to pregnancy and is designated as corpus luteum of second early pregnancy. This corpus luteum shows the nuclear structure characteristic of an active steroidogenic cell. The first lactation phase overlaps with the second early pregnancy phase, both of which show active steroidogenic cells thereby suggesting the release of progesterone from both the corpora lutea. This also correlates with the high level of progesterone recorded by Date (1996) during this phase.

This newly formed corpus luteum of second early pregnancy is ultrastructurally different from that of the first early pregnancy. The polyhedral shape of active steroidogenic cells of this phase is similar to that of the active corpus luteum of human during high progesterone secretion (Crisp et al. 1970; Niswender et al. 1972; Paavola 1977).

The nuclear morphology of cells of corpus luteum in second midterm pregnancy resembled the first midterm pregnancy in some of the common aspects, such as round nuclei with the presence of perichromatin and interchromatin granules, suggesting them as nuclei of active steroidogenic cells. This view is in accordance with Crisp et al. (1970). The corpus luteum of first midterm pregnancy, however, did not show the presence of intranuclear inclusion as seen during second mid-pregnancy.

The steroidogenic cells of second lactation did not show any signs of revival as shown by the cells of previous lactation phase. These cells instead, showed more degenerative changes such as distorted heterochromatic nucleus and sparse perichromatin granules. This presence of sparse perichromatin granules suggests decreased progesterone synthesis, which is in agreement with Ghadially and Roy (1969) and Belt and Pease (1956). These changes perfectly match with the hormonal assay carried out by Date (1996), in *Rousettus leschenaulti.*

Mitochondria are present abundantly in different forms, in the cells of corpus luteum, asserting their importance in steroid synthesis. Fine structure of mitochondrial morphology of steroidogenic cells indicates the true physiological status of a cell because mitochondria are the rate-limiting cell organelle since they possess side-chain cleaving enzymes for conversion of cholesterol to pregnenolone. The mitochondria show severe myelination in persisted corpus luteum of second pregnancy after ovulation in the contralateral ovary, which distinctly indicates these cells are in degenerative state, after completing extensive steroid synthesis or after completing their role as energy suppliers. Swollen mitochondria during the same stage indicate reduced metabolic activity at the end of steroidogenic activity marking the beginning of corpus luteum regression (Ghadially 1988).

Fine structure of corpus luteum during anestrous phase show a vesiculated rough endoplasmic reticulum which is indicative of exhaustion of the organelle due to over-activity in the previous phase as mentioned by Ghadially (1988), who described the vesiculation of rough endoplasmic reticulum during arrested protein synthesis, which is a step towards atrophy. Well-developed rough endoplasmic reticulum during persisted corpus luteum of second pregnancy after ovulation in the contralateral ovary is suggestive of increased proteinaceous secretion as a requisite for stimulation of progesterone synthesis by luteinizing hormone and for the presence of cyclic AMP in the corpus luteum (Kumar et al. 1978). The scattering of ribosomes during the same phase supports our above statement, which is in correlation with the observations of Ghadially (1988).

The new corpus luteum that is formed simultaneously in the contralateral ovary, shows sparse single cisternae of rough endoplasmic reticulum suggesting, it to be in developing stage. Rough endoplasmic reticulum in second mid-pregnancy is an indistinct feature supported by absence of proteinaceous secretion during this phase as suggested by Brar et al. (1994b).

An abundance of smooth endoplasmic reticulum has been reported in all lutein cells as a consistent feature of steroidogenically active cells (Christensen and Gillim 1969). It is difficult to isolate smooth endoplasmic reticulum from rough endoplasmic reticulum and study it independently, as there are continuities between the cisternae of both types of endoplasmic reticulum. An extensive development of smooth endoplasmic reticulum is a characteristic feature of active steroid secretors (Enders 1973). Smooth endoplasmic reticulum usually shows a compartmentalized arrangement in steroidogenic cells, occupying the cell periphery during active stage (Niswender and Nett 1988; Niswender et al. 1994; Meyer 1991). Unlike rough endoplasmic reticulum, smooth endoplasmic reticulum is more persistent, and it does not disappear even during inactive phases. Smooth endoplasmic reticulum changes its forms during various phases of reproductive cycle, confusing many workers, for which some of them who have tried getting some solutions are Hashimoto and Wiest (1969); Wiest and Kidwell (1969); Guraya (1971a) and Strauss et al. (1972).

The significant presence of lipid droplets could be due to the extensive utilization of lipids as precursors to progesterone synthesis, which is in accordance with Crisp and Francis; Everett (1947) and Crisp and Browning (1968). Presence of numerous secretory granules during this phase supports our above statement that, steroids are being synthesized by utilizing lipids. Neill et al. (1969) have reported presence of such secretory granules along with decreased lipid droplets.

Autophagic vacuoles are the structures that bring about self-destruction while lysosomal action brings about heterophagy, both of which are seen during luteal cell regression (Moor et al. 1970; Gemmell et al. 1976). Thus, abundant autophagic vacuoles present in the persisted corpus luteum of second pregnancy during anestrous stage, persisting corpus luteum after ovulation in contralateral ovary and second full-term pregnancy are suggestive of regressive changes as recorded by Quatacker (1971) in human corpus luteum and by Levine et al. (1979), in cycling mares. The coexistence of autophagic vacuoles along with large lipid droplets confirm the progression of regression of corpus luteum, which is also observed by Paavola (1977), in corpus luteum of Guinea pig. The presence of autophagic vacuoles in corpus luteum of first lactation and in corpus luteum of early second pregnancy during lactation is difficult to understand as the progesterone level during these phases is high and the other cell organelles do not show any sign of regression.

Fine structure of corpus luteum of *Scotophilus heathi*, during various phases of reproductive cycle, resembles well with the observations of the corpus luteum for corresponding phases in other mammalian species. However, some of the structures or organelles in the corpus luteum of *Scotophilus heathi* are unusual for a particular phase of reproductive cycle in comparison to other mammals, hence unique to *Scotophilus heathi*.

Our study confirms the previous reports that the corpus luteum volume is relatively constant during early pregnancy but increases during mid-pregnancy followed by a decrease in the volume towards term (Bassett 1949; Cavazos et al. 1969; Uchida et al. 1984; Meyer and Bruce 1979). Our ultrastructural observations show that the steroidogenic cells are the largest cells in the corpus luteum of *Scotophilus heathi* as compared to the rest of the cell types. Steroidogenic cells of only one type have been observed in corpus luteum of *Scotophilus heathi*. This is similar to observations by Enders and Lyons (1964), in rats; Goodman et al. (1968), in sow; Long (1973), in rats and mares; Harrison (1946), and Paavola (1977), in guinea pig. In contrast to our observations, however, Corner, (1921) in sow; Belt et al. (1970), in pig; Crisp et al. (1970), in human; Sinha et al. (1971b), in white-tailed deer; Mossman and Duke (1973), in horse, llama, deer, goat and sheep; O'Shea et al. (1979), in sheep; O'Shea, (1987) in sheep; goat cow and buffalo; HildPetito, et al. (1989), in monkey; mentioned the presence of two types of cells based on their origin as theca lutein cells and theca granulosa cells. Some of the investigators have made a mention of not one or two types of luteal cells but five types (Foley and Greenstein 1958 in cow, Thwaites and Edey 1969 in ewe).

Unlike in *Rousettus leschenaulti*, in *Scotophilus heathi*, the round shape of the nucleus is maintained all throughout pregnancy except at term when the nuclear envelope undergoes invagination. The invagination of nuclear envelope results in the formation of a distorted and crenated nucleus, which is a characteristic feature of pycnotic nuclei that marks the beginning of regression of corpus luteum. This corroborated by necrotic changes observed by Ghadially and Roy (1969), in necrotic chondrocytes from articular cartilage of rabbits.

Fragmentation of nucleus is observed before mid-pregnancy and is a feature unusual when compared to corpus luteum of other mammals of that stage. Weakly (1968), has reported the presence of lobulation of nuclei and pinocytotic vesicles, in hamster ovary during full-term pregnancy. Nuclear fragmentation is suggestive of high activity of cell, which is supported by the presence of abundant secretory granules in the cell. It might be one of the means to overcome the high activity of cells (Ghadially 1988).

Mitochondria of early pregnancy are abundant and distinctly show varied shapes. There is no particular significance linked to the variations seen in mitochondrial shape in steroidogenic cells, according to Crisp, et al., (1970). Similar mitochondrial variations have been observed by Priedkalns and Weber (1968a); Adams and Hertig (1969a, b); Cavazos et al. (1969); Belt et al. (1970) and Koering et al. (1973). The mitochondria of active phase show tubular cristae, which are similar to the findings of Belt and Pease (1956), who have noted it as a characteristic feature of steroid secretors.

Mitochondrial number shows a slow decline towards termination of pregnancy so also it shows variation in its positions and association with other cell organelles. The close association of mitochondria with Golgi complex during early pregnancy justifies its function as a major organelle participating in steroidogenesis (Christensen and Gillim 1969; Flint and Armstrong 1971). The decline in the number of mitochondria after early pregnancy through term correlates with the increased and decreased progesterone level during early pregnancy and term pregnancy supporting its participation in progesterone secretion. This is because the mitochondrial enzymes that bring about cleavage of side chains of cholesterol form a rate-limiting step in the biosynthesis of steroid (Kortiz 1962; Garren et al. 1971; Savard 1973).

The distribution of smooth endoplasmic reticulum all throughout the pregnancy is random unlike in rat, rabbit and human where it is compartmentalized and present only in the peripheral region of the cell (Green and Maqueo 1965; Koizumi 1965; Gillim et al. 1969). This also is different from our observations in *Rousettus leschenaulti* where smooth endoplasmic reticulum shows compartmentalized arrangement. The tubular form of smooth endoplasmic reticulum observed by us in *Scotophilus heathi* is a normal component of the luteal cells as discussed by Enders, (1962). Similar observations have been recorded in corpus luteum in mouse by Enders (1962) & Yamada and Ishikawa (1960); Crisp and Browning (1968) in rat; Blanchette (1966) in rabbit; Yates et al. (1967) in hamster; Bjersing (1967) in pig; Priedkalns and Weber (1968a, b) in cow and Adams and Hertig, (1969a, b) in human. The typical honeycomb pattern of smooth endoplasmic reticulum observed in *Scotophilus heathi* during midterm pregnancy is also reported by Goodman et al. (1968) in sow and Crisp et al. (1970) in human.

The most distinct feature seen in the *Scotophilus heathi* is the concentrically whorled smooth endoplasmic reticulum seen from mid-pregnancy till term. Similar findings have been reported in corpus luteum of many mammals. (Corner 1919 in pig; Carr and Carr 1962 in interstitial cells of mouse; Christensen 1965; Christensen and Fawcett 1966 in testicular interstitial cells of guinea pig; Blanchette 1966 and Bjersing 1967 in rabbit interstitial cells of guinea pig; Vacek 1967 and Crisp et al. 1970 in granulosa luteal cell of human and Enders 1973 in luteal cells of ferret). Bjersing (1967) related the presence of whorls of smooth endoplasmic reticulum, to increased steroidogenesis in luteal cells of sow. Our observations of whorled smooth endoplasmic reticulum during mid-pregnancy to term pregnancy are not in a similar line of thought to that of Bjersing (1967), as progesterone levels begin to decline showing other ultrastructural changes indicating regression of corpus luteum. Thus, our observations are not in line with the observations of Bjersing (1967) and Paavola (1977), who observed whorls during maximum progesterone secretion. Our observations, on the contrary, emphasize the presence of whorled smooth endoplasmic reticulum during post active phase of corpus luteum. This is in agreement with Leavitt et al. (1973), who described the presence of such whorls as indicators of initiation of luteolysis in hamster corpus luteum. Similar observations have been recorded by Neill et al. (1967), who have mentioned the failure to observe whorled smooth endoplasmic reticulum during high progesterone secreting phase of menstrual cycle. Adams and Hertig (1969a) have made a mention of such whorled smooth endoplasmic reticulum only during regression which is in complete agreement with our observations. Enders (1973) has suggested that the whorled arrangement of smooth endoplasmic reticulum along with mitochondria is an inefficient system for interchange of steroid molecules between the two organelles which results in reduced steroid synthesis.

At term pregnancy in *Scotophilus heathi* the presence of rarefied mitochondria with smooth endoplasmic reticulum probably enables the termination of progesterone synthesis. This is caused by blocking the conversion of cholesterol to pregnenolone, due to the absence of the enzyme 20 alpha hydroxylase in mitochondria which forms the rate-limiting factor (Kortiz 1962; Sulimovici and Boyd 1969; Garren et al. 1971; Savard 1973). The correlation of increased ribosome population with polypeptide synthesis has been reported in the adrenal cortical steroid secreting cells of rats and dogs (Fiala et al. 1956; Bransome Jr and Reddy 1961).

Lipids share a major role in the steroidogenic cells of corpus luteum as they store the steroid hormone precursors namely cholesterol and cholesterol esters. Lipid droplets form a characteristic feature of many steroid secreting cells (Christensen and Gillim 1969), and their presence in granulosa lutein cells is well documented in humans (Van Lennep and Madden 1965; Adams and Hertig 1969a, b), in cow (Priedkalns and Weber 1968b), in pig (Bjersing 1967; Cavazos et al. 1969; Belt et al. 1970) and in rabbit (Blanchette 1966). The absence or presence of lipid droplets, however, is strongly influenced by various hormonal stimuli (Everett 1947).

Ultrastructural studies on the corpus luteum at various phases of reproduction in two cyclical breeders (*Rousettus leschenaulti* and *Scotophilus heathi*) clearly show the involvement of the tissue in steroidogenesis. The changes observed in our studies correlate with those reported in other animals. Some of the unique ultrastructural characteristics do not correlate with the hormonal levels during the respective stages of gestation. These need further evaluation to better understand their physiological and functional roles.

References

Adams EC, Hertig AT (1969a) Studies on human corpus luteum; observations on ultrastructure of development and regression of the luteal cells during menstrual cycle. J Cell Biol 41:696–715

Adams EC, Hertig AT (1969b) Studies on human corpus luteum; observations on ultrastructure of the luteal cells during pregnancy. J Cell Biol 41:716–735

Alila HW, Hansel W (1984) Origin of different cell types in the bovine corpus luteum as characterized by specific monoclonal antibodies. Biol Reprod 31:1015–1025

Allen WM, Wintersteiner O (1934) Crystalline progestin–Sc 80:190–191

Anderson JW (1969) Ultrastructure of the placenta and fetal membranes of the dog. I the placental labyrinth. Anat Rec 165:15–36

Azmi TI, O'Shea JD, Bruce NW, Rodgers RJ (1984) Morphometry of the functional and regressing corpus luteum of Guinea pig. Anat Rec 210:33–40

Bassett DL (1949) The lutein cell population and mitotic activity in the corpus luteum of pregnancy in the albino rat. Anat Rec 103:597–609

Belt WD, Cavazos LE, Anderson LL, Kraeling RR (1970) Fine structure and progesterone levels in the corpus luteum of pig during pregnancy and after hysterectomy. Biol Repd 2:98–113

Belt WD, Pease D (1956) Mitochondiral structure in sites of steroid secretion. J Biophy Biochem Cytol 2:369–374

Bhiwgade DA (1977) Development of the fetal membranes in the Indian horse-shoe bat, *Rhinolophus rouxi* (Temminck). Proc Ind Acad Sci 86:61–72

Bhiwgade DA (1989) Ultrastructural studies: the chorio-allantoic placental barrier and anterior pituitary in some Indian chiroptera and the effect of cyproterone acetate on testis and epididymis of rat: DSc thesis Nagpur

Bhiwgade DA (1990) Comparative electron microscopy of chorio-allantoic placental barrier in some Indian Chiroptera. Acta Anat 138:302–317

Bhiwgade DA (1991) The development of the chorioallantoic placental barrier in the Indian fruit bat. *Rousettus leschenaulti* (Desmarest). Proc Ind Acad Sci. (submitted)

Bhiwgade DA, Singh AB, Manekar AR, Menon SN (1992) Ultrastructural development of chorioallantoic placenta in the Indian Miniopterus bat, *Miniopterus schreibersii fuliginosus* (Hodgson). Acta Anat 145: 248–264

Bhiwgade DA, Taskar AS, Bhatnagar KP (2000) The interhemal membrane of the bats, Ronsettus leschenaultia and Cynopterus sphinx (megachiroptera): an ultrastructural investigation. Acta Chiropterologica 2(1):107–122

Bhiwgade DA, Thakur VA (1992) The fine structure of term placenta in Indian molossid bat, *Chaerophon plicata* IXth Intematioanl Bat Research conference Madurai, Abst. No. 13

Bjersing L (1967) On the ultrastructure of granulosa lutein cells in porcine corpus luteum with special reference to endoplasmic reticulum and steroid hormone synthesis. Z Zellforsch Mikrosk Anat 82:187–211

Bjorkman NH, Wimsatt WA (1968) The allantoic placenta of the vampire bat (*Desmodus rotundus murinus*): a reinterpretation of its structure based on electron microscopic observations. Anat Rec 162:83–97

Blanchette EJ (1966) Ovarian steroid cells: differentiation of lutein cell. J Cell Biol 31:501–516

Bodley HD (1974) Ultrastructural development of the chorioallantoic placental barrier in the bat, *Macrotus waterhousii*. Anat Rec 180:351–368

Body JW, Hamilton WJ (1966) Electron microscopic observations on the cytotrophoblast contribution to the syncytium of the human placenta. J Anat 100: 535–548

Bouteille M, Kalifat SR, Delarue J (1967) Ultrastructural variations of nuclear bodies in human diseases. J Ultrastruct Res 19:474

Branca A, Recherchcs sur la placentation des chdiropteres (1927) Archs Zool Exp 66:291–450

Bransome ED Jr, Reddy WJ (1961) Endocrinology 69:997

Brar AS, Guraya SS, Khera KS (1994a) Large luteal cell polymorphism and ultrastructure of corpus luteum of buffalo (*Bubalus bubalis*). In: National Symposium of reproductive health care. University of Rajasthan, Jaipur, India, p 147

Brar AS, Guraya SS, Khera KS (1994b) Histomorphology of corpus luteum of goat. Small Ruminant Res

Brar AS, Khera KS, Guraya SS (1994) Immunohistochemical localization of inhibin, relaxin and oxytocin in the corpus luteum of goat. In: Ovulation Inductions: Basic Science and Clinical Advances. January 20–22, 1994, Plam Beach, Florida, U. S. A. Abstract 36

Bronstein RD, Holt JA, Zirkin BR (1984) Estradiol induced changes in progesterone secretion by rabbit corpora lutea are associated with quantitative ultrastructural changes in luteal cells. Endocrinology 115: 342–348

Carpenter SJ (1972) Light and electron microscopic observations on the morphogenesis of the chorioallantoic placenta of the golden hamster (*Cricetus auratus*). Days seven through nine of gestation. Am J Anat 135: 445–476

Carr, I. And Carr, J., 1962. Membranous whorls in the testicular interstitial cells. Anat Rec, 144: 143–147

Cavazos LF, Anderson LL, Belt WD, Henricks DM, Kraeling RR, Melampy RM (1969) Fine structure and progesterone levels in the corpus luteum of the pig during the estrus cycle. Biol Reprod 1:83–106

Cavazos LF, Melampy RM, Anderson LL, Wehl SK (1968) Fine structure of luteal cells of pig. Anat Rec 160:328–329

Chari A, Gopalakrishna A (1984) Morphogenesis of the fetal membranes and placentation in the bat. *Miniopterus schreibersii fuliginosus* (hod- son). Proc Ind Acad Sci (Anim) 93:463–483

Christensen AK (1965) The fine structure of testicular interstitial cells in Guinea pigs. J Cell Biol 26:91935

Christensen AK, Fawcett DW (1966) The fine structure of testicular interstitial cells in mice. Am J Anat 118:551–572

Christensen AK, Gillim SW (1969) The correlation of fine structure and function in the steroid–secreting cells with emphasis on those of the gonads. The Gonads-K. W. Mc. Kerns, eds., 415–488

Clellan MC, Diekman MA, Abel JH Jr, Niswender GD (1975) Luteinizing hormone, progesterone and the morphological development of normal and superovulated corpora lutea in sheep. Cell Tiss Res 164:291–307

Corner GW (1919) The corpus luteum of sow from both granulosa and theca interna. Am J Anat 26:117–183

Corner GW (1921) Cyclic charges in the ovaries uterus of the sow and their relation to the mechanism of implantation. Contr Embryol Carngie Inst 13:117–146

Creed RFS, Biggers JD (1963) Development of the raccoon placenta. Am J Anat 113:417–431

Crisp TM, Browning HC (1968) The fine structure of corpora lutea in ovarian transplants of mice following luteotrophin stimulation. Am J Anat 122:162–192

Crisp TM, Dessouky DA, Denys FR (1970) The fine structure of the human corpus luteum of early pregnancy and during progestational phase of mentrual cycle. Am J Anat 127:3770

Csapo AI, Pulkkinen MO, Ruttner B, Sauvage JP, Wiesk WG (1972) The significance of human corpus luteum in pregnancy maintenance. Am J Obstet Gynecol 112: 1061–1067

Cukierski MA (1987) Synthesis and transport studies of the intrasyncytial lamina: an unusual placental basement membrane in the little brown bat *Myotis lucifugus lucifugus*. Am J Anat 178:387–409

Danell B (1987) Oestrous behaviour, ovarian morphology and cyclical variation in follicular system and endocrine pattern in water buffalo heifers, Ph.D. Thesis pp. 1–124. Swedish University of Agricultural Sciences. Uppsala, Sweden

Deane HW, Rubin BL, Driks EC, Lobel BL, Leipsner G (1962) Trophoblast giant cells in placentas of rats and mice and their probable role in steroid-hormone production. Endocrinology 70:407–419

Dempsey EW, Wislocki GB (1956) Electron microscopic observations on the placenta of the cat. J Biophys Biochem Cytol 2:734–754

Donaldson L, Hansel W (1965) Histological study of bovine corpora luteal. J Dairy Sci 48:905–909

Enders AC (1960) Electron microscopic observations on the villous haemochorial placenta of the ninebanded armadillo (*Dasypus novemcinctus*). J Anat 94:205–215

Enders AC (1962) Observations on the fine structure of lutein cells. J Cell Biol 12:101–113

Enders AC (1965) A comparative study of the fine structure of the trophoblast in several haemochorial placentas. Am J Anat 116:29–68

Enders AC (1973) Cytology of the corpus luteum. Biol Reprod 8:158–182

Enders AC, Lyons WR (1964) Observation on the fine structure of lutein cells. II. The effect of hypophysectomy and mammotrophic hormone in rat. J Cell Biol 22:127–141

Enders AC, Wimsatt WA (1968) Formation and structure of the hemodichorial chorioallantoic placenta of the bat (*Myotis lucifugus lucifugus*). Am J Anat 122:453–490

Everett JW (1947) Endocrinology of Reproduction: Endocrine interaction during the different phases of female reproductive cycle in the Indian fruit bat, Rousettus leschenaultia. Ph. D. thesis (Mumbai University). 88: 805

Farin CE, Moeller CL, Sawyer HR, Niswender GD (1989) Analysis of cell types in corpus luteum of the sheep. J Reprod Fert 37:181–187

Fiala S, Sproul EE, Fiala AE (1956) J Biophys Biochem Cytol 2:115

Fields PA (1991) Relaxin and other luteal secretory peptides: cell localization and function in the ovary. Ultrastructure of ovary–Famillari, G; Makabe, S and Netta, P.N. Eds. 177–198

Fields MJ, Barros CM, Watkins WB, Fields PA (1992) Characterization of large luteal cells and their secretory granules during the estrous cycle of the cow. Biol Reprod 46:535–545

Fitz TA, Mayan MH, Sawyer HR, Niswender GD (1982) Characterization of two steroidogenic cell types in ovine corpus luteum. Biol Repord 27:703–711

Flint APF, Armstrong DT (1971) Nat New Biol 231:60

Foley RC, Greenstein JS (1958) Cytological changes in the bovine corpus luteum during early pregnancy. In: Gassner FX (ed) Reproduction and infertility. Pergammon Press, New York, pp 88–96

Fuller GB, Hansel W (1971) Proc Soc Exp Biol Med 137: 539

Gammal GB (1985) Syncytial channels in the villous trophoblast of the macaque. J Anat 141:181–191

Garren LD, Gill GN, Masui H, Walton GM (1971) On the mechanism of action of adreno cortico trophic hormone. Recent Prog, Hormone Res 27:443

Garverick HA, Smith MF (1986) Mechanisms associated with subnormal luteal function. J Anim Sci 62:92–105

Gemmell RT, Stacy BD, Thorburn GD (1976) Morphology of the regressing corpus luteum in ewe. Biol Reprod 14:270–279

Ghadially FN (1988) Ultrastructural pathology of the cell and matrix. Volume I & II

Ghadially FN, Roy S (1969) Ultrastructure of synovial joints in health and disease. Butterworths, London

Gillim SW, Christensen AK, Mclennan CE (1969) Fine structure of human menstrual corpus luteum of its stage of maximum secretory activity. Am J Anat 126:409–428

Goodman P, Latta JS, Wilson RB, Kadis B (1968) The fine structure of sow lutein cells. Anat Rec 161:77–90

Gopalakrishna A, Chari GC (1983) A review of the taxonomic position of Miniopterus based on embryological characters. Curr Sci 50:152–154

Gopalakrishna A, Karim KB (1979) Fetal membranes and placentation in Chiroplera. J Reprod Fertil 56:417–429

Gopalakrishna A, Karim KB (1980) Female genital anatomy and morphogenesis of fetal membranes of Chiroptera and their bearing on the phylogenetic relationships of the group. Golden Jubilee commemoration volume. Nat Acad Sci, Allahabad, India, pp 379–428

Gopalakrishna A, Khaparde MS (1978) Development of the foetal membranes and placentation in the Indian false vampire bat, Megaderma lyra lyra (Geoffroy). Proc Ind Acad Sci 87:179–194

Gopalakrishna A, Moghe MA (1960) Development of the foetal membranes in the Indian leaf-nosed bat, Hipposideros bicolor pallidus. Zeitschrift fur Anatomie und Entwicklungsgeschichte 122:137–149

Gopalakrishna A, Pendharkar YP, Badwaik NB (1989) Morphogenesis of the fetal membranes and placentation in the Indian molossid bat Chaerephon plicata (Buchanan). Proc Ind Acad Sci (Anim Sci) 98:149–166

Green JA, Maqueo M (1965) Ultrastructure of human ovary. Am J Obstet Gynec 92:946–957

Grosser O (1927) Friihentwicklung, Eihautbildung und Placentation des Menschen und der Saiigetiere. Bergman, Munich

Gulamhussein AP, Beck F (1975) Development and structure of the extraembryonic membranes of the ferret. Alight microscopic and ultrastructural study. J Anat 120:349–365

Guraya SS (1968) Histochemical study of human corpus luteum at term. Am J Obst Gynec 102:219221

Guraya SS (1971a) Morphology, histochemistry and biochemistry of human ovarian compartments and steriod hormone synthesis. Physiol Rev 51:785–807

Guraya SS (1971b) Histochemical changes of nucleic acids in the follicles of rabbit ovary before and during ovulation. J Reprod Fert 24:107–108

Guraya SS (1972) Function of the human ovary during pregnancy as revealed by the histochemical biochemical and electron microscope techniques. Acta Endocr (kbh) 69:107–118

Guraya SS (1978) Recent advances in the morphology, histochemistry, biochemistry and physiology of bovine ovarian compartments and steroid hormone synthesis. J Anim Morph Physiol 86:103

Guraya SS (1991) Interstitial cells. In: Famillari S, Motta PM (eds) Ultrastructure of the ovary. Kluwer academic publishers, Boston, pp 199–224

Hansel W, Dowed JP (1986) New concepts of the control of corpus luteum function. J Reprod Fertil 78:755–768

Harrison RJ (1946) Early development of corpus luteum in mare. J Anat 80:160–166

Harrison RJ, Young BA (1961) Specialization in the pinniped placenta. J Anat 95:450

Hashimoto, I. And Wiest, W. G., 1969. Luteotrophic and luteolytic mechanisms in rat corpora lutea. Endocrinology, 84: 886–892

Hilliard J (1973) Corpus luteum function in Guinea pigs, hamsters, rats, mice and rabbits. Biol Reprod 8:203–221

Hillier SG (1994) Hormonal control of folliculogenesis and luteinization. In: Findlay JK (ed) Molecular biology of human female reproductive system. Academic Press, San Diego, pp 1–37

Horta H, Gordillo FJ, Soto de Leon B, Cortex-Gallegoz V (1977) Direct evidence of luteal insufficiency in women with habitual abortion. Obstet Gynecol 49(705):708

Horvath E, Kovacs K (1988) Fine structural cytology of the adenohypophysis in rat and man. J Electron Micro Tech 8:401–432

Jollie WP (1973) Fine structural changes in the placental membrane of the marmoset with increasing gestational age. Anat Rec 176:307–320

Jones GS (1949) Some newer aspect of management of infertility. Med Asso:1123–1129

Karim KB, Bhatnagar KP (1996) Observations on the chorioallantoic placenta of the Indian flying fox, Pteropus giganteus giganteus. Ann Anat 178:523–530

Karsch FJ, Roche JF, Noveroske JW, Foster DL, Norton HW, Nalbondov AV (1971) Prolonged maintenance of corpus luteum of ewe by continuous infusion of luteinizing hormone. Biol Reprod 4:129–136

Kempcrmann CT (1929) Die Placenta der Fledermaus Miniopterus schreibersii, und ihre functionelle Bedeutung. Z Anat Entw Gesch 91:292–303

Keyes PL, Nalbandov AV (1968) Endocrine function of the ovarian interstitial gland of the rabbits. Endocrinology 82:799–804

Kimura K, Uchida TA (1984) Development of the main and accessory placentae in the Japanese long-fingered bat, Miniopterus schreibersii fuliginosus. J Reprod Fertil 71:119–126

King BF, Mossman HW (1974) The fetal membranes and unusual giant cell placenta of the jeroba (Jaculus) and jumping mouse (Zapus). Am J Anat 140:405–432

King BF, Tibbitts FD (1976) The fine structure of the chinchilla placenta am. J Anat 145(1):33–56

Kirby DRS, Bilington WD, Bradsbury S, Goldstein BJ (1964) Abtigen barrier in mouse. Nature 204:548

Knobil E (1973) On the regulation of the primate cycle. Biol Reprod 8:222–245

Koering MJ (1974) Comparative morphology of primate ovary. In: Luckett WP (ed) Contribution to primatology, vol 3. Kagel Basel, pp 33–81

Koering MJ, Wolf RC, Meyer RK (1973) Morphological changes in the corpus luteum corelated with progestin levels in the rhesus monkey during early pregnancy. Biol Reprod 9:254–271

Koizumi T (1965) Electron microscopic study of corpus luteum in rabbits and rats. Folia Endocrinol Jop 41: 994–1009

Koopman KF (1993) Order Chiroptera. pp. 137–241. In: Wilson DE, Reeder DM (eds) Mammal species of the world: a taxonomic and geographic reference, 2nd edn. Smithsonian Institution Press, Washington, D.C, p 1207

Kortiz SB (1962) The effect of calcium ions and freezing on the in vitro synthesis of pregnendone by rat adrenal preparations. Biochem Biophys Acta 56:63

Kumar A, Das M, Manchanda SK (1978) Effect of PGF and cyclic AMP on production HCGstimulated progesterone by bovine luteal cells. Ind J Med Res 85:665–669

Lauderdale JW (1986) A review of patterns of change in luteal function. J Anim Sci 62:79–91

Lawn AM, Chiquoine AD (1965) Ultrastructure of the placental labyrinth of ferret (Mustella putorius furo). J Anat 99:47–69

Leavitt WW, Basom CR, Bagwell JN, Blaha GC (1973) Structure and function of the hamster corpus luteum during pregnancy. Am J Anat 136:235–250

Leduc EH, Wilson JW (1959) An electron microscopic study of intra nuclear inclusions in mouse liver and hepatoma. J Histochem Cytochem 7:8

Levine H, Wight T, Squires E (1979) Ultrastructure of the corpus luteum of the cycling mare. Biol Reprod 20: 492–504

Long JA (1973) Corpus luteum of pregnancy in the rat-ultrastructural and cytochemical observation. Biol Reprod 8:87–99

Luckett WD (1970) The fine structure of the placental villi of rhesus monkey (Macaca mulata). Anat Rec 167: 141–164

Mahaiey J, Manekar AP, Panse C, Bhiwgade DA (1995) The chorioallantoic placental barrier in the Indian Vespertilionidae bat, Tylonycteris pachypus. Bat Res News 36:85

Mahaley JA (1998) In plantation, development of the fetal membranes and ultrastructure of the placenta of bat, Tylonycteris pachypus (Tem).: Ph.D. thesis,. University of Mumbai

Malassine A (1970) Etude histologique et ultra- structurale du disque placentaire de Miniooptere, Miiniopterus schreibersii (chiroptere): sa nature endothelio-choriale, son caractere endocrine. Archives d'Anatomie Microscopique et de Morphologie Experimentale 59: 99–112

Mandal A (1991) Ultrastructural studies on the chorioallantoic placenta and pituitary in the Indian mouse tailed bat, Rhinopoma microphyllum. Ph.D. thesis,. University of Mumbai

Meyer GT (1991) Ultrastructural dynamics during corpus luteum development and growth. In: Famillari G, Makabe S, Motta FM (eds) Ultrastructure of the ovary. Kluwer Academic Publishers, Boston, pp 161–176

Meyer GT, Bruce NW (1979) The cellular pattern of corpus luteal growth during pregnancy in rat. Anat Rec 193:823–830

Midgley AP, Pierce GB, Denaue GA, Gosling JRG (1963) Morphogenesis of syncytiotrophoblast in vivo: an autoradiographic demonstration. Science 141:349–350

Moor RM, Hay MF, Short RV, Rowson LE (1970) The corpus luteum of sheep: effect of uterine removal during luteal regres. J Reprod Fertil 21:319–326

Mossman HW (1987) Vertebrate fetal membranes: comparative ontogeny and morphology; evolution; phylogenetic significance; basic functions; research opportunities. Rutgers University Press, New Jersey, p 383

Mossman HW, Duke KL (1973) Comparative morphology of the mammalian ovary. University of Wisconsin Press. 1973: 117–230

Neill JD, Johansson EDB, Datta JK, Knobil E (1967) Relationship between plasma levels of luteinizing hormone and progesterone during the normal menstrual cycle. J Clin Endo Metab 27:1167

Neill JD, Johansson EDB, Knobil E (1969) Failure of hysterectomy to influence the normal pattern of cyclic progesterone secretion in the rhesus monkey. Endocrinology 84:464–465

Niswender GD, Juengel JL, Guire WJ, Belfiore CJ, Wiltbank MC (1994) Luteal function: the estrous cycle and early pregnancy. Biol Reprod 50:239–247

Niswender GD, Menon KMJ, JafFe RB (1972) Regulation of corpus luteum during the menstrual cycle and early pregnancy. Fert Ster 23(6):432–439

Niswender GD, Nett TM (1988) Corpus luteum and its control in intraprismate species. In: Knobil E, Neill JD (eds) Physiology of reproduction, vol 1, pp 781–816

Niswender GD, Schwall RH, Fitz TA, Farm CE, Sawyer HR (1983) Regulation of luteal function in domestic ruminants: new concepts. Rec Prog Horm Res 41:101–152

O'Shea JD (1987) Heterogeneous cell types in the corpus luteum of sheep, goats and cattle. J Reprod Fert 34:71–85

O'Shea JD, Cran DJ, Hay MF (1979) The small luteal cell of the sheep. J Anat 128:239–251

Oduor-Okelo D, Musewe VO, Gombe S (1983) Electron microscope study of the Chorio allantoic placenta of the rock-hyrax (Heterohyrax brucei). J Reprod Fertil 68:311–316

Paavola LG (1977) The corpus luteum of Guinea pig. Fine structure at the time of maximum progesterone secretion and during regression. Am J Anat 150:565–604

Paavola LG (1979) The corpus luteum of the Guinea pig IV: fine structure of macrophages during pregnancy and postpartum luteolysis and the phagocytosis of luteal cells. Am J Anat 154:337–364

Parry DM, Wilcox DL, Thorburn GD (1980) Ultrastructural and cytochemical study of the bovine corpus luteum. J Reprod Fert 60:349–357

Pedersen PH, Larsen JF (1968) The ultrastructure of human granlosal lutein cell of the lst trimester of gestation. Acta Endocrinol 58:481–496

Priedkalns J, Weber AF (1968a) Ultrastructure studies of the bovine graffian follicle and corpus luteum. Z Zellforsch 91:554–573

Priedkalns J, Weber AF (1968b) Quantitative ultrastructural analysis of the follicular and luteal cells of the bovine ovary. Z Zellforsch 91:554–573

Quatacker (1971) Formation of autophagic vacuoles during human corpus luteum involution. Z Zellforsch Mikrosk Anat 122:479–487

Rasweiler JJ IV (1991) Development of the discoidal haemochorial placenta in the black mastiff bat, Molossus ater evidence for a role of maternal endothelial cells in control of trophoblastic growth. Am J Anat 191:186–207

Rhodin JA, Terzakis J (1962) The ultrastructure of the human full term placenta. J Ultra Res 6:88–106

Richardson DW, Goldsmith LT, Pohl CR, Schallenberger E, Knobil E (1985) The role of prolactin in the regulation of the primate corpus luteum. J Clin Endocrinol Metab 60:501

Rodgers RJ, O'Shea JD, Bruce NW (1984) Morphometric analysis of the cellular composition of the ovine corpus luteum. J Anat 138(Pt 4):757–770

Rothchild I (1981) The regulation of the mammalian corpus luteum. Recent Prog Horm Res 37:183–298

Saeger W (1992) Effect of drugs on pituitary ultrastructure. Microscopy Res Tech 20:162–176

Sandhu SK (1986) Studies on the embryology of some Indian Chiroptera. PhD dissertation, Nagpur University, India

Savard K (1973) The biochemistry of the corpus luteum. Biol Reprod 8:183–202

Schwall RH, Sawyer HR, Niswender GD (1986) Differential regulation by LH and prostaglandins of steroidogenesis in small luteal cells of the ewe. J Reprod Fert 76:821–829

Shomita SB, Bhiwgade DA (1995) Light and electron microscope study–reexamination of the placental barrier in Rousettus leschenaulti and Cynopterus sphinx gangeticus. Bat Res News 36:110

Short RV, (1977) The discovery of the ovaries (Zuckerman and Wess eds.). The Ovary. 1–39

Short RV, Buss IO (1965) Biochemical and histological observation on the corpora lutea of the African elephant, Loxondonta ajricana. J Repd Fert 9:61

Sinha AA, Seal US, Doe RP (1971a) Fine structure of the corpus luteum of raccoon during pregnancy. Z Zellforsch Mikrosk Anat 117:35–45

Sinha AA, Seal US, Doe RP (1971b) Ultrastructure of corpus luteum of the while tailed deer during pregnancy. Am J Anat 132:189–205

Spies HG, Warren DR, Grier HT (1967) Endocrinology 81:1435

Spies HG, Hilliard J, Sawyer CH (1968) Endocrinology 83:291

Stacy BD, Gemell RT, Thorburn GD (1976) Morphology of the corpus luteum in the sheep during regression induced by prostaglandin F2a. Biol Reprod 14:280–291

Stephens RJ (1969) The development and fine structure of the allantoic placental barrier in the bat, Tadarida brasiliensis cynocephali. J Ultrastruct Res 28:371–398
Stephens RJ, Cabral L (1970) Direct contribution of the cytotrophoblast to the syncytiotrophoblast in the diffuse labyrinthine Endothelio, chorial placenta of bat. Anat Rec 169:243–252
Strauss JF, Foley B, Stambaugh R (1972) 20-a-hydroxysteroid dehydrogenase activity in the rabbit ovary. Biol Reprod 6:78–86
Sulimovici SI, Boyd GS (1969) The cholesterol side chain cleavage enzymes in steroid hormone producing tissue. Vit Horm 27:199–234
Thwaites CJ, Edey TN (1969) Histology of the corpus luteum in ewe: changes during estrous cycle, early pregnancy and in response to some experimental treatments. Am J Anat 129:439–438
Uchida TA, Inoue C, Kimura K (1984) Effects of elevated temperatures on the embryonic development and corpus luteum activity in the Japanese long. Fingered bat, Miniopterus Schreibersii fuliginosus. J Repord Fert 71: 439–444
Vacek Z (1967) Ultrastructure and enzyme histochemistry of the corpus luteum graviditalis and its correlation to the decidual transformation of the endometrium. Folia morph (Praha) 15:375–383
Van Lennep EW, Madden LM (1965) Electron microscopic observations on the involution of the human corpus luteum of menstruation. Z Zellforsch 66:365–380
Weakly BS (1968) Comparison of cytoplasmic lamellae and membranous elements in the oocytes of five mammalian species. Z Zellforsch Mikrosk Anat 85:109–123
Wiest WG, Kidwell WR (1969) The regulation of progesterone secretion by ovarian dehydrogenases. In: McKerns KW (ed) The gonads. Appleton-century-crofts, New York, pp 295–320
Wimsatt WA (1945a) Notes on breeding behavior, pregnancy and parturition in some vespertilionid bats of eastern United States. J Mammal 26:23–33
Wimsatt WA (1945b) The placentation in the vespertilionid bat, Myotis lucifugus lucifugus. Am J Anat 77:1–151
Wimsatt WA (1954) The fetal membranes and placentation of the tropical American vampire bat, Desmodus rotundus murinus. Acta Anat 21:285–341
Wimsatt WA (1958) The allantoic placental barrier in Chiroptera: a new concept of its organization and histochemistry. Acta Anat 32:141–186
Wimsatt WA, Enders AC (1980) Structure and morphogenesis of the uterus, placenta and paraplacental organs of the neotropical disc-winged bat, Thyroptera tricolor spix (Microchiroptera: Thyropteridae). Am J Anat 159: 209–243
Wimsatt WA, Enders AC, Mossman HW (1973) A reexamination of the chorio-allantoic placental membrane of a shrew, Blarina brevicauda: resolution of a controversy. Am J Anat 138:207–238
Wislocki GB, Dempsey EW (1955) Electron microscopy of the placenta of the rat. Am J Anat 123:33–63
Wislocki GB, Wimsatt WA (1947) Chemical cytology of the placenta in two north American shrew (Blarina brevicauda and Sorex fumeus). Am J Anat 81:269–308
Wooding FBP (1982a) Structure and function of placental binucleate ("giant cells"). Bibl Anat 22:134–139
Wooding FBP (1982b) The role of the binucleate cells in ruminant placental structure. J Repro Fertil 31:31–39
Wooding FBP (1984) Role of binucleate cells in fetomaternal cell fusion and implantation in the sheep. Am J Anat 170:233–250
Wooding FBP, Flint APF (1994) Placentation. In: Lemming GE (ed) Marshall's physiology of reproduction, fourth ed. vol. 3: pregnancy and lactation, part 1: ovulation and early pregnancy. Chapman and Hall, London, pp 235–460
Wynn RM, Bjorkman NH (1968) Ultrastructure of the feline placental membrane. Am J Obstet Gynec 102: 34–43
Wynn RM, Davies J (1965) Comparative electron microscopy of the haemochorial placenta. Am J Obstet Gynec 91:533–549
Yamada E, Ishikawa JM (1960) The fine structure of the corpus luteum in the mouse ovary as revealed by election microscopy. Kyushu. J Med Sci 11:235–259
Yates RD, Arai K, Rappoport DA (1967) Fine structure and chemical composition of opaque cytoplasmic bodies of triparanol-treated Syrian hamsters. Exp Cell Res 47:459–478

7 Fine Structure of Placenta in Two Myomorph Rodents

The placenta is the site where the maternal blood (uterine vasculature) and the foetal blood (foetal capillaries) exchange materials including gases. The efficiency of this exchange is a vital factor that influences the development of embryo. The two blood circulations are separated by the placental barrier or the interhemal membrane. We studied the placental structure in two myomorph rodents (*Rattus norvegicus* (Common street rat) and *Bondicoota bengalensis* (Indian mole rat)) at term to make a comparative evaluation of the fine structure of the interhemal membrane.

The placenta in *Rattus norvegicus* shows there are three trophoblast layers; the outer, the middle and the inner layers that separate the maternal blood spaces from foetal blood vessels and from other extraembryonic connective tissues. The placenta can be classified as hemotrichorial in nature. The placenta in *Bandicoota bengalensis* has essentially the same organization as the trophoblast in the placental labyrinth. The placental layers that are encountered can be sequentially listed as follows:

1. Outer cellular layer of trophoblast lining the maternal blood space.
2. Middle syncytial layer of trophoblast.
3. Inner syncytial layer of trophoblast.
4. Basal lamina.
5. Foetal endothelium.
6. Foetal mesenchyme.

7.1 Outer Layer of Trophoblast

The Placental labyrinth clearly shows trilaminar nature of the trophoblasts. The outer trophoblast layer, which is adjacent to maternal blood, is evidently cellular in nature (i.e. cytotrophoblastic) since there are desmosomes that connect these cells to each other and to the plasma membrane of the middle layer.

In advanced pregnancy, the ultrastructural of placenta indicates the gradually progressing thinning of various layers in the placenta along with the appearance of various structural modifications that effectively reduces the diffusion distance between maternal and foetal bloodstreams (Figs. 7.1 and 7.2). In both, rat and Bandicoota, several gaps appear at sites where the layers have shrunken very significantly. These gaps become numerous and wider at this stage of gestation (Figs. 7.3 and 7.4). In many places, it is porous with the pores closed by a diaphragm. This layer of the placental barrier shows signs of degeneration. Some areas of peripheral cytoplasm appear swollen and protruded into maternal blood spaces and the cytoplasm shows hypertrophied mitochondria with dilated cisternae endoplasmic reticulum and Golgi complex.

D. A. Bhiwgade, S. Menon, *Ultrastructural Investigations on the Pituitary-Gonadal Axis*,
https://doi.org/10.1007/978-981-99-3276-4_7

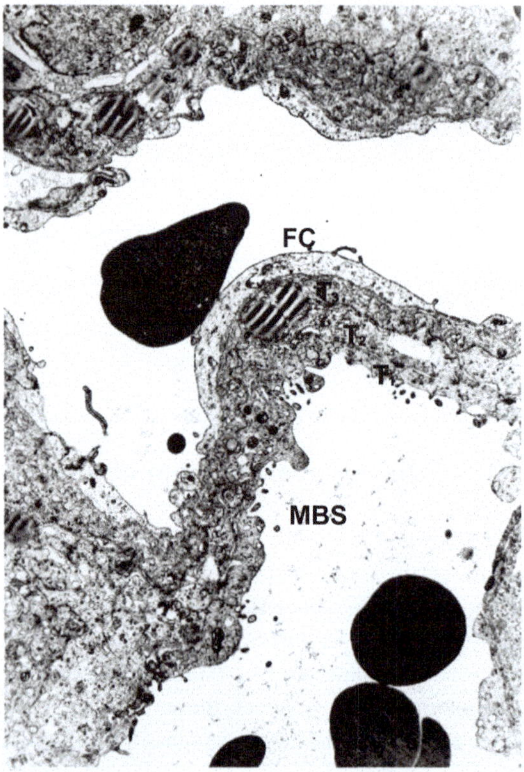

Fig. 7.1 Low-power electron micrograph to show the general structure of the interhemal membrane of rat at advanced pregnancy. Three trophoblastic layers, outer (T1), 1 middle (T2) and inner (T3) separate maternal blood (MBS) from foetal capillary (FC). In the upper left corner, the nucleus (N) is seen. Note sparsely distributed stubby microvilli and large lipid 1 droplets (L) in T3. Note irregularly developed foetal endothelium (FE) [X 3500]. Unpublished electron micrograph from Bhiwgade and Thakur

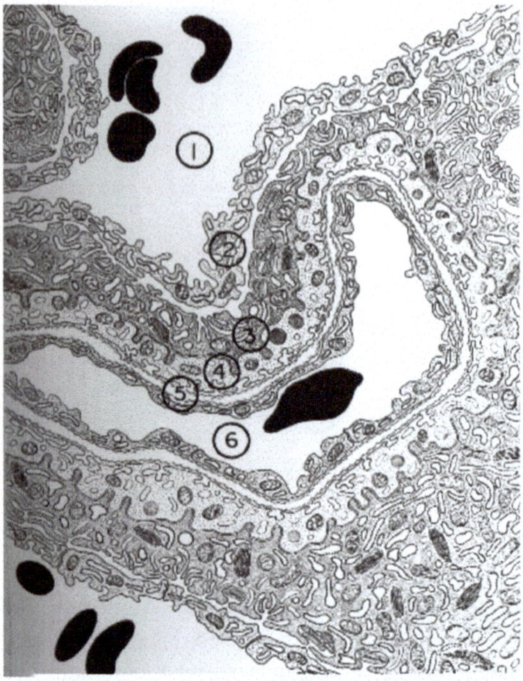

7.2 Middle Layer of Trophoblast

In advanced pregnancy, the middle trophoblastic layer is the thickest and the least electron dense of the three layers of the trophoblast. This layer of the labyrinth shows no evidence of being cellular and assumes to be syncytial in character.

It is very thin and shrunken in many places at this stage of gestation (Figs. 7.1 and 7.2). The outer side of this layer is uneven with many irregular microvilli and invaginations. At some sites, the folding is so deep that it spans the entire width of this layer (Figs. 7.3 and 7.4). The intercellular gap between outer and middle trophoblast layers is filled with electron-dense material. Gap junctions and maculae are more observed between middle and inner layers of trophoblast than in earlier stages of gestation (Figs. 7.1 and 7.2).

7.3 Innermost Layer of Trophoblast

The innermost layer of trophoblast is variable in width with an outer undulating surface closely attached to the middle layer of trophoblast. The basal plasma membrane of the inner layer is with frequent infoldings extending into the interior of the cell. This layer is related to a thin basal lamina which does not show the convolutions of the plasma membrane. Presence of large lipid

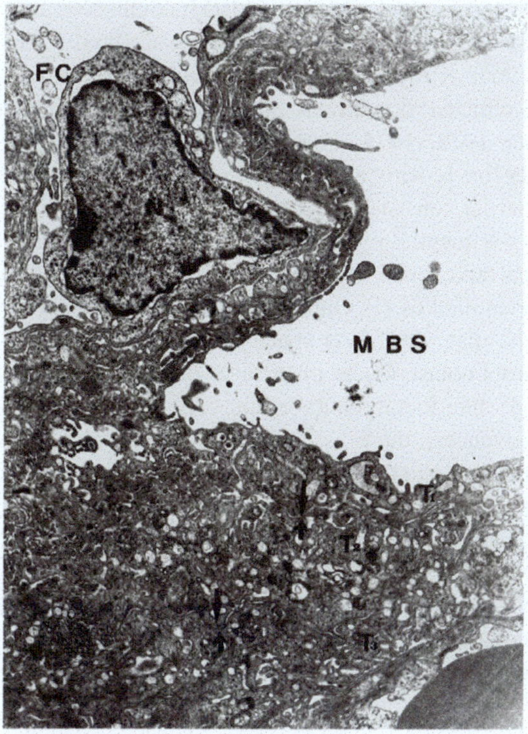

Fig. 7.3 Low-power electron micrograph of interhemal membrane at advanced gestation in Bandicoota. The section passing through the membrane shows three layered trophoblastic epithelium, outer (T1), middle ($T2_2$) and inner (T3) with an intervening maternal blood space (MBS). The outer surface of T2 and inner surface of T3 show deep infoldings. Distinct basal lamina (curved arrow) closely abuts the foetal endothelium (FE) of foetal capillary (FC). A large nucleus of foetal endothelium is seen at the top. Arrows indicate separation between T1 and T2 and T2–T3 [X 5000]. Unpublished electron micrograph from Bhiwgade and Thakur

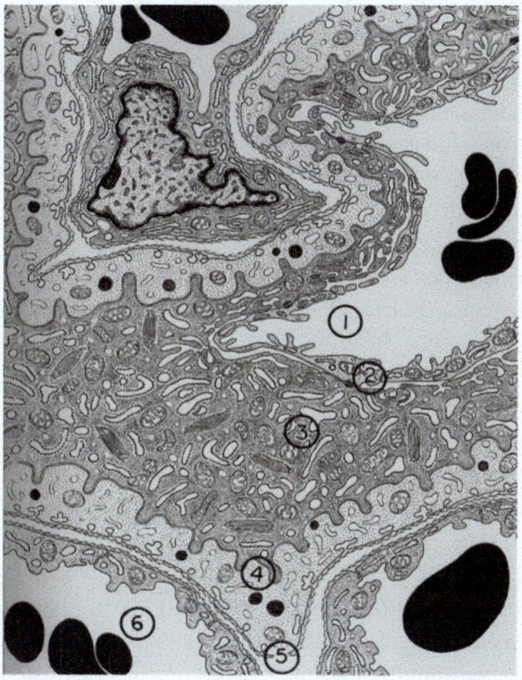

Fig. 7.4 Schematic representation of the interrelationship between the maternal and foetal bloodstreams. Note that the interhemal membrane in Bandicoota shows hemotrichorial condition with the following sequential layers: (1) Maternal blood space (2) Outer layer of trophoblast (3) Middle layer of trophoblast (4) Inner layer of trophoblast (5) Basal lamina (6) Foetal capillary with foetal endothelium. Unpublished electron micrograph from Bhiwgade

droplets is a prominent feature of this layer and are more commonly found than in any other layers (Figs. 7.1 and 7.2). Although intracellular desmosomes are found in this layer, a close examination reveals that it is also syncytial in nature (Figs. 7.3 and 7.4).

During advanced Pregnancy, obvious degenerative changes are observed in component layers of the labyrinthine trabeculae. In rat, the inner layer is distinctly thinner than the middle layer and has a characteristic somewhat beaded appearance (Figs. 7.1 and 7.2). In Bandicoota, this layer is reduced in width due to deep infoldings into the interior of the cytoplasm and its inner plasma membrane is no longer parallel to the endothelial basal lamina. The relationship of three layers of the trophoblast in hemotrichorial placenta of Rat and Bandicoota is depicted schematically in Figs. 7.3 and 7.4. In both the animals, the basal lamina is distinct and closely abut against the foetal endothelium. The foetal endothelium is separated from the inner layer of trophoblast by a distinct basal lamina.

The most striking feature observed in both the animals, during advanced pregnancy is the peripheral endothelium. The cytoplasm in the peripheral endothelium becomes highly reduced, with distinct endothelial openings (pores) which are closed by a diaphragm. Luminal surface of the endothelium exhibits blunt projections and elongated projections.

7.4 Cytoplasmic Organelles

The rough endoplasmic reticulum is the most abundant organelle seen in the cytoplasm of the trophoblast. Cisternae are well developed with various profiles such as tubular, partially dilated and fully dilated with irregular outline. In some places, cisternae are heavily studded with ribosomes whereas in other places ribosome distribution is irregular. In most of the cisternae, a moderately electron-dense substance is present. There is no specific directional orientation of the cisternae, and are scattered throughout the cytoplasm. Free ribosomes are abundant throughout the cytoplasm and the ribosomes are not seen associated with any membrane.

The mitochondria seen in all three trophoblast layers are approximately the same size in both species. Mitochondria are scanty in the inner layers but are comparatively more in abundance in the middle and outer layers. Some elongated mitochondria are observed in the outer cellular layer than in the two syncytial layers.

The characteristic feature of the trophospongium of Bandicoota, in the term placenta, is the abundance of myelinosomes towards the surface. The myelinosomes or myelinoid bodies are secretory granules of characteristic morphology (Ghadially 1988).

In both species, sieve-like zones are found in some of the endothelial cells. In many places, a thin central membrane is seen stretching across individual 'Pores'. Caveolae and sieve areas are seen within the same region of endothelium. Irregular folds are seen projecting into the lumen of the capillaries, especially at all junctions between endothelial cells.

7.5 Discussion

Ultrastructural studies on the placenta of two murid rodents at progressive stages of gestational, confirm the observations reported earlier that, the trophoblast layer lies between maternal blood and foetal endothelium throughout the pregnancy as in other rodents (Dempsey and Wislocki 1953; Wislocki and Dempsey 1955; Schiebler and Knoop 1960; Owers and Mossman 1963; Jollie 1964; Kirby and Bradbury 1965; Toro Jr and Rohlich 1966; Davies and Classer 1968; Carpenter 1972, 1975; Forssmann et al. 1975; Jollie 1976a, b; King and Hastings 1977). The placental barrier consists of discrete constituents structured as sequential cytoplasmic layers between maternal and foetal blood. These constituents can be identified as (1) Outer layer of trophoblast lining the maternal blood space, (2) a middle layer of trophoblast, (3) an inner layer of trophoblast and (4) the foetal endothelium. As the pregnancy advances, there is a progressive shrinking or reduction in these layers, but the four constituents persist throughout the gestation. Though exceptions may be found, our observations together with observations reported earlier, provide ample evidence to conclude that, hemotrichorial pattern is the typical nature of placenta in murid rodents.

Wislocki and Dempsey (1955) have reported that there is no overlapping of cells within the placental barrier. This could be attributed to the limitation of facilities available at that time which could account for their failure to differentiate between different placental elements.

Though numerous physiological studies have been conducted on the placental membrane of rodents, there is paucity of studies to correlate placental ultrastructural features with their functionality. The rates of placental transport of many substances change with the progression of gestation. Electron microscopic studies do not show active transport, but indirect evidences like abundance of organelles, signs of pinocytosis, exocytosis, and modifications of membrane structures are discernible. In our observations, clear changes in ultrastructure placental barrier have been recorded with the progression in gestation. The changes that we have observed, include shrinkage of membrane as the gestation progresses, the trophoblast layers developing pores or clear discontinuities which clearly lead to easier exchange of material between the maternal and foetal environments. Similar reduction in the barrier between maternal and foetal bloodstreams with the progression gestation has been reported earlier in rat placental labyrinth by Jollie (1964), in hamster by Carpenter (1975) and in six myomorph rodents by King and Hastings (1977). We have observed the presence of thinner areas, pores and gaps in the term placental barrier, especially in the outer layer trophoblast, in all myomorph rodents. Studies using tracers have demonstrated that the outer trophoblastic layer (probably because of the pores) offers little hindrance to the movement of macromolecules across the barrier into the space between outer and middle layers (Jollie 1964; Tillack 1966; Robertson et al. 1971; Fels and Themann 1971). These studies have shown that the macromolecules reach the middle layer by pinocytosis but there is hardly any transport evident across the middle trophoblastic layer.

7.6 Functional Characteristics of some Ultrastructural Features

1. Outer Layer of the Trophoblast
   (a) Gaps and Discontinuities

In our study, many discontinuities or gaps are seen in the outer layer of the trophoblast in both the rodents. These gaps become more numerous and wider in succeeding gestational days. According to Carpenter (1975) because of development of such gaps, subcellular components of the maternal blood, can easily enter the large area between the outer and the middle layer of the placenta. Both Enders (1965) and Davies and Classer (1968) have shown that once the space between the outer and middle trophoblastic layers is in communication with the maternal blood compartment, the space between the trophoblastic layer becomes an 'area of relative stasis' enabling uptake of substances from the maternal plasma by the middle trophoblastic layer.

In the hamster labyrinth, gaps in the outer layer appear on the eleventh day of gestation (Carpenter 1975) and according to Jollie (1964) it appears in the rat placental labyrinth on the twentieth day of gestation. According to Carpenter (1975), gaps in the outer layer permits direct contact with the outer surface of the middle layer. In our study, gaps are discernible in the outer trophoblastic layer of both, the rat and the Bandicoot placenta. These gaps are seen from early pregnancy and during advanced pregnancy, and the gaps become sieve like with fenestrae. Such gaps have been described in various other tissue systems like kidney (Longley et al. 1960; Rhodin 1962), thyroid (Dempsey and Peterson 1955; Ekholm 1957; Ekholm and Sjostrand 1957), adrenal cortex (Ross 1962; Zelander 1959), anterior pituitary (Enders and Schlafke 1971), choroid plexus (Maxwell and Pease 1956) and exocrine pancreas (Ekholm and Edlund 1959). The gaps as observed by us in myomorph rodents, are similar to gaps seen on endothelium. Such gaps are reported to facilitate transendothelial transport to various tissues and such fenestrations effectively enhance trans cellular transport.

According to Tillack (1966), the outer trophoblastic layer in the chorioallantoic placenta of the rat is highly permeable to large molecules like ferritin. The real barrier that checks the exchange of materials between the maternal foetal environment is probably the middle and inner trophoblastic layers (Forssmann et al. 1975). Enders (1965) and King and Hastings (1977) have noted that the outer cellular trophoblastic layer in myomorph placenta has abundant endoplasmic reticulum. Out studies and observations on myomorph placenta confirm their findings.

2. Middle Layer of the Trophoblast
   (b) Cap Junction Between Middle and Inner Layer

Some macromolecules are transported across the chorioallantoic placenta but the presence of gap junctions between the middle and inner layers of trophoblast enhances the passage of small molecules between the trophoblast layers. It was Ender (1965) who brought out the significance of the close proximity between cell membranes of the inner and middle trophoblast layers. Carpenter (1975) demonstrated such close association between these two layers in hamster placenta. He categorized them as close junctions and intermediate junctions. Forssmann et al. (1975) used electron microscopy and freeze–fracture techniques while studying rat placenta and showed the presence of gap junctions and maculae adherences between the middle and inner layers of the trophoblast layers in the placenta. King and Hastings (1977) have also shown gap junctions as a common feature between middle and inner layers of trophoblast in all hemotrichorial placentas. Although we did not use specialized freeze-fracture techniques, our studies with high-power electron micrographs undoubtedly establish the presence of that gap junctions and maculae adherences between the two trophoblast layers. As per observations from studies reported earlier, one can presume that a low-resistance electrical coupling exits within the inner and middle layers of the placenta. Such an association of layers could create a 'functional double syncytium' (Forssmann et al. 1975). As suggested by Barr et al. (1965 & 1968), Dreifuss et al. (1966) and Loewenstein and Ann (1966), gap junctions could be a conduit for intercellular exchange of small molecules. Along with the electrical coupling sites, gap junctions also serve as points of metabolic coupling (Cilula et al. 1970). Since these conduits extend across the foeto–maternal barrier, they would function like a sieve for the passage of small molecules.

(c) Micropinocytic Vesicles

Ultrastructural characteristics that indicate active micropinocytosis occurring in cells include the presence of coated vesicles and caveolae. Enders (1965) reported the greatest number of caveolae in the middle trophoblastic layer of the placenta. In rat placenta, Jollie (1976a, b) reported the presence of micropinocytosis and observed strings of coated vesicles spanning the width of trophoblastic layer. We have also observed numerous coated vesicles and caveolae in the middle and inner layers of trophoblast in both rat and bandicoot placentae. The ultrastructural features like deep infoldings, caveolae and coated vesicles in the trophoblastic layer suggest their active involvement in micropinocytosis.

(d) Glomerular Body

We observed glomerular-like body in the cytoplasm of the middle layer of *Bandicoot* placental labyrinth. King and Hastings (1977) had already reported a filamentous structure in the cytoplasm of the middle trophoblast layer which was similar to that reported by Enders (1965) in the mouse placenta. Toro Jr and Rohlich (1966), named these structures 'glomerular bodies' and noted that these bodies were always observed in the middle layer. Wynn et al. (1971) have reported the same in cytotrophoblast of the baboon and in the human placenta by King (Unpublished observations). The chemical composition or function of glomerular bodies is yet to be explained. In hamster placenta, however, Carpenter (1975) did not encounter these structures.

3. The Inner Layer of the trophoblast
   (e) Lipid Droplets

Large lipid droplets are a characteristic of inner layer of trophoblast in both rat and Bandicoota. Similar type of lipid droplets has been observed by Dempsey and Wislocki (1953); Wislocki and Dempsey (1955); Jollie (1964); Enders (1965); King and Hastings (1977); Carpenter (1975). According to Enders (1965) in rat and mouse, large lipid droplets are uniform and of lighter intensity in the inner layer and few fine droplets may be found elsewhere. In the hamster placenta, lipid droplets are generally denser and irregularly distributed in the middle trophoblast layer than in the inner layer. On 11th–13th gestational days, in hamster placenta, extensive accumulation of lipid droplets in the cytoplasm is seen for the first time in the trophoblastic middle layer. Similarly large membranous whorls (myelin figures) are seen often associated with lipid droplets in the cytoplasm of the inner layer. King and Hastings (1977) have observed occasional lipid droplets in the innermost trophoblastic layer of mice. They also found partially or wholly extracted lipid droplets in the cytoplasm of the middle layer in mice placenta, but in general, it is not common. Jollie (1976a, b) has shown that the main difference in the inner trophoblast of the aged placenta is the absence of lipids but they are present frequently till term. Despite the wealth of information concerning the structure of the inner layer of trophoblast, the significance of the presence of large number of lipid droplets in it has remained somewhat obscure. According to Luckett (1970) lipid droplets in the syncytiotrophoblast of the monkey are nutritive lipids rather than involved in steroid synthesis.

(f) Membranous Whorls

Membranous whorls in the inner layer of trophoblast were observed in placentas of rabbit (Enders 1965; Sinha 1968); rat (Carlson and Ollerich 1969); human (Wynn and Davies 1965); hamster (Carpenter 1975) and mice (King and Hastings 1977). Membranous whorls were not recorded in our study in the two myomorph rodents.

4. Fenestrations in Foetal Endothelium

Fenestrations in the foetal endothelium were observed by us during advanced pregnancy stage in both myomorph rodents, in our study. Jollie (1964) has remarked that micropinocytotic activity is evident in endothelium and the outer trophoblast layer though, there is fenestration. In the case of renal glomerulus, similar fenestrations have been observed on the endothelium. It is postulated that the fenestration on the renal endothelium allows the passage of small molecules like water, glucose, urea and various ions, but it is impermeable to large protein molecules (Rhodin 1962). One can apply the same principle to placental labyrinth of rat and Bandicoota and suggest the existence of both micropinocytosis and fenestrations facilitates separate placental exchange routes for different transportable materials. Micropinocytosis enables transport of serum proteins and fenestrations allow transport of small molecules (including respiratory gases) by simple diffusion.

5. Foetal Mesenchyme

The ultrastructural observations on foetal mesenchyme of rat placenta indicate the presence of granular endoplasmic reticulum and hypertrophied mitochondria during early pregnancy. This points to the extensive synthetic activity occurring in the fatal mesenchyme.

7.7 Trophospongium

1. Secretory activity

The trophospongium of *Bandicoota bengalensis* under an electron microscope shows well-developed granular endoplasmic reticulum and Golgi apparatus. These characteristics in a cell are usually regarded as signs of protein synthesis and secretion (King and Tibbits 1976). In almost all cisternae of granular endoplasmic reticulum moderately dense substances are found to be present. Similar types of intra-cisternal granules have been reported in trophospongium of chinchilla (King and Tibbits 1976), acinar cells of the pancreas in guinea pig (Palade 1956) and in dog (Ichikawa 1965). Similar observations have also been made in cells of anterior pituitary after thyroidectomy (Farqwhar 1971; Pelletier and Puviani 1973). The trophoblast cells of blastocyst in rabbit at the time of implantation also show such signs of secretory activity (Enders and Schlafke 1971; Enders 1971). King and Tibbits (1976) observed granules in dilated endoplasmic reticulum cisternae in the coarse syncytium at 1–5 weeks of gestation and suggested that during the early weeks of gestation, the syncytial layer probably is a major site of active secretion. The composition of the secretion of the interlobular syncytium is not known. Mossman and Fisher (1969) have suggested that areas like the interlobular syncytium may be involved in the secretion of gonadotrophic hormones into the maternal circulation. This could be facilitated by the proximity of the syncytial trophoblast to maternal blood sinuses which drain away from the labyrinth. Therefore, such secretions would proceed directly into the general maternal circulation circumventing the initial labyrinthine circulation. Others have suggested that the trophoblast of this region in the guinea pig (Burgess and Tam 1974), or in the homologous region of the rat placenta (Chan and Leathem 1975) may be synthesizing and secreting steroid hormones. Thus, it seems evident that trophospongium must play an important synthetic role during gestation. Presence of abundant granular endoplasmic reticulum, Golgi complexes and large mitochondria in Bandicoota support the suggestion that trophospongium is actively synthetic in function.

7.8 Active Transport

The presence of coated vesicles, multivesicular bodies and large membrane bound structures containing membranous whorls and numerous vesicles in the present report indicate that trophospongium provides active transport for the products of absorption, secretion and excretion through it. Similar types of multivesicular bodies have been observed by King and Tibbits (1976) in placental labyrinth of the chinchilla and by King and Enders (1971) in the placental labyrinth of the guinea pig.

7.9 Myelinosomes

One of the striking features of the trophospongium of the Bandicoota placenta is the presence of myelinosomes. As suggested by Ghadially (1988) initially, the multivesicular bodies that are less electron dense later get converted into the more electron dense type of vesicles. Then, a few osmiophilic membranous materials (myelinoid membranes) appear in the matrix of dense multivesicular body. As the amount of membranous material increases, the number of vesicles inside the body reduces until the fully mature myelinoid body packed with stacks or whorls of electron-dense membranes (myelin figure) is formed.

We observed myelinosomes near the surface of the trophospongium of the Bandicoota placenta during early pregnancy. Large membrane bound structures containing only myelinoid figures were seen during the late limb bud stage. Similar myeloid figures were observed by Wynn and Davies (1965) in the villi of human placenta at full term. According to them, these myeloid figures are formed either from endoplasmic reticulum, mitochondria or possibly Glogi complex which increases in number and size. These myelinosomes then moves to the surface of the cell and their contents (lipids, mainly phospholipids) are discharged like other merocrine secretion (Ghadially 1988).

References

Barr L, Berger W, Dewey MM (1968) J Gen Physiol 51: 347

Barr L, Dewey MM, Berger W (1965) Propogation of action. Potentials and the structure of the nexus in cardiac muscle. J Gen Physiol 48:797

Burgess SM, Tam WH (1974) The development of biochemistry and electron microscopy of the steroid producing tissue in the Guinea pig placenta. Seventh Ann. Meeting Soc. for the Study of Reproduction, Ottawa, Canada (Abstract)

Carlson EC, Ollerich DA (1969) Intranuclear tubules in trophoblast III of rat and mouse chorioallantoic placenta. J Ultra Res 28:150–160

Carpenter SJ (1972) Light and electron microscopic observations on the morphogenesis of the chor ioallantoic placenta of the golden hamster. (Cricetus auratus). Days seven through nine of gestation. Am J Anat 133:445–476

Carpenter SJ (1975) Ultrastructural observations on the maturation of the placental labyrinth of the golden hamster (days 10 to 16 of gestation). Am J Anat 143(31):5–348

Chan S, Leathem J (1975) Placental steroidogenesis in the rat: progesterone production by tissue of the basal zone. Endocrinology 96:298–303

Cilula NB, Reeves OR, Steinbach A (1970) Nature (London) 235

Davies J, Classer SR (1968) Histological and fine structural observations on the placenta of the rat. Acta Anat 69:542–608

Dempsey EW, Peterson RR (1955) Endocrinology 56:46

Dempsey EW, Wislocki CB (1953) Electron microscopy of the rat's placenta. Anat Rec 117:581–582

Dreifuss JJ, Cirardier L, Forssmann WC (1966) Pfligers Arch 292:13

Ekholm RZ (1957) Zel Iforsch Mikroskop. Anat 46:139

Ekholm RZ, Edlund YJ (1959) Ultrastruct 2:453

Ekholm R, Sjostrand FS (1957) J Ultrastruct Res 1:178

Enders AC (1965) Comparative study of the fine structure of the trophoblast in sevral hemochorial placentas. Am J Anat 116:29–68

Enders AC (1971) The fine structure of the blastocyst. In: Blandau RJ (ed) Biology of the blastocyst, vol 71. University of Chicago Press, Chicago, p 94

Enders AC, Schlafke SJ (1971) Penetration of the uterine epithelium during implantation in the rabbit. Am J Anat 132(21):9–240

Farqwhar MC (1971) Processing of secretory products by cells of the anterior pituitary gland. In: Heller H, Lederis K (eds) (Memoirs Soc. Endocrinol. No. 19) Subcellular organization and function in endocrine tissues. Cambridge University Press, London

Fels R, Themann H (1971) Die permeabilat der Mause placenta fur Meerrettich peroxidase. Eine elektonenmkroskopische Untersuchung. Cytobiologie 3:171

Forssmann WC, Metz J, Heinrich. (1975) Cap junctions in the hemotrichori placenta of the rat. J Ultra Res S3: 374–381

Ichikawa A (1965) Fine structural changes in response to hormonal stimulation of the perfused canine pancreas. J Cell Biol 24:369–385

Jollie WP (1964) Fine structural changes in placental labyrinth of the rat with increasing gestational age. J Ultra Res 10:27–47

Jollie WP (1976a) The fine structure of the interhemal membrane of the rat chorio- allanotic placenta during prolonged pregnancy. Anat Rec 184:73–90

Jollie WP (1976b) Vascularization of a block to placental transport of protein and dextrin in the rat. Bull Tulane Univ Med Fac 24(21):3–224

King BF, Enders AC (1971) Protein absorption by the Guinea pig chorioallantoic placenta. Am J Anat 130: 409–430

King BF, Hastings RA (1977) The comparative fine structure of the interhemal membrane of chorioallantoic placentas from six genera of myomorph rodents. J Anat 149:165–180

King BF, Tibbits DF (1976) The fine structure of the Chinchilla placenta. J Anat 145:33–56

Kirby DRS, Bradbury S (1965) The hemochorial mouse lacenta. Anat Rec 152:279–282

Loewenstein WR, Ann NY (1966) Acad Sci 137:441

Longley JB, Banfield WC, Brindley DC (1960) J Biophys Biochem Cytol 7:103

Maxwell DS, Pease DC (1956) J Biophys Biochem Cytol 2:467

Mossman HW, Fisher TV (1969) The preplacenta of pedetes, the trager, and the maternal circulatory pattern in rodent placentae. J Reprod Fert Suppl 6:175–184

Owers N, Mossman HW (1963) New observsations by electron microscopy on the structure of the placental barrier of the white rat–Musnorvegicus. Anat Rec 145: 269. (Abstract)

Palade GE (1956) Intracisternal granules in the exocrine cells of the pancreas. J Biochem Biophys Cytol 2:417–422

Pelletier G, Puviani R (1973) Detection of glycoproteins and autoradiographic localizations of (3H) fucose in the thyroidectory cells of the rat pituitary gland. J Cell Biol 56:600–605

Rhodin JAC (1962) J Ultrastruct Res 6:171

Robertson TA, Archer JM, Papadimitriou JM, Walters MNI (1971) Transport of horse radish peroxidase in the murine placenta. J Pathol 103:141–147

Ross MH (1962) Cited by Rhodin JAG. J Ultrastruct Res 6:171

Schiebler TH, Knoop A (1960) Verhandl Anat Ces 82

Sinha AA (1968) The interlobular cleft and membraneous whorl in the rabbit placenta. Anat Rec 160:187–200

Tillack TW (1966) The transport of ferritin across the placenta of the rat. Lab Investig 15:896–909

Toro I Jr, Rohlich PA (1966) New cytoplasmic component in the trophoblast cells of the rat and mouse. Anat Rec 155:385–400

Wislocki GB, Dempsey EW (1955) Electron Microscopy of the placenta of the rat. Anat Rec 123:33–63

Wynn RM, Davies J (1965) Comparative electron microscopy of the hemochorial placenta. Am J Obstet Gynec 91:533–549

Wynn RN, Panigel M, Maclennan AH (1971) Fine structure of the placenta and fetal membrane of the baboon. Am J Obstet Gynecol 109:638–648

Zelander T (1959) J Ultrastruct Res Suppl 2:1

8 Ultrastructure of Placental Barrier and Chorion Laeve of Bonnet Macaque

In this study major emphasis is given to the ultrastructure of the placental barrier and the chorion laeve during early and late pregnancy in the monkey, *Macaca radiata* (Bonnet macaque).

8.1 Early Placenta

(a) The Syncytiotrophoblast

It is a true syncytium, because the small villus portion in cross section, shows many nuclei without any separating cell boundaries (Fig. 8.1). The syncytiotrophoblast thus, forms a continuous cover over each villus and probably could be continuous from villus to villus.

The syncytiotrophoblastic surface is thrown up into irregularly distributed cytoplasmic promontories and the surface is more irregular due to numerous depressions between the promontories (Fig. 8.1). These promontories are further profusely branched and bordered by numerous, long, slender microvilli (Fig. 8.1). The tips of the some of the villi are slightly bulbous in appearance. The surface of the syncytium in some places invaginates to form pinocytotic vesicles. The tip of the villi contains large, light multivesicular bodies enclosing numerous vesicles (Fig. 8.1).

During early pregnancy, the syncytiotrophoblast was considerably thick and its basal surface rests upon distinct but thin basal lamina. The basal surface although exhibits irregular contour does not show any specialized features (Fig. 8.1). The nuclei of the early syncytiotrophoblast appears in clusters and are oval to irregular in shape. It is with distinct nucleolus and clumped chromatin and nuclei are arranged more towards the basal side (Fig. 8.1). The cytoplasm of the syncytiotrophoblast is well equipped with various organelles. One of the striking features is the rough endoplasmic reticulum seen in abundance. The cytoplasm is crowded with two distinct morphological forms of rough endoplasmic reticulum. Towards the apical surface, it is mainly in the vesicular form, whereas at the basal side two-third of the cytoplasm, the cisternae are dilated giving it a cribriform appearance (Fig. 8.1). The cisternae are filled with material of light electron density and with uneven distribution of ribosomes on the wall of the cisternae (Fig. 8.1). Golgi zones are scattered throughout the cytoplasm (Fig. 8.1) and it consists of stacks of 3–4 flattened sacs which are slightly curved with slightly enlarged ends associated with numerous vesicles (Fig. 8.1). Mitochondria generally are seen at tips of the villi and are present in large aggregations. They are hypertrophied, bounded by a double membrane with degenerated lamellar cristae, filled with a dense intramitochondrial matrix. Large electron-dense bodies can be seen scattered throughout the cytoplasm.

D. A. Bhiwgade, S. Menon, *Ultrastructural Investigations on the Pituitary-Gonadal Axis*,
https://doi.org/10.1007/978-981-99-3276-4_8

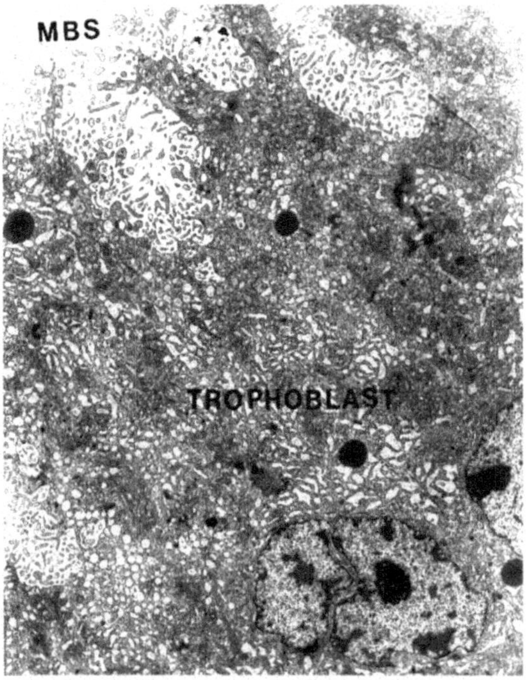

Fig. 8.1 This is a cross section of a free villus in the early placenta of Bonnet macaque, under low-power electron micrograph. The syncytial trophoblast (S) shows its surface, thrown up into cytoplasmic promontories (P) which display multitude of microvilli extending into the maternal blood space (MBS). The cytoplasm shows abundance of rough endoplasmic reticulum. Two varieties of endoplasmic reticulum are seen, the vesicular form (VrEr) and the tubular form (T-rER) and numerous scattered Golgi zones (G). Mitochondria (M) occupy tips of the villus and electron-dense bodies (arrows) are scattered. In the lower portion, dense nuclei (N) are seen [X 3500]. Unpublished electron micrograph from Bhiwgade and Thakur

(b) Villous Stroma

In the mesodermal core of the young villi, undifferentiated mesenchymal cells can be seen scattered in the loose textured matrix which shows a delicate network of collagen fibres. Mesenchymal cells are irregular with distinct contour with some projections. Prominent feature is the abundance of vesicular form of rough endoplasmic reticulum (Fig. 8.1).

8.2 Term Placenta

As the placenta ages, the width of the syncytiotrophoblast reduces and it becomes thinnest in the place, where it is overlying the fetal capillary forming syncytio-capillary membrane.

(a) The Syncytiotrophoblast

The syncytium, at the term placenta covers the villi completely. Its thickness varies considerably in different regions. Over foetal capillary, the thickness is the least (Fig. 8.2), especially at advanced gestation. The surface of the syncytiotrophoblast is thrown up into small promontories which bear irregularly distributed microvilli. These promontories, however, are not further branched as in the early placenta. The apical surface is irregular due to numerous pinocytotic invaginations. Just beneath the surface the cytoplasm shows a number of coated vesicles (Fig. 8.2). The tips of some villi are slightly bulbous but no multivesicular body is found in

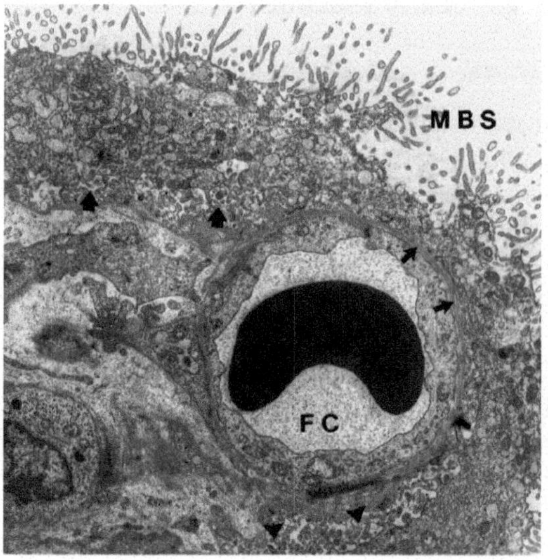

Fig. 8.2 Overview of the villus at term of Bonnet macaque, showing thinning of layer of the syncytium with Nuclei. At many places, the apical surface invaginates to form pinocytotic invaginations (small arrow) and long, slender microvilli project into the maternal blood space (MBS). The basal surface of the syncytium shows nymerous foot like processes (thick arrow), where it comes in contact with the thick, amorphous basal lamina. Basal lamina is followed by fetal endothelium. Extracellular space is with densefilamentous fibrinoid material. Also seen are rough endoplasmic reticulum, Golgi zone, and mitochondria. [X 5,000]. Unpublished electron micrograph from Bhiwgade and Thakur

it. The basal surface of the syncytium at some places shows the cytotrophoblast cell (Fig. 8.3) but is generally bound by the basal lamina. The surface where the syncytium comes in contact with the basal lamina there are numerous cytoplasmic infoldings which may be called as foot processes. In some sites, the folds move away from the adjacent surface and intervening extracellular space is seen (Fig. 8.3). Nuclei are big in size and are seen in groups in the syncytium (Fig. 8.3). The nuclei are more electron dense than earlier stage where it is uniformly granular with one or more nucleoli. The cytoplasm is crowded with rough endoplasmic reticulum. At this stage of gestation, vesicular form of rough endoplasmic reticulum predominates (Fig. 8.3). As in early pregnancy, no distinct zones of rough endoplasmic reticulum are observed. The cisternae are filled with moderately electron-dense material. Golgi complex is like the one observed in the early placenta. Mitochondria are dispersed throughout the cytoplasm, but they are less in number as compared to the earlier stage. They are hypertrophied with degenerated lamellar cristae. Electron dense granules are rarely observed within the cytoplasm.

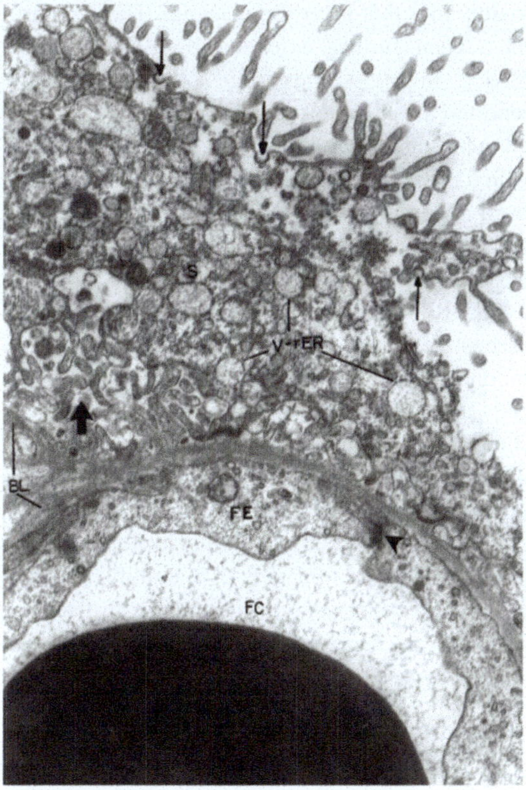

Fig. 8.3 High power electron micrograph showing the syncytiotrophoblast, during late pregnancy in Bonnet macaque . The thick amorphous basal lamina (BL) splits into two, one underlying the syncytiotrophoblast and the other surrounding fetal capillary (FC). The basal surface of the syncytium shows numerous folding (thick arrow), whereasthe apical surface shows numerous pinocytotic invaginations (thin arrow). Fetal endothelial cells (FE) show tight junctions (arrow head), V–rER, vesicular form of rough endoplasmic reticulum. [X 13,000]. Unpublished electron micrograph from Bhiwgade and Thakur

(b) Cytotrophoblast [Langhans Cell]

Cytotrophoblast cells are easily distinguishable because of their lighter staining (Fig. 8.2). These cells are infrequently seen in the villi and occurred singly. They rest on the basal lamina and do not reach the surface of the villus, since they are constantly overlayed by the syncytial trophoblast. The cell body is oval or elongated in shape with the plasma membrane thrown up into folds having tiny projections. These folds are, however, lesser than those in the plasma membrane of the syncytium (Fig. 8.3). The cells cohesion is established via desmosomes and the base of the syncytium shows short filamentous extensions of the plasma membrane in the direction of Langhans cell (Fig. 8.3). The cytoplasm shows a significant number of dense particles of ribonucleoprotein [RNP] scattered in clusters. Rough endoplasmic reticulum is of vesicular type and mitochondria are oval to elongated in shape and are hypertrophied. Both rough endoplasmic reticulum and mitochondria do not form dominant features in the ultrastructural picture. The nucleus is oval in shape and shows finely granular contents of moderate electron density but they are smaller than those seen in the syncytium.

(c) Villus Stoma

The stroma of the villus shows many capillaries that are lined by endothelial cells which are resting upon basal lamina. The basal lamina is thick and amorphous in nature. It is split into two, the one related to the trophoblast and the other surrounding the capillary (Fig. 8.2). The site of attachment between endothelial cells shows electron-dense material in which adjacent cell membranes can be seen. Luminal surface of the endothelial cell appears smooth but in some places it shows blunt projections (Fig. 8.3). The prominent feature of the cytoplasm is the presence of hypertrophied mitochondria and many coated vesicles (Fig. 8.2).

The core of villus shows connective tissue with dense collagen fibres and well-differentiated mesenchymal cells. Mesenchymal cells are oval with a distinct smooth contour. The nucleus is oval with peripherally arranged chromatin without nucleolus (Fig. 8.2). The cytoplasm contains scanty fragments of rough endoplasmic reticulum.

Difference between early and late placenta in Bonnet Macaque.

Early placenta	Late placenta
1. Cytoplasm promontories are further divided into branches bearing numerous microvilli.	1. Cytoplasmic promontories are small and are undivided. Microvilli although long and slender are comparatively few in number.
2. Pinocytotic invaginations are rarely observed at the apical surface of the syncytium.	2. Pinocytotic invaginations are numerous at the apical surface of the syncytium.
3. At the basal surface of the syncytium, foot-like processes are not seen.	3. The basal surface of the syncytium, where it comes in contact with basal lamina, shows distinct foot-like processes.
4. In syncytial cytoplasm, the rough endoplasmic reticulum is in two distinct morphological forms—Vesicular cisternae and dilated cisternae which are distributed in two distinct zones.	4. The vesicular form of rough endoplasmic reticulum predominates and no two distribution zones are observed.
5. Mitochondria are aggregated in large numbers at the tip of villi.	5. Mitochondria are scattered throughout the cytoplasm and comparatively few in number.
6. Basal lamina is thin.	6. Basal lamina is thick and amorphous.
7. Syncytiotrophoblast is considerably reduced	7. The syncytiotrophoblast is distinctively thicker
8. Mesenchymal cells are undifferentiated and irregular with distinct contour.	8. Mesenchymal cells are well differentiated, with smooth contour.

8.2.1 Discussion (Placenta)

Since the publication of Wislocki and Dempsey (1955), reporting the rich variety of membranes in the placenta, electron microscopic studies on several other species confirmed the complexity of membranes that layer the placenta. These reports have been further extended by Schiebler and Knoop in 1959 and Jollie (1973) who also reported the changes that take place in the placental membranes with the advancement of gestation which result in increased permeability. The layers from maternal blood to foetal mesenchyme in the placenta were designated as trophoblast I, trophoblast II and element II. Wislocki and Dempsey 1955, Rhodin and Terzakis 1962 made the earliest electron microscopic studies on the human placenta. The contribution of Langhans cells in the formation of syncytium was not clear (Enders 1965a, b). Other than these studies, studies on the true structure of placental layers in other species were very scanty.

In this context our study on placenta of bonnet monkey takes significance. In our study, we have observed a consistency in the layers of placenta. The class of hemochorial placenta is very large and the examples studied by may not be the most comprehensive ones. We may find different placenta which may be hemodichorial, hemotrichorial or even hemomonochorial. It is interesting to point out here that in molossid bats, C*haerephon plicata* (Bhiwgade et al. 1997) and in *Tadarida brasiliensis cynocephala* the placenta is of cellular hemomonochorial type. Both in labyrinthine and villous hemomonochorial placenta, it is the syncytial trophoblast that forms the continuous layer. In myomorph rodents, there is evidence of two syncytial layers which do not fuse but are very close to each other with tight junctions between them. The origin of this tri-layered pattern, in rodents, is probably from three sources namely, the giant cells, the trigger from epithelium, and the true chorionic epithelium. The hemotrichorial rodent placenta has been further investigated and its functional significance in glucose transport is also discussed (Takata et al. 1997). They have also confirmed the presence of gap junction between the syncytial trophoblast layers. It can be presumed that several variations in layering of placenta can be expected to be found in rodents.

In all placentae studies, it has been observed that caveolae are more frequent in areas where the trophoblast layers are less directly in contact with maternal blood. This indicates that areas where there are least changes in maternal plasma, are the areas where maximum absorption takes place. In a hemotrichorial placenta, caveolae are more abundant in the second layer of trophoblast. The space between the first and second layers with maternal blood spaces would be the area with comparatively low exchange of materials. In the villous placenta of monkey, unlike the labyrinthine placenta, the maternal blood spaces are randomly placed giving a network appearance. This gives rise to a countercurrent flow within the placenta. Ramsey in 1962, suggested that in the human placenta, the blood enters and leaves through the basal plate region. The microvilli are swollen at tips and branched giving rise to an irregular arrangement of blood spaces. This arrangement provides impediments to the flow of plasma giving rise to areas with static composition of plasma due to low exchange of materials. Caveolae are abundant in such regions of low material exchange.

In our study, in all placentae that we examined, it is observed that the granular endoplasmic reticulum is maximum developed at trophoblast sites that are in close proximity to maternal blood. This is a common observation in all placentae with different numbers of trophoblast layers. Since granular endoplasmic reticulum is involved with protein synthesis, it is presumable that proximity to maternal blood ensures better exchange of gasses and nutrients which would also optimize energy utilization.

8.2.2 Chorion Laeve

In the early monkey placenta, the villous chorionic surface entirely covers the placenta. With the advancement of gestation, the villi degenerate and a smooth structure is identifiable, which is recognized as the chorion leave (smooth chorion).

1. Early chorion Laeve

The chorion laeve during early gestation, under electron microscope, shows numerous desmosomes joining adjacent epithelial cells (Fig. 8.4). Numerous microvilli are seen extending from the apical surface of the trophoblast cells (Figs. 8.4 and 8.5). A thin and indistinct basal lamina underlines the foetal tissue which is followed by numerous filamentous features (Fig. 8.4). The nucleus of trophoblast cells is irregular with a distinct nucleolus and peripherally arranged chromatin.

2. Late Chorion Laeve

The cells comprising the chorion laeve during late gestation are more irregular in shape than earlier. Their apical surfaces exhibit blunt microvilli extending into the uterine lumen (Fig. 8.6). Adjacent cell surfaces are joined by numerous desmosomes, tight junctions and at some places, wide gap is found in between the cell junctions which is occupied by blunt microvilli (Fig. 8.6). Numerous coated vesicles and caveolae are found just beneath the surface. Trophoblastic cells are resting on distinct basal lamina which is followed by abundant filaments. Extracellular space is packed with collagen fibrils with loosely scattered mesenchymal cells. The cytoplasm of mesenchymal cells is poorly differentiated with scanty cell organelles.

3. Parietal Decidua in Early Pregnancy

Cells of the parietal decidua are arranged in a stratified manner and readily distinguished from the trophoblastic cells in having numerous electron-dense bodies scattered throughout the cytoplasm (Fig. 8.3). Most of the cells are oval in shape with moderately well-differentiated

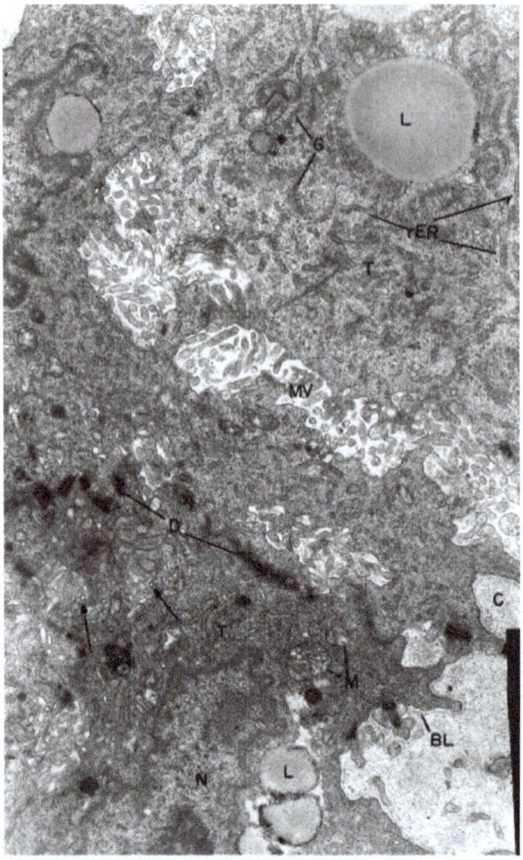

Fig. 8.4 Trophoblastic cells (T) of the smooth chorion during early pregnancy in Bonnet macaque. Note the deep recesses between the cells where numerous microvilli (MV) extend. The junction between adjacent cells is with numerous desmosomes (D). The lower surface of the trophoblast cells rests on indistinct basal lamina (BL) and the matrix is packed with abundant collagen (C). The nuclei (N) of the trophoblast are irregular and the cytoplasm exhibits large lipid droplets (L), hypertrophied mitochondria (M) and Golgi zones (G). Numerous coated vesicles (arrow) are observed near the intercellular junction [X 8000]. Unpublished electron micrograph from Bhiwgade and Thakur

cytoplasm. The cells are loosely arranged and attached by desmosomes at some places. Abundant filaments, collagen and fibrinoid material are present in the extracellular space. Fibrinoid material also surrounds each cell like a basement membrane (Fig. 8.3). Nuclei are almost oval in shape with few indentations with peripherally arranged chromatin materials.

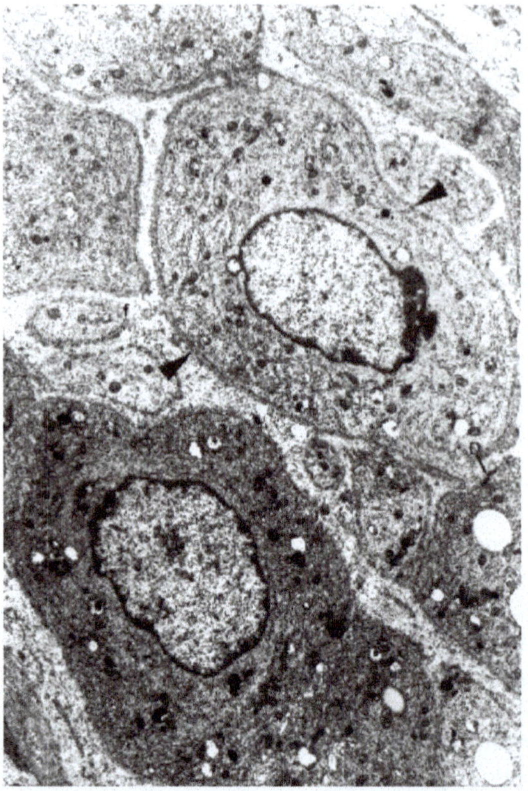

Fig. 8.5 Population of decidual cells in placenta during early pregnancy in Bonnet macaque, which are scattered within the matrix. The cells are packed with filaments (f) and fibrinoid material. Fibrinoid material which appears like a basement membrane (arrow head) surrounded each cell. Cells are loosely scattered but in some places, adjacent cells are joined by desmosomes (D) [X 8300]. Unpublished electron micrograph from Bhiwgade and Thakur

4. Parietal Decidua During Late Pregnancy

The decidual cells at this stage are more irregular in shape with long cytoplasmic flanges, scattered loosely in the matrix. Extracellular space is packed with collagen and fibrinoid material. Nuclei are irregular, cytoplasm contains few electron-dense granules (Fig. 8.6).

5. Cytoplasmic Organelles

During early pregnancy, mitochondria of Chorion Laeve are few in number. They are oval to balloon shaped with transverse lamellar cristae separated by considerable intermitochondrial matrix. During late pregnancy, however, they

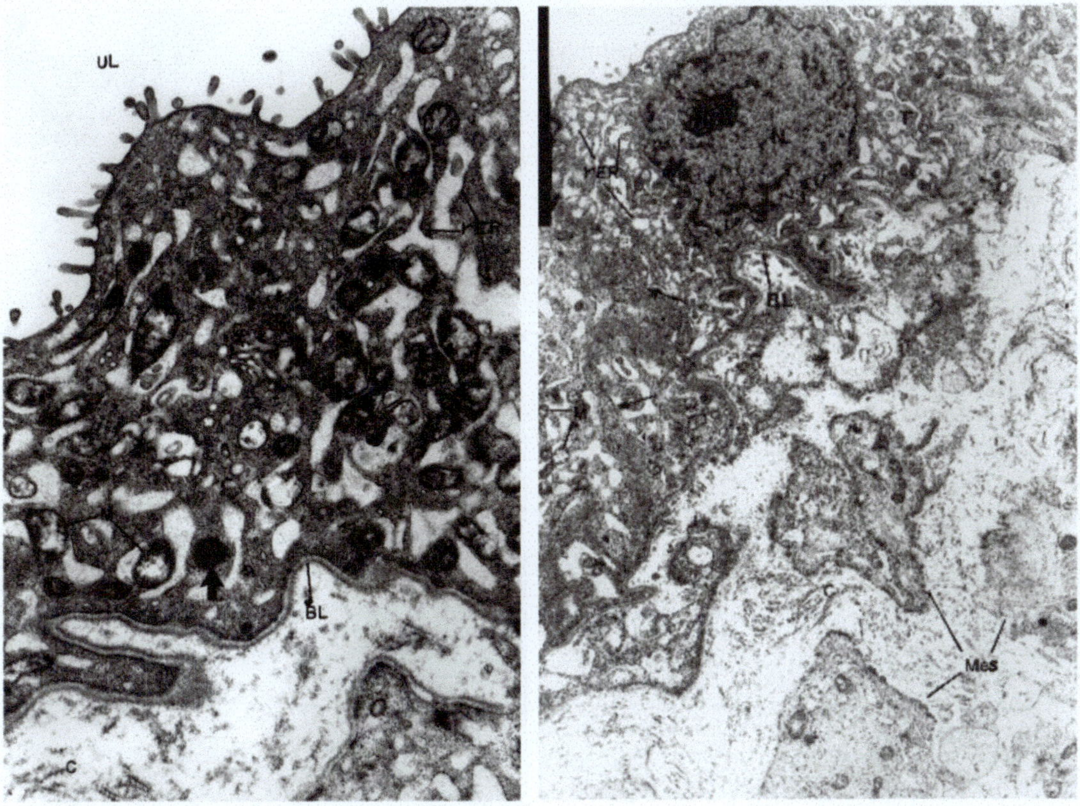

Fig. 8.6 Electron micrograph (low power) of chorion laeve and associated mesenchymal cells (Mes) of placenta in Bonnet macaque, near term. Irregular trophoblastic cells (T) with many cytoplasmic processes are seen. Abundant rough endoplasmic reticulum (rER) and mitochondria (M) are seen in the cytoplasm. Intracellular gap (arrow) is seen at some places filled with microvilli. Mesenchymal cells are loosely scattered in collagen (C) packed matrix. Nucleus of trophoblast cells (N) and basal lamina (BL) are also seen [X 8300]. Unpublished electron micrograph from Bhiwgade and Thakur

become many in number, hypertrophied with cavities within the matrix. They are found in close association with dilated cisternae of rough endoplasmic reticulum. Small, circular and oval-shaped mitochondria are scattered throughout the cytoplasm of the decidua cells during both, early and late pregnancy (Fig. 8.6).

In trophoblast cells during early pregnancy, the rough endoplasmic reticulum is tubular in shape and are arranged singly and loosely. The membranes are provided with uneven distribution of ribosomes, whereas during late pregnancy, the sacs of the rough endoplasmic reticulum are dilated and filled with amorphous material. The part of the membrane, which is in close contact with mitochondria, is devoid of ribosomes. During early pregnancy, rough endoplasmic reticulum in the cytoplasm of decidual cells are tubular in shape and scattered throughout the cytoplasm. In late pregnancy, however, the sacs of rough endoplasmic reticulum are partially dilated filled with amorphous material and its membranes show uneven distribution of ribosomes.

In trophoblast cells during early pregnancy, Golgi consists of a stack of flattened sacs which are slightly curved, associated with numerous vesicles on the outer concave, immature surface. Inner, concave and mature surface is generally directed towards the base of trophoblastic cells. During late pregnancy, it shows hypertrophy with dilated sacs and numerous vacuoles on inner surface. Golgi complex of the decidual cells during both the stages of gestation, exhibit hypertrophy.

Large lipid droplets are found scattered throughout the cytoplasm of the trophoblast cells and decidual cells during early pregnancy, but they are not observed in both cells during late pregnancy.

Numerous coated vesicles are found at the inter-cellular junction of the trophoblast cells during early pregnancy whereas during late pregnancy, they are comparatively few in number.

Large electron lucent vacuoles are found in the cytoplasm of the decidual cells during early pregnancy whereas during late pregnancy, they are comparatively smaller in size (Fig. 8.7).

Difference between chorionic cells and decidual cells during Early and late gestation are briefly summarized.

	Early chorion laeve		Late chorion laeve
1	Cells are irregular and considerably Thicker	1	Cells were more irregular and reduced in thickness.
2	Numerous, long microvilli are seen.	2	Microvilli are sparsely distributed and are blunt.
3	Intercellular gap is long and narrow.	3	Intercellular gap is small and wide.
4	Mitochondria are small and few in number.	4	Mitochondria are hypertrophied and large in number.
5	Rough endoplasmic reticulum is in the tubular form.	5	Rough endoplasmic reticulum is in the form of dilated cristernae.
6	Golgi consists of flattened sacs.	6	Golgi sacs are dilated.
7	Lipid droplets are observed prominently.	7	Lipid droplets are not prominent.
8	Numerous coated vesicles are seen.	8	Coated vesicles are few in number.
	Early decidual cells		Late decidual cells
1	Cells are arranged in layers.	1	Cells are sparse in population.
2	Cells are oval to fusiform in shape.	2	Cells are irregular.
3	Numerous electron-dense granules are seen.	3	Electron-dense granules are few in number.

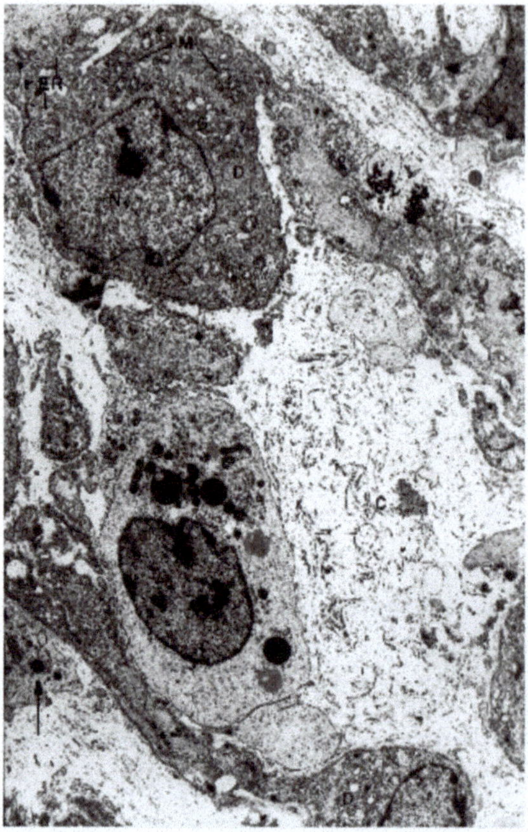

Fig. 8.7 Well-developed decidual cells (D) with long cytoplasmic processes in Bonnet macaque during late pregnancy. They are seen scattered in the collagen (C) packed matrix. Nucleus (N) of the decidual cells is irregular and the cytoplasm exhibits well-developed Golgi (G), mitochondria (M) and rough endoplasmic reticulum (rER). Electron-dense granules (arrow) are less numerous than earlier [X 5800]. Unpublished electron micrograph from Bhiwgade and Thakur

	Early chorion laeve		Late chorion laeve
4	Rough endoplasmic reticulum is in tubular form.	4	Cisternae of rough endoplasmic reticulum are partially dilated.

8.2.3 Discussion (Chorion Laeve)

Observations in our study confirm that chorion laeve undergoes significant changes with the progression of gestation. During the early gestation when the chorion laeve is yet to fuse with the deciduas, the trophoblast appears as a continuous layer around the placenta. The trophoblast cells are connected by desmosomes and tight junctions. The outermost trophoblast cells are differentiated mainly for phagocytic role of absorption with the presence of villi at their ends. A similar role of trophoblast has been suggested in earlier studies (Sinha and Erickson 1974; Gulamhusein and Beck 1975; Burton et al. 1976; Myagkaya and Vreeling-sindelarova 1976; Malassine 1977; King et al. 1978; King 1981). This is indicative that the main function of chorionic layer in the primate placenta, during early stages of gestation, is histotrophic nutrition.

If our observations in placenta of bonnet monkey are compared with studies on human chorion laeve, at term, it is interesting to note that similar observations have been reported in the ultrastructural observation of trophoblast in humans. The trophoblastic cells in human chorion laeve have been reported to have well-developed granular endoplasmic reticulum, a large number of glycogen granules and desmosomes at cell junctions. Similar findings have been reported by Anderson and McKay (1966) in normal and toxemic placentas of humans and by Thliveris and Speroff (1977) in normal and hypertensive pregnant women.

With the progression of gestation, the changes observed in the interrelationships and structural features of the foetal membranes have a definitive relationship with their functional significance in the transmembrane transport of materials. Since, during early gestation, the chorionic epithelium lines the uterine lumen directly, materials from the uterine lumen could be directly transferred to the fetus via the amniotic fluid through the process of histotrophic nutrition. It is significant to note that earlier studies in ferret and rat have shown that interference with the functions of extraembryonic membranes can result in teratogenesis during early gestation. Similar studies in primates are not available and such studies on the role of Para placental chorion in transfer of nutrients and teratogens would be very significant. Hendrickx and Binkerd in 1979 have shown that the time when chorion laeve becomes absorptive in function overlaps the critical period of teratogenesis in monkey.

In our observations on the chorion laeve in bonnet monkey, during the early gestation period, we have recorded the presence of tight junctions between the cells of chorionic epithelium. At late gestation, however, there are a lot of intercellular space seen between adjacent cells. This the junctions between cells are unobstructed by junctional complexes. These changes evidently suggest the increased permeability of chorionic epithelium during late pregnancy. There have been several in vitro experimental evidences that are available from studies on monkey and human Chorion Laeve or amniochorion, in support of this increased permeability during late gestation.

These studies have shown that through large, aqueous, extracellular channels, many substances can pass these layers by diffusion. It has been found that since the thickness of the membrane increases at term, the permeability of the tissue per unit thickness increases. The increased permeability could be attributed to the increase in size of the extracellular space and to the change in the composition of the extracellular space. The widened gaps between the trophoblast cells seen in the late placenta, in our observation, could be one of the factors that contribute to the increased permeability.

Though the increase in permeability of chorion laeve in both human and non-human primates is reported, the limit on the size of molecules that could be exchanged across the layers is not very clear.

Battalgia and co-workers in 1964 and Moore and co-workers in 1966 suggested that permeability decreased with molecular weights below 1000 Daltons. Chez and co workers in 1970 have shown that lactogen with a molecular weight of 22,000 Dalton could readily diffuse across the amnion-chorion layer of human placenta. On the basis of our observations, large intercellular spaces between many of the trophoblastic cells could easily permit the transfer of macromolecules across the chorion laeve during term.

In Bonnet monkey, the decidual reaction akin to that seen in humans is not observed. The stomal cells, however, show similarities to the decidual cells in humans due to the presence of granular ER and Golgi complexes. The unusual "secretory bodies" containing small vesicles seen in the bonnet monkey are very similar to those reported in human uterine and ovarian decidual cells (Lawn et al. 1971; Wynn 1974; Herr et al. 1978).

Therefore, even though bonnet monkey shows a different decidual reaction to that seen in humans, the functional similarities seen in the decidual cells of bonnet monkey and human are very useful in understanding the significant role that these cells play in pregnancy and parturition.

References

Anderson WA, McKay DC (1966) Electron microscope study of the trophoblast in normal and toxemic placentas. Am J Obstet Gynecol 95:1134–1148

Bhiwgade DA, Mahaley JA, Banergee SS (1997) J Univ Mumbai 3:80–81

Burton CJ, Samuel CA, Steven DH (1976) Ultrastructural studies of the placenta of the ewe: phagocytosis of erythrocytes by the chorionic epithelium at the central depression of the cotyledon. Quart J Exp Physiol 61: 275–286

Enders AC (1965a) Comparative study of the fine structure of the trophoblast in several hemochorial placentas. Am J Anat 116:29–68

Enders AC (1965b) Formation of syncytium from cytotrophoblast in the human placenta. Obstet Cyn 25:378–386

Gulamhusein AP, Beck F (1975) Development and structure of the extraembryonic membranes of the ferret. A light microscopic and ultrastructural study. J Anat 120: 349–365

Herr JC, Heidger PM, Scott JR, Anderson JW, Curet LB, Mossman HW (1978) Decidual cells in the human ovary at term. 1. Incidence, gross anatomy and ultrastructural features of merocrine secretion. Am J Anat 152:7–28

Jollie WP (1973) Fine structural changes in the placental membrane of the marmoset with increasing gestational age. Anat Rec 176:307–320

King BF (1981) Developmental changes in the fine structure of the chorion laeve (smooth chorion) of the rhesus monkey placenta. Anat Rec 200:163–175

King BF, Enders AC, Wimsatt WA (1978) The annular hematoma of the shrew yolk sac placenta. Am J Anat 152:45–58

Lawn AM, Wilson EW, Finn CA (1971) The ultrastructure of human decidual and predeciferal cells. J Reprod Fert 26:85–90

Malassine A (1977) Etude ultrastructurale do paraplacenta de chatte: mechanisme de I'erthrophagocytose par la cellule chorionique. Anat Embryol 151:267–283

Myagkaya G, Vreeling-sindelarova H (1976) Erythrophagocytosis by cells of the trophoblast epithelium in the sheep placenta in different stages of gestation. Acta Anat 95:234–248

Rhodin JA, Terzakis J (1962) The ultrastructure of the human full term placenta. J Ultra Res 6:88–106

Sinha AA, Erickson AW (1974) Ultrastructure of the placenta of antartic seals during the first third of pregnancy. Am J Anat 141:263–279

Takata K, Fujikura K, Shin B (1997) Ultrastructure of the rodent placental labyrinth: a site of barrier and transport. J Reprod Dev 43(1):13

Thliveris JA, Speroff L (1977) Ultrastructure of the placental villi, chorion laeve and decidua parietal is in normal and hypertensive pregnant women. Am J Obstet Cynec 129:492–498

Wislocki GB, Dempsey EW (1955) Electron microscopy of the human placenta. Anat Rec 123(2):133–167

Wynn RM (1974) Ultrastructural development of the human decidua. Am J Obstet Gynecol 118:652–670

GPSR Compliance

The European Union's (EU) General Product Safety Regulation (GPSR) is a set of rules that requires consumer products to be safe and our obligations to ensure this.

If you have any concerns about our products, you can contact us on ProductSafety@springernature.com

In case Publisher is established outside the EU, the EU authorized representative is:

Springer Nature Customer Service Center GmbH
Europaplatz 3
69115 Heidelberg, Germany

The manufacturer's authorised representative in the EU is Springer Nature Customer Service Centre GmbH, Europaplatz 3, 69115 Heidelberg, Germany. If you have any concerns regarding our products, please contact ProductSafety@springernature.com

Printed and bound by CPI Group (UK) Ltd, Croydon, CR0 4YY
01/10/2026
02225855-0019